27th Annual Conference of the International Society of Exposure Science (ISES 2017)

Integrating Exposure Science Across Diverse Communities

Book of Abstracts

Research Triangle Park, North Carolina, USA
15 - 19 October 2017

ISBN: 978-1-5108-9754-0

Printed from e-media with permission by:

Curran Associates, Inc.
57 Morehouse Lane
Red Hook, NY 12571

Some format issues inherent in the e-media version may also appear in this print version.

Printed by Curran Associates, Inc. (2019)

For permission requests, please contact International Society of Exposure Science (ISES) at the address below.

International Society of Exposure Science (ISES)
c/o Infinity Conference Group
1035 Sterling Road, Suite 202
Herndon VA 20170
USA

Phone: (703) 925-9620
Fax: (703) 925-9453

info@intlexposurescience.org

Additional copies of this publication are available from:

Curran Associates, Inc.
57 Morehouse Lane
Red Hook, NY 12571 USA
Phone: 845-758-0400
Fax: 845-758-2633
Email: curran@proceedings.com
Web: www.proceedings.com

2017 Annual Meeting

ISES2017

ABSTRACT BOOK

Integrating Exposure Science Across Diverse Communities

Research Triangle Park, NC, USA

October 15-19, 2017

Date: 10.6.17

INTERNATIONAL SOCIETY OF EXPOSURE SCIENCE

Table of Contents

Arranged by Day and Time

Note: Sessions are numbered as follows:

- First two letters indicate day (Su=Sunday, Mo=Monday, Tu=Tuesday, We=Wednesday, Th=Thursday)
- Next two letters indicate the type of presentation (PO=Poster; SY=Symposium; PL=Platform)
- Next letter indicates room (A=Bull Durham A/B, B=Crown A/B, C=Royal A/B, D=Crystal Ballroom, E=Imperial 1, F=Imperial 2, G=Auditorium Theater)
- First numeric digit indicates order of session in the day (1=early morning, 2=late morning, 3=early afternoon, 4=late afternoon)
- Numeric digits after the dash are unique abstract numbers indicating order within the session

For example, Mo-S-A1-01 is the first talk in a Monday, early morning, symposium session will be held in Modena 1.

SUNDAY, OCTOBER 15, 2017

MONDAY, OCTOBER 16, 2017

TUESDAY, OCTOBER 17, 2017

WEDNESDAY, OCTOBER 18, 2017

THURSDAY, OCTOBER 19, 2017

Student Poster Competition

SU-PO-01

Urinary Concentrations of Organophosphate Flame Retardants and Fertility Outcomes among Couples Undergoing in Vitro Fertilization

C. Carignan[1], L. Mínguez-Alarcón[1], J. Meeker[2], P. Williams[3], H. M. Stapleton[4], C. Butt[4], T. Toth[5], J. Ford[1], R. Hauser[1]; [1]Harvard T.H. Chan School of Public Health, Boston, MA, [2]University of Michigan, Ann Arbor, MI, [3]Harvard T.H. Chan School of Public Health, Boston, MA, [4]Duke University, Durham, NC, [5]Massachusetts General Hospital, Boston, MA

Abstract: Use of organophosphate flame retardants (PFRs) has increased over the past decade with the phase out of some brominated flame retardants. We recently reported associations of some urinary PFR metabolites with decreased proportions of fertilization, implantation, clinical pregnancy and live birth among women recruited from an academic fertility clinic. In this analysis, we report on urinary concentrations for the male partners, examine predictors of these concentrations, and examine associations between urinary concentrations of PFR metabolites and outcomes of *in vitro* fertilization in their partner. Our analysis included 209 couples enrolled in the Environment and Reproductive Health (EARTH) prospective cohort study (2005-2015). We measured five urinary PFR metabolites using negative electrospray ionization liquid chromatography tandem mass spectrometry. We used multivariable generalized linear mixed models to evaluate the association of demographic characteristics with the PFR metabolites and PFR metabolites with IVF outcomes, accounting for multiple IVF cycles per couple. Detection frequencies were high (>75%) for BDCIPP, DPHPand ip-PPP but low (<15%) for tb-PPP and BCIPP. Some PFR urinary metabolites were associated with race, body mass index, year of treatment cycle, and season. An 8% decline was observed for the highest compared to lowest quartile of urinary BDCIPP in adjusted means for cycles resulting in successful fertilization (adjusted absolute difference=0.06 (0.01, 0.12), p-trend=0.06), after adjusting for maternal exposure. We conclude that male partner urinary concentrations of BDCIPP may be associated with fertilization whereas *maternal* urinary DPHP, ip-PPP and ΣPFR account for negative associations previously observed for proportions of successful fertilization, implantation, clinical pregnancy and live birth. These results highlight potential reproductive effects of exposure to PFRs at levels currently common among the general population.

Keywords: A-biomonitoring, A-chemical alternatives, A-epidemiology, A-indoor environment, B-flame retardants

SU-PO-02

Dietary and inhalation exposure to polycyclic aromatic hydrocarbons (PAHs) and monohydroxy metabolites in urine: A panel study for the elderly in Tianjin,

B. Han[1], P. Li[2], X. Qin[3], L. Zhang[3], T. Ni[4], J. Fan[3], N. Zhang[1], F. He[5], J. Xu[6], W. Yang[1], W. Zhang[1], X. Wang[1], Z. Bai[1]; [1]Chinese Research Academy of Environmental Sciences, Beijing, China, [2]Tianjin University of Technology, Tianjin, China, [3]Tianjin Medical University, Tianjin, China, [4]Tianjin University of Sport, Tianjin, China, [5]Hubei Meteorological Service Center, Wuhan, China, [6]University of Washington, Seattle, WA

Abstract: Objective: Our study aimed to analyze the levels of PAHs species from both respiration and diet pathway, as well as the levels of monohydroxy metabolites in the urine. Methods: GC-MS was used to analyze the concentrations of PAHs in personal exposed PM2.5 and food samples as well as the levels of OH-PAHs in urine collected from a elderly panel. We preliminarily analyzed the characterizations of PAHs in PM2.5 and food, and OH-PAHs in urine. Results 1. Season variation had significant influence on the concentrations of all PAHs species associated with PM2.5, and some PAHs species associated with food. Respiration pathway contributed a small fraction to the total exposure amount of PAHs. 2. 7 and 10 OH-PAH analytes were well detected respectively in summer and winter. Low molecular weight PAHs hydroxyl metabolites were the most abundant compounds. The concentrations of 1- and 2-OH-Naphthalene were much higher than other analytes, accounting for over 46% of the total. 3. Good correlations were found between 1-OHNaP, 2-OHNaP, ΣOHNaP and PM2.5-bound NaP, indicating the PM2.5 contribution to elderly exposure. However, no other significant correlations were observed between other urine metabolites and PM2.5-bound PAHs, indicating that ingestion could be the more exposure route for the general population. Conclusion: The contents of OH-PAHs in urine are easily to be affected by some factors, such as smoking, cooking, etc. Therefore, in this study, we found that simultaneous determination of multiple OH-PAHs species (1-OHPyr, 1-OHNap, 2-OHNap, 2-OHFlu and 2-OHBcPhe) should be implemented to accurately and comprehensively reflect the respiration and diet PAHs internal exposure level.

Keywords: B-particulate matter, A-biomarkers,

SU-PO-03

Systematic Review and Meta-Analysis of Occupational Styrene-Induced Dyschromatopsia

*A. R. Choi[1], J. M. Braun[2], G. D. Papandonatos[3], P. B. Greenberg[1]; [1]The Warren Alpert Medical School of Brown University, Providence, RI, [2]Brown University School of Public Health, Providence, RI, [3]Brown University School of Public Health, Providence, R*I

Abstract: AIM: Styrene is an industrial solvent commonly utilized in the manufacture of plastics-based products. Observational human studies have indicated that prolonged occupational exposure to this chemical may result in acquired dyschromatopsia, or defective color discrimination. However, reviews and meta-analyses on this topic are conflicting and exclude studies from the past decade. To address this information gap, we systematically reviewed all studies of occupational styrene-induced dyschromatopsia. Using a subset of identified studies, we estimated the styrene-induced effect through meta-analysis. METHODS: PubMed, EMBASE, Web of Science electronic databases were systematically queried for eligible studies. We employed a random effects model to compare measures of dyschromatopsia between exposed and non-exposed workers, from which we calculated the standardized mean difference (Hedges' *g*). Assessments of between-study heterogeneity and publication bias were also conducted. RESULTS: All but one study presented findings in support of styrene-induced dyschromatopsia. Meta-analysis demonstrated significantly greater dyschromatopsia among exposed workers relative to their non-exposed counterparts (standardized mean = 0.56; 95% CI: 0.37, 0.76; $p < 0.0001$). Furthermore, a non-significant Cochran's Q test result ($Q = 23.2$; df = 10; $p = 0.171$) and an I^2 of 32.2% (95% CI: 0.0%, 69.9%) indicated low-to-moderate heterogeneity between studies. Funnel plot and trim-and-fill analyses showed evidence of publication bias. CONCLUSIONS: Qualitative synthesis of the evidence supports the putative exposure-outcome relationship, at levels well below regulatory agency-prescribed exposure limits. Additionally, meta-analysis indicates a moderate effect size with low-to-moderate heterogeneity. Our review suggests a need to re-assess current styrene exposure limits through further investigation. To this end, current work conditions and practices at high-exposure sites should be better elucidated.

Keywords: D-occupational, A-workplace

SU-PO-04

Characterizing persistent EDC mixtures in a diverse pregnancy cohort

S. Mehta[1], K. Coleman-Phox[2], N. E. Adler[3], B. Laraia[4], E. Epel[3], A. R. Zota[1]; [1]The George Washington University, Washington, DC, [2]University of California, San Francisco, San Francisco, CA, [3]University of California, San Francisco, San Francisco, CA, [4]University of California, Berkeley, Berkeley, CA

Abstract: Background: Exposure to persistent halogenated endocrine disrupting chemicals (EDCs) during pregnancy can impact fetal and maternal health. Despite an overall phase-out in use, these synthetic compounds are still measured concurrently in pregnant women at detectable concentrations in the US. In our current study, we aim to characterize exposure to EDC mixtures during early pregnancy by utilizing multiple aggregation techniques. **Methods:** Single spot serum samples of polybrominated diphenyl ethers (PBDEs), hydroxylated PBDEs (OH-PBDEs), polychlorinated biphenyls (PCBs), and perfluorinated chemicals (PFCs) were measured in a cohort of 104 pregnant women living in Northern

California. Cross-sectional associations between EDCs and predictors were examined using linear regression. To aggregate exposure, a chemical burden metric was calculated for each participant by counting the number of chemicals with concentrations in the 25th percentile of each chemical's distribution. Spearman correlation and principal components analysis were also used to understand grouping and to overcome the issue of multicollinearity. **Results:** 15 EDCs were detected in >50% of serum samples; PBDE-47, PFNA, and PFOS were detected in all participants. There was no consistency in the direction or magnitude of predictors in single pollutant and within-class models. Age was cross-sectionally associated with PCB-138, 153, and 180 (p<0.0001). Food insecurity was associated with maternal PFOS concentrations (p=0.03). Of the 15 EDCs, a range of 3-11 chemicals (median=6 chemicals) were detected at greater than the 75th percentile concentrations in all participants. Overall, strong with-in class and weak across-class correlations were seen. The first and second principal components, which explained 43% of the total variance, were highly correlated with PBDEs and PCBs, respectively. **Conclusions:** By utilizing various methods to aggregate exposure, we are better able to characterize EDC mixtures during pregnancy.

Keywords: A-aggregate exposure, B-mixtures, B-POPs, D-susceptible/vulnerable

SU-PO-05

Characteristics of exposure factors for consumer products in Korean women and children

M. Lim[1], J. Park[2], K. Kim[3], K. Lee[1]; [1]Graduate School of Public Health, Seoul National University, Seoul, Korea (the Republic of), [2]Seoul National University, Seoul, Korea (the Republic of), [3]Seoul National University of Science and Technology, Seoul, Korea (the Republic of)

Abstract: Concerns about potential health risks of chemicals in consumer products like cosmetic, personal care products, food containers are growing. Especially infant and children are more vulnerable to chemical exposure compared to adults. Exposure to these chemical could be determined by consumer products usage pattern. An accurate measurement of the usage patterns of consumer products is important for realistic exposure assessment. The mother's usage pattern of consumer products and children's use of consumer products are interrelated. The aims of this study were to determine exposure factors of consumer products for child and mother and analyze the relationship between consumer exposures of caregiver and child. We determined the exposure factors of 12 kinds of cosmetics (3 basic cosmetics, 1 UV protection products, 3 hair products, 3 body products and 2 cleansing products) for adults and 10 kinds of consumer products (2 cosmetics, 3 oral supplies and 7 household products) for children and 11 kinds of food containers for household. Survey was conducted on 505 mother-infant pairs from Oct. to Dec. 2015 in Seoul and the metropolitan area by using structured questionnaires. The number of subjects were determined by proportionate quota sampling based on the population composition ratio in children's sex and age distribution from 0 to 4. All cosmetics investigated in this study were used on a daily basis and usage rates ranged from 52.1% to 98.0%, except 9.9% for hair styling product and 7.7% for deodorant. The frequency of food intake by food containers ranged 2.52 to 17.39 times a month. The use of children's oral supplies varies according to the age of the child. There were a significant difference in the mother's usage rates of lotion, hair products and vinyl package food by age of children. These exposure characteristics and factor data would be useful input data for exposure and risk assessment for chemical regulation.

Keywords: A-exposure factors, C-consumer/personal care products, D-children

SU-PO-06

The joint effects of metals on general cognitive ability in adolescents living near ferromanganese industry

J. A. Bauer[1], B. Coull[2], J. Bobb[3], S. Guazzetti[4], S. Zoni[5], C. Fedrighi[5], D. Placidi[5], G. Cagna[5], R. White[1], M. Arora[6], R. O. Wright[6], D. R. Smith[7], R. G. Lucchini[6], B. Claus Henn[1]; [1]Boston University School of Public Health, Boston, MA, [2]Harvard T.H. Chan School of Public Health, Boston, MA, [3]University of Washington, Seattle, WA, [4]Azienda Unita Sanitaria Locale, Reggio Emilia, Italy, [5]University of Brescia, Brescia, Italy, [6]Icahn School of Medicine at Mount Sinai, New York, NY, [7]University of California, Santa Cruz, CA

Abstract: Introduction: Although mixtures research is increasingly popular, there are virtually no data on chemical mixture effects in adolescents. We explored associations of exposure to a metal mixture with general cognitive ability in adolescents living proximal to ferromanganese industry, a source of multiple airborne metals emissions. **Methods**: We measured manganese (Mn), lead (Pb), copper (Cu), and chromium (Cr) in hair and blood samples collected from 608 Italian adolescents ages 10-14 years. Concurrent with biosampling, trained neuropsychologists administered the Wechsler Intelligence Scale for Children, 3rd edition (WISC-III) to measure full-scale, verbal and performance IQ (FSIQ, VIQ, PIQ) scores. Bayesian kernel machine regression, which allows for higher order interactions and nonlinear effects, was used to estimate associations of FSIQ, VIQ, or PIQ scores with the metal mixture and its individual components. **Results:** Median metal concentrations were: hair Mn, 0.08 ug/g; hair Cu, 9.6 ug/g; hair Cr, 0.05 ug/g; blood Pb, 1.3 ug/dL. Adjusting for sex, age and socioeconomic status, we observed an inverse association between the metal mixture and VIQ scores (e.g. when all metals concentrations were at their 55th percentile compared to 50th percentile: -0.3 [95% credible interval: -0.5, -0.1]). Within the mixture, when other metals were fixed at their 25th percentile, VIQ scores were inversely associated with an IQR increase in hair Mn (-1.9 [-3.6, -0.2]) and with an IQR increase in blood Pb (-1.5 [-3.0, -0.1]). Inverted u-shaped associations between hair Cu and FSIQ and VIQ were found, consistent with Cu as an essential nutrient. There was suggestive evidence of interactions between Mn and Cu. **Conclusion**: Adolescent exposure to a mixture of metals was inversely associated with verbal cognitive ability, suggesting that metals from ferromanganese emissions jointly affect neurobehavior.

Keywords: A-epidemiology, B-metals, D-children

SU-PO-07

Top-Down Toxicology: an Experimental Application of the Pesticide Exposome in Honey Bee Queens

J. Milone, D. Tarpy; North Carolina State University, Raleigh, NC

Abstract: The exposome concept was developed for human exposure science and is often limited to observational research for exposure assessment. By adapting this framework for the investigation of pesticide exposure in honey bees (*Apis mellifera*) we are able to experimentally apply the exposome paradigm to a model species for social behavior and an essential agricultural pollinator. Studies of pesticide exposure in honey bees have historically focused on single pesticide dose-responses and have failed to accurately characterize exposure conditions faced by developing queens. Realistically, honey bees are exposed to numerous chemicals across multiple pathways during sensitive periods of development. In this study we reared queens in realistic exposure environments simulating elevated and reduced oral and contact exposures via pollen and wax. Pesticide treatments consisted of a combined mixture of insecticides, fungicides, and herbicides that have been commonly detected within commercial colonies and the toxicity of each mixture was based on field relevant Hazard Quotients. We sampled queens as newly emerged adults and at the onset of reproductive activity to quantify direct effects of developmental exposure, as well as sampling the offspring of mature queens to identify downstream effects on colony function. Our analysis includes measures of queen morphology, viral load intensity, mating number, insemination quality, and colony-level measures of health and disease prevalence. By understanding the effects of multi pesticide exposure across queen life history we have a more thorough understanding of honey bee exposome and multi-stressor interactions driving pollinator losses.

Keywords: B-pesticides, D-wildlife

SU-PO-08

Health effect of different gaseous formaldehyde fluctuation forms on mice: a preliminary study

X. Zhang[1], Y. Zhao[2], X. Yang[2], Y. Zhang[1], R. Li[2]; [1]Tsinghua University, Beijing, China, [2]Central China Normal University, Wuhan, China

Abstract: Formaldehyde (FA), among the most hazardous air pollutants in indoor environment, will result in diseases such as asthma, lung cancer and leukemia at certain dose. However, so far very few experiments covering its health effects of different fluctuation forms have been reported, especially those existing in the actual residence environment. The aim of this study is to compare the health damage difference between FA at fluctuant exposure (based on the real monitoring result) and at constant exposure.Mice were divided into 3 groups for fluctuant exposure (12 hours 1 mg/m^3 and 12 hours 0 mg/m^3), constant exposure (24 hours 0.5 mg/m^3) and zero exposure (no formaldehyde) per day for 7, 14 and 28 days, respectively. After exposure, the bronchoalveolar lavage fluid (BALF), lung tissue and its homogenate as well as marrow samples were prepared to examine the relevant biomarkers. It was shown that exposure to FA could induce elevated level of both oxidative stress and inflammation in the lung tissue of mice. Additionally, the inflammatory and pro-inflammatory cytokines like IL-4, IL-5, IL-13, IL-6, IL-17 and IL-1β were found to be up-regulated by FA exposure. What's more, total number of inflammatory cells and the number of some specific inflammatory cells as well both exhibited an increase in the BALF when exposed to fluctuating FA. Interestingly, our study has demonstrated that the fluctuating FA exposure rather than the constant FA exposureplayed a more intensified role in promoting the above biological alterations and even the apoptosis. This study has confirmed from several aspects that fluctuating FA exposure could be more likely to exert an adverse influence on the health of mice, leading to some guidance information for the construction of healthy indoor environment.

Keywords: C-air, A-epidemiology, B-VOCs, C-indoor

SU-PO-09

Predicting Daily $PM_{2.5}$ Concentrations in Texas Using High-Resolution Satellite Aerosol Optical Depth

X. Zhang[1], Y. Chu[2], K. Zhang[1]; [1]The University of Texas Health Science Center at Houston School of Public Health, Houston, TX, [2]The University of Texas Health Science Center at Houston School of Public Health, Houston, TX

Abstract: Background: The state of Texas contains many of most populous areas, prevalent diesel vehicles, and numerous industrial facilities that foreshadow a potential public health impact associated with particulate matter with aerodynamic diameter less than 2.5 μm ($PM_{2.5}$). Understanding the relevant public health impact requires a detailed knowledge of $PM_{2.5}$ exposure, but the current monitoring system is not capable of providing adequate spatial or temporal resolution for large cohort studies to conduct exposure assessments. **Objective**: The purpose of this study is to estimate ground-level $PM_{2.5}$ concentrations using satellite-retrieved Aerosol Optical Depth (AOD) in the state of Texas. **Methods**: We used gridded AOD values generated from the Moderate Resolution Imaging Spectroradiometer (MODIS) satellites data based on the Multi-angle Implementation of Atmospheric Correction (MAIAC) algorithm, which can provide 1km spatial and daily temporal resolution AOD data at a global coverage. In this study, we developed a mixed-effect regression model to estimate the ground-level $PM_{2.5}$ concentrations using MAIAC AOD, land use features, geographic characteristics, and weather conditions in the state of Texas for a study period of 2008-2013. **Results**: The results show high model fitting R^2 (0.83 - 0.88). The model performance was strongly influenced by dusts from unpaved roads and sand storms in the west of Texas, then yielded reasonable cross-validated R^2 (0.63 - 0.71). The multi-year predictions illustrated an unbalanced decrease of $PM_{2.5}$ concentrations between urban areas and rural areas from 2008 to 2013. Urban areas tended to have higher levels of reduction in air pollution. **Conclusions**: The MAIAC AOD was firstly used in estimating ground-level $PM_{2.5}$ concentrations in the state of Texas. The multi-year estimates enabled us to study the spatial variation of $PM_{2.5}$ levels at fine scale,, which can provide exposure metrics for epidemiological studies.

Keywords: C-air, A-exposure models, A-geospatial analysis/GIS,

SU-PO-10

Associations Between First Trimester Blood Metal Levels and Childhood Obesity in the NEST Cohort

S. S. Park[1], Y. Luo[2], T. Darrah[3], R. L. Maguire[1], J. Tzeng[2], C. Hoyo[1]; [1]North Carolina State University, Raleigh, NC, [2]North Carolina State University, Raleigh, NC, [3]The Ohio State University, Columbus, OH

Withdrawn

SU-PO-11

Analysis and Identification of Ozone-Squalene Particulate Phase By-Products

B. Coffaro, C. Weisel; Rutgers, Highland Park, NJ

Abstract: Ambient ozone penetrates into homes exposing residents to ozone and its by-products. Recently, squalene, a natural component of skin oil, has been shown to be highly reactive with ozone resulting in both volatile compounds and particulate phase by-products. The latter is likely to include a series of carbonyls and organic acids that may be respiratory and ocular irritants. While a number of the volatile ozone-squalene by-products have been identified, the composition of the particles formed has not been well characterized. This work identifies and quantifies the particulate products formed from the reactions of ozone and squalene in a controlled setting that is representative of indoor conditions. Squalene was loaded onto a surface at a concentration of 0.252 mg/cm^2 to simulate a human forehead and 100 ppb ozone was passed over the surface and allowed to react within the vessel. Produced particles were extracted from Teflon filters and followed by derivatization of the compounds with BSTFA. Subsequent analysis and identification by GC-MS was conducted where succinic acid and levulinic acid were identified as the likely unique squalene-ozone reaction products, as well as nine other alcohols and carboxylic acids. The effect of varying concentrations of ozone in the air, duration of reactions, humidity levels and whether ozone continues to react with the by-products produced on the filter are being determined. This research is meant to identify compounds that people may be exposed to inside their home from this reaction taking place on the surface of their skin.

Keywords: A-indoor environment, C-air

SU-PO-12

Impacts of Cold Weather on Emergency Hospital Admission in Texas, 2004-2013

T. Chen, K. Zhang; University of Texas Health Science Center at Houston School of Public Health, Houston, TX

Abstract: Background: Cold weather has been identified as a major cause of weather-related deaths in the U.S. While the effects of cold weather on mortality has been investigated extensively, cold-morbidity associations are less well studied. The purpose is to examine impacts of cold weather on emergency hospital admission (HA) in 12 major Texas Metropolitan Areas (MSAs) for the 10-year period, 2004-2013. Methods: Our study used a two-stage approach to examine the cold-HA association. We first applied distributed lag non-linear models (DLNM) to estimate cold effects for each MSA. A random effects meta-analysis was then used to estimate pooled effects. Percent increase in risk and corresponding 95% confidence intervals (CIs) were estimated as with a 1 °C decrease in temperature below a MSA-specific threshold. Age-stratified and cause-specific HA were modeled separately. Results: The majority of Texas MSAs were associated with an increased risk in HA ranging from 0.1% to 3.3% with a 1 °C decrease in temperature below cold thresholds. Pooled effect estimate was 1.4% (95% CI: 0.9%, 1.9%) increase in all-cause HA risk with 1 °C decrease in temperature. Cold wave effects in Texas were also examined and observed in most eastern and southern MSAs. Effects of cold on all-cause HA were highest among people over 75 years old (2.0%, 95% CI: 1.1%, 3.0%). Pooled estimates for cause-specific HA association were strongest in pneumonia (3.4%, 95% CI: 2.8%, 4.0%), followed by chronic obstructive pulmonary disease (3.2%, 95% CI: 2.1%, 4.5%) and respiratory diseases (2.4%, 95% CI: 2.0%, 2.9%). Conclusion: Cold weather generally increases hospital admission risk significantly in Texas, and cold effects were spatially heterogeneous across Texas. Our findings can provide insights to design better intervention strategies for targeted vulnerable populations towards reducing adverse health effects of cold weather.

Keywords: A-epidemiology, A - population exposure, A-climate change, D-susceptible/vulnerable

SU-PO-13

Relationships of Ambient Fine Particle Air Pollution, Endothelial Function, and Blood DNA Methylation Age: Implications for Renal Health

J. Nwanaji-Enwerem[1], M. Weisskopf[2], A. Baccarelli[3], J. Schwartz[2]; [1]Harvard Medical School & Harvard School of Public Health, Boston, MA, [2]Harvard T.H. Chan School of Public Health, Boston, MA, [3]Columbia Mailman School of Public Health, New York, NY

Abstract: Background: Research has demonstrated associations of long-term ambient fine particulate matter ($PM_{2.5}$) with renal function and DNA methylation (DNAm) age, an epigenome-wide DNA methylation-based predictor of chronological age. It has also been shown that the association of $PM_{2.5}$ component species with DNAm-age can be modified by normal variation of genes involved in endothelial physiology. **Objective**: Given that renal and endothelial physiology are closely related, we determined the relationship of DNAm-age with measures of renal function and examined the modifying role of DNAm-age on the relationship of $PM_{2.5}$ with renal function. **Methods**: This analysis included 559 participants from the Normative Aging Study with multiple visits between 2000 and 2011 (n=943 visits). We estimated one-year $PM_{2.5}$ and component species (ammonium and sulfate) levels at participants' addresses using the GEOS-chem model. Participants' blood, collected at each study visit, was used to derive DNAm-age and four renal function markers (serum creatinine, serum BUN, eGFR, and BUN-to-creatinine ratio). Blood DNAm-age was calculated using CpG sites on the Illumina HumanMethylation450 BeadChip. We estimated associations of DNAm-age with renal function using linear mixed-effects models adjusted for chronological age, lifestyle/environmental factors, and aging-related diseases. **Results**: A one-year increase in DNAm-age was positively associated with serum creatinine (β = 0.006; 95%CI 0.001, 0.010). DNAm-age was negatively associated with both eGFR (β = -0.34 mL/min/1.73 m^2; 95%CI -0.51, -0.18) and BUN-to-creatinine ratio (β = -0.07; 95%CI -0.13, -0.02). The data also suggested a modifying effect of DNAm-age on the association of one-year $PM_{2.5}$ and Ammonium levels with eGFR. **Conclusions**: DNAm-age is significantly associated with measures of renal function. Additional work is necessary to understand these relationships particularly in the contexts of $PM_{2.5}$ exposure, effect modification, and mediation.

Keywords: B-particulate matter, A-biomarkers, C-air, A - population exposure, A-epidemiology

SU-PO-14

Exposure to Air Pollutants in Selected High-rise Buildings in High Density Urban Areas in Hong Kong

T. Li[1], W. Che[1], A. K. Lau[1], H. Frey[2]; [1]Hong Kong University of Science and Techonology, Hong Kong, Hong Kong, [2]North Carolina State University, Raleigh, CA

Abstract: Air pollution is identified as the largest single global environmental health risk, contributing to 7 million annually. Half of the world's population lives in urban areas. Dense urban development characterized by high-rise buildings is becoming common, especially in Asian cities, to cope with population growth. Such urban morphology may trap air pollution in street canyons, resulting in elevated air pollutant concentrations. Urban citizens spend substantial time in high-rise buildings located in canyon-like areas; however, little is known about their exposure to air pollutants and how they are affected by outdoor air pollution. A field study was designed to quantify human exposure to fine particulate matter ($PM_{2.5}$) and carbon monoxide (CO) in three high-rise buildings in Hong Kong. Simultaneous indoor and outdoor concentrations were measured on multiple floors using portable air quality monitors. Variability in the $PM_{2.5}$ and CO concentrations was quantified. Factors affecting variability in exposure concentrations were identified. The associations between indoor and outdoor concentrations were examined. Preliminary results indicate that indoor $PM_{2.5}$ concentrations are sensitive to outdoor concentration, ventilation mode, height of air intake, and intensity of indoor sources. Based on linear regression, the infiltration of outdoor particles varies from 0.2 to 0.6 among selected locations, related to ventilation practice, choice of fresh air handling and maintenance of filtering system. Outdoor $PM_{2.5}$ concentrations are highly correlated (R^2 = 0.7~0.9) among multiple floors at the same building, but less correlated (R^2 =0.2~0.5) with $PM_{2.5}$ concentrations recorded at nearby fixed site monitors (FSMs), suggesting that FSMs may not be an appropriate surrogate for outdoor concentration in high-rise buildings. Such field studies may help better characterize exposure to air pollutants in high-rise buildings and parameterize exposure estimates in urban areas.

Keywords: A-exposure factors, A-built environment, A-sampling methods, C-air, B-particulate matter

SU-PO-15

Public's Risk Perception and Willingness-to-Pay for Air Pollution

D. Bi[2], S. Pu[1], J. Liu[2], S. Wu[2], J. Ding[2], M. Fang[1], X. Lu[2], M. Liu[1]; [1]Nanjing University, Nanjing, China, [2]Nanjing foreign language school, Nanjing, China

Abstract: Frequent haze event in China raised great public concerns on air pollution issue. This work analyzed temporal and spatial difference of four risk perception factors (risk cognition, worry, perceived benefit and trust), risk acceptance and willingness to pay (WTP) (shorted as "five variables and WTP"), so as to provide decision-makers science-based evidence. Specifically, Likert scale and contingent value method (CVM) were applied to quantitatively measure five variables and WTP. We carried out 3-years questionnaire survey from November to December in Nanjing from 2014 to 2016, and collected 570, 598 and 595 valid questionnaires respectively. To better capture spatial dynamics, we collected another 10,646 valid questionnaires in 2016 from 31 provinces in China through Internet platform. T-test and one-way ANOVA were used to examine significant difference of five variables among different demographical groups, years and provinces to explore temporal and spatial variations. In addition, average value of WTP in Nanjing and China were calculated. Result show that females , environmental professionals, highly educated people, and medium to high income groups are sensitive population. Risk cognition in 2015 was significantly higher than that in 2014($p<.000$) and 2016($p<.000$), degree of worry in 2016 was significantly higher than that in 2014($p<.000$) and 2015($p<.000$), perceived benefit was increasing year by year

(4.14<4.18<4.45), while trust was declining year by year (3.11>3.00>2.84). Moreover, we didn't observe any significant heterogeneity among regions in terms of risk perception. This result indicate that in a society with increasing air pollution risk, government should pay attention to guide public' risk perception, especially for sensitive population. Additionally, public's WTP for 50% reduction of heavy air pollution days within two years in Nanjing and China are 364 and 283 RMB per year, respectively. When making policy, government should also take it into account.

Keywords: A-behavior, C-air, A-statistical methods

SU-PO-16

An exposure assessment of occupational exposures among small quantity pesticides users in amenity horticulture.

A. Connolly[1], K. Jones[2], K. Galea[3], I. Basinas[3], L. kenny[2], P. McGowan[4], M. Coggins[1]; [1]National University of Ireland Galway, Galway, Ireland, [2]Health and Safety Laboratory (HSL), Buxton, United Kingdom, [3]Institute of Occupational Medicine (IOM), Edinburgh, United Kingdom, [4]Irish Commissioners for Public Works, Co. Meath, Ireland

Abstract: An occupational biomonitoring study to assess gardeners and amenity horticulturalist exposure to pesticides was completed over the summer period of 2015. Spot urine samples were collected pre and post work tasks involving small volumes of pesticides products (ranging from 100 ml to 2 L of concentrated product used), samples were analysed for pesticide content. The active ingredients of interest in this study included glyphosate and fluroxypyr. Study results (GM (GSD) 0.66(1.11) µg L^{-1} for glyphosate and GM (GSD) 0.29 (1.69) µg L^{-1} for fluroxypyr) showed an exposure potential during amenity horticulture work, which warrants further investigation. Results also suggested that factors such as the re-use of PPE and contact with contaminated objects used during spraying could result in increased exposure concentrations. A larger study involving 24-hour biological monitoring, dermal exposure assessment and an assessment for the potential for inadvertent ingestion is now planned for the gardening season of 2017. Workers will be grouped into four similar exposure groups based on the pesticide application method used and whether they were working with glyphosate or fluroxypyr. Dermal exposure assessment will involve collecting wipe samples from the worker hands, gloves and potentially contaminated objects such as mobile phones, product containers and vehicle steering wheels. Wipe samples from the worker perioral region will also be collected before and after completion of work tasks alongside detailed contextual information regarding the worker, task and environmental conditions. Data collection and analysis will be completed by August 2017. Study results will be used to estimate total body burden and the contribution of dermal and inadvertent ingestion as exposure routes.

Keywords: A-biomonitoring, A - exposure measurement, A-industrial hygiene, B-pesticides, D-occupational

SU-PO-17

Patterns and Predictors of Environmental Chemical Mixtures Among Pregnant Women: The HOME Study

G. Kalloo[1], A. M. Calafat[2], A. Sjodin[2], A. Chen[3], K. Yolton[4], B. Lanphear[5], J. M. Braun[1]; [1]Brown University School of Public Health, Providence, RI, [2]Centers for Disease Control and Prevention, Atlanta, GA, [3]University of Cincinnati, Cincinnati, OH, [4]Cincinnati Children's Hospital Medical Center, Cincinnati, OH, [5]Simon Fraser University, Burnaby, Canada

Withdrawn

Withdrawn

SU-PO-18

Early-life Manganese Exposure and Intrinsic Functional Connectivity of the Developing Brain

E. de Water[1], E. Proal[3], V. Wang[2], S. Martínez Medina[3], L. Schnaas[3], M. M. Téllez-Rojo[4], R. O. Wright[1], C. Y. Tang[2], M. K. Horton[1]; [1]Icahn School of Medicine at Mount Sinai, New York, NY, [2]Icahn School of Medicine at Mount Sinai, New York, NY, [3]National Institute of Perinatology (INPer), Mexico City, Mexico, [4]National Institute of Public Health (INSP), Mexico City, Mexico

Abstract: Manganese (Mn) is an essential trace metal that is neurotoxic at high levels of exposure. Disruption of brain maturation processes during the prenatal period may have lasting consequences. During this critical period, the developing human brain is uniquely vulnerable to exposure to environmental toxicants such as Mn, and prenatal Mn exposure has been associated with changes in brain areas involved in emotion processing and regulation. The goal of the present study was to examine whether prenatal Mn exposure is associated with changes in the intrinsic functional connectivity (iFC) of the brain in childhood, focusing on changes in emotional brain areas. In this pilot study, 15 subjects (aged 6-7 years) were selected from an ongoing longitudinal birth cohort study to participate in a resting state functional magnetic resonance imaging (fMRI) study. Prenatal Mn exposure was determined from maternal blood collected during the 2nd and 3rd trimesters of pregnancy. We used seed-based correlation analyses and independent component analyses to examine whether prenatal Mn exposure was associated with the iFC of the brain in children. We found that the bilateral anterior cingulate cortex and right globus pallidus showed reduced iFC with medial and lateral prefrontal areas in children who were exposed to higher prenatal Mn levels, and these children further showed reduced iFC between the bilateral insula and occipito-temporal areas. These findings indicate that prenatal Mn exposure is associated with reduced iFC of brain areas involved in emotion processing and regulation in children. Future studies should investigate whether this reduced iFC mediates the association between prenatal Mn exposure and emotional dysfunction in childhood.

Keywords: B-metals, D-children

SU-PO-19

Comparing chemical exposures across diverse communities using silicone wristbands

H. Dixon, C. Donald, A. Bergmann, G. Points III, R. Scott, B. Smith, K. A. Anderson; Oregon State University, Corvallis, OR

Abstract: Assessing personal chemical exposure is critical to linking environmental contaminants with adverse health effects. Yet, for most environmental chemicals, there is little information about the frequency of personal exposures between different human populations. In this study, silicone wristbands were used to capture chemical detection frequencies in seven distinct populations from Africa, South America, and North America. Wristbands were worn by residents in Senegal (n=70), the Alto Mayo region of Peru (n=69), in Eugene, Oregon (n=25), in rural Ohio (n=26), near the East Coast of the U.S. (n=29), in western Washington (n=10), and children in Corvallis and Bend, Oregon (n=20). Gender and age was also recorded. All 204 wristband extracts were analyzed for the presence/absence of >1,400 chemicals. Combining all locations, the average number of chemical detections was not significantly different for participants of ages 11-20, 21-40, 41-60, and >61 (Tukey-Kramer HSD, p>0.05). However, the children in Oregon, making up the <10 age bracket, had a significantly higher average number of chemical detections than all six other populations investigated (p<0.05), largely due to flame retardant and plasticizer chemicals. Examining location further, Senegal participants had significantly lower average number of chemical detections than all six other populations (p<0.05). In the U.S., Oregon adults had significantly higher average number of chemical detections than both Ohio and East Coast residents (p<0.001). There were no significant differences in the number of chemical detections for males compared to females even when stratified by continent or age (p>0.05). These results suggest that variables such as age and geographic location may contribute to differences in personal chemical exposures.

Keywords: B-SVOCs, B-flame retardants, A - exposure measurement, A - population exposure, B-phthalates

SU-PO-20

A Field Study to Validate Inhalation Exposure Factors Used to Create the Particle Inhalation Rate Metric

L. Corlin[1], H. Amaravadi[2], N. Henderson[1], J. Cordova[2], S. Ball[2], M. Woodin[1], J. L. Durant[1], D. Gute[1], D. Brugge[2]; [1]Tufts University, Boston, MA, [2]Tufts University, Boston, MA

Abstract: Exposure assessment in air pollution epidemiology usually focuses on ambient concentrations, ignoring inhalation as part of the exposure pathway. To account for inhalation patterns, we developed a new exposure metric that multiples ambient particle concentrations by minute respiratory volume (MRV). The MRV estimates were derived from fairly homogenous, healthy populations. We sought to test whether the published MRV estimates were valid among a vulnerable population representative of participants in the Boston Puerto Rican Health Study (BPRHS). We recruited 38 adults (79% Puerto Rican, 68% female, mean age=60.9 years) to participate in a Spanish-English bilingual health fair/research day. Participants reported health and demographic characteristics that affect inhalation patterns. We measured participants' height, weight, and peak expiratory volume. We also measured participants' pulse, oxygen saturation, and inhalation patterns while they walked, stood, lied down, and sat. From participants' respiration rate (RR) and inspiratory volume (IV), we calculated the average age, sex, and weight-adjusted MRV for each activity level. The effect of using these calculated values instead of published values on the PIR was assessed. Demographic and health characteristics were similar between our sample and the larger BPRHS cohort (34% of our participants had also participated in the BPRHS). About one-third of our participants were smokers, had asthma, and had low peak expiratory volume. Pulse, RR, IV, and MRV were significantly greater when participants walked or stood compared to sat or lied down. The MRVs were 1.9-4.9 times higher than the published estimates. Applying these coefficients within the BPRHS cohort increased the variability of the PIR distribution and doubled the mean and median PIR. Using the new values strengthened the effect estimates for PIR on blood pressure. Population-specific estimates of inhalation metrics are important for accurate exposure assessment.

Keywords: A-exposure factors, D-susceptible/vulnerable, C-air, B-particulate matter

SU-PO-21

Comparison of Bioaerosol Samplers and Media for the Collection of Aerosolized Norovirus

C. L. Boles, M. W. Nonnenmann; University of Iowa, Iowa City, IA

Abstract: The United States (US) has an estimated 179 million cases of acute gastroenteritis (AGE) each year. Of the total cases of AGE more than 39 million are caused be an infectious agent, such as bacteria and virus. One of the most common infectious agents that result in AGE (iAGE) is norovirus. Norovirus causes greater than 90% of iAGE outbreaks, and results in 56,000-71,000 hospitalizations in the US each year. Route of transmission is known to occur by direct and indirect contact with contaminated surfaces; however, evidence has suggested that the virus can be transmitted via aerosol. Determining routes of exposure is important as the infectious dose for norovirus has been reported as low as 10 virions. Little research has been performed measuring aerosolized norovirus in the field and laboratory, which may be a significant route of transmission. Therefore, a developed methodology is needed for norovirus aerosol sampling so that inhalation exposure to norovirus can be determined. This study compared two samplers and two liquid sampling medias in the collection of airborne norovirus. The SKC BioSampler and NIOSH Bioaerosol sampler containing either Phosphate-Buffered Saline or Hanks Balanced Salt Solution were compared for collection of murine norovirus (MNV:CW3). Ten 30-minute trials were completed by aerosolizing MNV:CW3 in a laboratory bioaerosol chamber to compare concentrations of norovirus collected across two types of bioaerosol samplers using two liquid sampling medias. Samples were analyzed using RT-qPCR, as well as PMA:RT-qPCR to evaluate virus capsid integrity. The mean total virus particles per cubic meter of sampled air were compared between samplers and liquid sampling medias. Preliminary results (n=40) indicate no difference between media type, however there was a difference between the percentage of virus with intact capsids before and after sampling. Future investigations targeting aerosolized norovirus can incorporate the findings of this study.

Keywords: A-industrial hygiene, A-Infectious disease, D-occupational, B-microbial agents, A-workplace

SU-PO-22

Assessing the potential impact of global warming on infiltration of outdoor air pollutants into residential indoor environments

D. Liang[1], K. Ono[1], W. Lee[2], J. Liao[1], J. Lawrence[3], S. E. Sarnat[1], J. A. Sarnat[1], P. Koutrakis[3]; [1]Emory University, Decatur, GA, [2]National Taiwan University, Taipei, Taiwan, [3]Harvard University, Boston, MA

Abstract: Aim. Rising temperatures due to climate change are expected to impact human adaptive response, including changes to home cooling and ventilation patterns. These change may affect potential air pollution exposures via alteration in residential air exchange rates, which impact contributions from indoor and outdoor particle sources to indoor air. We conducted a field study examining associations between particle infiltration and temperature to inform future studies of air pollution health effects. **Methods.** We measured indoor and outdoor fine particulate matter ($PM_{2.5}$) in Atlanta in 60 homes totaling 840 sampling-days. Indoor-outdoor sulfur ratios were used to estimate particle infiltration. Linear mixed-effects models were used to examine sulfur ratio-temperature relationships. Projected meteorological values were incorporated in the models to predict sulfur ratios in a 20 year future scenario (2046-2065) and past scenario (1981-2000). **Results.** The average sulfur ratio in Atlanta was 0.70 ± 0.30, with a 0.21 lower sulfur ratio in summer compared to transition seasons. Sulfur ratios were also 0.19 lower in houses using air conditioning (AC) compared to those without AC usage. We observed a negative linear relationship between temperature and sulfur ratio; a 1 degree Celsius increase in temperature was associated with a decrease in the sulfur ratio of 0.008. Future temperature was projected to increase by 2.1 degrees Celsius compared to the past, which would result in a corresponding 0.023 increase in sulfur ratio during cooler months and a 0.037 decrease during warmer months. **Conclusion.** We observed substantial variability in sulfur ratios in Atlanta by temperature, season, and AC usage, and projected changes to sulfur ratios in the future under a warming climate. Ongoing analyses will compare these relationships to a similar study in Boston. These analyses can help to provide a better understanding of the potential influence of climate change on $PM_{2.5}$ health effects.

Keywords: A-climate change, A - exposure measurement, B-particulate matter, A-indoor environment, C-air

SU-PO-23

Performance of Consumer-Grade Air Pollution Measurement Devices in Residential Environments

G. Mainelis, S. N. Manibusan; Rutgers University, Edison, NJ

Abstract: A developing trend in consumer electronics is production and use of inexpensive (<$300) devices to monitor airborne contaminants of health concern. Most of these devices monitor the mass concentration of $PM_{2.5}$ and PM_{10} particles, temperature, relative humidity, and some include volatile organic compounds or other pollutants. This study focused on four consumer devices: the Air Quality Egg 2, Blueair Aware, Foobot, and Speck that use optical sensors to measure airborne particulate matter concentrations. The devices were collocated in three residences over 7 consecutive days in each residence. Household measurements were compared against established optical sensing devices including two Personal Data RAMs (passive pDR-1000, Thermo Fisher), a DustTrak DRX (TSI Inc.), and gravimetric mass measurements using two Personal Modular Impactors (PMI, SKC, Inc.). A follow-up study in one residence included a collocation of Foobot, Speck, and the pDR-1000 in the kitchen (near source cooking exposure) and bedroom (residential exposure away from a source) combined with time-activity behaviors in the residence over 14-days. Overall, the results show inconsistent performance of consumer grade devices relative to reference methods. For example, in one household, the Foobot registered a Pearson's R^2 of 0.565 when compared to the DustTrak DRX for 1-hour averages, while the Speck showed poor associations (R^2=0.206). Similar results were shown in the cooking study with poor associations from the Speck compared with the pDR-1000 (R^2=0.003), while the Foobot had a strong association with the pDR (R^2=0.830). Among the investigated sensors, the Foobot was most correlated with the reference devices, but the cooking study indicated its limitations regarding maximum loading capacity. Overall, the data show the utility of the consumer-grade air quality measurement devices, allowing for wide-scale deployment; however, further testing is needed to determine their applicability in field studies.

Keywords: A-sensor technology, B-particulate matter, A-indoor environment, A-sampling methods

SU-PO-24

Evaluating effects of global change on patterns of aeroallergens and ground-level ozone across the contiguous US

T. Cai[1], Z. Mi[1], X. Ren[2], Y. Zhang[2], P. Georgopoulos[1]; [1]Rutgers University, Piscataway, NJ, [2]Rutgers University, Piscataway, NJ

Abstract: The prevalence of Allergic Airway Disease (AAD) is growing globally, resulting in increased numbers of emergency department visits and hospitalizations. Clinical studies show that AAD can be exacerbated by the synergistic action of aeroallergens such as pollen and fungi, and atmospheric pollutants such as ozone. Furthermore, climate change has critically affected atmospheric processes involved in the dynamics of air pollution systems and emissions of natural pollutants such as pollen and spores. Previous studies, involving data from nationwide observations of airborne pollen counts of selected plant species in conjunction with climatic factors, indicated that the start date and length of pollen season, the average peak value and annual total of daily counted airborne pollen have been affected substantially by the changing climate. The present study investigates co-occurrences of peak ozone concentrations and peak pollen counts across the contiguous United States (ConUS). Analysis of observed pollen counts and ozone concentrations at 58 pollen monitor stations were conducted. Concentrations of pollen in base years (2001-2004) and future years (2047-2050) are simulated with a customized version of CMAQ (the Community Multiscale Air Quality model) employing a grid with 36 km by 36 km horizontal resolution, while corresponding ozone concentrations at similar resolutions for the above timeframes were conducted by a multi-university/agency consortium including USEPA. The study employs spatiotemporal correlation analysis to examine patterns of co-occurring ozone and pollen concentrations. The outcomes of this study provide information that could support development of strategies for managing health-impacts of co-occurring photochemical pollutants and aeroallergens.

Keywords: A-climate change, A - ambient monitoring, B - ozone

SU-PO-25

Heat and Hydration Assessment of Migrant Grape-Workers in Sonora, Mexico

R. Wagoner; University of Arizona, Tucson, AZ

Abstract: The demanding physical labor that makes agricultural production possible in Northern Mexico is exacerbated by extreme temperatures, migration, and lack of health and safety oversight on a national level. This study addressed heat and hydration in a commercial grape farm in Northern Mexico where 95% of the land is classified as arid or semi-arid. Notably, one-third of Mexico's heat-related deaths reported from 2002 to 2010 occurred in agricultural workers in this region. Migration status, low socio-economic ranking, and poor literacy rates contribute to the vulnerability of workers in this region to heat exposure. Furthermore, there is growing concern about the direct and indirect health effects of climate change, especially in regions such as Sonora that have limited access to resources and occupational health control methods. A total of 38 participants were recruited for three sampling periods during spring and summer 2016. For each sampling period, an oral questionnaire was administered in Spanish; core body temperature was measured using ingestible thermometers; urine was collected and analyzed for specific gravity; and the effective Wet-Bulb Globe Temperature (WBGT) and metabolic rate were calculated. The majority of participants listed Chiapas as their home state, nearly half spoke an indigenous language as well as Spanish, most were between the ages of 18 and 24, and none had completed high school. The effective WBGT was higher during the summer. However, the core body temperature of workers in the spring were not significantly different than the core body temperatures of workers in June and August. As indicated by urine specific gravity, the majority of workers in all months were either mildly or clinically dehydrated. These results indicate the need for enhanced administrative and engineering controls, policies, and binational collaborations to reduce levels of heat stress and dehydration in agricultural workers in Northern Mexico.

Keywords: D-occupational, D-susceptible/vulnerable, D-environmental justice, A-climate change, A-workplace

SU-PO-26

Health Risk Ranking Assessment of Human Exposure to Multiple Air Pollutants Emitted from Municipal Solid Waste Incineration in China

Q. Zhou, M. Liu, J. Bi; Nanjing University, Nanjing, China

Abstract: Municipal Solid Waste Incineration (MSWI) has developed rapidly with doubled increase in China during 2010-2015. However, risk level of MSWI as well as its spatial distribution were still unknown. A systematic risk identification and ranking assessment is essential to improve effectiveness and pertinence of risk management. In this study, we developed a health risk ranking framework and applied it to assess the MSWI related risk across China. Firstly, a bottom-up multiple-pollutant emission inventory was calculated based on detailed information of 222 MSWIs in mainland China and localized emission factors from literature investigation. Then a Gaussian Plume Model was used to simulate the dispersion of pollutants emitted by each MSWI, considering the perennial dominant wind direction and mean wind speed. Finally, spatialized and population-weighted Hazard Index (HI) and Relative Risk (RR) were calculated to evaluate both non-carcinogenic and carcinogenic health risks caused by individual respiratory exposure from heavy metals and PCDD/Fs. Results show that, in 2015, 59651.6 t of NOx, 20072.2 t of SO_2, 13215.3 t of CO, 7240.0 t of HCl, 4392.5 t of particulate matter (PM), 26.0 t of Pb, 25.8 t of Cr, 15.5 g of Hg, 5.0 t of Cd, 3.4 t of Ni, 273.2 kg of As, and 73.2 g-TEQ of polychlorinated dibenzo-p-dioxins and polychlorinated dibenzofurans (PCDD/Fs) were emitted by MSWI in mainland China. The biggest HI and RR at county-level were 2.75E-3 (<1) and 9.11E-7 (<1E-6), and the relatively high-risk regions mainly concentrated in the eastern coastal areas (Shanghai, Jiangsu), and southwestern inland (Sichuan, Chongqing). Besides, the greatest contributor to HI and RR were Pb and PCDD/Fs separately. The identified relatively high-risk areas and pollutants above need to be set as the priority control areas or pollutants in the risk management. Certainly, more field tests are suggested to be conducted in these places to provide more exact information.

Keywords: A-risk assessment, A-geospatial analysis/GIS, C-air, A-environmental policy

SU-PO-27

Spatial-temporal Characteristics of Metal Contents in Ambient Fine Particles in Nanjing, China

D. Bi[1], H. Cheng[1], X. Liu[2], M. Liu[2], J. Chen[1], X. Cui[2], Z. Zhu[1]; [1]Nanjing Foreign Language School, Nanjing, China, [2]Nanjing University, Nanjing, China

Abstract: Recently, frequent hazy weather across China rose great public concern about health consequences of air pollution. Among different air pollutants, ambient fine particles ($PM_{2.5}$) became the focus partly due to its strong adsorption of nocuous substances like Pb and Cd, which will do serious harm to human health through affecting body's nervous, immune, cardiovascular system, etc. Many studies have analyzed the composition and sources of metal elements in $PM_{2.5}$. However, few of them have explored spatial-temporal variations, specifically for a school in China. With the support from STEM (Science, Technology, Engineering and Mathematics) program directed by Nanjing Foreign Language School, this study measured the metals in $PM_{2.5}$. Specifically, we selected the playground and classroom in Nanjing Foreign Language School as sample sites for outdoor and indoor environment. Samples were collected by the portable particulate monitor (MicroPEM, RTI International) for four seasons. For each site in each season, there were three samples with 24-hours continuous monitoring $PM_{2.5}$ data. Then, all samples were analyzed by inductively coupled plasma mass spectrometry (ICP-MS) to characterize six typical metal elements, including Pb, Mn, Cu, Cr, Sb and Ni. Results showed that spatial and temporal distribution of metals in $PM_{2.5}$ followed some rules. The concentrations of most metals were higher in spring and winter, lower in summer and autumn. Outdoor metal contents were significantly higher than indoor in spring and winter, probably because closed indoor environment in cold days significantly hindered penetration of outdoor pollutants. In addition, Pb is the dominated metals in $PM_{2.5}$ in Nanjing mainly due to vehicle emissions. From the results, the government should pay more attention to pollutant sources like motor vehicle. And it is recommended for individuals to take action like wearing masks or staying indoors so as to reduce damage to their health in spring and winter.

Keywords: B-particulate matter, B-metals, A - ambient monitoring

SU-PO-28

Health Risk Assessment of Endocrine Disruptor Organophosphorus Pesticides Exposure through Dietary Intake of Fresh Vegetables for children in Tehran, Iran

M. Zare Jeddi[1], Y. Bruinen de Bruin[2], V. Mahdavi[3]; [1]Institute for Environmental Research (IER), Tehran University of Medical Sciences, Tehran, Iran (the Islamic Republic of), [2]Joint Research Centre of the European Commission, Ispra, Italy, [3]Laboratory of Pesticide Research,, Tehran, Iran (the Islamic Republic of)

Withdrawn

SU-PO-29

Individual's Exposure and Emotional Response to Visual Impact of Particulate Matter Pollution: A Psychophysiological Study

J. Yang[1], X. Liu[1], J. Chen[2], M. Liu[1]; [1]Nanjing University, Nanjing, China, [2]Nanjing Foreign Language School, Nanjing, China

Abstract: As a leading risk factor for respiratory and coronary disease in China, air pollution has been studied extensively from physical health perspective. However, it is gradually recognized that exposure to air pollution also exert massive adverse impacts on individual's mental health. Particulate matter pollution (e.g. $PM_{2.5}$) is perceived as one of the most harmful air pollutants by Chinese lay public. High-concentration and large-scale PM pollution can damage visibility seriously, thus evoke individual's negative emotions. In this study, we administered a psychophysiological study to quantify association between individual exposure and emotional response to PM pollution. Participants were instructed to view a series of photos taken at the same site and occasion but with different $PM_{2.5}$ concentrations. The images were mixed with pictures selected from IAPS (International Affective Picture System) in case of fatigue. Three self-report scales on basic dimension of emotion (i.e. valence, arousal and motivation) were rated by participants after viewing each photo. Moreover, their psychophysiological signals, including skin conductance responses, electrocardiogram, electromyogram, were also recorded during the whole experiment. Obvious but diverse trends of different emotion indicators were observed. A 10% increase of $PM_{2.5}$ concentrations was associated with a 0.07 decrease of valence scores. Arousal scores, however, dropped down sharply when $PM_{2.5}$ concentration increased from 11 to 50μg/m^3, and tended to be quite stable when the number exceeded 50μg/m^3. Physiological responses indicating negative emotion status were significantly stronger if $PM_{2.5}$ concentrations were higher than 50μg/m^3. We conclude the policy implications revealed from our study, and discuss the feasibility for further applying this novel method to measure individual's emotional reactions towards various environmental risks.

Keywords: B-particulate matter, A-exposure models, A-environmental policy

SU-PO-30

Perfluoroalkyl substances in the Fernald Community Cohort: Exposure distributions over time and associations with thyroid and kidney function and obesity

B. E. Blake[1], S. M. Pinney[2], S. E. Fenton[1], K. Ferguson[3]; [1]National Institute of Environmental Health Sciences, Durham, NC, [2]University of Cincinnati, Cincinnati, OH, [3]National Institute of Environmental Health Sciences, Durham, NC

Abstract: Perfluoroalkyl substances (PFAS) are a diverse class of chemicals used in industrial settings and consumer products to make materials resistant to heat, water, and oil. PFAS exposure in the general population is common, and PFAS have been associated with altered kidney function, disrupted thyroid hormones, and are suspected obesogens. We investigated a panel of PFAS in participants of the Fernald Community Cohort (FCC), who were identified as at risk for high PFOA exposure based on living in zip codes bordering the Ohio River. Subjects had up to 3 repeated measures of serum PFAS (N=210 participants, N=517 total measurements). We compared changes in serum PFAS concentrations over time with those observed in the National Health and Nutrition Examination Survey (NHANES) at similar time points. We tested the association between PFAS measures and concurrent outcome measurements of obesity and kidney and thyroid function, including body mass index (BMI), kidney function (estimated glomerular filtration rate, eGFR), and thyroid stimulating hormone (TSH). We used linear mixed effects models to examine each association, adjusting for participant age, sex, and year of sample collection among other confounders. No associations were observed in models of PFAS and BMI or TSH. We observed significant positive associations between 2-(N-methyl-perfluorooctane sulfonamido) acetate (β=1.49, SE=0.70, p=0.04) and 2-(N-ethyl-perfluorooctane sulfonamido) acetate and eGFR (β=1.82, SE=0.71, p=0.01) and significant negative associations between perfluorononanoate (β=-2.96, SE=1.30, p=0.02), perfluorohexane sulfonate (β=-2.26, SE=0.86, p=0.01), and perfluorodecanoate and eGFR (β=-2.40, SE=1.00, p=0.02). Geometric means for serum perfluorooctanoic acid (PFOA) were threefold higher in the FCC compared to NHANES from 1999 to 2008 while perfluorooctanesulfonate (PFOS) levels did not differ between the groups. These results suggest PFAS may affect kidney function in a compound-specific manner.

Keywords: A - exposure measurement, A-biomarkers

SU-PO-31

Exposure to Short Lived Air Pollutants on Public Transit in Brazil

E. Patrick[1], J. F. Segura[2], A. C. Targino[2], M. Gibson[1]; [1]Dalhousie University, Halifax, Canada, [2]Universidade Tecnológica Federal do Paraná, Londrina, Brazil

Abstract: Air quality is of particular concern in urban areas where high traffic and city design lead to accumulation of toxic air pollutants within street corridors. Urban aerosols are a heterogeneous mixture of solid particles and liquid droplets of varying physiochemical characteristics. Their concentration and composition are influenced by local and long range anthropogenic, geogenic, and biogenic emissions. Brazil in particular has grown rapidly in population, car ownership, and industry; it is currently the fifth most populous country in the world. Public transit is often chosen as the 'green' alternative to driving and helps alleviate congestion in cities. However, there is a gap in knowledge with regard to air pollution exposure on public transport in Brazil. The majority of buses in Brazil run on high sulfur diesel, do not have particle traps, and many cities contain old vehicle fleets. This contributes to higher emissions than buses that use cleaner fuels and have emission control. This study aims to quantify the bus passengers' exposure to air pollution in three varying sized cities in southern Brazil. Two different bus routes from each city were surveyed over a period of one week. Black carbon and particulate matter concentrations were measured using hand-held devices (an aethalometer model AE51, Aethlabs and a photometer model DustTrak 8520, TSI). Data between the cities will be compared to determine if city size and traffic variation affect the amount of air pollution aboard public transit. Highly temporally-resolved air pollution data will be linked with GPS data to create maps of air pollution observed during the bus trips. Preliminary results show that the average mass concentrations of fine particulate matter ≤ 2.5 μm ($PM_{2.5}$) and black carbon in a megacity, 32 μg/m^3 and 10 μg/m^3 respectively, are double the concentrations in a small city, 15 μg/m^3 and 4.5 μg/m^3 respectively.

Keywords: A - exposure measurement, A-geospatial analysis/GIS, B-particulate matter, C-air, D-environmental justice

SU-PO-32

Evaluating the synergistic effects of cyanotoxic mixtures on the ALS pathway using targeted proteomics and statistical design of experiments

R. Martin, M. Bereman; NCSU, Raleigh, NC

Abstract: Amyotrophic lateral sclerosis (ALS) is a devastating fatal disease characterized by the death of motor neurons which leads to the loss of voluntary muscle control - ultimately robbing the individual of the capability to walk, eat and eventually breathe. ALS is classified in to two categories: idiopathic (sporadic ALS) and heritable (familial ALS). Familial forms of ALS have autosomal inheritance patterns and account for about 10% of all ALS cases. The cause for the other 90% of sporadic ALS cases remains poorly understood. Current research suggests that environmental stressors may contribute to the development of sporadic ALS. β-methylamino-L-alanine (BMAA) is a neurotoxin produced ubiquitously by cyanobacteria and has been linked to various neurodegenerative disorders, including sporadic ALS. In addition to BMAA, cyanobacteria produce other known neurotoxins such as anatoxin, mycrocystins, and 2,4 diaminobutyric acid (2,4 DAB). To further investigate if cyanotoxins mixtures could potentially act in a synergistic manner to enhance neurodegeneration and promote the onset of sporadic ALS, we performed targeted mass spectrometry (MS) based proteomics of a preselected group of proteins that are associated with motor neuron degeneration in the ALS pathway. We used the NSC-34 cell line, which is an increasingly used in vitro model for neurodegenerative diseases, as it shares many properties of motor neurons cells. We employ statistical design of experiments, a powerful approach in process optimization, to screen for synergistic effects of mixtures of cyanotoxins. Taken together, we intend to develop new methodologies to screen for interactions amongst chemicals and apply these methods to elucidate the mechanisms at which the environment initiates motor neuron degeneration.

Keywords: A - exposure measurement, A-analytical methods, A-statistical methods, B-mixtures, B-microbial agents

SU-PO-33

Improving ambient ultrafine particle concentration predictions at residences by adding central-site monitoring to a mobile monitoring campaign

M. C. Simon[1], A. P. Patton[2], D. Brugge[3], J. L. Durant[1]; [1]Tufts University, Medford, MA, [2]Health Effects Institute, Boston, MA, [3]Tufts University, Boston, MA

Abstract: Aim: Accurate exposure estimates of traffic-related ultrafine particles (UFP; <100 nm diameter) are needed to reduce exposure misclassification in health studies; however, because UFP concentrations are highly variable in space and time, accurate estimation is a considerable challenge. Our aim was to determine if hourly particle number concentration (PNC; a proxy for UFP) models built from intermittent, continuous mobile-monitoring data could be improved with the addition of data from a central-site monitor. **Methods:** PNC was measured in Boston (Dec 2011-Nov 2013) and Chelsea (Dec 2013-May 2015), Massachusetts (USA) by (1) mobile monitoring on ~45 days (3-6 hours/day), (2) continuous monitoring at central sites, and (3) continuous monitoring outside homes (6 weeks at 9 homes) (per city). Spatial-temporal models were built from the mobile-monitoring-only data set in each city. To incorporate central-site data, we built a time-series model from that data set and a spatial model based on the mobile-monitoring data standardized to the central site. Traffic, meteorology, indicators of secondary particle formation, and land use (e.g., distance from roads) were used as predictors in regression models of log-transformed PNC [ln(PNC)]. Models were evaluated by root-mean-square-error (RMSE) and validated by 10-fold cross-validation, and assessed for their ability to predict ambient concentrations made outside of the 18 homes using Pearson correlations. **Results:** Models showed generally good agreement with measurements, but overestimated PNC at homes 52-83% of hours. The addition of a central monitor increased the mean Pearson correlations between modeled and measured ln(PNC) at individual homes from 0.33 to 0.54 and decreased the mean RMSE by 7%. **Conclusions:** PNC models based on both mobile-monitoring and central-site data improved model predictions and reduced error of estimated ambient PNC at residences in two urban areas compared to models based on mobile-monitoring data alone.

Keywords: A-exposure models, B-particulate matter, C-air, A - ambient monitoring

SU-PO-34

Method development for the detection of pyrethroid metabolites in saliva

M. Wren, I. Yang, B. Buckley; Rutgers University, Piscataway, NJ

Abstract: Saliva may be used as a less-invasive alternative to urine or blood in pesticide biomonitoring. Salivary concentrations of lipophilic pesticides have been correlated to the unbound fraction level present in blood, making it a useful biomarker for occupational and environmental exposures. No method has previously been developed to quantify biomarkers of pyrethroid insecticides in saliva. The purpose of this study was to develop a method to detect three pyrethroid metabolites in saliva: *cis*-3-(2,2-dibromovinyl)-2,2-dimethylcyclopropane carboxylic acid (DBCA), 3-phenoxybenzoic acid (3-PBA) and 4-fluoro-3-phenoxybenzoic acid (FPBA). Two different sample collection methods were assessed: 1) passive flow and 2) the Salivette® collection device. Two solvents, toluene (MePh) and dichloromethane (DCM), and three time points, 20, 40 and 60 min, were evaluated for extraction efficiency. Liquid-liquid extraction (2 x 3 mL solvent) was performed to separate analytes from the aqueous layer. Metabolites were derivatized with hexa-2-fluoroisopropanol and N,N'-Diisopropylcarbodiimide, cleaned up with $NaHCO_3$ and isolated in hexane, then concentrated to 40 µL hexane prior to analysis by ion trap gas chromatography/mass spectrometry using the Agilent 7890B Gas Chromatograph and Agilent 240 Mass Spectrometer. MePh extracted all pyrethroid metabolites with 30% - 70% greater efficiency per one mL than DCM. The 40 min extraction period yielded greatest recoveries for both MePh and DCM: 109%, 88% and 104% recovery and 93%, 77% and 88% recovery for DBCA, FPBA and 3-PBA, respectively. Collection from the Salivette® cotton roll required three diH_2O washes to extract metabolites. Recoveries of over 150% for all analytes from the Salivette suggest interfering components from the collection device. A correlative study to quantify salivary, blood and urine concentrations of pyrethroid metabolites is underway.

Keywords: B-pesticides, A-biomonitoring, A-analytical methods, A - exposure measurement

SU-PO-35

Intraurban distributions of NO_2 and $PM_{2.5}$ at three dimensions in Lanzhou, China

L. Jin[1], Y. Zhang[1], Y. Zhang[2], M. L. Bell[1]; [1]Yale University, New Haven, CT, [2]Gansu Academy of Environmental Sciences, Lanzhou, China

Abstract: Previous research linked maternal exposures to NO_2 or PM_{10} with increased risks of adverse birth outcomes in Lanzhou. To improve exposure assessment, a land use regression study at ground level with consideration of vertical variation by building height was conducted in Lanzhou. In each of four seasons in 2016-2017, NO_2 was measured using Ogawa badges for 2 weeks at 75 ground-level sites. $PM_{2.5}$ was measured using DataRAM at a subset (N=38) of the 75 sites. The $PM_{2.5}$ monitor was systematically moved in pre-determined patterns around each site. The measurements were adjusted for temporal trend using government monitoring data. Vertical profile measurements were conducted at 18 sites evenly distributed at increasing floors of 2 buildings: one facing traffic and the other facing away from traffic. The annual and seasonal concentrations of NO_2 and $PM_{2.5}$ at ground level were regressed against spatial predictors, including elevation, population, road network, land cover, and land use with buffers 100-2000 m around each site. The vertical variations were investigated using polynomial models. NO_2 concentrations were associated with lengths of major roads and of all roads, slope, and cultivated land (adjusted R^2: 0.61). $PM_{2.5}$ concentrations were higher in certain districts, and associated with population density, lengths of roads with vehicles, and elevation (adjusted R^2: 0.56). From the 1st to 32nd floors in the building facing traffic, NO_2 and $PM_{2.5}$ concentrations decreased 22% and 26%, respectively. These distributions showed seasonal and diurnal variations. The ground-level NO_2 and $PM_{2.5}$ showed different patterns, which was not captured by the 4 government monitors in Lanzhou. Concentrations of these pollutants decreased substantially with height. Therefore, characterizing 3-dimensional distributions of multiple pollutants might be crucial for exposure and risk assessment in cities with mixed pollution sources and tall buildings, and potentially have policy implications.

Keywords: A-exposure models, A - ambient monitoring, A - exposure measurement, C-air

SU-PO-36

Spatial point pattern analysis of congenital heart defects in Lanzhou, China

L. Jin[1], J. Reuning-Scherer[1], J. Qiu[2], Q. Liu[2], Y. Zhang[1], M. L. Bell[1]; [1]Yale University, New Haven, CT, [2]Gansu Provincial Maternity and Child Care Hospital, Lanzhou, China

Abstract: Analysis of spatial point processes has been used to investigate spatial variation of disease risks, and their relationships with environmental factors; however, few studies have been conducted in China, where emissions from industries and heavy traffic pose serious public health concerns. The objective of this work is to investigate the impacts of major point sources and road networks on congenital heart defects (CHD) risks in Lanzhou, China, where maternal exposure to ambient air pollution was linked to increased risks of CHD. From a Lanzhou birth cohort during 2010-2012, 8,227 singleton live births with home addresses in the city urban area were included in this study. K, L, and Kcross functions were used to detect clustering tendency of the CHD (n=65) and healthy infants (n=8162). A Kernel density ratio was used to identify potential clusters of CHD cases adjusting for the distribution of healthy infants. A Poisson point process model was used to model the intensity of CHD cases as a function of emission-weighted distance to major point sources (power plants and cement factories), road length density within 100m buffers, the estimated intensity of healthy infants, maternal income, and education. Similar distributional patterns were identified for CHD cases and healthy infants. Adjusting for the distribution of healthy infants, maternal income and education, CHD risks were significantly associated with increased road length density within 100 m buffer (RR: 1.07) and decreased emission-weighted distance from major point sources (RR: 0.07). Chi-square tests of quadrat counts and Kolmogorov-Smirnov tests validated the model and showed no spatial clustering of residuals. Results indicate proximity to major point sources and road networks have adverse impacts on newborn's health in Lanzhou, China. The identified clusters of CHD cases may allow policy makers or future mothers to make informed decisions regarding exposure control and risk management.

Keywords: A-geospatial analysis/GIS, A-epidemiology, D-prenatal

SU-PO-37

Assessing the Role of Caregiver Stress and Cockroach Allergen in Asthma Related Healthcare Utilization

D. Werthmann, F. Rabito; Tulane University SPHTM, New Orleans, LA

Withdrawn

SU-PO-38

A Pilot Study on Migrant Grape Workers Exposure to Pesticides in Sonora, Mexico

N. I. Lopez-Galvez; University of Arizona, Tucson, AZ

Abstract: Expanding agribusiness in Sonora, a state in Northern Mexico known for its cattle, tomato, and table grape production, has increased the demand for temporary migrant agricultural workers from the poorest regions of Southern Mexico. These migrant agricultural workers participate in strenuous tasks while exposed to a wide variety of occupational risks and hazards including pesticides. A cross-sectional observational study was conducted to assess the pesticide exposure of migrant grape workers. This is the first study to characterize exposure of migrant grape workers to pesticides in this region. A convenience sample of 20 participants were recruited from a large commercial grape farm employing approximately 2,000 workers during the harvest season near Hermosillo- Sonora, Mexico. A questionnaire was used to obtain information on working activities and demographics. Morning void urine samples were collected to assess pesticide exposure. Most participants were originally from the state of Chiapas-Mexico, none had completed high school, and half spoke an indigenous language. Pyrethroid and organophosphate urine metabolites were detected in the majority of workers. The creatinine-adjusted concentration for cyfluthrin, chlorpyrifos, and parathion metabolites in urine obtained in this study (geometric means: 0.942 µg/g, 3.559 µg/g and 1.630 µg/g, respectively) were higher than in the Mexican American population included in NHANES. Unfortunately, there is no study of the general population in Mexico for comparison. Our results suggest that grape workers in this region are exposed to high levels of pesticides, which may, eventually, affect their health. Additional research is needed to confirm these findings and to evaluate health outcomes associated to pesticide exposure in this region. Results from this pilot study can be used to conduct a larger pesticide study, create binational partnerships between researchers, and develop occupational health training resources in Mexico.

Keywords: A-biomonitoring, D-occupational, B-pesticides, A-environmental justice, A - population exposure

SU-PO-39

Household dust is a good predictor of children's lead exposure

S. Cao[1], X. Duan[2], X. Zhao[1], C. Sun[3], F. Wei[4]; [1]Chinese Research Academy of Environmental Sciences, Beijing, China, [2]University of Science and Technology Beijing, Beijing, China, [3]Chinese Center for Disease Control and Prevention, Beijing, China, [4]China National Environmental Monitoring Center, Beijing, China

Abstract: China is one of the main producer and consumer of lead in the world. As one of the typical lead-related activities, lead-bearing ore mining and refining influences the environmental quality and poses high risks to human health, due to large amounts of pollutants are emitted into the environment, particularly into the surrounding dust. Children's lead exposure depends heavily on multi-pathway and source-specific exposure, which have been rarely studied around typical ore mining and refining area. The objectives of this study are to 1) quantify the lead levels in the blood of children living around typical ore mining and refining area, China, as well as 2) lead levels in their household dust, and 3) estimate the contribution of household dust to the children's blood lead. Thus, lead contents and lead isotopic ratios ($^{207}Pb/^{206}Pb$ and $^{208}Pb/^{206}Pb$) in blood and household dust were studied. Lead contents in indoor and outdoor dust were 1358.08 and 2116.18 mg kg^{-1}, respectively, which showed a significant variance and were ~ 10 times higher than the thresholds of China (300 mg kg^{-1}).The children's blood lead level was 22.26 µg dL^{-1}, quite higher than the threshold in the criteria of China (10 µg dL^{-1}). Lead contents in indoor and outdoor were quite correlated ($p<0.01$, $r=0.77$), and the household dust lead showed a significant correlation to the children' blood lead ($p<0.01$, $r=0.60$). The isotopic ratios in the blood were 2.117±0.022 for $^{208}Pb/^{206}Pb$ and 0.851±0.021 for $^{207}Pb/^{206}Pb$, similar to those of the slag (2.121±0.018, 0.849±0.001 for $^{208}Pb/^{206}Pb$ and $^{207}Pb/^{206}Pb$) and dust (2.122±0.048, 0.852±0.027 for $^{208}Pb/^{206}Pb$ and $^{207}Pb/^{206}Pb$, respectively), suggesting the slag was the main pollution sources and household dust was the important exposure pathway of children's lead exposure. The study further indicated the slag in the mining and refining activities is the dominate pollution source of lead in children's blood, and highlight the household dust to children's lead exposure.

Keywords: A - exposure measurement, A-biomarkers, B-metals, D-children

SU-PO-40

Personal exposure to black carbon among women in semi-rural Mozambique

A. Curto Tirado[1], D. Donaire-Gonzalez[1], M. Manaca[2], R. Gonzalez[4], C. Sacoor[2], I. Rivas[1], G. Wellenius[3], J. Sunyer[1], C. Menéndez[4], C. Tonne[1]; [1]ISGlobal, Centre for Research in Environmental Epidemiology (CREAL), Barcelona, Spain, [2]Centro de Investigação em Saúde da Manhiça (CISM), Manhiça, Mozambique, [3]Brown University School of Public Health, Providence, RI, [4]ISGlobal, Centre for International Health Research (CRESIB), Hospital Clínic- UB, Barcelona, Spain

Abstract: Background: Sub-Saharan Africa has the highest prevalence of people using solid fuels for household energy, which can result in products of incomplete combustion that are damaging for health. Black carbon (BC) is a useful marker of combustion-related particles; however air quality data and temporal patterns of personal exposure to BC in sub-Saharan Africa are scarce. We characterized personal exposure to BC in women from a semi-rural area of southern Mozambique. **Methods:** A total of 204 non-smoking women were randomly selected from the demographic surveillance system of the Manhiça Health Research Centre between 2014-15. We measured 24h personal exposure to BC using a portable MicroAeth worn in a waist pouch and ambient elemental carbon (EC, highly comparable to BC) with a centrally-located stationary monitor. We used a questionnaire to obtain information on socio-demographic, household characteristics, and cooking/lighting use. We used backward linear regression to identify predictors of log-transformed 24-h mean BC exposure. **Results:** After excluding participants with poor data quality, we analyzed data from 142 (70%) women with a mean age of 30 years. Participants were largely housekeepers (88%) who principally used firewood (85%) and kerosene (52%) for cooking and lighting, respectively. Mean personal BC was 3.69µg/m^3 (SD ±2.04) and 24h mean ambient EC was 0.85µg/m^3 (SD ±0.59). Cooking fuel used during sampling (firewood vs charcoal) and same-day ambient EC were the determinants of exposure that were kept in the model ($p<0.005$). Plots of BC for all participants combined revealed a peak between 6-7pm, suggesting the contribution of lighting sources since the last cooking episode was 3-5pm. **Conclusions:** This study provides temporal patterns of BC exposure and suggests the important contribution of lighting, in addition to cooking, sources to personal exposure to combustion particles in populations that lack access to clean household energy.

Keywords: A - exposure measurement, A-indoor environment, C-air, A - ambient monitoring

SU-PO-41

NO_2 Air Pollution Exposure Assessment in Urban Mysore, India

A. A. Nori-Sarma[1], R. K. Thimulappa[2], V. G. Venkatareddy[3], A. K. Fauzie[3], S. K. Dey[2], M. L. Bell[1]; [1]Yale University, New Haven, CT, [2]JSS Medical College, Mysore, India, [3]University of Mysore, Mysore, India

Abstract: *Introduction*: In India, rapid urbanization has lead to a decrease in urban air quality. Few ongoing studies apply spatially heterogeneous sampling to assess ambient air pollution levels in urban India. Understanding air pollution exposure differentials is particularly important in cities, where socio-economic disparity by neighborhood may increase health impacts of varying exposures. *Methods*: Seasonal sampling campaigns have been conducted at 150 sampling sites throughout urban Mysore, Karnataka, India. Sampling sites were selected using a 3-step combination of purposeful site selection, systematic random sampling, and purposeful site selection to fill in remaining gaps of exposure measurement. Nitrogen dioxide (NO_2) levels were assessed using passive Palmes tube and Ogawa badge technologies at these sites, to develop a spatial interpolation of ambient air pollution exposure in these cities. *Results*: Mysore, considered a "clean city" of India, has minimum air pollution levels during the post-monsoon season in excess of 20 ppb averaged over a 2-week period. Maximum air pollution levels measured in the city around 50-60ppb over a 2-week period, indicating that pollution levels may exceed the health-protective thresholds established by the World Health Organization in highly populated urban areas. Strong spatial and temporal patterns indicate a seasonal trend in air pollution levels, as well as the importance of point sources of pollution. *Conclusions*: While traffic pollution has been indicated in the past as a major contributor to ambient air pollution levels in urbanizing centers of Asia, our results indicate that air pollution levels do not follow major roadway patterns. Other pollution sources (e.g., building density), as well as highly localized environmental factors (such as waste burning and small trash fires), might be critically influencing air pollution exposure patterns in urban India.

Keywords: A - ambient monitoring, A - exposure measurement, A-exposure models, A-geospatial analysis/GIS, A-sampling methods

SU-PO-42

Mercury Co-Benefits of Climate Policy in China

K. Mulvaney[1], N. E. Selin[2], A. Giang[3], S. Y. Kwon[3], M. Muntean[4], M. Li[5], C. Li[6], D. Zhang[7], V. J. Karplus[8]; [1]University of North Carolina at Chapel Hill, Chapel Hill, NC, [2]Massachusetts Institue of Technology, Cambridge, MA, [3]Massachusetts Institue of Technology, Cambridge, MA, [4]European Commission, Joint Research Centre, Institute for Environment and Sustainability, Ispra, Italy, [5]Massachusetts Institute of Technology, Cambridge, MA, [6]Massachusetts Institute of Technology, Cambridge, MA, [7]Massachusetts Institute of Technology and Tsinghua University, Cambridge, MA, [8]Massachusetts Institute of Technology, Cambridge, MA

Abstract: National commitments on the Minamata Convention on Mercury interact with other global environmental objectives, such as the Paris Agreement on climate change. Understanding the interactions between mercury and other policy interventions can help decision-makers identify more effective policies that can address multiple environmental issues simultaneously. Here, we examine how China's national climate policy focused on decarbonization affects the country's ability to meet its environmental goals related to mercury. To address environmental goals, we assess how mercury co-benefits (positive side effects that are peripheral to a policy's main goal) of a national climate policy in China could contribute to the country's commitments under the Minamata Convention. We examine climate policy scenarios in 2030 corresponding to various levels of carbon intensity reductions in addition to a business-as-usual scenario and a scenario that implements end-of-pipe controls aligned with the requirements of the Minamata Convention. Economic analysis from a computable general equilibrium model of China's economy (C-REM) provides information on changes in economic activity resulting from the climate policy scenarios. Using this economic data, we scale 2007 mercury emissions from the Emissions Database for Global Atmospheric Research (EDGAR) in a variety of sectors to 2030. We then use a global atmospheric transport model (GEOS-Chem) to project changes in mercury deposition at the regional scale in China for each policy scenario, and evaluate the resulting spatial distribution of mercury co-benefits. We find that mercury co-benefits of climate policy can lead to reductions in emissions comparable to end-of-pipe controls. We also find that regions exhibiting the greatest reductions in emissions across the policy scenarios also exhibit the greatest reductions in total mercury deposition.

Keywords: A - population exposure, A-climate change, A-environmental policy, A-geospatial analysis/GIS, B-metals

SU-PO-43

Can exposure surfaces and GPS data predict personal exposures to air pollution and noise? Findings from a panel study

L. Minet[1], R. Liu[1], M. Valois[2], J. Xu[1], M. Shekarrizfard[1], S. Weichenthal[2], M. Hatzopoulou[1]; [1]University of Toronto, Toronto, Canada, [2]McGill University, Montreal, Canada

Abstract: Exposure surfaces developed from land-use regression (LUR) models are commonly used to quantify the risks of various air pollutants and noise on public health. Typically, exposures are associated with the home location of participants in a cohort. While a few studies have begun to incorporate data on the daily mobility to refine exposure estimates, often, the "mobility-based" exposures are not validated against personal data. During summer 2016, we conducted a panel study in the city of Toronto, Canada, which involved 46 participants who participated on two days. Participants were provided with instruments measuring levels of Ultrafine Particles (UFP), Black Carbon (BC) and noise, as well as a GPS. Over a 6-hours duration, they were free to pursue their daily activities with the constraint of walking for 2 hours outdoors. During the same summer a data collection campaign took place whereby the levels of UFP, BC and noise were measured in Toronto with the same devices via two different protocols. A mobile monitoring campaign using bicycles enabled the coverage of 3,895 unique road segments visited 5 to 6 times. In addition, 92 fixed points were also sampled 5 to 6 times. These simultaneous monitoring campaigns enabled the development of LUR models and associated exposure surfaces. The GPS coordinates of all participants were synchronised with the air pollution and noise data to generate estimates of personal exposures. These estimates were analyzed in terms of spatial variability throughout the city, and differences between indoor and outdoor levels. In addition, GPS data were intersected with the surfaces to derive "mobility-based" exposures. We are currently comparing the personal and "mobility-based" exposure estimates and identifying locations, and trajectories with strong and poor agreement between the two measures. We are also developing models that can correct "mobility-based" exposures using information on daily levels measured at a central location.

Keywords: A - exposure measurement, A-activity patterns, A - population exposure, A-exposure models, A-epidemiology

SU-PO-44

A New Approach for Deriving Numerical Relationships Between Different Elongate Mineral Particles (EMPs) Definitions

Y. Shao[1], G. Ramachandran[2]; [1]University of Minnesota, Minneapolis, MN, [2]Johns Hopkins University, Baltimore, MD

Abstract: Different dimensions of elongate mineral particles (EMPs) have been proposed as being relevant to respiratory health end-points but industriies still monitor the NIOSH-regulated EMP only.The goal of this study is to develop a mathematical approach for deriving numerical conversion factors between these exposure metrics for each mine-similar exposure group (SEG) combination of the Minnesota Taconite Worker Health Study. These factors could allow us to create job exposure matrices (JEMs) for any EMP metric from our existing NIOSH EMP data. There are a total of 162 mine-SEG combinations in our study, and in each combination, we assume that the overall EMP dimensions (lengths and widths) follow a bivariate lognormal distribution. A Bayesian approach was used to predict these distributions based on priors for EMP dimensions derived from a limited number (N=92) of area samples that were then updated using the SEG-specific personal EMP measurements (1267 personal samples). All EMP samples were analyzed using ISO-TEM methods. Once the EMP size distribution was ready, each of the selected EMP definitions ('NIOSH EMP', 'Chatfield asbestiform EMP', 'Chatfield non-asbestiform EMP' and 'Suzuki EMP') were then applied to the bivariate lognormal distribution, and the volume under the curve corresponding to the length and width for that definition is the percentage of the total number of EMPs for that definition. We then derived numerical conversion factors between any two EMP definitions, which would be the ratios of the volumes under the bivariate posterior EMP distribution for those two definitions. There are a total of 2,791 EMPs identified from the area samples, and 11,190 EMPs identified from the personal samples. For the final predicted conversion factors, the median [IQR] of the conversion factors of "Chatfield asbestiform EMP", "Suzuki EMP", and "Chatfield non-asbestiform EMP" to the "NIOSH EMP" were 0.38 [0.10-1.17], 8.59 [5.05-20.15], and 13.6 [8.3-25.4] respectively.

Keywords: A - exposure measurement, A-statistical methods, D-occupational, A-industrial hygiene, A-exposure models

SU-PO-45

A novel air quality model fusion method for improving spatial resolution (250m) of ambient air pollutant exposure

J. Bates[1], A. F. Pennington[2], X. Zhai[1], M. Friberg[1], F. Metcalf[1], M. Strickland[3], L. Darrow[3], J. Mulholland[1], A. Russell[1]; [1]Georgia Institute of Technology, Atlanta, GA, [2]Emory University, Atlanta, GA, [3]University of Nevada, Reno, Reno, NV

Abstract: Improving spatial resolution of air pollutant concentration estimates is critical for epidemiologic studies with spatially resolved health data. Air quality exposure assessment is often limited by spatially sparse monitoring data or model estimates that are either spatially coarse (chemical transport models) or lack chemistry (dispersion models). Therefore, two novel model fusion methods were developed to create spatially resolved, chemically comprehensive air pollutant concentration estimates, one for gaseous pollutants and one for particulate matter. The methods use computationally efficient linear combinations and mass conservative bilinear interpolation to fuse model results at two scales: the chemical transport model CMAQ at a 12km resolution, which accounts for numerous emission sources and complex chemistry, and the line dispersion model RLINE at a 250m resolution, which simulates the dispersion of inert, primary roadway pollutants. These methods are applied to Atlanta, GA to obtain daily 24-hr averaged $PM_{2.5}$ and 1-hour maximum CO and NOx from 2002-2011 at a 250m grid resolution, but these methods can be applied at different spatial scales and for other pollutants. Resulting estimates are spatially and temporally resolved air pollutant concentrations that include contributions from secondary processes and primary emissions from a comprehensive set of regional and local sources. Fused results show an improvement in both spatial R and temporal R^2 compared to CMAQ and RLINE, providing more accurate concentration estimates for improved exposure assessment. Results also capture steep spatial gradients in pollutant concentration near roadways, reducing exposure error in intraurban studies on the health effects of vehicle emissions. Capturing steep spatial gradients near roadways while also simulating effects of regional sources that can influence daily variation in pollutant concentrations can potentially improve exposure estimates in epidemiologic studies.

Keywords: C-air, B-particulate matter

SU-PO-46

Health risks of heavy metals from drinking water exposure near a typical river basin area in China, after pollution control measurement

X. Zhao[1], S. Cao[1], X. Duan[2], B. Zheng[1], Y. Ma[1]; [1]Chinese Research Academy of Environmental Sciences, Beijing, China, [2]University of Science and Technology Beijing, Beijing, China

Abstract: The metal pollution still is a great concern due to the effects from urbanization and industrialization. While, the health risks from the toxic metal could decrease if strict pollution control measures were adopted. However, few studies to date investigate the health risks of heavy metal in a systematic river basin for the dependent residents, after taking pollution control measures. Thus, the contents of metal(loid)s (Cu, Pb, Zn, Cd, Mn, As) in surface water along a typical river basin were investigated in this study, and the potential non-carcinogenic and carcinogenic health risks posed to the residents were assessed. Although existing mining activity in the upstream, the soluble concentrations of all the metal were within the relevant thresholds. However, the closer to the mining area, the higher the pollution levels of metal. The total hazard index for non-carcinogenic risks of metal were far lower than the threshold (1) for the local population. Whereas, although the content of metal were low (such as As and Cd), they could pose relative higher non-carcinogenic health risks. The result illustrated that pollution levels, toxicity of the contaminants and exposure behavior patterns all could contribute to the potential detrimental health risks. Additionally, the non-carcinogenic and carcinogenic risks from ingestion exposure were ~2- ~4 orders of magnitude higher than those from dermal contact. The total carcinogenic risks were ~2- ~4 orders of magnitude lower than the maximum tolerable levels (1.0×10^{-4}), indicating carcinogenic risks could also be neglected. Among different population groups, heavy metal posed relative higher non-carcinogenic and carcinogenic risks to the children in 0~5 years old. Fortunately, the surface water in this basin is safe in usage for the local population and the health risks were acceptable in case exposed to the target metal, after the river basin was in the charge of strict pollution control measures.

Keywords: A - exposure measurement, A - population exposure, A-environmental policy, C-water

SU-PO-47

Reconstructing Historical Exposures to Respirable Dust and Respirable Silica in the Taconite Mining Industry for 1956-2010

Y. Shao[1], G. Ramachandran[2]; [1]University of Minnesota, Minneapolis, MN, [2]Johns Hopkins University, Baltimore, MD

Abstract: As part of ongoing epidemiological studies for assessing the association between exposure to taconite dusts and the development of respiratory diseases, the goal of our study is to reconstruct the historical silica (RS) and respirable dust (RD) exposures of workers in Minnesota taconite industry from 1955-2010 by developing a job exposure matrix (JEM) that uses 9,128 RS and 19,408 RD industrial hygiene monitoring data. Historical RS and RD data were obtained for seven taconite mines from three sources:(1) the Mine Safety and Health Administration (MSHA) has online database records for all inspection results since 1978 for 13 MSHA Mine IDs with 4,303 RD monitoring records;(2) the mining companies' internal monitoring reports provided 14,417 RD records, most of which date from the late 1970s.(3) University of Minnesota in 2010 conducted 688 pairs of RS and RD measurements covering six active mines. Unlike RD data which can be directly read from IH reports, many historical RS data have to be calculated using RD and silica percentage information. 8,840 RS data were calculated using available silica percentage information. After data treatment, all these data were grouped into 7 mines and then into 8 departments. Within each department, we applied a two-level random-intercept model which assumes that the natural log of Y (RD or RS) changes over time at a constant rate. Among the 56 mine-department combinations, 13 combinations show a significant (p-value<0.05) decreasing RS and RD trend with the maximum rate of decrease of 3.3% per year;12 combinations show a significant increasing trend with the maximum annual increase rate of 8.7%. The estimated percent silica varies by department, with the maximum value usually occurring in the crushing department (8.5%-27.6%) and the minimum value occurring in the pelletizing department (2.6%-8.5%).The result of this study is a JEM by mine, department, and year for RD and RS that will form the basis for future epidemiological studies.

Keywords: A-exposure models, A-cumulative exposure, D-occupational, B-particulate matter, A - population exposure

SU-PO-48

Applications of Passive Silicone Wristband Samplers: Childhood Para-Occupational Exposures to Pesticide Mixtures

C. Poutasse[1], T. Arcury[2], S. Quandt[3], P. Laurienti[4], K. A. Anderson[1]; [1]Oregon State University, Corvallis, OR, [2]Wake Forest School of Medicine, Winston-Salem, NC, [3]Wake Forest School of Medicine, Winston-Salem, NC, [4]Wake Forest School of Medicine, Winston-Salem, NC

Abstract: Agricultural workers are exposed to a variety of pesticides at work and may put household members at risk through para-occupational exposures. However, self-reporting and biomonitoring samples may not fully capture total pesticide exposure. For a complementary assessment of external chemical exposure, silicone wristbands were used as passive sampling devices (PSDs) to sample complex environmental mixtures. In this pilot study, 20 children wore PSDs, which were subsequently analyzed for 72 pesticides by gas chromatography (GC) electron capture detection (ECD). It was hypothesized that farmworker household children (FHC) experience different pesticide exposures compared to non-farmworker household children (NFHC), which could lead to distinguishable neurological differences between study groups. FHC wristbands detected twice as many pesticides as the NFHC wristbands, and only FHC wristbands detected agricultural organophosphate pesticides. These preliminary data suggest an association between FHC group membership and an increased number of pesticide detections from para-occupational exposures. Future directions include investigating pesticide exposures as related to the agricultural season, incorporating data on participant neurological development, and expanding the study by increasing the sample size.

Keywords: B-pesticides, D-children, A - exposure measurement, A-epidemiology, A-sampling methods

SU-PO-49

Relationship between Reactive Oxygen Species (ROS) activity and Cytotoxicity of Ambient Particles

Y. Wang[1], M. Plewa[2], V. Verma[1]; [1]University of Illinois at Urbana-Champaign, Urbana, IL, [2]University of Illinois at Urbana-Champaign, Urbana, IL

Abstract: Associations of particulate matter (PM) concentration with cardiopulmonary disease and mortality have been reported in several studies. Threat of PM to public health is likely due to reactive oxygen species (ROS) generation by PM. However, only a handful of studies could demonstrate a consistent correlation between biological toxicity endpoints and ROS formation measured in acellular assays. In this study, we targeted to investigate these associations by measuring the acellular ROS generation and cytotoxicity of ambient particulate matter (PM) samples collected from an urban site. The cytotoxicity of water-soluble PM is assessed by exposing aerosol extracts to Chinese hamster ovary (CHO) cells. Two methods were used to assess the ROS generation ability - 1. dithiothreitol (DTT) loss (conventional DTT assay); and 2. hydroxyl radical generation through redox cycling in the DTT system. Inhibition of CHO cells by soluble PM was observed, and half maximal inhibitory concentrations (IC_{50}) were calculated for the PM samples, which range from 43.6% to 74.0% (percentage of soluble PM extract). Interestingly, we found strong and significant correlation between ROS generation in the DTT system and IC_{50} ($R^2 = 0.60$, $p<0.05$). To further understand the chemical components, which might drive the response of both DTT and cytotoxicity assays, we measured the concentrations of metals and water-soluble organic carbon (WSOC) in PM extracts. Fe, Cu, Zn and WSOC were significantly correlated with both IC_{50} and ROS generation in DTT assay. Further chemical analysis of the collected samples in underway and would be presented in this talk. However, Our preliminary results indicate that measuring the ROS generation from PM could be used as a biologically relevant metric to assess the toxicity and health impacts from ambient aerosol pollution.

Keywords: C-air, A - exposure measurement, B-metals

SU-PO-50

The Pregnancy Exposome: Multiple Environmental Exposures in the Iranian Environment and Neurodevelopmental Disorders (TEND) Birth Cohort

M. Zare Jeddi[1], M. Shamsipour[1], M. Yunesian[1], Y. Bruinen de Bruin[2], S. Hassanvand[1]; [1]Institute for Environmental Research (IER), Tehran University of Medical Sciences, Tehran, Iran (the Islamic Republic of), [2]Joint Research Centre of the European Commission, Ispra, Italy

Withdrawn

SU-PO-51

STATISTICAL FUSION OF PRESENT-DAY OBSERVED GLOBAL OZONE CONCENTRATIONS AND CCMI-1 MULTI-MODEL SURFACE OZONE ESTIMATES

M. Buechlein, M. Serre, J. West, Y. Zhang; UNC, Chapel Hill, NC

Abstract: With over 9000 surface monitoring stations globally, the Tropospheric Ozone Assessment Report (TOAR) has compiled the largest database of global ozone observations using metrics relevant for human health impacts, vegetation impacts, and climate impacts. Despite the effort to assemble a comprehensive global ozone observation database, the TOAR database still contains large regions with sparse data. Chemistry-climate models (CCMs) and chemical-transport models (CTMs) can provide estimates of air pollution concentration in regions where observed concentrations are sparse. Phase 1 of the Chemistry-Climate Model Initiative (CCMI-1) includes 20 CCMs and CTMs that use simulated meteorological conditions and chemistry to estimate air pollutant concentrations globally. We conducted the first global Bayesian Maximum Entropy (BME) data fusion of ozone observations and CCMI-1 model predictions. We used a Constant Air Quality Model Performance (CAMP) evaluation of the CCMI-1 model outputs to account for nonlinear and heteroscedastic performance of global models and to correct for consistent bias within the models. The observations and CAMP corrected global model outputs were then integrated in the BME framework to provide a comprehensive and robust global estimate of "present-day" surface ozone that can be used to estimate the global health burden of ambient ozone.

Keywords: B - ozone, A-geospatial analysis/GIS, A-statistical methods

SU-PO-52

Citywide validation and improvement of the MAIAC aerosol product in Lima, Peru

J. Bi[1], B. Vu[1], A. Wildani[2], Y. Wang[3], A. Lyapustin[3], Y. Liu[1]; [1]Emory University, Atlanta, GA, [2]Emory University, Atlanta, GA, [3]National Aeronautics and Space Administration, Greenbelt, MD

Abstract: Background: Satellite-observed aerosol optical depth (AOD) have been widely applied to estimate the ground-level concentrations of fine particulate matters (PM2.5). One of the state-of-the-art AOD products is the Multi-Angle Implementation of Atmospheric Correction (MAIAC). While MAIAC AOD has been shown in North America and Europe to be a robust parameter of ground-level PM2.5 estimation, its quality in other parts of the world has not been extensively evaluated. **Objectives:** We conducted a validation analysis on Terra and Aqua MAIAC AOD in Peru and developed a gap-filling model to improve its coverage in Lima, Peru. **Methods:** We validated the accuracy of MAIAC AOD in Lima by comparing it with the AOD measurements from the aerosol robotic network (AERONET). A moving average (MA) model was then applied to impute the missing AOD observations in order to enhance the dataset's spatial coverage and quality. **Results:** The proportion of available MAIAC AOD data in Lima was ~20%. MAIAC AOD had a moderate correlation ($r = 0.60$) with the AERONET measurements. After applying an MA model with fixed window sizes (5 km for the spatial coverage and 7 days for the time span) on the time series of MAIAC AOD, the proportion of available data raised to 31% with a slightly decreased correlation ($r = 0.53$). **Conclusions:** MAIAC AOD has a reasonable but still limited quality in Lima in terms of its accuracy and spatial coverage. Improvements in the MA model by introducing flexible window sizes and new gap-filling strategies and the application of machine learning techniques are expected to further increase the quality of the dataset in Lima.

Keywords: A - exposure measurement, B-particulate matter, C-air

SU-PO-53

Harmonization of Sensor Metadata and Measurements to Support Exposomic Research

N. Burnett[1], R. Gouripeddi[2], J. Wen[2], P. Mo[3], R. Madsen[3], R. Butcher[3], K. A. Sward[4], J. Facelli[2]; [1]University of Utah, Salt Lake City, UT, [2]University of Utah, Salt Lake City, UT, [3]University of Utah, Salt Lake City, UT, [4]University of Utah, Salt Lake City, UT

Abstract: The concept of the exposome includes assessing exposures to different environmental factors and understanding endogenous processes within the body. Quantifying exposures and their effects therefore requires sensors that measure the general and personal environments for different physical, chemical and biological species. Using data from these different sensors requires them to be harmonized into a common data representation for ease of use in different data analytic approaches, and for integration for generation as comprehensive, high-resolution spatio-temporal profiles of exposomes. Towards developing such a common representation, we started with modeling air quality (AQ) data which has well-documented effects on health. We performed a literature review using PubMed with the search criterion "Pediatric Asthma Sensor Studies'. We manually extracted a list of metadata elements describing AQ sensors from the literature, and developed a first draft of a conceptual data model (CDM). Next we collected sample data from different research studies using different types of sensors including Environmental Protection Agency monitors, personal, in-home, outdoor and citizen networks. We evaluated the CDM with the collected and revised it with fields found in the data. We then met with AQ experts in Utah and The Pediatric Research using Integrated Sensor Monitoring Systems (PRISMS) group to review the model and modified it further based on their inputs. We now have a complete CDM that accommodate different types of sensors and is available for community review at goo.gl/DgvzGz. This model consists of Instrument, Deployment and Measurement Output domains that describe characteristics of a sensor device, its deployment in research settings, and its measurements respectively. We have deployed this model with different databases harmonizing vast amounts of air quality data from different sensors. This model will now be used in the PRISMS project to study and effects of AQ on health.

Keywords: C-air,

SU-PO-54

A Novel Method for Characterizing Resident Behaviors and Housing Attributes using Photo Survey

P. Fabian, Z. Petropoulos, C. Scollaert, M. Scammell, J. I. Levy; Boston University School of Public Health, Boston, MA

Abstract: Exposure to air pollutants such as $PM_{2.5}$ is associated with a variety of adverse health outcomes including cardiovascular and respiratory diseases, preterm birth, and low birth weight. Accurately capturing an individual's exposure to ambient air pollutants requires focusing on the indoor environment, as this is where the average person spends the majority of their day. The impact of resident behaviors (i.e. window opening or air conditioning use) on infiltration of air pollutants into a home has been well-characterized. However, efforts to accurately predict these behaviors and their distribution (through field campaigns and mail-in surveys) have many limitations, including the ability to characterize these behaviors across entire communities. Using the ArcGIS Photo Survey tool and a GPS-enabled camera, we collected photos of homes in a community in Massachusetts, and created an online survey which was used to classify open windows and installation of window air conditioning units via crowdsourcing. Photos are also linked to parcel-level housing data and neighborhood-level sociodemographic data which will be used to develop predictive models of these behaviors. In our winter 2016-17 survey, we captured photos of 1,100 homes and classified the necessary parameters using crowdsourcing for image classification. These data will be used to identify the significant predictors—including housing characteristics, sociodemographic variables, and meteorological parameters—of window-opening and ownership of window air conditioning units. This approach to data collection can be used to concurrently capture information relevant to community stakeholders—which may include data on recycling and trash bins, trees, blighted or damaged homes, and graffiti—and can be implemented in any community to collect a wide variety of information. Photo survey techniques can complement parcel data and other geospatial information to enhance residential exposure assessment.

Keywords: A-behavior, A-activity patterns, A-built environment, A-indoor environment, C-air

SU-PO-55

Assessing Indoor $PM_{2.5}$ Concentrations in Households on the Hopi Reservation

M. K. O'Rourke[1], S. Hadeed[1], G. Honanie[2], A. Mahkewa[2], M. Paukgana[2], C. Kelley[1], M. A. Alshammari[1], R. Canales[1], R. Harris[1]; [1]The University of Arizona, Tucson, AZ, [2]The Hopi Tribe, Kykotsmovi, AZ

Abstract: Introduction: A significant proportion of the Hopi Tribe burn coal, wood and other organic combustible materials for heating, lighting and cooking in homes. An estimated 37% and 33% of Hopi household report burning coal and wood for heating, respectively. The Hopi Tribe has expressed their concerns about indoor air quality from the combustion of these fuels. The purpose of this pilot study is to measure indoor $PM_{2.5}$ concentrations in homes and assess the effects of fuel type and housing structure on indoor levels. Method: Pilot data was collected during the 2017 heating season in four households on the Hopi Reservation. Indoor concentration of $PM_{2.5}$ was measured over a 24-hour period using real time monitors (pDR-1500) set at 1-minute logging intervals. Subsequent collection of primary fuel type and housing characteristics was recorded by field team members at the time monitors were placed indoors. Average $PM_{2.5}$ concentration was calculated by fuel and housing type. Results: In the four households sampled to date, the indoor mean (SD) $PM_{2.5}$ concentration was 18.7 (37.3) μg/m^3. Two modern households that used a combination of coal, wood and electricity for fuel had an indoor mean $PM_{2.5}$

concentration of 17.9 (36.0) μg/m^3 and 1-minute peak concentration of 332.8 μg/m^3. Average $PM_{2.5}$ concentrations for a hybrid modern-traditional home with coal and wood fuel was 33.0 (50.6) μg/m^3 and 6.1 (6.8) μg/m^3 for an electrically powered mobile home. Discussion: From this pilot study, households using coal and wood as primary fuel sources were found to have elevated indoor $PM_{2.5}$ concentrations compared to electrically powered homes. These initial results are similar to previous studies that found increased exposures and indoor $PM_{2.5}$ concentrations in households using combustible fuels as a primary fuel source. This project will continue to assess indoor air quality of households on the Hopi Reservation by various fuel sources and housing types during heating and non-heating seasons.

Keywords: B-particulate matter, A-indoor environment

SU-PO-56

Distributions of real time black carbon concentration and its association to PM2.5 concentration at an urban hotspot in Seoul, South Korea : Preliminary study results

S. Kim[1], S. Lee[1], H. KIM[1], J. Lee[1], S. Park[1], J. Park[2], B. Park[3]; [1]Soonchunhyang University, Asan, Korea (the Republic of), [2]KOREA TECH, Cheonan, Korea (the Republic of), [3]Korea Environment Corporation, Incheon, Korea (the Republic of)

Abstract: Among the many airborne contaminants, the black carbon (BC) has been recently identified as a carcinogenic compounds, and new scientific evidence has led to a recognition of its representativeness as indicator of exposure to combustion PM, i.e., diesel traffic sources and the significant role of BC as one of the short-lived climate forcers. This study conducted to obtained distributions of BC concentrations and its association to PM.2.5 concentration at an urban hotspot of Seoul, South Korea. Real-time PM2.5 measurement was conducted using real-time portable monitors of Sidepak AM510 personal Aerosol Monitor (TSI Inc., St. Paul, MN) then results were compared with those obtained from national monitoring site. Comparative measurement of BC concentrations was conducted using single wave length (SWL) light absorption devices and multi channel wave length (MWL) light absorption devices. Descriptive analyses were conducted for continuous variables of hourly mean values of BC and PM2.5. Associations between BC and PM2.5 concentrations in different sampling time (morning, afternoon and evening) were evaluated with multivariate regression models after adjusting for temperature, relative humidity, wind direction and wind speed. The SAS statistical package (version 21.0) was used for statistical analyses. BC concentration increased 7.5 % by unit increase of PM2.5 concentration in morning and evening rush hour period of the weekend whereas we obtained 5.5% increase in morning rush hour period of weekdays ($p<0.05$). Degree of the increase rate was different depending on sampling time of weekdays or weekend. Our study provides compelling evidence supporting necessity of immediate monitoring of BC concentration at urban areas for emission source control and management as BC is known as a better indicator of harmful particulate substances from combustion sources (i.e., traffic or power plant) than undifferentiated PM mass.

Keywords: A - ambient monitoring, B-particulate matter, A - population exposure, C-air

SU-PO-57

Characterization of Organophosphate Pesticides in Urine and Home Environment Dust in an Agricultural Community

C. Tamaro[1], M. Smith[1], T. Workman[1], W. Griffith[1], B. Thompson[2], E. Faustman[1]; [1]University of Washington, Seattle, WA, [2]Cancer Prevention Program, Fred Hutchinson Cancer Research Center, Seattle, WA

Abstract: Organophosphorus insecticides (OPs) are used on agricultural crops to control pests. Farmworkers (FW) have higher exposure to OPs than non-farmworkers (NFW), and FW children may have higher exposure than NFW children. A community-based participatory research strategy was used in the lower Yakima Valley of Washington state to identify 100 FW and 100 NFW adults, each with a referent child 2-6 years old. FW adults worked as thinners or harvesters in pome fruit orchards. Parents and children participated in three data collection periods over the course of one year. Urine samples were evaluated for the dialkylphosphate metabolites (DAP) characteristic of OP exposure, and dust from homes and vehicles was evaluated for intact OP residues. Geometric mean (GM) concentrations of OPs in house and vehicle dust were higher in FW households than NFW households across all three agricultural seasons. GM concentration of urinary DAPs was higher for children in FW households than NFW households. We found significant seasonal and occupational associations between urinary DAPs and intact OP residues in house dust among this cohort of FW and NFW adults and children. We found higher concentrations of intact OP residues in FW house dust and higher concentrations of OP metabolites in FW family urine. Results of our regression analysis showed a significant positive association between the concentrations of dimethyl OPs in household dust and dimethyl OP metabolites in urine, most notably during thinning season but also during harvest season when OPs are applied less frequently. These results provide support for the occupational take-home pathway of exposure by which children are exposed to agricultural pesticides used in the workplace. This project was made possible by The University of Washington Center for Child Environmental Health Risks Research supported by grant PO1 ES009601 from NIEHS, grants RD826886, RD83451401 and RD-83273301 from EPA, and NIH contract HHSN267200700023C.

Keywords: A - exposure measurement, B-pesticides, C-indoor, D-children, D-occupational

SU-PO-58

Modeling Historic Air Pollution Concentrations with Land Use Regression in Tucson, AZ

N. Lothrop, M. Bell, H. Brown, M. Furlong, S. Guerra, M. K. O'Rourke, P. I. Beamer; University of Arizona, Tucson, AZ

Abstract: Numerous studies have linked air pollution exposure to respiratory health problems. However, air pollutant sampling is seldom completed at participant homes in studies investigating respiratory health outcomes due to costs, resulting in spatially limited data on air pollution exposures. When no or limited air pollution measurements are taken, land use regression (LUR) modelling is a common method for predicting air pollution levels. Using LUR, we retrospectively predicted outdoor annual average traffic-related air pollution (nitrogen dioxide or NO_2) levels, which were measured with Palmes diffusion tubes at 146 sites in the Tucson, AZ area in 1987. Multiple samples collected at each site were corrected for seasonal variation to obtain the annual average outcome (mean=15.5 ppb, std. deviation=5.04 ppb). Predictors were selected from the European Study for Cohorts for Air Pollution Effects (ESCAPE) protocols (e.g., roadways, population density, land uses), with Tucson-specific variables added (e.g., surface mines). We tested the efficacy of the ESCAPE and a Vancouver, Canada model building approaches, both of which use supervised forward stepwise variable selection procedures. The Vancouver approach performed marginally better (increased R^2 leave-one-out cross validation) than the ESCAPE approach (R^2 = 0.61 vs 0.58, 8 vs 5 predictors). Regardless of approach, several predictors were common, including: elevation, commercial land use, nearby (major) roadway length, and distance to the nearest active surface mine. The relatively large number of samples taken in 1987, the spatial representation of the data set, and the historic nature of the outdoor NO_2 data make this study a unique opportunity. Predicted outdoor NO_2 concentrations will be matched to study participant homes for various respiratory disease cohorts with almost 40 years of health data from Tucson, AZ to better understand the relationship between traffic-related air pollution and respiratory health outcomes.

Keywords: D-children, A-exposure models, A-geospatial analysis/GIS, A-built environment, D-Southwest-specific

SU-PO-59

Integrating Spatiotemporal Information System Approaches with Agent-Based Modeling for Studies of Human Exposures to Traffic Related Air Pollution (TRAP)

Z. Mi, R. Laumbach, H. Kipen, P. Georgopoulos; Environmental and Occupational Health Sciences Institute, Rutgers University, Piscataway, NJ

Abstract: Exposure to Traffic Related Air Pollution (TRAP) during commuting has been associated with various adverse cardiovascular and respiratory effects. A prototype Spatiotemporal Information System (STIS) was designed and implemented to support analysis of information collected through the Rutgers Commuter Community Cohort (RC3) study, providing data management, analytics, visualization, and screening exposure modeling capabilities. Using information collected via the RC3 study (i.e. spatiotemporal commuting trajectories and associated measured personal in-vehicle concentration profiles), the STIS was designed and implemented for direct integration with a comprehensive multiscale environmental and microenvironmental quality and human exposure modeling system, under continuing development at the Computational Chemodynamics Laboratory (CCL) of EOHSI. This integration can be used to inform and enhance the design of future exposure-relevant studies, while providing access to the field information collected via the RC3 study. Furthermore, an Agent-Based Model of Human Exposures has been under development as an option within the PRoTEGE (Prioritization and Ranking of Toxic Exposures with GIS Extension) modeling framework of CCL, to support further analyses of commuting behavior and of associated exposures, under both real and hypothetical scenarios that extend beyond the situations assessed and data collected via the RC3 study. The ABM presented here is being built on the REPAST (Recursive Porous Agent Simulation Toolkit) modeling platform (customizable via Python 2.7 scripts) and is used to simulate traveling patterns and behaviors of "virtual subjects" (represented as "agents" in the model) that allow the study of situations involving altered traffic patterns that reflect conditions corresponding to alternative planning options considered in urban and regional development plans.

Keywords: A-exposure models, A-geospatial analysis/GIS, A-risk assessment, C-air

SU-PO-60

Assessing Indoor $PM_{2.5}$ Concentrations in Households on the Hopi Reservation

S. Hadeed[1], M. K. O'Rourke[1], G. Honanie[2], A. Mahkewa[2], M. Paukgana[2], C. Kelley[1], M. A. Alshammari[1], R. Canales[1], R. Harris[1]; [1]The University of Arizona, Tucson, AZ, [2]The Hopi Tribe, Kykotsmovi, AZ

Abstract: Introduction: A significant proportion of the Hopi Tribe burn coal, wood and other organic combustible materials for heating, lighting and cooking in homes. An estimated 37% and 33% of Hopi household report burning coal and wood for heating, respectively. The Hopi Tribe has expressed their concerns about indoor air quality from the combustion of these fuels. The purpose of this pilot study is to measure indoor $PM_{2.5}$ concentrations in homes and assess the effects of fuel type and housing structure on indoor levels. Method: Pilot data was collected during the 2017 heating season in four households on the Hopi Reservation. Indoor concentration of $PM_{2.5}$ was measured over a 24-hour period using real time monitors (pDR-1500) set at 1-minute logging intervals. Subsequent collection of primary fuel type and housing characteristics was recorded by field team members at the time monitors were placed indoors. Average $PM_{2.5}$ concentration was calculated by fuel and housing type. Results: In the four households sampled to date, the indoor mean (SD) $PM_{2.5}$ concentration was 18.7 (37.3) μg/m^3. Two modern households that used a combination of coal, wood and electricity for fuel had an indoor mean $PM_{2.5}$ concentration of 17.9 (36.0) μg/m^3 and 1-minute peak concentration of 332.8 μg/m^3. Average $PM_{2.5}$ concentrations for a hybrid modern-traditional home with coal and wood fuel was 33.0 (50.6) μg/m^3 and 6.1 (6.8) μg/m^3 for an electrically powered mobile home. Discussion: From this pilot study, households using coal and wood as primary fuel sources were found to have elevated indoor $PM_{2.5}$ concentrations compared to electrically powered homes. These initial results are similar to previous studies that found increased exposures and indoor $PM_{2.5}$ concentrations in households using combustible fuels as a primary fuel source. This project will continue to assess indoor air quality of households on the Hopi Reservation by various fuel sources and housing types during heating and non-heating seasons.

Keywords: A-indoor environment, B-particulate matter

SU-PO-61

Spatial variation of secondary inorganic $PM_{2.5}$ exposure: from exposure magnitude to exposure distance

K. Stylianou, O. Jolliet; University of Michigan School of Public Health, Ann Arbor, MI

Abstract: Secondary $PM_{2.5}$ exposure, particles formed from precursor emissions, is currently poorly characterized. We estimate secondary $PM_{2.5}$ exposure and exposure travel distances due to precursors such as NH_3, NOx, and SO_2 and evaluate the respective spatial variability within the US. We characterize $PM_{2.5}$ exposure due to NH_3, NOx, and SO_2 using intake fractions (iF), the fraction of $PM_{2.5}$ inhaled by population from a unitary precursor emission. We use a source-receptor (S-R) matrix approach based on the InMAP (Intervention Model for Air Pollution), a reduced-complexity air pollution model, to characterize exposure from ground level emissions in >40,000 source locations in the US. To investigate exposure travel distance we estimate *intake travel distance* of x (ITDx), distance from the source at which we reach x% of total cumulative iF, for each source. Results indicate substantial spatial variability for $PM_{2.5}$ exposure due to NH_3 emissions in the US. Estimates of $iF_{PM2.5,NH3}$ range between 0.01 and 41.5 ppm. Sources locations in close proximity to big cities (e.g. NY, LA) result in the highest $PM_{2.5}$ exposure from NH_3 emissions. Similar trends are observed for SO_2 and NOx emissions, with $iF_{PM2.5,SO2}$ and $iF_{PM2.5,NOx}$ estimates spanning between 0.009-3.0 ppm and 0.002-1.3 ppm, respectively. Our analysis produced higher secondary $iF_{PM2.5,NH3}$ estimates with higher variability compared to published estimates. In addition, our analysis suggest that exposure radius is precursor dependent with NH_3 having on average the shortest ($ITD_{50,NH3}$=70 km, 95% CI: 0-450) and SO_2 having on average the largest ($ITD_{50,SO2}$=200 km, 95% CI: 0-740) travel distance. Our results improve the exposure characterization of secondary $PM_{2.5}$, especially from NH_3 emissions. This S-R exposure assessment approach enables us to calculate sector specific (e.g. agriculture, electricity production) average iF at local, regional or nation level that can be used in alternative chemical assessment or life cycle assessment.

Keywords: A-exposure models, A-geospatial analysis/GIS, B-particulate matter, C-air

SU-PO-62

Amyotrophic lateral sclerosis and exposure to diesel exhaust in a Danish cohort

A. S. Dickerson[1], J. Hansen[2], O. Gredal[2], M. Weisskopf[1]; [1]Harvard T.H. Chan School of Public Health, Boston, MA, [2]Danish Cancer Society Research Center, Copenhagen, Denmark

Abstract: Background: Previous studies have suggested relationships between occupational exposures and risk of ALS. Specifically, some manuscripts have reported increased risk of ALS and other motor neuron diseases in occupations commonly exposed to diesel exhaust, such as truck and bus drivers. Additionally, studies have implied that residential exposure to air pollutants may increase risk of ALS. **Objective:** To investigate the association between occupational exposures to diesel exhaust and odds of ALS. **Methods:** 4600 ALS cases were identified from the Danish National Patient Registry with 459,906 sex- and birth year-matched controls from 1982 to 2013. We acquired occupational history since 1964 from the Danish Pension Fund. Diesel exhaust exposures and probability were estimated for each job and incorporated into a job exposure matrix, which was used to calculate cumulative and average exposures for jobs held at least 3 years before diagnosis/index dates. Adjusted odds ratios (aOR) and 95% confidence intervals (CI) for quartiles of cumulative and average exposure were obtained using conditional logistic regression analyses and stratified by sex. **Results:** 1090 (23.70%) cases and 110,988 (24.13%) controls were ever exposed to diesel exhaust at work. No significant results were seen in our investigation of cumulative exposures. However, we did observe a significant increase in ALS for those with the highest quartile for average daily exposure compared to no exposure (aOR = 1.17; 95% CI: 1.02, 1.34). Stratified analyses revealed that this association was only in males (aOR = 1.19; 95% CI: 1.03, 1.38). **Conclusion:** Our study suggests an association between consistently higher exposures to diesel exhaust. These findings support those of previously reported associations between ALS and occupations commonly exposed to diesel exhaust.

Keywords: D-occupational, A-epidemiology, A-cumulative exposure, B-particulate matter, A-workplace

SU-PO-63

Investigation of Arsenic and co-occurring metals near abandoned mine wastes in Cheyenne River, South Dakota (CRST)

C. DeVore, L. Rodriguez Freire, J. Cerrato; University of New Mexico, Albuquerque, NM

Abstract: The contamination of arsenic (As), chromium (Cr) and co-occurring metals (*i.e.*, vanadium and mercury) in surface water and sediment has occurred due to mining legacy conducted across the Western United States near Native American lands. The Cheyenne River Sioux Tribe (CRST) in South Dakota has expressed their concerns for the millions of tons of mine waste released into the Cheyenne River between the 1870's and 1980's from the Homestake Mine, the largest and deepest gold mine in North America. Because of the known health impacts of As, Hg, Cr and other co-occurring heavy metals, we need to investigate their concentration and speciation within stream sediment, stream water and ubiquitous plants (fruit trees, medicinal herbs, burning wood); to determine the extent of heavy metals contamination in areas used by Native American communities exposed to mining legacy waste material. This effort is part of the Center for Native American Environmental Health Equity Research's environmental monitoring core's specific aim to quantify environmental exposures to metals resulting from traditional and cultural practices. The goal of this research is to determine the chemical composition and mobility of As in sediment and surface water near abandoned mine wastes from sites located along the Cheyenne River in South Dakota. Furthermore, this study seeks to examine the uptake and chemical characteristics of these metals and co- constituents in plants. This study serves as a foundation to build an interdisciplinary partnership with tribal community members, and to understand the broader impacts of mining on human health and the environment. Characterizing the spatial distribution of metals in the CRST environment may help to address community concerns about exposure and subsequent risk reduction strategies.

Keywords: B-metals

SU-PO-64

Analyzing NO2 concentration variations from 2005 to 2016 over the atmosphere of Kazakhstan using Satellite data

M. Amouei Torkmahalleh[1], Z. Darynova[1], T. Holloway[2]; [1]Nazarbayev University, Astana, Kazakhstan, [2]University of Wisconsin Madison, Madison, WI

Abstract: NO_x is the major precursor of ozone and particulate matter (PM), which in turn affect the air quality and climate. NO_x concentrations in the tropospheric layer are highly variable in both space and time, reflecting its short chemical lifetime in the atmosphere, and its nonuniform distribution. NO_x is released to the troposphere as a result of anthropogenic activities, such as fossil fuel combustion, fertilizer application, and prescribed burning. Major sources of energy in Kazakhstan include coal, petroleum, biomass and natural gas. Combustion of these fuels is the main source of pollution in the natural environment. It was reported that In 2014, the highest per capita household coal consumption occurred in Poland (165 kgoe/cap), followed by Kazakhstan (157 kgoe/cap), and Mongolia (104 kgoe/cap). Given the rapid increment in ozone antecedent outflows in East Asia, investigating NOx concentration trends over the Central Asia, particularly Kazakhstan is critical. Particularly, so far no study was found in the literature investigating NO_2 concentrations variability in Kazakhstan, where significant amount of NO_2 is produced due to local combustion. The objective of this study was to investigate the temporal evolution of tropospheric NO_2 in three polluted cities of Kazakhstan including Shymkent, Almaty and Ekibastuz from 2005 to 2016 using the data from Ozone Monitoring Instrument (OMI) launched on the NASA Aura satellite in July 2004. The trends are, for the most part, associated with technological changes in energy use, as well as regional regulatory policies. Over the Almaty, NO_2 levels increased significantly from 2013 to 2015 by 30%. NO_2 concentrations over 3 cities and overall Kazakhstan decreased significantly after 2015 due to implementation of Euro 4 standard for automobiles. For all three cities, concentration of NO_2 reached a maximum during winter season and a minimum during summer, between 2005 and 2016.

Keywords: A-climate change, A - population exposure, C-air

SU-PO-65

Use of satellite imagery to identify a target population for recruitment of households within a large rural tribal area.

C. Kelley[1], R. Harris[1], J. Tabor[1], M. K. O'Rourke[1], G. S. Honanie[2], L. Joshweseoma[2], S. Hadeed[1], M. A. Alshammari[1]; [1]University of Arizona, Tucson, AZ, [2]Hopi Tribe, Kykotsmovi, AZ

Abstract: Introduction: Study recruitment is increasingly challenging. In rural and tribal communities, which may lack traditional street addresses. Strategies are needed to identify and randomly select participants. The Hopi Environmental Health Project is seeking recruitment of households representative of housing type, fuel source and village. Hopi lands are >1.5 million acres, with 12 villages. **Methods:** Census block vectors for tribal lands were overlaid onto satellite imagery in Google Earth Pro 7.1. Potential household structures were marked with a geotag to record coordinates and an identifier describing block number, housing type, and region. Metadata were copied to an associated .KLM file and included block number, housing type, region, latitude, and longitude. Recruitment goals were set to be proportional to population size, based on the number of houses identified within village vectors. 'Pins' were randomized by village/region; latitude and longitude were used to generate web-addresses which were used by Hopi to locate dwellings and print maps. **Results:** The process identified 2512 potential 'houses' within 400 block vectors. This number is similar to US Census 2010. Initial recruitment of the first 30 target households identified several issues: some 'houses' were multi-family dwellings, or vacant, outbuildings of an occupied house, public building or a kiva, all of which were ineligible. Recruiters are tracking all households for eligibility; reasons for ineligibility are type of structure, probable fuel type, and response. **Discussion:** Some researchers suggest that a random recruitment process will not be effective on native lands. Our approach provides a baseline list that is "field-truthed" prior to recruitment. Recruitment statistics generated using this approach will provide insight to response rates by region and housing type. The Hopi project provides an opportunity to evaluate the described approach.

Keywords: A-epidemiology, D-Southwest-specific, D-First Nation, C-multimedia, A-geospatial analysis/GIS

SU-PO-66

Daily and diurnal trends in $PM_{10 \& 2.5}$ collected by a TEOM 1405-DF on the Hopi Reservation of Arizona.

M. A. Alshammari[1], S. Hadeed[1], G. S. Honanie[2], A. Mahkewa[2], M. Paukgana[1], M. K. O'Rourke[1]; [1]University of Arizona, Tucson, AZ, [2]Hopi Tribe, Kykotsmovi, AZ

Abstract: Air quality data from Native American reservations in Northern Arizona is limited. The Hopi Tribe is interested in air quality assessment to evaluate the impact of regional coal burning power plants and local household burning of wood and coal on ambient air quality. The combination of regional, local and indoor combustion may influence the self-reported asthma rate (25.5%) experienced by tribal members. As part of our larger study examining health disparities, we have installed a TEOM 1405-DF to monitor regional PM_{10} and $PM_{2.5}$ at the Hopi Mission School in Kykotsmovi, Arizona. The Hopi value knowledge although some have concerns about the impact of air quality information on traditional practices. Following approval of the tribal resolution granting permission for the study, construction of the site began with leveling the ground and pouring the concrete pad in August 2016. In September and October the container, electricity, back-up power and TEOM were placed at the site. Operation of the site began on October 28, 2016. Adjustments were made in November of 2016 and again in February of 2017. Preliminary winter data indicate that ambient concentrations of $PM_{2.5}$ range from 1.9 to 5.3 µg/m^3. PM_{10} ranges from 0 (during rainstorms) to 8.0 µg/m^3. Diurnal patterns of typical clear days show the highest concentrations of both PM_{10} and $PM_{2.5}$ during night time and lowest values during the late afternoon. As expected days with high relative humidity have low PM_{10} concentrations. $PM_{2.5}$ may remain unaffected during the winter since home heating with coal and wood may be more prevalent during cold conditions. These are only preliminary results.

Keywords: A - ambient monitoring, A - population exposure, D-Southwest-specific, D-First Nation, B-particulate matter

SU-PO-67

Risk Assessment of Polluted Soils in Relation to Transfer of Metals to Human Food Chain

D. Golui, S. Datta, B. Dwivedi, M. Meena; ICAR-Indian Agricultural Research Institute, New Delhi, India

Abstract: The success of risk assessment of metal contaminated soils depends on how precisely one can predict the solubility of metals in soil and subsequent transfer of these metals from soil to human food chain via plant. The specific objectives of the present investigation were to predict the uptake of Zn, Cu, Ni, Cd and Pb by Indian spinach (*Beta Vulgaris* L. var. All green) grown on metal contaminated soil using free ion activity model (FIAM) and to assess the risk to human health from intake of these metals from consumption of this leafy vegetable. For this purpose, twenty eight bulk surface soil samples with diverse soil properties such as pH, organic carbon and metal content were collected from smelter effluent, industrial sewage, polluted river water, domestic sewage, solid waste and cycle factory effluent deposited sites across the country. Free metal ion activity was estimated in soil solution as extracted by rhizon samplers using version 7 of the 'Windermere Humic Aqueous Model' (WHAM-VII). Results indicated that 94, 70, 75, 81 and 91% variation in Zn, Cu, Ni, Pb and Cd content, respectively, of Indian spinach could be explained by free ion activity model based on WHAM VII (Model I). Predicted free ion activity derived from solubility model could also significantly predict Zn, Cu, Ni, Pb, and Cd content in spinach to the extent of 85, 64, 66, 78 and 95%, respectively. Risk assessment of metal contaminated soil in terms of hazard quotient (HQ) indicated that Indian spinach grown on solid waste and industrial effluent irrigated soils were not safe (>0.5) to be consumed by human being as far as Zn, Cd, Pb and Ni content in this leafy vegetable are concerned. A ready reckoner was developed for computing safe limit of extractable metal in soil based on the predicted HQ by Model I for intake of metal by human through consumption of spinach. Novel approach as used in assessing the risk of metal polluted soils in relation to human health proved to be useful and promising.

Keywords: A-risk assessment, B-metals, C-soil

MONDAY, OCTOBER 16, 2017

MO-PL-A1: Population Biomonitoring

MO-PL-A1-67

California biomonitoring leads to region specific environmental exposure findings

K. Attfield, N. Wu; California Department of Public Health, Richmond, CA

Abstract: The California Environmental Contaminant Biomonitoring Program (Biomonitoring California) was established by legislation in 2006 to determine levels of environmental chemicals in a representative sample of Californians, establish trends over time, and help assess the effectiveness of public health efforts to decrease exposures. Biomonitoring California uncovers regional and state differences in chemical levels due to source and population differences which cannot be provided by the nationally representative National Health and Nutrition Examination Survey (NHANES). Biomonitoring California seeks to investigate exposures in populations potentially at risk in California. For example, a study of southern California firefighters has identified relatively high levels of polybrominated diphenyl ethers. Benzophenone 3 was also found to be elevated, and further state-based studies will help elucidate if this is an occupational effect or if Californians overall are exposed at higher levels. NHANES data has shown elevated levels of metals and per-and polyfluroalkyl substances (PFAS) in Asians. In 2016, we initiated a study investigating metals and PFAS exposures in Asian and Pacific Islander sub-populations in the San Francisco Bay Area. Among the 96 people of Chinese background participating, the blood mercury geometric mean was 3.6 ug/L, 4.5 times higher than the level found in all 2013-2014 NHANES adults and 2.1 times higher than non-Hispanic Asians. Lead and cadmium were also elevated in our population (30% and 81% higher, respectively) relative to NHANES, and inversely associated with time lived in the US. Urinary metals and PFAS data are pending, and recruitment of Vietnamese participants is currently on-going. Biomonitoring California is using these results to inform upcoming statewide surveillance in which we will examine differences among geographic regions and demographic groups and identify the differences in exposures between Californians and the overall US.

Keywords: A-biomonitoring, B-flame retardants, A-longitudinal metrics, A-biomarkers

MO-PL-A1-68

Contemporary Flame Retardants Exposures: Firefighters vs. Non-Occupationally Exposed Groups

M. Ospina, N. Jayatilaka, P. Restrepo, A. M. Calafat; Centers for Disease Control and Prevention, Atlanta, GA

Abstract: Introduction: Because of the phase-out of polybrominated diphenyl ethers (PBDEs), flame retardant formulations containing mixtures of chlorinated and non-chlorinated organophosphates, and other non-BDE brominated flame retardant chemicals have entered consumers markets. Understanding exposure to these contemporary flame retardants is of public health interest as components of these formulations are frequently detected in many consumer products. **Methods:** We assessed exposure to nine contemporary flame retardants by measuring the following urinary biomarkers: diphenyl phosphate (DPhP), bis(1,3-dichloro-2-propyl) phosphate (BDCPP), bis(1-chloro-2-propyl) phosphate (BCPP), bis(2-chloroethyl) phosphate (BCEtP), di-p-cresylphosphate (DpCP), di-o-cresylphosphate (DoCP), dibutyl phosphate (DBuP), dibenzyl phosphate (DBzP), and 2,3,4,5-tetrabromobenzoic acid (TBBA). We measured the concentrations of these metabolites in 76 urine samples collected in Atlanta in 2015 from adults with no known flame retardant occupational exposure and compared them to those from a group of 146 firefighters collected in 2010-2011 after performing structural firefighting while wearing full protective clothing and a SCBA respirator. **Results:** DPhP (median: 0.89; range: 0.26-5.6 ng/mL) and BDCPP (median: 0.69; range: 0.31-6.8 ng/mL) were detected in all of the non-occupationally exposed samples and all of the firefighters samples (DPhP [median: 2.9; range: 0.24-28 ng/mL], BDCPP [median: 3.4; range: 0.30-44 ng/mL]) with median concentrations of DBCPP and DPhP about five and three times higher in the firefighters, respectively. DBzP and DoCP were not detected in any of the samples. **Conclusions:** The results suggest exposure to several contemporary flame retardants in the general population and firefighters having occupational exposures higher than background exposures. We plan to obtain reference range concentrations of these biomarkers from the National Health and Nutrition Examination Survey samples.

Keywords: A-biomonitoring, A-biomarkers, B-flame retardants, D-occupational

MO-PL-A1-69

Characterization of Pregnant Women's Phthalate Exposure using Multiple Urine Samples

H. Shin[1], D. Bennett[2], X. Ye[3], A. M. Calafat[3], I. Hertz-Picciotto[2]; [1]University of Texas, Arlington, Arlington, TX, [2]University of California, Davis, Davis, CA, [3]Centers for Disease Control and Prevention, Atlanta, GA

Abstract: Background: Epidemiologic studies have reported associations between prenatal exposure to some phthalates and adverse health effects in children. Although phthalates quickly metabolize and excrete in urine, most epidemiologic studies have relied on phthalate metabolite concentrations obtained from one or a few spot maternal samples to characterize gestational exposures. **Objective:** We aimed to better characterize pregnant women's phthalate exposure of (e.g., within-subject variability, temporal trends) using multiple spot and 24-hour urine samples available in the MARBLES (***M***arkers of ***A***utism ***R***isk in ***B***abies - ***L***earning ***E***arly ***S***igns) study. **Methods:** We quantified 14 phthalate metabolites in 666 urine samples collected from 208 pregnancies in MARBLES. To reduce analytical cost while maintaining information about temporal variability, for each woman, we analyzed separately the first spot sample collected each trimester and a pooled sample, prepared from all remaining samples for that trimester. To evaluate within-subject variability, we calculated intraclass correlation coefficients (ICCs) using samples collected from 9 randomly selected mothers who provided 8 or 9 individual samples over 8 weeks. **Results:** Except for monoethyl phthalate, biomarkers of phthalates known to be used in personal care products with relatively consistent day-to-day exposure sources have relatively higher ICCs than biomarkers of phthalates with predominantly dietary sources. Biomarker concentrations varied over two orders of magnitude for some subjects, indicating that selecting one individual sample to reflect average exposures may bias exposure characterization in epidemiologic studies. **Conclusions:** Analysis of multiple samples improved women's characterization of phthalate exposure throughout pregnancy. When collecting multiple samples, pooling samples can reduce analytical cost and minimize exposure misclassification in epidemiologic studies

Keywords: A-biomarkers, B-phthalates, C-consumer/personal care products, D-prenatal

MO-PL-A1-70

Variation in glucuronidation of phthalate monoesters: Different metabolism or degradation?

J. W. Brock[1], J. Bell[2], D. Hagendorfer[1]; [1]University of North Carolina at Asheville, Asheville, NC, [2]Warren Wilson College, Asheville, NC

Abstract: Many recent epidemiologic studies have examined the relationship between adverse health outcomes and exposure to phthalates diesters. These studies predominantly use urinary metabolites as measures of internal doses. In these studies, urine samples are treated with glucuronidase converting the glucuronidated metabolites to the monoester. In a few studies examining urine samples without enzyme treatment, free monoester levels have been shown to vary temporal among and within individuals. This variation could arise from differences in metabolism or from differential degradation of the glucuronidated monoesters from different sampling conditions. One recent paper examined the temporal stability of phthalate monoester glucuronides but lacked some clinically relevant time points and temperatures. We extend this work to examine temporal stability at additional clinically relevant time points and temperatures. Urine sample were stored at 20 or 37°C for 4, 8 or 12 hours. After treatment, samples were spiked with ^{13}C-labeled phthalate monoester standards and half were treated with glucuronidase. Urinary phthalate monoesters (MiBP, MBP, MEHP, MEHHP, MEOHP, MBzP) were extracted using solid phase extraction and analyzed by liquid chromatography couple to a tandem mass spectrometer. Free and glucuronidated levels were obtained by difference. The ranges of percent glucuronide remaining at 20°C and 4, 8, or 12 hours were: MiBP 99.6-99.8%, MBP 101.5-101.9%, MEHP 99.4-102.7%, MEHHP 99.3-

100.1%, MEOHP 99.6-99.0%, and for MBzP 99.9-100.2%. The ranges of percent glucuronide remaining at 37°C and 4, 8, and 12 hours were: MiBP 97.1-99.2%, MBP 96.3-99.6%, MEHP 95.8-100.1%, MEHHP 99.3-100.1%, MEOHP 97.5-99.4%, and for MBzP 97.2-99.3%. These results suggest urinary phthalate glucuronides are relatively stable and the previously reported larger variations are due to other factors.

Keywords: A-biomarkers, B-phthalates

MO-PL-A1-71

Elaborating exposure reference values from the French national human biomonitoring program

L. RAMBAUD, A. Zeghnoun, A. SAOUDI, C. DEREUMEAUX, C. FILLOL; Santé publique France, Saint-Maurice, France

Abstract: The French human biomonitoring program has started in 2010 in order to provide a representative estimation of the French population's exposure to multiple environmental contaminants. It currently consists in two national surveys: a perinatal component including 4145 pregnant women from the Elfe cohort (French Longitudinal Study since Childhood) and a general population survey (Esteban) including 3021 adults (18-74 years) and 1355 children (6-17 years). These surveys provide a particularly good material to produce exposure reference values (ERV) to describe the ranges of environmental contaminants among the national population. ERV are a useful tool for public health actors who need to compare individual or population's level of exposure to a reference population in order to reduce highest exposures to toxic substances. Santé publique France, the French national public health agency, defines a national strategy to derive ERV following the recommendations of the International Union of Pure and Applied Chemistry (IUPAC) and other international experts. This strategy is based on the use of the 95th percentile as the ERV for the measured pollutant concentration levels in the relevant matrix. Age, sex and body mass index are used as partitioning criteria to derivate separate ERV for sub-group population when it seems relevant. A first step will take place in 2017 when ERV will be derived for French pregnant women for metals, organophosphorus and organochlorine compounds, pyrethroids, bisphenol A and phthalates. In a second step, in 2018, ERV will be derived for French adults and children included in the Esteban survey for more than a hundred environmental pollutants. Beyond the help that these ERV will provide to public health actors in France, their production under a harmonized procedure will allow international comparisons on levels of exposure for environmental pollutants and help public health policies to set a target in the reduction of exposure at national scale.

Keywords: A-biomonitoring, A-biomarkers, A-environmental policy, B-metals, B-POPs

MO-SY-B1: Diverse Applications of Exposure Science for Big Impact in Global Health - Domestic hazards

MO-SY-B1-72

Exposure to Lead from Used Lead-Acid Battery Recycling in Low- and Moderate-Income Countries

J. S. Keith, B. Ericson; Blacksmith Institute, Gardiner, NY

Abstract: Exposure to lead impacts the health of millions of people in low- and moderate-income countries (LMICs), with children often most impacted. One of the largest greatest sources of lead exposure is recycling of used lead-acid batteries (ULABs) from cars, trucks and motorbikes. Because lead isDue to lead's intrinsic valuable, almost all ULABs are recycled. In low- and moderate-income countries (LMICs), this is often done in small-scale shops in the informal economy. Large numbers of people are exposed to lead from these shops because there are thousands of them, they are commonly located in residential areas, and they they most often have poor controls for smelting smoke, dust, worker hygiene and waste disposal. Pure Earth has evaluated over 500 active and former ULAB sites in 33 LMIC countries around the world, with a particular focus on Bangladesh, Ghana, India, Indonesia and , Kenya. and the Philippines. Pure Earth has also developed estimates for the number of ULAB shops based on the number of batteries motor vehicle in use and the average battery life in LMICs. Based on this work, the total number of active or closed ULAB shops in LMICs is estimated to be in excess of 102,5000, and there are thousands more now-closed ULAB sites that continue to present lead exposure risks. Over 6nearly 2 million. Over 7,000,000 people are estimated to be at risk due to lead exposure from these shops. Many of these the affected people are children, and blood lead levels for people living in children around such shops is are often >40 ug/dLl , compared to a WHO CDC actionable level of 5 ug/dl. This paper will review the data known about ULAB shops and people whose health is impacted by them. Examples of ULAB recycling sites in Asia, Africa and Latin America will be shown, along with a brief review and examples of how such sites can be efficiently evaluated and remediated.

Keywords: A - population exposure, B-metals, C-soil, D-community, batteries

MO-SY-B1-73

Soil Ingestion Among Children 0-3 Years Old in Rural Bangladesh

L. Kwong[1], A. Ercumen[2], A. Pickering[3], L. Unicomb[4], J. Davis[1], J. O. Leckie[1], S. Luby[3]; [1]Stanford University, Palo Alto, CA, [2]UC Berkeley, Berkeley, CA, [3]Stanford University, Stanford, CA, [4]International Centre of Diarrhoeal Disease Research, Bangladesh, Dhaka, Bangladesh

Abstract: Ingestion of soil and dust is a primary pathway of children's exposure to several environmental contaminants, including lead and pesticides. Previous estimates of soil consumption by children in high-income countries range from 3.8-448 mg/day, with the most recent studies estimating 32-62 mg/day. The estimates for children in low-income countries range from 2-21 g/day. However, the sophisticated modeling that has been applied in high-income countries has not previously been applied in low-income countries. We modeled the mass of soil ingested per day by rural Bengali children using a Monte Carlo simulation. We combined anthropometric and observational data of children <3 years old in rural Bangladesh with measurements of soil mass on the hands of children and their mothers. Unlike previously published activity pattern simulations of soil ingestion, which estimated ingestion based on hand-to-mouth and object-to-mouth contacts, our model also includes ingestion of soil placed directly into the mouth. Among children <6 months old, mean ingestion of soil was 66 mg/day (standard deviation [sd] = 262). Values were similar among children 6-11 months old (mean = 82 mg/day [sd = 201]). Compared to young children, children >12 months old ingested significantly greater amounts of soil (p = 0.01), with a mean of 129 mg/day (sd = 212) among children 12-23 months old and a mean of 196 mg/day (sd = 335) among children 24-35 months old. The most influential parameters affecting the estimates were the frequency with which children put their own hands in their mouths and the efficiency of soil removal during hand-to-mouth contacts. Due to greater incorporation of empirical evidence, our estimates are 100-1000 times lower than previous estimates for rural children in low-income countries. Nonetheless, comparing our results to the most recent estimates for children in high-income countries, rural children in low-income countries may ingest more soil.

Keywords: C-soil, D-children, A-activity patterns, A-exposure factors, A-environmental justice

MO-SY-B1-74

Human exposure to by-products from anthrax infected animals in Bangladesh

M. Islam[1], S. Hasan[1], F. Haque[1], N. Haider[1], M. B. Hossain[1], S. U. Ahmed[1], M. Rahman[2], J. S. Salzer[3], M. Kadzik[3], E. Kennedy[3], E. Gurley[1]; [1]icddr,b, Dhaka, Bangladesh, [2]Institute of Epidemiology Disease Control and Research, Dhaka, Bangladesh, [3]Centers for Disease Control and Prevention, Atlanta, GA

Abstract: Anthrax is an acute infectious zoonotic disease caused by the spore-forming bacteria, *Bacillus anthracis*. In Bangladesh, anthrax is endemic in ruminants and animal cases of anthrax are closely associated with human outbreaks. Humans may become infected with *B. anthracis* through slaughtering of infected animals or processing of infected animal by-products. This study aimed to explore the use of by-products from ruminants slaughtered during suspected anthrax outbreaks. Between 2013 and 2015, we conducted a qualitative study in 28 suspected anthrax outbreak villages in 4 districts in Bangladesh. We explored the processing, distribution and use of by-products in these areas by conducting 65 in-depth interviews with cattle owners and animal by-product businessmen, and 10 group discussions with village residents. We found that slaughtering sick ruminants was common in anthrax affected areas. Ruminant owners reportedly sold meat from slaughtered sick ruminants to village residents, hides to local businessmen, and discarded butchering waste, namely bones, into nearby ditches or open fields. Businessmen reportedly washed hides with water and removed the fat and meat. The animal hides were reported to be used to make wallets, belts, shoes, balls and clothing. Bone collectors visited these villages occasionally to collect discarded bones. The bone collectors dried the bones naturally before grinding into granular and powder forms, and distributing them within Bangladesh and internationally. Ground bones were reportedly used for fertilizer, poultry and fish feed. We identified numerous uses of animal by-products. These by-products may be contaminated with *B. anthracis* spores and therefore pose a health risk to the people involved in the processing and distribution of these by-products. The use of fertilizers containing bone meal may also pose a risk of *B. anthracis* exposure to ruminants. Routine animal anthrax vaccination would reduce anthrax cases and contamination of by-products and therefore reduce the risk of human exposures to *B. anthracis*.

Keywords: B-microbial agents, D-occupational, A-epidemiology, A-Infectious disease, D-susceptible/vulnerable

MO-SY-B1-75

Advances in measuring household air pollution: Findings from India and beyond

K. R. Smith[1], A. Pillarisetti[1], K. Balakrishnan[2], C. R. Garland[3], M. A. Johnson[3], S. Madhav[2], R. Piedrahita[3], S. Sambandam[2]; [1]University of California, Berkeley, Oakland, CA, [2]Sri Ramachandra University, Chennai, India, [3]Berkeley Air Monitoring Group, Berkeley, CA

Abstract: Exposure to household air pollution (HAP) is a leading risk factor for ill-health, resulting in 3-4 million deaths globally per year. A critical challenge in assessing the impacts of household air pollution on health is obtaining accurate exposure estimates. A new generation of sensors have enabled improved measurement and modeling of exposure and exposure-related parameters, including appliance usage and ventilation. Technologies must optimize against the conflicting constraints of accuracy, replicability, weight, battery-life, noise, and comfort for participants and their family members. When used in rural areas, they must additionally be robust to harsh environments and difficult transit conditions. In this presentation, we provide a broad survey of the findings from field-testing of new sensing technologies in India and Nepal, with a short discussion of previous measurement methods. Specifically, we will discuss the following: (1) advances in particle monitoring technologies, including inter-device comparisons and participant/fieldworker feedback from field tests in two Indian states; (2) advances in stove use monitoring, including a novel event-counting stove use sensor deployed in Maharashtra and a new Bluetooth-enabled data-logging thermocouple and related analysis platform; (3) findings from the deployment of a quadcopter-based particle and temperature sensor to detect temperature inversions in North India; (4) results from field evaluations of a CO_2 decay-based estimate of household air exchange rates (AERs) in Nepal, including a short demonstration of an analysis tool to determine AERs from timestamped area measurements of particle and gas concentrations. We conclude with a discussion of future steps and challenges in using the data generated from these sensors to better model exposure, ideally with fewer intrusive measurements.

Keywords: A - exposure measurement, A-global health, B-particulate matter, A-sampling methods, C-air

MO-SY-B1-76

Dietary Exposure Analyses for Central and South America using Creme Global's Food Safety Model

C. O' Mahony, S. Pigat; Creme Global, Dublin, Ireland

Abstract: Creme Food is a scientific cloud based software service used for estimating dietary intakes of foods, chemicals and nutrients consumer populations. This is achieved by linking food consumption studies to the appropriate food composition and chemical occurrence data using a number of validated and published models. The system supports both deterministic and probabilistic input data, and these data sets are then combined model using Monte Carlo simulation. A key input for the model is food consumption data, in the form of dietary surveys typically carried out at a country level. Two such surveys from Central and South America currently installed in the Creme Food model are the and the Brazil National Dietary Survey 2008-2009 and the Mexican Encuesta Nacional de Salud y Nutrici6n (ENSANUT) 2006 and 2012. Both dietary surveys were adapted to the Creme Food system, despite having different methodologies; the Mexican data in the form of a seven day FFQ and the Brazilian data in the form of a two day 24-hour recall. This presentation will discuss how the data was sourced, formatted and analysed and can be used to produce calculation types include daily average intakes, acute exposures, as well other population health statistics, standard errors and confidence intervals.

Keywords: A - population exposure, A-aggregate exposure, A-exposure models, C-food

MO-SY-B1-77

Panel Discussion on Applications of Exposure Science for Global Health

C. V. Muianga[5], J. S. Keith[2], C. O' Mahony[6], A. Pillarisetti[3], M. Islam[4], L. Kwong[1]; [1]Stanford University, Stanford, CA, [2]Blacksmith Institute, New York, NY, [3]UC Berkeley, Berkeley, CA, [4]International Centre of Diarrhoeal Disease Research, Bangladesh, Dhaka, Bangladesh, [5]Centers for Disease Control and Prevention, Atlanta, GA, [6]Creme Global, Dublin, Ireland

Abstract: How can we expand the impact of exposure science to settings where it matters most? In this session, a panel of speakers who have applied exposure science to global health issues will discuss themes related to the following questions: 1) What are the beneftis of and challenges to conducting exposure research in low-income country settings? 2) Which global health issues would most benefit from further incorporation of exposure science? 3) What are exposure science mehtods used in high-income countries that are infeasbile in low-income settings and why? How can these methods become more accessible for exposure research in global health? 4) How can exposure scientists most effectively engage with global health research to help create effective interventions?

Keywords: A-environmental justice, A - population exposure, B-metals, A-behavior

MO-PL-C1: Risk Assessment Models

MO-PL-C1-78

A case study to illustrate the utility of the Aggregate Exposure Pathway and Adverse Outcome Pathway frameworks for integrating human health and ecological data into cumulative risk assessment

D. E. Hines, S. W. Edwards, R. B. Conolly, A. M. Jarabek; U.S. Environmental Protection Agency, Research Triangle Park, NC

Abstract: Cumulative risk assessment (CRA) methods, which evaluate the risk of multiple adverse outcomes (AOs) from multiple chemicals, promote the use of a conceptual site model (CSM) to integrate risk from relevant stressors. The Adverse Outcome Pathway (AOP) framework can inform CRAs by describing biological mechanisms of action for chemicals from molecular initiating events (MIEs), through key events, to AOs. However, AOPs do not consider the exposures that an organism may encounter. The Aggregate Exposure Pathway (AEP) framework was created to track stressors from sources, through key exposure states, to a target site exposure. Together, the joint AEP-AOP construct can act as a mechanistic, source-to-outcome summary, which can provide the basis for a CSM. The resulting CSM can allow for consideration of multiple stressors and sources that impact multiple outcomes across several species. We demonstrate how this construct can support the mechanistic integration of human health and ecological endpoints into CRA using a case study of perchlorate, an environmental contaminant found in multiple media that affects the sodium-iodide symporter (NIS) inhibition MIE. Computational models and dose-response data were used to evaluate responses to perchlorate exposure in eight vertebrates and four invertebrates. We observed a dose-response concordance across key events and species, reflecting the conserved nature of the NIS inhibition AOP. Results suggested that amphibian (*Xenopus laevis*) and mammal (*Rattus* sp.) AOs may be more sensitive to perchlorate exposure than AOs in fish species (*Danio rerio* and *Gambusia holbrooki*), but also highlighted gaps in the fish data. The AEP-AOP construct can help integrate human health and ecological endpoints into CRA by providing a common framework for evaluating risk in multiple species. *The views expressed in this abstract are those of the authors and do not necessarily reflect the views or policies of the U.S. Environmental Protection Agency.*

Keywords: A-risk assessment, A-cumulative exposure, A-ecological exposure

MO-PL-C1-79

The U.S. EPA RTR Program's Tiered Screening Approach to Evaluating Ingestion Risks from Emissions of PBHAPs

C. Holder[1], G. Carter[1], A. Varghese[1], M. McVey[3], S. Kumar[1], T. Palma[2], J. Hirtz[2]; [1]ICF, Durham, NC, [2]U.S. Environmental Protection Agency, Research Triangle Park, NC, [3]ICF, Blue Hill, ME

Abstract: U.S. EPA's Risk and Technology Review (RTR) assesses risks remaining from emissions of hazardous air pollutants (HAP) after implementing maximum achievable control technologies. Risk characterization includes a tired screen run with custom software tools to evaluate potential risks from ingestion of fish, farm foods, soil, and breastmilk (for infants) contaminated from persistent and bioaccumulative HAP (PBHAP) emissions. The screens are based on pre-run scenarios in the Total Risk Integrated Methodology's Fate, Transport, and Ecological Exposure module (TRIM.FaTE), producing PBHAP threshold emission rates below which a facility's emissions are unlikely to pose risks above benchmark levels. Each tier uses additional site-specific data to reduce instances of false-positive estimates of such risk. Tier 1 uses settings that maximize media contamination and ingestion. EPA uses a software tool to compare facility emissions to Tier 1 threshold rates. For facilities not passing Tier 1, Tier 2 allows EPA to efficiently incorporate more site-specificity. EPA ran many TRIM.FaTE simulations with a variety of lake locations and meteorology to produce a variety of Tier 2 threshold rates. EPA uses a tool with national databases of lakes and local meteorology to identify appropriate threshold rates per facility. While Tier 1 combines high-end ingestion of fish and farm foods, Tier 2 evaluates them separately, allows fish to come from multiple lakes, and lowers the farm-food ingestion to home-gardener levels. Using Tier 3, EPA continues to refine the evaluation by evaluating local lakes for fishability, estimating the exposure-mitigating effects of plume rise, and better accounts for the effects of local hourly meteorology on chemical fate and transport. After Tier 3, if the potential remains for risks above benchmark levels, EPA may conduct a site-specific risk assessment using TRIM.FaTE and local data on watersheds and soil and water characteristics.

Keywords: A-risk assessment, C-multimedia, A-environmental regulation, A-exposure factors, A-exposure models

MO-PL-C1-80

Leveraging Global Air Sampling Data - Validation for Regulatory Chemical Risk Assessment Models

R. Tibaldi[1], N. Barone[1], C. Money[2]; [1]ExxonMobil Biomedical Sciences, Annandale, NJ, [2]Cynara Consulting, Hampshire, United Kingdom

Abstract: Exposure risk assessment has grown in recent years due to new regulatory requirements. Models to predict occupational exposures have been published to support these assessments. An exposure estimation tool used for occupational risk assessment for European chemical registrations (REACH) is the ECETOC Targeted Risk Assessment (TRA) model. The TRA is intended to provide conservative estimates of exposure. A study was made with in house data to examine and validate how well the model matched real exposures under field conditions. Industrial hygiene air sampling data collected during work activities across company operations were retrieved from historical databases. A data set of 24,759 sample measurements for various substances and tasks spanning 1999 to 2015 were compared to the TRA model estimates. This required mapping of company task codes to TRA task definitions and substances measured to TRA vapor pressure or dustiness categories. The 75th percentile of the measurements was calculated for comparison to TRA 75th percentile exposure estimates. The key objective was to validate if the TRA model performed conservatively with estimates higher than task exposure measurments from our operations. A methodology was developed to analyze the dataset of historical air sample results. The 24,759 sample measurements were grouped into 86 cases that aligned with distinct TRA model inputs. While we had a large number of data points, many were below the Limit Of Detection, i.e., censored. In 65 cases more than 50% of the data was censored; preventing any firm statistical conclusion. For the remaining 21 robust data cases, the TRA estimate was validated as higher than measurement results. This project compared measured air sampling data from occupational exposure assessments to an exposure model (ECETOC TRA) used for risk assessments in the EU for chemical registration regulations. The evaluation concluded to some extent that the model performed conservatively as intended.

Keywords: D-occupational, A-exposure models, A-industrial hygiene, A-risk assessment

MO-PL-C1-81

Human Exposure Model (HEM): A Modular, Web-based Application to Characterize Near-field Chemical Exposures and Releases

K. Dionisio[1], K. Isaacs[1], K. Phillips[1], D. Lyons[1], N. Brandon[1], G. Glen[2], T. Hong[2], A. Varghese[2], J. Levasseur[2], H. Hubbard[2], D. A. Vallero[1], P. P. Egeghy[1], P. Price[1]; [1]Environmental Protection Agency, Research Triangle Park, NC, [2]ICF International, RTP, NC

Abstract: The U.S. EPA's Chemical Safety and Sustainability research program is developing the Human Exposure Model (HEM) to assess near-field exposures to chemicals that occur in various populations over the entire life cycle of a consumer product. The model will be implemented as a web-based, modular system that will produce estimates of population distributions of chemical exposure by route and releases to the environment, with one intended use to support human health impact assessments in Life Cycle Impact Assessments (LCIAs). The model determines aggregate doses of thousands of chemicals from the use of over 300 categories of consumer products over one year. Fundamental components of the model included in the beta-HEM release (September 2017) are the Residential-Population Generator (RPGen), Human Behavior, product formulation, and Source-to-Dose modules. The RPGen module generates synthetic populations, together with characteristics that may drive product use such as detailed demographic and residential information for each household. The Human Behavior module uses agent-based modeling to determine longitudinal patterns of macro activities (eating, sleeping, commuting, working, and 'idle' time) and a product usage scheduler to define the use of consumer products on each day. CPDat, a database of product composition and chemical functional use data, is used to generate empirical product formulations. The Source-to-Dose module uses these outputs and data on chemical properties to calculate doses (by chemical, route and product category) for each individual, and corresponding environmental releases. This model allows for evaluation of near-field chemical exposures and environmental releases from the use of consumer products for incorporation into LCIA and other chemical assessments.

Keywords: A - population exposure, A-chemical prioritization, C-consumer products, A-life cycle analysis

MO-PL-C1-82

The DustEx model for calculating human exposure to semivolatile organic chemicals (SVOC) in dust: Evaluation and validation

V. Sukiene[1], N. von Goetz[1], A. C. Gerecke[2], M. Bakker[3], C. Delmaar[3], H. M. Koch[4], C. Paelmke[4], K. Hungerbuehler[1]; [1]ETH Zurich, Zurich, Switzerland, [2]EMPA, Dübendorf, Switzerland, [3]RIVM, Bilthoven, Netherlands, [4]DGUV, Bochum, Germany

Abstract: For substances used in consumer products, human exposure via dust is often suggested to be a potentially important exposure pathway. In the DustEx project the relative contribution of the dust ingestion pathway was systematically evaluated by using a model framework consisting of a combination of published models, which describe the transfer of substances from consumer products into the indoor environment (DustEx model). On the basis of the calculated indoor air and dust concentrations and further conservative parameters the exposure of consumers via the dust pathway was calculated and compared to other exposure pathways. This inter-pathway comparison was performed (1) in a generic way by exploring the chemical property space and (2) by selecting realistic case studies for different types of consumer products. The development and evaluation of the DustEx model was supported by a small-scale field study under controlled conditions. In this study both environmental monitoring and biomonitoring were conducted. Deuterium-labelled semi-volatile organic compounds (SVOCs) were embedded in artificial consumer products and placed into regular inhabited apartments. At regular time intervals indoor air and settled dust was sampled to assess transfer into these media under realistic conditions. Altogether two measurement campaigns were conducted in 5 (first campaign) or 3 inhabited apartments (second campaign). During the second campaign, at regular intervals spot urine samples were taken from at least one inhabitant per apartment and analysed for two of the nine substances. For the evaluation and validation of the DustEx model the respective SVOCs were modelled and the estimates for environmental concentrations compared to the environmental monitoring. The biomonitoring results were compared to systemic human exposure estimates based on dust ingestion.

Keywords: A-exposure models, A-indoor environment, A-biomonitoring, B-SVOCs, C-consumer products

MO-SY-D1: US EPA STAR Program: Novel Research for 21st Century Exposure Science - Part 1

MO-SY-D1-83

Novel Research in 21st Century Exposure Science: Introduction and Welcome

M. Fry, S. Ramasamy; US Environmental Protection Agency, Washington, DC

Abstract: The US Environmental Protection Agency (US EPA) National Center for Environmental Research awarded Science to Achieve Results (STAR) grants from 2014 to 2017 for New Methods in 21st Century Exposure Science, and Susceptibility and Variability in Human Response to Chemical Exposure. Through these research grants, researchers have 1) developed innovative methods to better characterize indoor chemical exposures, and 2) improved our understanding of the variability in human response to chemicals and factors that influence susceptibility. This symposium will feature presentations that advance exposure assessment methods, models, and data. STAR grantees from Virginia Polytechnic Institute and State University, University of California Davis, and University of Michigan have developed new devices and methods for rapid, portable measurement of semivolatile organic compounds (SVOCs) indoors from consumer products. Duke University and University of California San Francisco STAR grantees will also share innovative techniques for measuring indoor chemical exposures in children and among a diverse population of pregnant women. Finally, Texas A&M University STAR grantee will present a novel mouse population-based model for characterizing the susceptibility and variability in human response to tetrachloroethylene exposure. These STAR research grants advance rapid assessment techniques for evaluating human exposure to chemicals, and expand the knowledge base needed for developing chemical exposure pathways and models.

Keywords: B-SVOCs, A-exposure models, A-indoor environment, A-sampling methods, D-susceptible/vulnerable

MO-SY-D1-84

Rapid Methods to Estimate Exposure to SVOCs in the Indoor Environment

J. Little, Y. Wu, C. Eichler, L. Marr; Virginia Tech, Blacksburg, VA

Abstract: A systematic and efficient strategy is needed to assess and manage the potential risks to human health that arise from the manufacture and use of thousands of chemicals. Among available tools for rapid assessment of large numbers of chemicals, significant gaps are associated with the capability to evaluate exposures to semi-volatile organic compounds (SVOCs) that occur indoors. We have developed simple methods to measure the key parameters that govern the emission and transport of SVOCs in the indoor environment. Then, accounting for product use, emission characteristics, and the properties of the SVOCs, we estimate exposure via inhalation of SVOCs in the gas-phase, inhalation of SVOCs adsorbed to airborne particles, ingestion of SVOCs adsorbed to dust, and dermal sorption of SVOCs from air into blood. Further development of a comprehensive set of models for estimating exposure to SVOCs in materials and products is needed. When combined with rapid toxicity estimates, screening-level exposure estimates for SVOCs can be used for health-risk-based prioritization of a wide range of chemicals of concern.

Keywords: A-exposure models, B-SVOCs, C-indoor

MO-SY-D1-85

Determining Indoor Source Strengths: Utilizing Measurements and Models

D. Bennett[1], C. Moschet[1], H. Shin[2], T. Young[1]; [1]University of California, Davis, Davis, CA, [2]University of Texas, Arlington, Arlington, TX

Abstract: A combination of modeling and measurements are used to determine the potential range of emission of compounds into homes. Dust samples were collected using a Mighty Mite sampler from the main living area of 38 homes. Samples were analyzed for a broad range of targeted compounds as well as non-targeted compounds. The targeted chemicals were from the following chemical classes: 17 flame retardants (FRs) (brominated diphenyl ethers (BDEs), organophosphates, novel brominated FRs), 14 phthalates and other plasticizers, 3 surfactants, 10 antimicrobials and biocides, 26 insecticides and insecticide transformation products, 10 PFCs, and 53 various (antimicrobials, fragrances, fatty acids, skin oils, PAHs, UV filters). We report dust concentrations for many ubiquitous compounds that have not been quantified as frequently in dust, including the fragrances galaxolide (mean concentration 1958 mg/kg) and 2-Benzylideneoctanal (2072 mg/kg), skin oils squalene (16,100 mg/kg) and cis-hexadec-6-enoic acid (112,000 mg/kg), surfactants Aacohol ethoxylated C10E6 (12,500 mg/kg) and didecyldimethylammonium (3,900 mg/kg), as well as typically measured flame retardants, phthalates, pesticides, and PAHs. Surprisingly, the BPA replacement, BPS (1445 mg/kg) had a concentration than BPA (534 mg/kg). In addition, we highlight several compounds identified in dust through non-targeted methods, including the plasticizer Toluene-2-sulfonamide (identified in 38 homes), the fungicides fludioxonil (37), Thiabendazole (37), and Propiconazole-II (31), the pigment physcion (38), the Package Additive Butylated Hydroxytoluene (38), as well as numerous food additives, uv filters, and pharmaceuticals. Finally, we use a fugacity based indoor model to estimate the source strength to air that would need to be present in the home to result in the measured concentrations. This approach takes into consideration the chemical properties and thus fate and transport in the home.

Keywords: A-indoor environment, A - exposure measurement, B-SVOCs, C-consumer products, A-exposure models

MO-SY-D1-86

On-site monitoring of occupational exposure to volatile organic compounds by a portable comprehensive 2-dimensional gas chromatography device

J. Lee[1], S. Sayler[2], M. Zhou[1], H. Zhu[1], R. J. Richardson[2], R. Neitzel[2], K. Kurabayashi[3], X. Fan[1]; [1]University of Michigan, Ann Arbor, MI, [2]University of Michigan, Ann Arbor, MI, [3]University of Michigan, Ann Arbor, MI

Abstract: Workers routinely encounter volatile organic compounds (VOCs) in their workplaces, which can exert various forms of acute and chronic health effects. Monitoring those VOCs on-site can help ensure the safe conditions for workers and guide an immediate first aid for workers. Here we report on using a fully automated portable comprehensive 2-dimensional (2-D) gas chromatography (GC) device (60 cm × 50 cm × 10 cm and <5 kg) for the field analysis of occupational VOCs at the Spray and Finishing Shop of the University of Michigan. The entire device consists of a micro-preconcentrator/injector, commercial columns, micro-Deans switches, micro-thermal injectors, and micro-photoionization detectors, as well as miniaturized valve, pump, helium cartridge, power supply, and computer control and user interface. The indoor air at the Shop was sampled and analyzed before, during, and after spraying the five most frequently used paints. The 2-D analysis yields enhanced peak capacity and shortened analysis time as compared to those achieved by portable GC instruments reported to date. Also, the portable system is able to perform on-site analysis of VOCs, providing the results every 5 to 15 min, which would allow for enhanced acquisition of information regarding workers' safety and health risks from airborne VOCs. The performance of the portable 2-D GC device was validated by comparing the quantitative results of various type of paints using traditional occupational VOCs monitoring techniques in accordance with Occupational Safety and Health Administration Method ORG-07.

Keywords: A-indoor environment, A-analytical methods, A-industrial hygiene, A-sensor technology, B-VOCs

MO-SY-D1-87

Young Children's Exposure to Phthalates in the Home Environment

H. M. Stapleton[1], K. Hoffman[1], A. Lorenzo[1], A. Killius[1], S. C. Hammel[1], A. Phillips[1], X. Ye[3], A. M. Calafat[3], T. F. Webster[2]; [1]Duke University, Durham, NC, [2]Boston University, Boston, MA, [3]Centers for Disease Control & Prevention, Atlanta, GA

Abstract: Phthalates, used in plastics to enhance flexibility and durability, are also used as solvents. The average home contains hundreds of products with phthalates, leading to widespread human exposure. We evaluated relationships between personal behavior and characteristics of the home environment, including air and dust measurements, and indicators of children's phthalate exposure. Central North Carolina families with children between 3 and 6 years of age were recruited in 2014-2016 from an ongoing pregnancy cohort. Home visits were conducted with each family during which detailed information on the child's behavior was collected. Children also provided three spot urine samples over 48 hours which were pooled for the measurement of phthalate metabolites and some of their alternatives (n=181 children). Handwipe and indoor dust samples were collected and analyzed for phthalates and terephthalates (n=181), and in a subset of 49 homes, passive air samplers were deployed and collected after 3 weeks. Phthalates were detected frequently in all matrices assessed (dust, air and handwipes) and their metabolites were detected in all urine samples. Concentrations varied by the demographic characteristics of study participants as well as characteristic of the home. For example, dioctyl terephthalate concentrations were 2.5 times higher in homes that had vinyl flooring compared to homes with no vinyl ($p=0.002$). Similar associations were observed for urinary metabolites, with characteristic of the home and family associated with higher urinary metabolite concentrations. In particular, maternal education attainment was strongly associated with children's urinary metabolite concentrations. Reported use of some personal care products was also significantly associated with children's urinary metabolite concentrations. Taken together, results suggest that the home environment and personal behavior are significant drivers of children's exposure to phthalates.

Keywords: D-children, A-epidemiology, A-activity patterns, B-phthalates

MO-PL-E1: Indoor Air Quality & Particulates

MO-PL-E1-88

Prevalence and timing of indoor PM emission events observed in a small cohort of homes using low-cost dust sensors

I. Longley, G. Olivares, A. Kachhara, S. Edwards; NIWA Ltd, Auckland, New Zealand

Abstract: Home occupants are exposed to airborne particulate matter from both indoor sources and the infiltration of outdoor sources. Mostly this differentiation has been achieved through compositional analysis. We conducted a pilot study using time series analysis of continuous data from a low-cost unobtrusive sensor package that would be suitable for mass deployment. 8 homes were equipped with three custom-built air quality monitors for a 2-week period in winter 2016. Two monitors (PACMAN) were placed in two rooms indoors and one (ODIN) outdoors on the same property. All homes but one used wood-burning appliances as the primary source of heat and were all located in a town where wood-burning dominates both heating and PM emissions. Householders were invited to keep time-activity diaries focussing on cooking, heating and other activities which may impact PM levels in the home, and fire-lighting was monitored with temperature sensors. Distinct particle 'events' were detectable in every home in which PM levels rapidly increased then decayed. Events were attributed to indoor or outdoor sources depending on whether a corresponding event was observed outdoors or not. The average prevalence of indoor sources was 1.2 per day (range 0.8 - 1.5). Events were attributed to outdoor sources in 6 out of 8 houses with an average prevalence of 0.4 events per day. Indoor events produced peak concentrations 3 times greater than outdoor events on average. Indoor source events were most common between 7 am and 8 am and between 5 pm and 6 pm, but with variations between homes. Outdoor source events appeared to have no temporal pattern. Completion of time-activity diaries, and the detail provided was highly variable. However, there was no clear correlation between events described and observed particle levels. Particle events did not, in general, correspond to the lighting of wood-burning fires. Data was visualised for presentation to, and discussion with, householders.

Keywords: A-sensor technology, B-particulate matter, C-air, C-indoor, A-activity patterns

MO-PL-E1-89

Assessing Indoor $PM_{2.5}$ Concentrations in Households on the Hopi Reservation

M. K. O'Rourke[1], S. Hadeed[1], G. Honanie[2], A. Mahkewa[2], M. Paukgana[2], C. Kelley[1], M. A. Alshammari[1], R. Canales[1], R. Harris[1]; [1]The University of Arizona, Tucson, AZ, [2]The Hopi Tribe, Kykotsmovi, AZ

Abstract: Introduction: A significant proportion of the Hopi Tribe burn coal, wood and other organic combustible materials for heating, lighting and cooking in homes. An estimated 37% and 33% of Hopi household report burning coal and wood for heating, respectively. The Hopi Tribe has expressed their concerns about indoor air quality from the combustion of these fuels. The purpose of this pilot study is to measure indoor $PM_{2.5}$ concentrations in homes and assess the effects of fuel type and housing structure on indoor levels. Method: Pilot data was collected during the 2017 heating season in four households on the Hopi Reservation. Indoor concentration of $PM_{2.5}$ was measured over a 24-hour period using real time monitors (pDR-1500) set at 1-minute logging intervals. Subsequent collection of primary fuel type and housing characteristics was recorded by field team members at the time monitors were placed indoors. Average $PM_{2.5}$ concentration was calculated by fuel and housing type. Results: In the four households sampled to date, the indoor mean (SD) $PM_{2.5}$ concentration was 18.7 (37.3) µg/m^3. Two modern households that used a combination of coal, wood and electricity for fuel had an indoor mean $PM_{2.5}$ concentration of 17.9 (36.0) µg/m^3 and 1-minute peak concentration of 332.8 µg/m^3. Average $PM_{2.5}$ concentrations for a hybrid modern-traditional home with coal and wood fuel was 33.0 (50.6) µg/m^3 and 6.1 (6.8) µg/m^3 for an electrically powered mobile home. Discussion: From this pilot study, households using coal and wood as primary fuel sources were found to have elevated indoor $PM_{2.5}$ concentrations compared to electrically powered homes. These initial results are similar to previous studies that found increased exposures and indoor $PM_{2.5}$ concentrations in households using combustible fuels as a primary fuel source. This project will continue to assess indoor air quality of households on the Hopi Reservation by various fuel sources and housing types during heating and non-heating seasons.

Keywords: B-particulate matter, A-indoor environment

MO-PL-E1-90

Assessment of air exchange rate in classrooms of middle schools with natural ventilation in Beijing, China

Z. Sun, C. Cai, R. Du, Y. Zhang; Tsinghua University, Beijing, China

Abstract: BACKGROUND: Air exchange rate (AER) greatly influences the indoor air quality in classrooms of middle schools where the pupil density is quite high. It is strongly related to indoor $PM_{2.5}$ concentrations in ambient $PM_{2.5}$ pollution period. In addition, it determines classroom CO_2 concentration which influences the learning ability of pupils. Therefore, it is important to accurately assess it. Build-up method and decay method based upon CO_2 are often used to determine the AERs during class and after school, respectively. However, the relative error of assessing the both AERs are not clear. OBJECTIVES: (1) Determine the relative error of assessing AERs by build-up method and decay method based upon CO_2; (2) Measure the AERs of one classroom in each of two middle schools; (3) Validate the assessed AER after school by comparing the indoor $PM_{2.5}$ concentrations with the predicted value based upon the assessed AER. METHODS: Concentration of $PM_{2.5}$ and CO_2 were measured per minute in two natural ventilated classrooms. Build-up method and decay method based upon CO_2 are used in the during-class and after school periods, respectively. The AERs and CO_2 emission rates during a class were calculated using least square method. The relative error of the former was estimated by taking account that the CO_2 sensors have an accuracy of 10%. RESULTS: The relative error for the AERs varied from 146% to 11% for the build-up method when they ranged from 0.1 to 5 h^{-1}, and from 6.1% to 1.4% for the decay method when they ranged from 0.1 to 0.5 h^{-1}. The AERs assessed varied from 0.2 h^{-1} to 4.5 h^{-1} during class and from 0.1 h^{-1} to 0.4 h^{-1} after school. CONCLUSION: The decay method has smaller relative error than the build-up method. The AER during class was usually higher than that after school. The AER can be used to estimate indoor $PM_{2.5}$ concentration. However, the precision relies heavily on the value of deposition rate and penetration factor.

Keywords: A-exposure factors, A-indoor environment, C-indoor, C-air

MO-PL-E1-91

Reductions in Personal $PM_{2.5}$ Exposure via Indoor Air Filtration in the Reducing Air Pollution in Detroit Intervention Study (RAPIDS)

M. M. Maestas[1], R. Brook[2], R. Bard[2], R. Ziemba[3], F. Li[1], R. Crane[1], K. Thompson[4], S. Adar[4], C. Spino[4], M. Morishita[1]; [1]Michigan State University, East Lansing, MI, [2]University of Michigan, Ann Arbor, MI, [3]Community Health Nursing, Ann Arbor, MI, [4]University of Michigan, Ann Arbor, MI

Abstract: The Reducing Air Pollution in Detroit Intervention Study (RAPIDS) was designed to evaluate $PM_{2.5}$ exposure reduction and cardiovascular health improvement with indoor air filtration among seniors. In this paper we evaluate the utility of indoor air filtration to reduce indoor $PM_{2.5}$ exposure, discuss whether the reduction of indoor home $PM_{2.5}$ concentrations is sufficient to measurably decrease personal $PM_{2.5}$ exposure, and describe how effectively economical low-efficiency (LE) filters reduce indoor $PM_{2.5}$ concentrations compared to high-efficiency (HE) filters. We enrolled 40 participants from a low-income senior-citizen residential facility in Midtown Detroit, Michigan for a blinded crossover trial randomized to order. Each participant was subjected to three intervention scenarios: HE, LE, or no filter (control) of three consecutive days each, separated by washout periods. Personal, indoor, and ambient $PM_{2.5}$ concentrations and cardiovascular health outcomes were measured daily during each scenario. Indoor $PM_{2.5}$ concentrations were significantly higher for control (17.5±17.0 μg m^{-3}) than for LE (8.4±5.4 μg m^{-3}) and HE (7.0±4.5 μg m^{-3}). The personal $PM_{2.5}$ concentrations were 15.7±13.8 μg m^{-3} for control, 10.9±9.6 μg m^{-3} for LE, and 7.4±4.2 μg m^{-3} for HE, and the distribution of personal $PM_{2.5}$ concentration based on generalized Friedman and post-hoc Wilcoxon signed ranks tests was significantly different for each scenario. These findings indicate that LE filters do significantly reduce indoor $PM_{2.5}$ compared to control, and personal $PM_{2.5}$ exposure are measurably reduced by using either HE or LE indoor air filtration in the home compared to no filtration. To our knowledge, this is the first study to show that both LE and HE indoor filtration systems significantly reduce not only indoor but also personal $PM_{2.5}$ exposure.

Keywords: A-indoor environment, B-particulate matter, D-community

MO-PL-E1-92

Indoor air quality in inner-city schools; Spatial and seasonal variability

E. Majd[1], M. McCormack[5], F. Connolly[2], P. Leaf[3], C. Gummerson[5], D. Clemons-Erby[1], F. Curriero[3], M. Davis[3], T. Webb[4], N. Price[4], K. Spicer[3], K. Koehler[1]; [1]Johns Hopkins University, Baltimore, MD, [2]Johns Hopkins University, Baltimore, MD, [3]Johns Hopkins University, Baltimore, MD, [4]Baltimore City Public Schools, Baltimore, MD, [5]Johns Hopkins University, Baltimore, MD

Abstract: Introduction: The adverse impacts of poor indoor air quality (IAQ) at school on students' health and academic performance have been the subject of several studies. To understand these impacts, appropriate characterization of IAQ in schools is essential, which requires accounting for the spatiotemporal variability of indoor pollutants concentrations in this microenvironment. **Methods:** This work aims to determine and compare the variability of indoor air pollutants within and across 15 inner-city schools. Indoor air pollutants including $PM_{2.5}$, CO, NO_2 and volatile organic compounds (expressed as TVOC) were monitored during a 2-week period at 4 locations within each school (classrooms, gyms, etc). Measurements were repeated in 3 seasons. Data analysis was limited to school days/hours. Between and within-school variability of these pollutants were estimated using mixed-effect models fitted for each with the random effect of location nested within school. The significance of season was verified by adding it as a fixed-effect variable to the models and using the likelihood ratio test. **Results:** Higher between-school than within-school variability was observed for CO and NO_2. For $PM_{2.5}$ and TVOC, the estimated variances were up to 9 and 7 times higher within schools compared to between schools. This may be due to different types and levels of indoor activities at each school. Season showed a significant effect on CO and NO_2 variability, but not on $PM_{2.5}$ and TVOC. **Conclusions:** Spatial and seasonal variabilities of IAQ in school were shown to be different by pollutant. While higher between-school variability was observed for gaseous pollutants, PM and VOCs showed higher within-school variability. This heterogeneity must be taken into account in future studies that aim to determine the adverse health and performance impacts of indoor IAQ.

Keywords: A - exposure measurement, A-exposure factors, A-indoor environment, B-particulate matter

MO-SY-F1: Pathways and mechanisms linking exposure to the natural environment with health and well-being benefits

MO-SY-F1-93

Green health pathways: a theoretical framework and applications

J. Roe; University of Virginia, Charlottesville, VA

Abstract: This presentation addresses why green space is important to public health, including the context of 'planetary health' and recent calls for a new research agenda that focus on the natural systems on which health depends (The Lancet, April 17, 2017). It briefly summarizes the contribution that green space can make to public health including improved life longevity and morbidity outcomes, improved pregnancy outcomes, improved mental health and cognitive functioning. It sets out four pathways by which green space impacts on health including air quality, physical activity, social contact and psychological restoration (Hartig et al., 2014). It considers what types of contact with nature support health and what is currently known about the dose-response relations for nature and health i.e. how much and how frequently people need to engage with nature - and what types of nature contact - makes a difference.

Keywords: planetary health, urbanisation, health and wellbeing, urban green space

MO-SY-F1-94

Properties of urban green spaces facilitating psychological restoration

M. van den Bosch[1], P. Währborg[2], J. Björk[3], P. Grahn[2], E. Skärbäck[2], P. Jönsson[4], M. Wallergård[3], G. Johansson[3], B. Karlson†[3], P. Östergren[3]; [1]The University of British Columbia, Vancouver, Canada, [2]Swedish University of Agricultural Sciences, Alnarp, Sweden, [3]Lund University, Lund, Sweden, [4]Högskolan Kristianstad, Kristianstad, Sweden

Abstract: Mental disorders contribute substantially to the global burden of disease. Research shows that natural environments may prevent mental illness, but it remains unclear which properties are optimal for psychological restoration. This presentation summarises findings from three studies that have attempted to refine and categorise type of nature exposure and the impact on mental wellbeing. Two studies used geocoded data from a Swedish longitudinal health survey linked to land use data, classified in GIS as perceived restoration qualities. The first study used multivariate logistic analyses, analysing interaction effects on population mental wellbeing (as measured by the General Health Questionnaire, GHQ-12) of restoration qualities, in combination with other health determinants, in a population remaining at same residence. The second study used a subsample of the same population, investigating effect on mental wellbeing of a move between areas of varying restorative qualities. Both studies suggest that access to serene nature, e.g. places of peace, silence, and tranquillity with sounds of wind, water, and birds, are particularly important for mental health (as measured by GHQ-12), in comparison to nature perceived as wild, lush, spacious, or historical. GIS-criteria of serene nature include broad-leaved and mixed forests, pastures, water courses, and ponds. The third study had an experimental design, monitoring parasympathetic activity during and after a virtual reality stress test and subsequent stress recovery in three different conditions - virtual natural environment with and without sounds and a control condition. This study showed that sounds of nature are crucial for efficient stress recovery. While these studies must be replicated and complemented with studies from different regions and populations, they provide preliminary support for using serene nature exposure, including nature sounds, as a means to psychological restoration and mental wellbeing.

Keywords: A-geospatial analysis/GIS, A-longitudinal metrics, green spaces, virtual reality, A-epidemiology, green spaces, virtual reality

MO-SY-F1-95

Linking urban green spaces with physical activity

J. A. Hipp; NC State University, Raleigh, NC

Abstract: There are a diversity of public open spaces (parks, greenways, plazas) under the surveillance of public, online, streaming webcameras. Since 2006, the Archive of Many Outdoor Scenes (AMOS; amos.cse.wustl.edu) has captured and archived images every 30 minutes from 36,000 international outdoor, public, webcams resulting in a repository of over 1.1 billion images of outdoor spaces. Ten percent of these webcams are systematically capturing images of public open spaces allowing for the audit of green spaces and built environments and annotation of human behaviors within the environments (e.g., physical activity). We identified 25 international webcams capturing images of parks, plazas, and squares, most with visible built environment or programmatic changes, and annotated the use of these public, green spaces. To compare change in human behavior within public spaces, randomly selected images from daylight hours across two time-matched weeks per year (e.g., June 1-14, 2012 and June 1-14, 2013) were chosen for annotation. One hundred images from year one and 100 images from year two, per the 25 webcams, were annotated (n=5,000 total images). The annotation process was completed by an online crowdsource (Amazon Mechanical Turk) and by machine learning with Research Assistant validation. Crowdsource and Research Assistant annotation included the digital drawing of polygons around each person captured in a scene, resulting in two forms of data; 1) number of people per image and 2) geospatial raster data on location of people in the image. The open source neural network, YOLO, was used for the machine learning annotation of images. Data was analyzed to determine change in number of people present pre and post green space and built environment change or programmatic change (e.g., addition of a farmer's market). Heat, or density, maps of scenes have been created from raster data to analyze density of use across time of day and day of week. Heat maps also serve the purpose of a visual dissemination tool to stakeholders interested in the behavior change associated with environments. Captured images from public webcams can be used for the public health surveillance of public green and open space utilization across time and space. Methodologies developed and presented here are scalable across webcams and environments and provide non-invasive and biased evaluation of human behavior in greenspaces.

Keywords: A-geospatial analysis/GIS, A-activity patterns, D-community, A-built environment, A-sensor technology

MO-SY-F1-96

Fine-scale health-related measures of exposure to urban green spaces

L. E. Jackson; U.S. EPA, Research Triangle Park, NC

Abstract: Nature's benefits to society, also known as ecosystem services, contribute to many aspects of public health and well-being. Green infrastructure routinely buffers both natural and man-made hazards. It is also a source of healthful exposures due to promoting physical activity, social interaction, and engagement with nature. The U.S. EPA's EnviroAtlas project maps neighborhood and individual-scale metrics of green infrastructure to indicate potential for hazard buffering and beneficial exposures. Example metrics include tree cover along roads, green window views, and proximity to parks. Based on literature review, these metrics are further tested and refined in EnviroAtlas communities with partners in health science research. Relevant health correlates include physical fitness, social connectedness, school performance, and longevity. Often, stronger beneficial associations with greenery are found in disproportionately vulnerable populations such as children, the elderly, and those of lower socioeconomic status. However, green infrastructure is often disproportionately lacking in low-income communities. EnviroAtlas maps and supporting information can help screen for neighborhoods most likely to benefit from interventions such as enhancing street trees in walkable areas and reducing impervious surfaces in urban heat islands. EPA and partners develop one-meter scale landcover data used in most metrics. Additional inputs include 30-meter resolution population estimates, and existing datasets such as roads and waterways. All EnviroAtlas data are available online. A suite of fine-scale metrics has been developed for 20 U.S. urbanized areas to date, using consistent methods to facilitate comparisons across neighborhoods and communities. Research is ongoing to increase sample size, environmental and population heterogeneity, and generalizability of results. Collaborators are welcome! This abstract has been approved by the U.S. EPA, but does not necessarily reflect Agency policy.

Keywords: A-ecological exposure, A-geospatial analysis/GIS, D-environmental justice, D-community, C-multimedia

MO-SY-F1-97

How vegetation affects local air quality and human exposure

R. Baldauf; US EPA, Durham, NC

Abstract: Debate has grown on the role of vegetation in improving local air quality. Many studies indicate that vegetation can have a positive, albeit minor, impact on regional air quality; however, studies are mixed on the impacts to local air quality. Interest has particularly expanded on how roadside vegetation affects population exposures to traffic-emitted air pollution, which numerous health studies have associated with adverse health effects. Air quality studies indicate that roadside vegetation impacts downwind, near-road air quality, with some studies identifying reductions in air pollution concentrations while others indicating increases in pollutant levels. These contradictory findings have resulted in confusion regarding the capability and utility of using vegetation to improve local air quality. These roadside vegetation studies have investigated the impact of many different types and characteristics of plants and urban forests, including preserved, existing vegetation stands usually consisting of mixtures of trees and shrubs or plantings of individual trees. This presentation will provide an overview of studies which have investigated how roadside vegetation alters near-road air quality and how these studies can be reconciled to highlight the characteristics and conditions of the vegetation which can lead to positive or negative impacts on local air quality. This presentation also provides recommendations on the design and location for preserving or planting roadside vegetation to maximize opportunities for pollution reduction and minimize potential unintended increases in near-road pollutant concentrations and exposures for nearby populations.

Keywords: A-sustainability, A-built environment, B-particulate matter, C-air, Green Infrastructure

MO-SY-G1: The Influence of Aviation Emissions on Exposures at Local, Regional and National Scales

MO-SY-G1-98

A Simple Dispersion Model of Ground Level Ultrafine Particle Concentrations Downwind of a Major Airport

T. V. Larson[1], N. Hudda[2], S. Fruin[3], S. Boonyarattaphan[1]; [1]University of Washington, Seattle, WA, [2]Tufts University, Boston, MA, [3]University of Southern California, Los Angeles, CA

Abstract: Introduction: Recent studies have observed a broad footprint of increased ultrafine particle number concentrations up to and possibly beyond 10 km downwind of major airports. Here we compare mobile monitoring measurements downwind of LAX with predictions from a set of relatively simple plume dispersion models. **Methods:** Mobile monitoring of ultrafine particle number concentrations took place on six days during daytime hours along seven N-S transects ranging from 2 to 9 km downwind of LAX. Background concentrations were first derived using the values obtained by applying a moving boxcar filter to the observations. Cross-wind integrated concentrations (CICs) of estimated ultrafine particle background concentrations were then computed each sampling day for each transect. ANCOVA was used to compare observations with dispersion model predictions with a random effect of sampling day. The Aermod dispersion model was run separately with hourly average wind data averaged from LAX one minute ASOS data. The source configuration was either a volume-line source to represent the descent path emissions, a surface area source to represent ground level sources at the airport, or a surface trapezoidal area representing cross-wind, ground level transport of the aircraft vortices generated near the ground during descent. The volume-line source geometry was tilted to represent the descent path of the aircraft at a 3 degree inclination. This tilted source was then lowered uniformly by 300 m (descent 300) to represent the effect of transport from descending vortices, resulting in a part of this source being placed at ground level near the airport. **Results:** Comparison of predicted versus observed CICs using ANCOVA indicate that either of the ground-level area source models performed better (adjusted R^2 was 0.57 and 0.58 for the trapezoid and the airport area source models, respectively) than the volume line source model (adjusted $R^2 = 0.39$). In two variable ANCOVA models (line and area source predictions), only the area source term was significant. **Conclusions:** Two different ground level area source models performed moderately well in predicting background-adjusted, cross-wind integrated concentrations obtained from a mobile monitoring platform at distances up to 9 km downwind of LAX. A descending volume line source model had less skill.

Keywords: C-air, A - ambient monitoring, A-exposure models

MO-SY-G1-99

The Impact of Aviation Emissions on Ultrafine Particulate Matter (UFP) Concentrations in Communities at Varying Distances from Flight Paths

C. Kim, J. I. Levy; Boston University School of Public Health, Boston, MA

Abstract: Multiple studies have concluded that aircraft arrivals contribute significantly to UFP concentrations over large regions, but few studies have had the necessary monitoring infrastructure or statistical approaches to formally evaluate the magnitude and spatial extent of impact. Real-time (1-second resolution) UFP measurements and concurrent meteorological data were collected at six monitoring sites in communities near an arrival flight path (4R) to Boston Logan International Airport at varying distances from the airport and lateral distances from 4R. Three UFP monitors were rotated among the sites for one week at a time from April to September of 2017 in order to capture the impact of aircraft arrivals under varying meteorological conditions in different communities. Real-time flight location data (latitude, longitude, and altitude) were obtained from the Federal Aviation Administration (FAA) for the purpose of source attribution. Due to airport construction, runway 4R had highly variable usage patterns across our monitoring campaign, with the runway being completely closed at times, closed for part of the day at times, and completely open at times. We also observed significant day-over-day variability in usage as a function of meteorological conditions. This provided a natural experiment that enhanced our ability to estimate aviation source contributions. We developed multivariable regression models associating real-time UFP concentrations with flight activity and meteorology in order to compare the patterns of arrival aircraft's contributions to UFPs under various meteorological conditions and runway configurations, taking account of lags between elevated emissions and concentrations at the surface. In addition, the long-term deployment over different seasons allowed us to investigate the diurnal and seasonal patterns of ambient UFPs. Our study results can be used to evaluate the impact of aircraft arriving on concentrated flight paths on population exposures.

Keywords: A-exposure models, A-analytical methods

MO-SY-G1-100

Model-based Assessment of Particle Mass and Number Concentrations due to Commercial Aircraft Activity in the U.S.

S. Arunachalam[1], J. Huang[1], L. P. Vennam[1], B. Murphy[2], F. Binkowski[2]; [1]University of North Carolina at Chapel Hill, Chapel Hill, NC, [2]U.S. Environmental Protection Agency, Research Triangle Park, NC

Abstract: Bulk of the emissions from aircraft during landing and takeoff (LTO) cycles are emitted in the size range of ultrafine particles (< 100 nm). To accurately characterize aircraft emissions, we enhanced the default approach for treating primary emissions of aircraft-emitted PM in the Community Multiscale Air

Quality (CMAQ) model. Currently, CMAQ treats aerosol formation without considering variability in particles size distribution (PSD)from individual emission sectors. Individual source sectors have varying PSD, and having a constant PSD for all source sectors might lead to uncertainties in predicted aerosol characteristics - number and mass concentrations. We developed a new CMAQ module to read particle emissions from a specific sector, and treat its PSD separately. We investigated ambient PM impacts due to commercial aircraft emissions in North America using this new module, and based upon information on aircraft emissions characteristics from engine measurement campaigns. We simulated three scenarios: background (all anthropogenic and biogenic emissions except aircraft emissions), base (background and aircraft emissions with uniform particle size distribution called the default configuration), and sensitivity (background and aircraft emissions with two different PSDs) scenarios. Total and aircraft attributed number concentrations of UFP (NUFP) are improved using this new CMAQ module. NUFP dramatically increase in the immediate vicinity of the airport locations due to changes in aircraft emitted PSD. NUFP at the Los Angeles international airport (LAX) from sensitivity scenario are 5.2 times higher than from background scenarios. Although mass concentrations in accumulation modes in airports were marginally lower, NUFP values were higher in sensitivity than in base scenarios. This study is the first CMAQ-based study to look into ambient particle matter impacted by PSD changes in emissions from an individual source sector. We will present results from this study focusing on NUFP impacts both from US-wide aircraft activity as well as at individual large airports.

Keywords: A-exposure models, B-particulate matter

MO-SY-G1-101

Local- and national-scale impacts of emissions from leaded aviation gasoline

P. J. Wolfe; Self, Ann Arbor, MA

Abstract: Lead is a persistent toxic pollutant that can impact development, impair children's IQ, and can generate lower earnings potentials among other physical and cognitive impacts. In the United States, general aviation piston-engine aircraft are now the largest source of anthropogenic lead emitted to the atmosphere. At the site of maximum concentration at an airport serving piston-engine aircraft, 3-month average atmospheric lead concentrations attributable to emissions from leaded avgas have the potential to exceed the National Ambient Air Quality Standard for Lead. Modeled atmospheric lead concentrations from aircraft landing and takeoff operations are distinguishable from the background up to 900 m downwind of the airport. On a national-scale, lead emissions from all phases of flight have the potential to contribute to small increases in ambient lead concentration. The impact of the emissions on atmospheric concentrations is modeled at the national-scale using the community multi-scale air quality model (CMAQ). These concentrations are used to quantify the impacts of annual aviation lead emissions on IQ loss of an annual cohort of 1-year old children in the United States. The average modeled aircraft-attributable lead impacts on potential earnings reductions to be $1.06 [0.16 - 10.1] billion 2006 USD in annual damages, and that dynamic economy-wide methods result in damage estimates that are 54% larger. Results on the local-and national-scale are sensitive to key modeling parameters and continued and future work is necessary to better characterize these parameters and assumptions. Thus, it is crucial to understand aircraft lead emissions as a multi-scale problem. Further, because the marginal costs of lead are dependent on background concentration, the costs of piston-driven aircraft lead emissions are expected to increase over time as regulations on other emissions sources are tightened.

Keywords: A - population exposure, B-metals, D-children

MO-SY-G1-102

Crossover and Distinction between Methods, Models, and Data to Assess Long-term Aviation-related Air Pollutant Exposures at Local, Regional, and National Spatial Scales

N. Hudda[1], T. Larson[2], P. J. Wolfe[3], S. Arunachalam[4], J. I. Levy[5]; [1]Tufts University, Medford, MA, [2]University of Washington, Seattle, WA, [3]MIT, Cambridge, MA, [4]University of North Carolina, Chapel Hill, NC, [5]Boston University, Boston, MA

Abstract: Aviation emissions impact the environment and human health at several scales; they contribute to climate change and the global burden of disease, impact regional air quality, and impact air quality and health for communities in the vicinity of the airports. Assessment of air pollution exposures at each of these scales requires distinct methods, models, and data. Moreover, insights from analyses at one scale may inform investigations at other scales. For example, while elevated ultrafine particulate matter (UFP) concentrations were noted for some time in the immediate vicinity of the airport, several studies now report elevated levels of aviation-attributable UFP and other air pollutants at larger spatial scales (~10 kilometers). Long-term exposure assessment tools at such spatial scales are sparse, with limitations in both monitoring-based approaches and atmospheric dispersion models. At larger spatial scales, it is also critical to distinguish aviation emissions impacts from those from surface transportation emissions. New data and methods are emerging that can quantify the contributions of aviation emissions at local and regional scales. The symposium aims to bring together researchers that are (a) measuring pollutant concentrations and applying statistical approaches to model the air pollution impacts of aviation activity, (b) developing methods to model long-term air pollution and noise exposures attributable to aviation at local and regional scales, and (c) assessing the national health and climate change impacts of aviation emissions. A panel will discuss crossover and distinction between methods for assessing long-term air pollution exposures resulting from aviation emissions at local, regional and national scales.

Keywords: A - ambient monitoring, A - exposure measurement, A - population exposure, A-environmental justice, A-built environment

MO-PL-A2: Wrist Bands As Sensors

MO-PL-A2-103

Passive Sampling-Silicone Wristbands Are Not Just A Fashion Statement

H. Rubenstein[1], K. A. Anderson[2], G. Wilson[2], R. Scott[2], F. Kathy[1]; [1]USAF, Beavercreek, OH, [2]Oregon State University, Corvallis, OR

Abstract: Military members are exposed to unique operational environments that include toxic industrial chemicals and materials. Capturing these exposures using current methods is often problematic due to supply logistics, power requirements, equipment ruggedness, and shipping to analytical labs outside of theater. Research to provide a low cost, rugged, sensitive, specific, small, lightweight dosimeter that can integrate doses over time will be presented. This sampler will provide information for exposure records. A specially prepared silicon wristband (SWB) was evaluated for 45 volatile organic compounds (VOCs) including those found in aircraft exhaust (i.e., JP8 and BTEX) and in aircraft lubricants/coolants (tri-organophosphates/poly alpha olefins). Previously, SWBs were infused with VOCs/semi-volatile organic compounds (SVOCs). SVOCs and VOCs were determined by thermal desorption. All analytes were analyzed by gas chromatography mass spectrometry. Studies of the effects of shipping were performed and SWBs were infused and stored at 4°C for 7 d. Stability studies were also performed. C-130 Aircraft were sampled on two separate occasions and the results will be discussed. Additional discussion will provide insight to analytical methods in development to discover a wide variety of analytes that military personnel are exposed. The goal of our approach is to have a sampler that is not limited to organic class and can be easily deployed.

Keywords: A-analytical methods, A-risk assessment, D-occupational, A-industrial hygiene, A-longitudinal metrics

MO-PL-A2-104

Comparing chemical exposures across diverse communities using silicone wristbands

H. Dixon, C. Donald, A. Bergmann, G. Points III, R. Scott, B. Smith, K. A. Anderson; Oregon State University, Corvallis, OR

Abstract: Assessing personal chemical exposure is critical to linking environmental contaminants with adverse health effects. Yet, for most environmental chemicals, there is little information about the frequency of personal exposures between different human populations. In this study, silicone wristbands were used to capture chemical detection frequencies in seven distinct populations from Africa, South America, and North America. Wristbands were worn by residents in Senegal (n=70), the Alto Mayo region of Peru (n=69), in Eugene, Oregon (n=25), in rural Ohio (n=26), near the East Coast of the U.S. (n=29), in western Washington (n=10), and children in Corvallis and Bend, Oregon (n=20). Gender and age was also recorded. All 204 wristband extracts were analyzed for the presence/absence of >1,400 chemicals. Combining all locations, the average number of chemical detections was not significantly different for participants of ages 11-20, 21-40, 41-60, and >61 (Tukey-Kramer HSD, $p>0.05$). However, the children in Oregon, making up the <10 age bracket, had a significantly higher average number of chemical detections than all six other populations investigated ($p<0.05$), largely due to flame retardant and plasticizer chemicals. Examining location further, Senegal participants had significantly lower average number of chemical detections than all six other populations ($p<0.05$). In the U.S., Oregon adults had significantly higher average number of chemical detections than both Ohio and East Coast residents ($p<0.001$). There were no significant differences in the number of chemical detections for males compared to females even when stratified by continent or age ($p>0.05$). These results suggest that variables such as age and geographic location may contribute to differences in personal chemical exposures.

Keywords: B-SVOCs, B-flame retardants, A - exposure measurement, A - population exposure, B-phthalates

MO-PL-A2-105

Passive wristband sampling of pesticide exposures among adolescent Latina girls in an agricultural community

K. Harley[1], R. Gunier[1], J. Nolan[1], K. A. Anderson[2], C. Poutasse[2], R. Scott[2], A. Bradman[1]; [1]UC Berkeley, Berkeley, CA, [2]Oregon State University, Corvallis, OR

Abstract: **Background:** Personal exposure to pesticides has not been well characterized, especially among adolescents. Passive sampling using silicone wristbands offers a promising method. **Methods:** Participants were 14-16 year old girls (N=96) living in the agricultural Salinas Valley, California and enrolled in the CHAMACOS cohort study. We measured cumulative pesticide concentrations (ng/g wristband/day) in silicone wristbands worn by girls for one week. Pesticides were extracted using ethyl acetate and concentrations were characterized using gas chromatography electron capture detection. Potential predictors of pesticide levels were collected during home visits. We calculated odds ratios (ORs) for detectable pesticide concentrations using logistic regression models and percentage change in continuous concentrations using Tobit regression models. **Results:** The most frequently detected pesticides in wristbands were fipronil (86%), 4,4'-dichlorodiphenyldichloroethylene (4,4'-DDE) (56%), cypermethrin (55%), dacthal (52%), and permethrin (52%). We observed higher odds of detection (OR =3.2, 95 % Confidence Interval (CI): 1.1-9.7) and five times higher concentrations of dacthal ($p=0.01$) in wristbands of participants living within 100m of agricultural crops. Permethrin concentrations were higher for participants with pesticide products stored in the home (GM: 3.8 vs. 0.6 ng/g/day, $p= 0.008$). Permethrin (GM: 15.1 vs. 1.2 ng/g/day, $p=0.01$) and fipronil (GM: 12.7 vs. 9.7 ng/g/day, $p<0.001$) were lower among participants with door mats in the entryway. We observed no differences comparing participants living in households with agricultural workers to those without. **Conclusions:** Silicone wristband samplers can be used to estimate personal pesticide exposures. While individual behaviors such as use of doormat and consumer pesticides may affect wristband concentrations, evidence indicates that nearby agricultural use of pesticides is also associated with some of the observed exposures.

Keywords: B-pesticides, A - exposure measurement, D-community

MO-PL-B2: Biomass Burning & Health

MO-PL-B2-106

Personal exposure to black carbon among women in semi-rural Mozambique

A. Curto Tirado[1], D. Donaire-Gonzalez[1], M. Manaca[2], R. Gonzalez[4], C. Sacoor[2], I. Rivas[1], G. Wellenius[3], J. Sunyer[1], C. Menéndez[4], C. Tonne[1]; [1]ISGlobal, Centre for Research in Environmental Epidemiology (CREAL), Barcelona, Spain, [2]Centro de Investigação em Saúde da Manhiça (CISM), Manhiça, Mozambique, [3]Brown University School of Public Health, Providence, RI, [4]ISGlobal, Centre for International Health Research (CRESIB), Hospital Clínic- UB, Barcelona, Spain

Abstract: Background: Sub-Saharan Africa has the highest prevalence of people using solid fuels for household energy, which can result in products of incomplete combustion that are damaging for health. Black carbon (BC) is a useful marker of combustion-related particles; however air quality data and temporal patterns of personal exposure to BC in sub-Saharan Africa are scarce. We characterized personal exposure to BC in women from a semi-rural area of southern Mozambique. **Methods:** A total of 204 non-smoking women were randomly selected from the demographic surveillance system of the Manhiça Health Research Centre between 2014-15. We measured 24h personal exposure to BC using a portable MicroAeth worn in a waist pouch and ambient elemental carbon (EC, highly comparable to BC) with a centrally-located stationary monitor. We used a questionnaire to obtain information on socio-demographic, household characteristics, and cooking/lighting use. We used backward linear regression to identify predictors of log-transformed 24-h mean BC exposure. **Results:** After excluding participants with poor data quality, we analyzed data from 142 (70%) women with a mean age of 30 years. Participants were largely housekeepers (88%) who principally used firewood (85%) and kerosene (52%) for cooking and lighting, respectively. Mean personal BC was 3.69$\mu g/m^3$ (SD ±2.04) and 24h mean ambient EC was 0.85$\mu g/m^3$ (SD ±0.59). Cooking fuel used during sampling (firewood vs charcoal) and same-day ambient EC were the determinants of exposure that were kept in the model ($p<0.005$). Plots of BC for all participants combined revealed a peak between 6-7pm, suggesting the contribution of lighting sources since the last cooking episode was 3-5pm. **Conclusions:** This study provides temporal patterns of BC exposure and suggests the important contribution of lighting, in addition to cooking, sources to personal exposure to combustion particles in populations that lack access to clean household energy.

Keywords: A - exposure measurement, A-indoor environment, C-air, A - ambient monitoring

MO-PL-B2-107

Biomonitoring Human Exposure to Household Air Pollution and Association with Self-reported Health Symptoms - A Stove Intervention Study in Peru

Z. Li[1], A. Commodore[2], S. Hartinger[3], M. Lewin[1], A. Sjodin[4], E. Pittman[4], D. Trinidad[4], K. Hubbard[4], C. Lanata[5], A. Gil[5], D. Mäusezahl[6], L. Naeher[7]; [1]Agency for Toxic Substances and Disease Registry, Atlanta, GA, [2]Medical University of South Carolina, Charleston, SC, [3]Universidad Peruana Cayetano Heredia, Lima, Peru, [4]Centers for Disease Control and Prevention, Atlanta, GA, [5]Instituto de Investigación Nutricional, Lima, Peru, [6]Swiss Tropical and Public Health Institute, Basel, Switzerland, [7]University of Georgia, Athens, GA

Abstract: BACKGROUND: Household air pollution (HAP) from indoor biomass stoves contains harmful pollutants, such as polycyclic aromatic hydrocarbons (PAHs), and is a leading risk factor for global disease burden. We used biomonitoring to assess HAP exposure and association with self-reported symptoms in 334 non-smoking Peruvian women to evaluate the efficacy of a stove intervention program. METHODS: We conducted a cross-sectional study within the framework of a community randomized control trial. Using urinary PAH metabolites (OH-PAHs) as the exposure biomarkers, we investigated whether the intervention group (n=155, with new chimney-equipped stoves) were less exposed to HAP compared to the control group (n=179, with mostly open-fire stoves). We also estimated associations between the exposure biomarkers, risk factors, and self-reported health symptoms, such as recent eye conditions, respiratory conditions, and headache. RESULTS: We observed reduced headache and ocular symptoms in the intervention group compared to the control group. Urinary 2-naphthol, a suggested biomarker for inhalation PAH exposure, was significantly lower in the intervention group (GM with 95% CI: 13.4 [12.3, 14.6] µg/g creatinine) compared to control group (16.5 [15.0, 18.0] µg/g creatinine). Most other OH-PAHs followed similar patterns (i.e. control > intervention), but the differences were smaller and were not statistically significant. Stove type and/or 2-naphthol was associated with a number of self-reported symptoms, such as red eye (adjusted OR with 95% CI: 3.80 [1.32, 10.9]) in the past 48h. CONCLUSIONS: Even with the improved stoves, median concentrations of the 10 OH-PAHs were equivalent or several times higher than the 95th percentile in the U.S. adults. In addition, median 1-hydroxypyrene was higher than a proposed no-observed-genotoxic-effect-level, indicating high exposure and a potential for increased cancer risk in the population.

Keywords: A-biomonitoring, A-epidemiology, A-indoor environment, A-biomarkers

MO-PL-B2-108

Elevated blood pressure and household solid fuel use in premenopausal women: Analysis of 12 Demographic and Health Surveys (DHS) from 10 countries

R. E. Arku[1], M. Ezzati[2], J. Baumgartner[3], G. Fink[4], P. Hystad[5], M. Brauer[1]; [1]University of British Columbia, Vancouver, Canada, [2]Imperial College London, London, United Kingdom, [3]McGill University, Montreal, Canada, [4]Harvard T.H. Chan School of Public Health, Boston, MA, [5]Oregon State University, Corvallis, OR

Abstract: Background/Aim Approximately three billion people are exposed to household air pollution (HAP) from solid fuel cookstoves. Studies from single settings have linked HAP with elevated blood pressure (BP), but no evidence exists from multi-country analyses. Using nationally representative and internationally comparable data, we examined the association between solid fuel use and BP in 77 605 largely premenopausal women (aged 15-49) from ten resource-poor countries. Methods We obtained data on systolic and diastolic BP, self-reported primary cooking fuel, health and socio-demographic characteristics from 12 Demographic and Health Surveys conducted in Albania, Armenia, Azerbaijan, Bangladesh, Benin, Ghana, Kyrgyzstan, Lesotho, Namibia, and Peru. We estimated associations between fuel use [solid fuel (i.e., coal or biomass) versus clean fuel (i.e., electricity or gas)] with systolic and diastolic BP and hypertension by a meta-analytical approach. Results Overall systolic and diastolic BP (SD) were 117 (16) and 74 (11) mmHg, respectively. Systolic BP was, higher by ~ 2 mmHg among electricity or gas users compared to those who cooked using solid fuels (118 mmHg vs. 116 mmHg; 95% CI: 1.4-2.0 mmHg), however, the relationship reversed in the multivariate analysis. After adjusting for age, BMI, ethnicity, education, occupation, wealth index, rural/urban location, and month of interview in country-specific multivariate analysis, use of solid fuels was associated with summary 0.58 mmHg higher systolic BP (95% CI: 0.23, 0.93), and higher (summary) odds of hypertension [OR=1.08 (95% CI: 0.99, 1.16)], especially among rural women (OR=1.16 [95% CI: 1.01, 1.35]), when compared with cleaner fuels. There was no significant statistical heterogeneity across country-specific estimates. Conclusions Cooking with solid fuels was associated with small BP increases. Use of cleaner fuels like gas or electricity may reduce cardiovascular risk in developing countries.

Keywords: A-global health, A-epidemiology, A-indoor environment, A-exposure factors, C-air

MO-SY-C2: Tribal Sovereign Rights, Knowledge and Data Protections and Protocols in the Environmental Health Sciences

MO-SY-C2-109

Native American Knowledge Systems: Sovereign Rights, Protections and Protocols

C. Lefthand-Begay; University of Washington, Seattle, WA

Abstract: Who owns, protects and disseminates the knowledge systems and data among tribal communities and their Nations? For several decades, tribal nations in the United States have established internal mechanisms and protocols to protect their knowledge from appropriation, exploitation and misuse. In part, this has occurred in response to ethical concerns regarding the handling of biological and environmental specimens, the exclusion of key leaders in the design and implementation of research and the erroneous dissemination of results. Similar apprehensions are echoed in conversations across local, national and international communities who have a history of marginalization and environmental injustice. Furthermore, academicians and institutions are taking an interest in the convergence between western science and indigenous ways of knowing, including traditional ecological knowledge (TEK). The popularity of TEK has increased as our global society grapples with defining and measuring sustainability factors. Both indigenous and non-indigenous groups promote such convergence and agree that research practices must be of value to stakeholders by aligning with tribal and indigenous values. In our research, we have worked with tribal nations who are faced with abandoned mine waste site and subsequent water pollution, and who are challenged by revealing unique exposure pathways due to subsistence food practices and ceremonial uses. In this conference proposal, we will contribute to the groundwork of others who work to develop ethical research practices in environmental health and exposure sciences. First, we will outline the complexities tribes face when protecting their communities, convey the work among tribal citizens to revitalize their knowledge systems, and the global and local protections and protocols designed by sovereign tribes in the United States. Second, we'll share our research methodologies to communicate data and results about water contamination tribal leaders and individual stakeholders. These practices are useful to those working with Indigenous and other diverse communities around the world.

Keywords: A-environmental justice, D-First Nation, C-water, A-ecological exposure, D-community

MO-SY-C2-110

Protections and communication for tribal entities involved in research

M. Gonzales[2], E. Erdei[1]; [1]University of New Mexico College of Pharmacy, Albuquerque, NM, [2]University of New Mexico School of Medicine, Albuquerque, NM

Abstract: Native American tribes are increasingly use of designated Institutional Review Boards to provide review, approval and ongoing oversight of research involving Native participants. In this capacity, tribal IRBs examine research protocols for potential risks to specific native groups to which investigators may be unaware. In addition to identifying these risks for future researchers, we will also provide examples of how Native concepts can be integrated into research protocols, the reporting of results back to participants and communities and to tribal IRBs. The first example examines the approval disaster research protocols for Native communities and first responders who will come into direct contact with environmental contaminants when natural disasters occur on or near Native lands. Recognizing the rapid nature of exposure to communities and during response work, scalable research protocols for health effects are increasingly available. We will examine the approval of such protocols to be both sensitive to Native perspectives and inclusive of health effects in Native workers and communities exposed to environmental contaminants in the aftermath of disasters. The second example examines the important role of Native research teams in translating research results back to communities and tribal IRBs drawing illustrations from the UNM Navajo Birth Cohort Study. With significant input from Navajo research team members, communication also emphasizes risk avoidance and prevention of exposures as an appropriate way to build environmental health knowledge base in uranium mine waste exposed communities and across Navajo Nation. We will summarize the format used to inform Tribal leadership, the Human Research Review Board and to the Navajo Area Indian Health Service medical providers.

Keywords: D-community, D-susceptible/vulnerable, A-emergency response, D-First Nation, D-occupational

MO-SY-C2-111

Communicating Home Well Water Quality Results to Families

J. Doyle[2], M. J. Eggers[1]; [1]Montana State University Bozeman, Bozeman, MT, [2]Little Big Horn College, Crow Agency, MT

Abstract: On the Crow Reservation in Montana, the Crow Water Quality Project is a community-engaged research endeavor led by the Crow Environmental Health Steering Committee (CEHSC), a group of Tribal stakeholders and academic partners. About half of all Tribal families are dependent on well water. One of the CEHSC's top priorities is to decrease people's exposures to uranium, manganese, arsenic and nitrate via drinking and/or cooking with home well water. The cumulative risk from these inorganic contaminants varies from 64% to 11% of wells, depending upon the watershed. Secondary contaminants such as iron, sulfate and hardness complicate water treatment and create additional challenges for families. We offer free home well water testing, interpretation of these results and discussions about options for safer water. We will highlight lessons learned about working with families to test for well water contaminants, sharing the test results in a way that is usable and understandable, and providing affordable mitigation options for coping with unsafe and/or unpalatable well water. Maintaining an open dialogue to share with and learn from community members about water is essential. We continue to learn more about water and health, and are committed to using this information to improve lives in our community.

Keywords: D-First Nation, C-water, B-metals, risk communication, A-cumulative exposure

MO-SY-D2: US EPA STAR Program: Novel Research for 21st Century Exposure Science - Part 2

MO-SY-D2-112

A Suspect Screening Method for Characterizing Multiple Chemical Exposures among a Demographically Diverse Population of Pregnant Women in San Francisco, CA

A. Wang[1], J. M. Schwartz[2], R. R. Gerona[3], T. Lin[3], M. Sirota[4], R. Morello-Frosch[5], D. Woodruff[2]; [1]University of California, San Francisco, San Francisco, CA, [2]University of California, San Francisco, San Francisco, CA, [3]University of California, San Francisco, San Francisco, CA, [4]University of California, San Francisco, San Francisco, CA, [5]University of California, Berkeley, Berkeley, CA

Abstract: *In utero* exposure to multiple environmental chemicals can adversely impact pregnancy outcomes and increase risks of disease throughout the lifespan. To better understand human exposure to potentially hundreds of environmental chemicals, we perform a suspect screen for the presence of 725 Environmental Organic Acids (EOAs) in maternal serum, to characterize the exposure profile of EOAs, and to identify novel chemicals, using liquid chromatography-quadrupole time-of-flight mass spectrometry (LC-QTOF/MS). We analyzed serum samples collected at delivery from 83 pregnant women receiving care at San Francisco General Hospital and UCSF Mission Bay Medical Center, using Agilent LC 1260-QTOF/MS 6550. We examined demographic differences in detection frequency (DF) and peak areas of candidate compounds using a Fisher's exact test, analysis of variance, or Kruskal-Wallis Rank Sum test. We confirmed prioritized suspect compounds by running their reference standards. We detected, on average, 60 (SD: 8) suspect EOAs in each sample (range: 34-87), the majority of which were matched to phenols and phenolic pesticide metabolites. There were 15 suspect EOAs with high DF (≥80%) in our sample population, over half of which were potential novel chemicals that have not been previously biomonitored. We found differences by race/ethnicity, household income, and education in DF for several suspect EOAs as well as in relative concentration (measured by peak areas) for some highly detected suspect EOAs. We confirmed the presence of 6 novel compounds maternal serum: 2,4-Di-tert-butylphenol, 2,4-Dinitrophenol, Pyrocatechol, 3,5-Di-tert-butylsalicylic acid, 4-Hydroxycoumarin, and 3'-Hydroxyacetophenone. Suspect screening approaches to biomonitoring for environmental chemicals provides a viable method to more holistically characterize a broad spectrum of environmental chemicals and to identify novel, ubiquitously present compounds and thus prioritize chemicals for targeted method development.

Keywords: A-analytical methods, A-biomonitoring, A-chemical prioritization, A - exposure measurement, D-prenatal

MO-SY-D2-113

Toxicogenetics of PERC Metabolism and Toxicity: Collaborative Cross Mouse Population Approach to Address Remaining Gaps in Human Health Assessments

I. RUSYN, J. Cichocki, W. Chiu, D. Threadgill; TEXAS A&M UNIVERSITY, College Station, TX

Abstract: Tetrachloroethylene (Perchloroethylene, PERC) is a high-production chemical of great concern to both risk assessors and public health officials worldwide. It has long been assumed that PERC metabolism and toxicity closely mimic those of trichloroethylene (TCE), a structurally similar chlorinated solvent. However, recent human health assessments of these chemicals determined that major differences in toxicokinetics and toxicodynamics exist between TCE and PERC. Furthermore, it was concluded that critical gaps remain in understanding of the human health hazard of PERC, including toxicokinetics, toxicodynamics and population variability. Thus, the long-term objective of our project was to uncover the mechanistic linkages between the genome (e.g., variation in DNA sequence among individuals), metabolism (e.g., formation of organ-specific toxic intermediates), and adverse molecular events (e.g., transcriptional changes associated with toxicity) in response to PERC. The central hypotheses of this proposal were (i) genetic variability-associated differences in PERC metabolism affect organ-specific toxicity of PERC; and (ii) a population-based experimental design utilizing Collaborative Cross (CC) can be used to exploit the variability in toxicity responses to better characterize uncertainties in human health assessments. We have successfully characterized variability in the toxicokinetics and toxicodynamics of PERC by using the Collaborative Cross mouse model of the human population. In addition we evaluate the effects of nonalcoholic fatty liver disease (NAFLD), an increasingly prevalent condition that can dramatically affect chemical metabolism, on PERC-associated liver and kidney effects.

Keywords: D-susceptible/vulnerable, A-analytical methods, A-environmental regulation, A-risk assessment, D-occupational

MO-SY-D2-114

Panel Discussion

M. Fry; US Environmental Protection Agency, Washington, DC

Abstract: This symposium showcases the findings from six US EPA STAR grantees who have developed novel methods for characterizing indoor chemical exposures and improved our understanding of the variability in human response to chemicals and susceptibility. This body of exposure science research advances exposure assessment methods, models, and data. STAR grantees from Virginia Polytechnic Institute and State University, University of California Davis, and University of Michigan have developed new devices and methods for rapid, portable measurement of SVOCs indoors from consumer products. Duke University and University of California San Francisco STAR grantees have provided innovative measurements of indoor chemical exposures among vulnerable subpopulations and diverse communities: children and pregnant women. Finally, Texas A&M University STAR grantee has successfully characterized the susceptibility and variability in human response to tetrachloroethylene exposure. The results from these STAR research grants advance rapid assessment methods for evaluating human exposure to chemicals and expand the knowledge base for further development of chemical exposure pathways and models.

Keywords: A-exposure models, A-indoor environment, A-sampling methods, B-SVOCs

MO-PL-E2: Food Packaging

MO-PL-E2-115

A Quantitative Property-Property Relationship for Estimating Packaging-Food Partition Coefficients of Organic Compounds

L. Huang[1], A. Ernstoff[2], H. Xu[1], S. Lu[1], P. Fantke[3], O. Jolliet[1]; [1]University of Michigan, Ann Arbor, MI, [2]Quantis Lausanne, Lausanne, Switzerland, [3]Technical University of Denmark, Lyngby, Denmark

Abstract: Organic chemicals encapsulated in beverage and food packaging can migrate to the food and lead to human exposures via ingestion. The packaging-food (K_{pf}) partition coefficient is a key parameter to estimate the chemical migration from packaging materials. Previous studies have simply set K_{pf} to 1 or 1000, or provided separate linear correlations for several discrete values of ethanol equivalencies of food simulants (EtOH-eq). The aim of the present study is to develop a single quantitative property-property relationship (QPPR) valid for different chemical-packaging combinations and for water or different EtOH-eq values. We compiled datasets of measured K_{pf} from 3 studies, which contained 302 data points of 152 chemicals in LDPE and HDPE (low and high density polyethylene) at 25 °C for EtOH-eq values ranging from 0% (water) to 95%. A multiple linear regression (MLR) model was developed to predict K_{pf} as a function of the chemical's K_{ow}, the EtOH-eq, the packaging type and an interaction term between K_{ow} and EtOH-eq. The model shows good fitting performance of the experimental datasets with adjusted R-square of 0.92. All predictors are highly significant except the packaging type, probably because only two packaging types are included. This preliminary QPPR demonstrates that the K_{pf} for various chemical-packaging-food combinations can be estimated by a single linear correlation. Based on more than 1000 collected K_{pf} in 15 materials, we will present extensive results for other packaging types and different temperatures. This QPPR provides a comprehensive correlation method to estimate the K_{pf} for a wide range of chemical-packaging-food combinations, and thus facilitate high-throughput estimates of human exposures to chemicals encapsulated in food contact materials.

Keywords: A-exposure models, C-consumer products, C-food, C-indoor

MO-PL-E2-116

Migration modeling to estimate exposure to chemicals in food packaging for application in high-throughput risk-based screening and Life Cycle Assessment

A. S. Ernstoff[1], O. Jolliet[2], L. Huang[2], P. Fantke[1]; [1]DTU, Kgs. Lyngby, Denmark, [2]University of Michigan, Ann Arbor, MI

Abstract: Specialty software and simplified models are often used to estimate "worst-case" migration of potentially toxic chemicals from packaging into food. Current approaches, however, cannot efficiently and accurately provide estimates of migration for emerging applications, e.g. in Life Cycle Assessment and risk prioritization and screening. To fulfill the need for a migration model flexibly suitable for such tools, we develop an accurate and rapid (high-throughput) approach. The developed model estimates the fraction of an organic chemical migrating from polymeric packaging into food for user-defined scenarios and requires limited parameters (i.e. physicochemical properties). Several hundred step-wise simulations optimized the coefficients of the model to cover a wide-range of scenarios (e.g. packaging thickness, food etc.). The developed model, implemented in a disseminatable spreadsheet, nearly instantaneously estimates migration from packaging into food for user-defined scenarios, and has improved performance over common model simplifications. The common practice of setting the package-food partition coefficient = 1 for specific "worst-case" scenarios is insufficient to predict the equilibrium concentration in food for diverse scenarios. Therefore a partition coefficient model, as a function of a chemical's octanol-water partition coefficient and a food's ethanol-equivalency, was also developed. When using measured diffusion coefficients the model accurately predicted ($R^2 = 0.9$, SE = 0.5) hundreds of empirical datapoints for various scenarios. Diffusion coefficient modeling, which determines the speed of chemical transfer from package to food, was found as a major contributor to uncertainty and decreased model performance ($R^2 = 0.5$, SE = 1). In all, this study provides a migration modeling approach that rapidly estimates the fraction migrated for emerging screening and prioritization approaches. To estimate exposure, chemical concentrations in packaging are essential.

Keywords: A-exposure models, C-food, A-life cycle analysis, A-chemical prioritization, C-consumer products

MO-PL-E2-117

Consumer Exposures to Engineered Nanomaterials through Dietary Sources

K. Grieger; RTI International, Research Triangle Park, NC

Abstract: Engineered nanomaterials (ENMs) are currently used in a wide range of consumer products, including food, dietary supplements, and food packaging. In addition, ENMs are also being developed for advanced water treatment technologies to help secure safe, sustainable water supplies. The use of ENMs in consumer products such as food and drinking water applications has been attractive to material scientists, engineers, and product developers given that ENMs often display unique, novel properties compared to their bulk counterparts. At the same time, there have also been safety concerns and questions raised regarding the use of ENMs in consumer products and other applications. The overall extent to which consumers are exposed to ENMs through dietary sources such as food and drinking water is still largely unknown at this time. This is largely due to the lack of robust analytical techniques for detection, characterization, and quantification of ENMs in food and drinking water matrices, as well as other challenges related to assessing consumer and environmental exposures to ENMs. This presentation provides an overview of consumer exposures to ENMs through dietary sources along with some concerns regarding these exposures. Recommendations are also made to stakeholder groups for the sustainable development and use of ENMs in food and drinking water treatment applications.

Keywords: A-nanotechnology, B-nanoparticles, C-consumer products, C-food

MO-SY-F2: Pathways and mechanisms linking exposure to the natural environment with health and well-being benefits - Part 2

MO-SY-F2-118

Using Health Effect Biomarkers to Characterize Benefits of Urban Green Spaces

A. I. Egorov, S. Griffin, R. Converse, J. Styles, E. Sams, A. Wilson, R. Baldauf, V. Isakov, L. E. Jackson, T. Wade; United States Environmental Protection Agency, RTP, NC

Abstract: Greater availability of urban green spaces has been linked to improved mental and physical health, and reduced mortality. Relative contributions of hypothesized pathways to improved health, such as relaxation and stress alleviation; enhanced physical activity; reduced exposure to air pollution, noise and heat; and beneficial exposure to natural allergens and microbes, remain to be characterized. The objective of an ongoing research project at U.S. EPA is to characterize subtle biological changes associated with long-term exposure to green space using a panel of biomarkers of metabolic, neuroendocrine and immune function, and to produce novel information on pathways to health. A pilot cross-sectional study in the Durham-Chapel Hill, NC area used 1-meter resolution data on trees and herbaceous vegetation within 500 m of residences derived from the U.S. EPA EnviroAtlas land cover dataset. Eighteen health effect biomarkers were measured in serum or saliva samples from more than 200 adults. Regression models controlled for demographic covariates and spatial autocorrelation. Increased proportion of vegetated land cover (trees and grass) was associated with: 1) a highly significant reduction of a composite biomarker-based measure of physiological dysregulation known as allostatic load; 2) reduced odds of having potentially unhealthy levels of individual biomarkers associated with chronic inflammation and chronic stress; and 3) reduced odds of previously diagnosed depression. Ongoing analysis aims to characterize residential exposures of study participants to air pollutants and noise from local traffic using the Community-LINE Source Model (C-LINE) developed by U.S. EPA and the University of North Carolina, and to assess potential effects of green barriers and green spaces in mitigating detrimental effects of gaseous and particulate air pollution on health biomarkers. Further research efforts aim to evaluate associations between various measures of exposure to urban greenery, stress, physical activity, exposure to air pollutants, human microbiome, allostatic load and susceptibility to infectious and non-communicable diseases in prospective settings. *This abstract does not represent EPA policy.*

Keywords: A-biomarkers, A-built environment, A-epidemiology, A-geospatial analysis/GIS

MO-SY-F2-119

The promises of the General Social Survey (GSS) as a research platform for evaluating well-being benefits from the natural environment

T. Smith; NORC, Chicago, IL

Abstract: The General Social Survey is the largest and longest-running project supported by the Sociology Program of the National Science Foundation. Since 1972 it has monitored societal change in the United States with annual or biennial national, full-probability in-person samples of adults living in households. Starting in 1982 it has had a cross-national component. In 1984 it co-founded the International Social Survey Program (ISSP) and the ISSP has conducted an annual cross-national studies since 1985. Sixty countries have participated in the ISSP. In recent years the GSS pioneered in developing the multi-level, multi-source (MLMS) study design. MLMS augments survey data by adding in information from other sources about the housing unit, neighborhood, and community that respondents reside in. This includes socio-demographic data from the census and ACS, socio-economic information from commercial databases, and much other data from a multitude of sources (e.g. election return, crime statistics, educational facilities, etc.). Existing MLMS information can be further augmented with data about the environment, land use, proximity to features in nature, etc. Both the wide ranging content of the GSS and the auxiliary data from MLMS can complement survey questions asking about the natural environment and assess both object and subjective factors that shape attitudes about the natural environment and the impact that the natural environment has on health and well-being.

Keywords: D-community

MO-SY-F2-120

Assessing the Effect of the Natural Environment on Subjective Well-being

R. A. Silva[1], S. Prince[2], F. Cochran[1], A. Neale[2], K. Rogers[2], T. J. Buckley[2]; [1]U.S. Environmental Protection Agency, Research Triangle Park, NC, [2]U.S. Environmental Protection Agency, Research Triangle Park, NC

Abstract: Exposure to the natural environment has a potentially large positive influence on human well-being. Thus, there has been growing interest in understanding the well-being benefits derived from that exposure and establishing causal pathways as they relate to physical health, psychological and social functioning, and subjective well-being. Although natural environments - from pristine natural areas to urban green infrastructure - have an objective impact on human beings related to the provision of life-supporting ecosystem services (e.g., air, water, food, etc.), additional effects are influenced by subjective factors, including individual perceptions and behavior, and social context. The objective of this presentation is to contribute to a better characterization of the size of the salutogenic effects of the natural environment in the US, as well as the mediators and modifiers that influence that association. We will present results from a study linking data from a nationwide survey on self-reported health and subjective well-being, as well as socio-demographic factors, such as age, sex, race/ethnicity, education, household income, social interaction, with measures of ecosystem services derived from the EPA's online tool EnviroAtlas. EnviroAtlas contains high-resolution geospatial data for many objective measures of natural environments (e.g. tree buffer around roadways, natural land cover, forest cover, access to parks, view of trees, view of water, presence of green space, climate, air quality, etc). This analysis will focus on benefits derived from the presence of natural environments. However, salutogenic effects depend on one's own experience of nature on a daily basis. Thus, we will discuss the importance of better assessing subjective factors such as connectedness to nature and past experiences, as well as perception of access, quality and safety of natural environments, motivation for exposure to natural environments and potential barriers.

Keywords: A-epidemiology, A-behavior, A-exposure models, Natural Environments, Individual Well-being

MO-SY-G2: Road Work Ahead: Progress in Assessing and Mitigating Exposure to Traffic-Related Air Pollution

MO-SY-G2-121

Strategies to reduce exposures to traffic-related air pollution at the local level

R. Baldauf; US EPA, Durham, NC

Abstract: As public health concerns for populations living, working and going to school near high-traffic roadways have increased, so has the need to identify and implement air pollution control strategies effective at the local level to protect the millions of people exposed to traffic-related emissions in these near-road microenvironments. While strategies implemented at the federal and state level, such as motor vehicle emission standards and vehicle activity reduction measures, are vital components of reducing exposures to traffic-related air pollutants, these control strategies can take a long time to implement, often requiring technology development, fleet turnover and infrastructure modifications. Air pollution mitigation options available to state and local transportation agencies, local community leaders, urban planners, and developers have also been identified and evaluated that can enhance reductions in air pollution exposures for near-road populations. These options can include improving indoor and ambient air quality. This presentation will summarize research and best practice recommendations for ambient air pollution mitigation strategies that can be implemented at the local level to reduce near-road air pollution exposures which can include improved land use and site layout designs, roadway configurations, and the use of roadside solid and vegetation barriers.

Keywords: A-built environment, B-particulate matter, C-air, D-environmental justice, Traffic

MO-SY-G2-122

Development of Internal Indicators of Primary Traffic Exposures Using Environmental Metabolomics in the Dorm Room Inhalation to Vehicle Emissions (DRIVE) Study

D. Liang[1], J. L. Moutinho[2], R. Golan[3], T. Yu[4], C. N. Ladva[1], M. Niedzwiecki[1], S. E. Sarnat[1], H. Chang[4], R. Greenwald[5], D. P. Jones[6], A. Russell[2], J. A. Sarnat[1]; [1]Emory University, Atlanta, GA, [2]Georgia Institute of Technology, Atlanta, GA, [3]Ben Gurion University of the Negev, Beer Sheva, Israel, [4]Emory University, Atlanta, GA, [5]Georgia State University, Atlanta, GA, [6]Emory University, Atlanta, GA

Abstract: Aim. High-resolution metabolomics (HRM) has recently emerged as a sensitive tool for measuring environmental exposures and corresponding biological response. Use of HRM in air pollution epidemiologic studies holds promise to improve internal exposure estimation to complex air pollution mixtures, including primary traffic emissions. **Methods.** The Dorm Room Inhalation to Vehicle Emission (DRIVE) study was conducted to measure an extensive suite of pollutants at multiple ambient and indoor sites along a major highway artery. In this panel-based study, 54 students living in dormitories either near (20 m) or far (1.4 km) from the highway conducted personal sampling and contributed bio samples (plasma and saliva). Untargeted HRM were used to identify potential metabolic pathways associated with traffic-related air pollutants in the panel. **Results.** Weekly levels of traffic pollutants were significantly higher at the near dorm compared to the far dorm ($p<0.05$). In total, 20,766 metabolites were reliably extracted from plasma samples. 847 features in a HILIC column and 444 features in a C18 column were significantly associated with at least one or more traffic indicator, including black carbon, carbon monoxide, nitrogen oxides and particulate matter ($p<0.05$), when controlling for covariates and false discovery rate. Pathway analysis indicated elicitation of several inflammatory pathways, including leukotriene metabolism and anti-inflammatory metabolites formation. In particular, 15 features were putatively matched with leukotriene B4 and leukotriene E4, all of which exhibited significant alteration of intensity associated with traffic air pollutant level. **Conclusions.** HRM is a sensitive, powerful tool for air pollution epidemiologic studies to examine the metabolic response to complex traffic exposures, and comprehensive chemical validation can help to further develop the biomarker of observed traffic pollution health effects.

Keywords: A-biomarkers, A - exposure measurement, A-biomonitoring, C-air, Exposome, Metabolomics

MO-SY-G2-123

Enhancing models and measurements of traffic-related air pollutants using a spatio-temporal Bayesian spatial model.

V. Berrocal[1], S. Batterman[2], O. Gilani[3]; [1]University of Michigan, Ann Arbor, MI, [2]University of Michigan, Ann Arbor, MI, [3]Bucknell University, Lewisburg, PA

Abstract: Measurements of traffic-related air pollutants concentrations are often limited, and spatially and temporally sparse. In this presentation, we discuss and present potential spatio-temporal statistical modeling approaches that can be employed to obtain enhanced estimates of traffic-related pollutants concentrations. The models considered include: (i) Bayesian hierarchical spatio-temporal models that model the non-stationary spatial dependence between traffic-related pollutants at any two sites as a function of their distance, difference in wind speed and wind direction; and (ii) Bayesian hierarchical data fusion models that leverage the information contained in the output of the dispersion model RLINE. Using observations of NO, NO_x, $PM_{2.5}$ and BC concentrations measured along nine transect areas that cross major highways in Detroit during seven consecutive days in December 2012, we evaluate the advantages and limitations of each model.

Keywords: A-geospatial analysis/GIS, A-statistical methods, A - exposure measurement, A - ambient monitoring, dispersion model

MO-PL-A3: Spatial Temporal Modeling of Air Pollution

MO-PL-A3-124

Modeling Historic Air Pollution Concentrations with Land Use Regression in Tucson, AZ

N. Lothrop, M. Bell, H. Brown, M. Furlong, S. Guerra, M. K. O'Rourke, P. I. Beamer; University of Arizona, Tucson, AZ

Abstract: Numerous studies have linked air pollution exposure to respiratory health problems. However, air pollutant sampling is seldom completed at participant homes in studies investigating respiratory health outcomes due to costs, resulting in spatially limited data on air pollution exposures. When no or limited air pollution measurements are taken, land use regression (LUR) modelling is a common method for predicting air pollution levels. Using LUR, we retrospectively predicted outdoor annual average traffic-related air pollution (nitrogen dioxide or NO_2) levels, which were measured with Palmes diffusion tubes at 146 sites in the Tucson, AZ area in 1987. Multiple samples collected at each site were corrected for seasonal variation to obtain the annual average outcome (mean=15.5 ppb, std. deviation=5.04 ppb). Predictors were selected from the European Study for Cohorts for Air Pollution Effects (ESCAPE) protocols (e.g., roadways, population density, land uses), with Tucson-specific variables added (e.g., surface mines). We tested the efficacy of the ESCAPE and a Vancouver, Canada model building approaches, both of which use supervised forward stepwise variable selection procedures. The Vancouver approach performed marginally better (increased R^2 leave-one-out cross validation) than the ESCAPE approach (R^2 = 0.61 vs 0.58, 8 vs 5 predictors). Regardless of approach, several predictors were common, including: elevation, commercial land use, nearby (major) roadway length, and distance to the nearest active surface mine. The relatively large number of samples taken in 1987, the spatial representation of the data set, and the historic nature of the outdoor NO_2 data make this study a unique opportunity. Predicted outdoor NO_2 concentrations will be matched to study participant homes for various respiratory disease cohorts with almost 40 years of health data from Tucson, AZ to better understand the relationship between traffic-related air pollution and respiratory health outcomes.

Keywords: D-children, A-exposure models, A-geospatial analysis/GIS, A-built environment, D-Southwest-specific

MO-PL-A3-125

Analyzing NO2 concentration variations from 2005 to 2016 over the atmosphere of Kazakhstan using Satellite data

M. Amouei Torkmahalleh[1], Z. Darynova[1], T. Holloway[2]; [1]Nazarbayev University, Astana, Kazakhstan, [2]University of Wisconsin Madison, Madison, WI

Abstract: NO_x is the major precursor of ozone and particulate matter (PM), which in turn affect the air quality and climate. NO_x concentrations in the tropospheric layer are highly variable in both space and time, reflecting its short chemical lifetime in the atmosphere, and its nonuniform distribution. NO_x is released to the troposphere as a result of anthropogenic activities, such as fossil fuel combustion, fertilizer application, and prescribed burning. Major sources of energy in Kazakhstan include coal, petroleum, biomass and natural gas. Combustion of these fuels is the main source of pollution in the natural environment. It was reported that In 2014, the highest per capita household coal consumption occurred in Poland (165 kgoe/cap), followed by Kazakhstan (157 kgoe/cap), and Mongolia (104 kgoe/cap). Given the rapid increment in ozone antecedent outflows in East Asia, investigating NOx concentration trends over the Central Asia, particularly Kazakhstan is critical. Particularly, so far no study was found in the literature investigating NO_2 concentrations variability in Kazakhstan, where significant amount of NO_2 is produced due to local combustion. The objective of this study was to investigate the temporal evolution of tropospheric NO_2 in three polluted cities of Kazakhstan including Shymkent, Almaty and Ekibastuz from 2005 to 2016 using the data from Ozone Monitoring Instrument (OMI) launched on the NASA Aura satellite in July 2004. The trends are, for the most part, associated with technological changes in energy use, as well as regional regulatory policies. Over the Almaty, NO_2 levels increased significantly from 2013 to 2015 by 30%. NO_2 concentrations over 3 cities and overall Kazakhstan decreased significantly after 2015 due to implementation of Euro 4 standard for automobiles. For all three cities, concentration of NO_2 reached a maximum during winter season and a minimum during summer, between 2005 and 2016.

Keywords: A-climate change, A - population exposure, C-air

MO-PL-A3-126

Spatial-temporal Characteristics of Metal Contents in Ambient Fine Particles in Nanjing, China

D. Bi[1], H. Cheng[1], X. Liu[2], M. Liu[2], J. Chen[1], X. Cui[2], Z. Zhu[1]; [1]Nanjing Foreign Language School, Nanjing, China, [2]Nanjing University, Nanjing, China

Abstract: Recently, frequent hazy weather across China rose great public concern about health consequences of air pollution. Among different air pollutants, ambient fine particles ($PM_{2.5}$) became the focus partly due to its strong adsorption of nocuous substances like Pb and Cd, which will do serious harm to human health through affecting body's nervous, immune, cardiovascular system, etc. Many studies have analyzed the composition and sources of metal elements in $PM_{2.5}$. However, few of them have explored spatial-temporal variations, specifically for a school in China. With the support from STEM (Science, Technology, Engineering and Mathematics) program directed by Nanjing Foreign Language School, this study measured the metals in $PM_{2.5}$. Specifically, we selected the playground and classroom in Nanjing Foreign Language School as sample sites for outdoor and indoor environment. Samples were collected by the portable particulate monitor (MicroPEM, RTI International) for four seasons. For each site in each season, there were three samples with 24-hours continuous monitoring $PM_{2.5}$ data. Then, all samples were analyzed by inductively coupled plasma mass spectrometry (ICP-MS) to characterize six typical metal elements, including Pb, Mn, Cu, Cr, Sb and Ni. Results showed that spatial and temporal distribution of metals in $PM_{2.5}$ followed some rules. The concentrations of most metals were higher in spring and winter, lower in summer and autumn. Outdoor metal contents were significantly higher than indoor in spring and winter, probably because closed indoor environment in cold days significantly hindered penetration of outdoor pollutants. In addition, Pb is the dominated metals in $PM_{2.5}$ in Nanjing mainly due to vehicle emissions. From the results, the government should pay more attention to pollutant sources like motor vehicle. And it is recommended for individuals to take action like wearing masks or staying indoors so as to reduce damage to their health in spring and winter.

Keywords: B-particulate matter, B-metals, A - ambient monitoring

MO-PL-A3-127

Intraurban distributions of NO_2 and $PM_{2.5}$ at three dimensions in Lanzhou, China

L. Jin[1], Y. Zhang[1], Y. Zhang[2], M. L. Bell[1]; [1]Yale University, New Haven, CT, [2]Gansu Academy of Environmental Sciences, Lanzhou, China

Abstract: Previous research linked maternal exposures to NO_2 or PM_{10} with increased risks of adverse birth outcomes in Lanzhou. To improve exposure assessment, a land use regression study at ground level with consideration of vertical variation by building height was conducted in Lanzhou. In each of four seasons in 2016-2017, NO_2 was measured using Ogawa badges for 2 weeks at 75 ground-level sites. $PM_{2.5}$ was measured using DataRAM at a subset (N=38) of the 75 sites. The $PM_{2.5}$ monitor was systematically moved in pre-determined patterns around each site. The measurements were adjusted for temporal trend using government monitoring data. Vertical profile measurements were conducted at 18 sites evenly distributed at increasing floors of 2 buildings: one facing traffic and the other facing away from traffic. The annual and seasonal concentrations of NO_2 and $PM_{2.5}$ at ground level were regressed against spatial predictors, including elevation, population, road network, land cover, and land use with buffers 100-2000 m around each site. The vertical variations were investigated using polynomial models. NO_2 concentrations were associated with lengths of major roads and of all roads, slope, and cultivated land (adjusted R^2: 0.61). $PM_{2.5}$ concentrations were higher in certain districts, and associated with population density, lengths of roads with vehicles, and elevation (adjusted R^2: 0.56). From the 1st to 32nd floors in the building facing traffic, NO_2 and $PM_{2.5}$ concentrations decreased 22% and 26%, respectively. These distributions showed seasonal and diurnal variations. The ground-level NO_2 and $PM_{2.5}$ showed different patterns, which was not captured by the 4 government monitors in Lanzhou. Concentrations of these pollutants decreased substantially with height. Therefore, characterizing 3-dimensional distributions of multiple pollutants might be crucial for exposure and risk assessment in cities with mixed pollution sources and tall buildings, and potentially have policy implications.

Keywords: A-exposure models, A - ambient monitoring, A - exposure measurement, C-air

MO-PL-A3-128

Spatial and Temporal Covariation of Ultrafine Particles and Nitrogen Oxides in Urban Air: Implications for Exposure Assessment

J. L. Durant[1], L. Corlin[1], M. C. Simon[1], M. Woodin[1], D. Gute[1], D. Brugge[2]; [1]Tufts University, Boston, MA, [2]Tufts University, Boston, MA

Abstract: Evaluating the effects of air pollutant mixtures requires an understanding of how pollutants covary. Our objective was to assess the spatial and temporal covariation of ultrafine particles and nitrogen oxides (NOx) in Boston and Chelsea (Massachusetts, USA) using measurements collected by a mobile platform. We collected particle number concentration (PNC), NOx, and GPS data on ~45 days in each study area during different seasons, days of the week, and times of day. We used one-minute meteorology data and different pollutant averaging times (10-second, minute, daily, and seasonal). Pollutant temporal covariation differed by study area. Averaging across longer time scales produced stronger Pearson correlations among the pollutants (e.g. r(lnPNC and lnNOx) = 0.51, 0.59, and 0.64 for 10-second, minute, and daily averages in Boston, respectively). The correlations of 10-second and minute-averaged PNC with NOx were 27-32% stronger in winter than summer in Chelsea; however, the correlation of minute-averaged PNC with NOx was 20% stronger in summer than winter in Boston. Correlations were 15-44% higher during morning rush hour than mid-day or evening rush hour in both study areas. Spatial covariation also differed by study area. In Boston, correlations between PNC and NOx tended to be higher closer to major roadways while in Chelsea, the opposite was true. For both study areas, correlations tended to be highest in areas where there is high density residential land use. Additionally, correlations were highest in both study areas when the wind came from the southwest and lowest when the wind came from the northeast. Despite overall high correlations, certain spatial and temporal factors partially differentiate PNC and NOx. Including these factors in exposure models could help disentangle the independent effects of highly correlated pollutants.

Keywords: C-air, B-particulate matter, B-mixtures, A-exposure models, A-geospatial analysis/GIS

MO-SY-B3: Toxic chemical water contaminants in low and middle income countries Part I: A global grand challenge for the WASH development sector

MO-SY-B3-129

Synthetic chemicals as agents of global change

E. Bernhardt; Duke University, Durham, NC

Abstract: Around 100 million unique synthetic chemicals have been produced over the past six decades, recently at a rate of about 10 million per year. Since 2000 there has been a clear increase in the rate of global production of chemical substances, with a particularly large increase in low- and middle- income countries. This presentation examines the rate of change in the production and variety of pesticides, pharmaceuticals, and other synthetic chemicals over the past four decades. We show that chemical production and diversification, particularly within the developing world, have outpaced other major drivers of global environmental change such as CO_2 emissions, nutrient pollution, habitat destruction and biodiversity loss. Sufficient evidence shows stresses on ecosystem and human health at local to global scales suggesting that conditions are transgressing the safe operating space delimited by a planetary boundary for chemical pollution. Though concerns about the proliferation of synthetic chemicals - including pesticides - gave rise to the modern environmental movement in the early 1960s, synthetic chemical pollution has not been included in most analyses of global change. However, despite these trends, mainstream ecological journals, ecological meetings, and ecological funding through the US

National Science Foundation devote less than 2% of their journal pages, meeting talks, and science funding, respectively, to the study of synthetic chemicals. Similarly, despite large implied environmental, economic, and human health costs, chemical pollutants have been overlooked in the international development agenda and pollution control currently receives <0.5% of global development spending.

Keywords: A-chemical prioritization, A-environmental justice, B-POPs, C-water, developing countries

MO-SY-B3-130

Shifting Exposures, Shifting Paradigms: Global Trends Warrant a Focus on Chemical Contaminants in the WASH Sector

A. M. Aceituno; RTI International, Research Triangle Park, NC

Abstract: WASH is the collective term for Water, Sanitation and Hygiene. Because access to, and quality of, water, sanitation, and hygiene are interrelated, they are grouped together in a collective sector. Traditionally, health impacts related to improvements in WASH infrastructure focused on the incidence of diarrhea in children under five years of age (under-5). This is because diarrhea is one of the major single causes of under-5 mortality, surpassed only by acute respiratory infections. As many international WASH organizations aim to reduce under-5 diarrhea and related mortality, WASH research and infrastructure improvements traditionally focus on stopping fecal-oral transmission of microbial pathogens via water, food and other exposure routes that cause diarrhea. However, over the past two decades, international efforts to improve WASH infrastructure and reduce under-5 mortality have been successful, and both the rate and number of child deaths were reduced more than half between 1900 and 2015. However, globalization and various development activities have contributed to a rise in chemical contamination of water sources during this same time period. A shift in the focus of the burden of disease related to WASH is warranted to include exposures to both microbial pathogens and chemical contaminants. This shift has begun, as evidenced by the United Nations' Sustainable Development Goal 6, to ensure availability and sustainable management of water and sanitation for all, which has targets including reducing pollution, eliminating dumping and minimizing release of hazardous chemicals and materials. It is critical that future research and innovation in the WASH sector shift to focusing on both chemical and microbial improvements in water quality. Low- and middle-income countries (LMICs) without resources to identify and mitigate chemical water quality problems need support to improve water quality and reduce acute and chronic morbidities from chemical exposures.

Keywords: A-global health, C-water, D-children, A - population exposure

MO-SY-B3-131

Minimizing potential groundwater and surface water exposures associated with agricultural practices

D. Womack, T. Lillys, J. H. Redmon; RTI International, RTP, NC

Abstract: Variability in agricultural practices in developing countries can pose unanticipated surface water and groundwater impacts. For example, these water sources can become impacted through leaching and runoff and erosion when soil amendments or fertilizers are misapplied to crops and fields. Human exposures can occur if these water sources are used for domestic proposes. This presentation will review typical agricultural practices and provide recommendations for minimizing water impacts and associated human exposures. Examples will be provided on how US applied techniques and models can be adapted to assess exposures and evaluate mitigation strategies. These examples will draw from best management practices and modeling techniques used to evaluate soil amendments and fertilizers.

Keywords: A-exposure models, C-water, C-multimedia

MO-SY-B3-132

Annual meeting of the WASH Toxics working group

J. Kearns; North Carolina State University, Raleigh, NC

Abstract: The annual meeting of the WASH Toxics working group and a discussion about our objectives for the coming year.

Keywords: A-sustainability, C-water, D-community

MO-SY-C3: ATSDR Methods for Assessing Exposure to Air Pollutants in Diverse Communities

MO-SY-C3-133

Estimating exposure to air pollutants from a pulp and paper mill in a coastal NW Washington community

B. Goodwin, J. Durant, D. Gable; Agency for Toxic Substances and Disease Registry, Atlanta, GA

Abstract: ATSDR developed a two-stage approach of using screening level modeling (human exposure model [HEM-3]) and refined CALPUFF modeling to estimate pollutant levels in a coastal community bounded by complex terrain. Emission estimates from many point and area sources for a variety of air pollutants were included in the screening model runs. Modeled concentrations were compared to carcinogenic, chronic, and acute health effect screening levels to determine which species should be included in refined modeling. Nine of the 35 pollutants evaluated in the screening model were retained in the refined modeling. Refined modeling using CALPUFF was conducted because of known complex wind fields in the area due to coastal and terrain effects. Gridded Weather Research and Forecasting (WRF) data were incorporated into the model to characterize the complex atmospheric flow in the region. Five year average concentrations were used to assess potential for carcinogenic effects, maximum one year averages were used to assess potential for chronic non-carcinogenic effects, and maximum 1- or 24-hour concentrations were used to assess potential for acute effects. An emission rate distribution approach for acute concentrations was employed to account for unknown short-term variability in emission rates. All nine of the species retained in the refined modeling were modeled to exceed screening levels for at least one of the three health effects considered. Spatial distributions of receptors exceeding screening levels were determined as well as estimates of how frequently acute screening levels might be exceeded at each model receptor. Results of this modeling study are being used to design an ambient air monitoring study that incorporates temporal and spatial variations by identifying pollutants of interest, locations where maximum air concentrations are expected, and times of year when meteorological conditions result in higher predicted concentrations.

Keywords: A-exposure models, B-particulate matter, B-VOCs, C-air, D-community

MO-SY-C3-134

Measuring Exposure to Hydrogen Sulfide and Particulate Matter in a Community Adjacent to a Construction and Demolition Debris Landfill in Louisiana

B. Goodwin, A. Young; ATSDR, Atlanta, GA

Abstract: Residents in Belle Chase, Louisiana, a city 30 minutes south of New Orleans, petitioned ATSDR for assistance in characterizing their exposure to contaminants in air. This underserved community has had longstanding complaints about odors they attribute to the construction and demolition (C&D) debris landfill adjacent to their neighborhood. Based on the community members' reports of odors and occasional smoke emitting from the C & D landfill, and due to the limited environmental data to evaluate this concern, ATSDR designed and executed an exposure investigation (EI) to measure hydrogen sulfide at three locations and particulate matter at one location near the landfill. Meteorological data were collected from one site in the community to help identify any related trends in measured concentrations. In addition to the environmental measurements, field staff completed odor logs to characterize odors present in the community during the sampling period. ATSDR EIs are targeted efforts to identify and estimate the exposure of the most-highly exposed individuals in a community. Although EI results are rarely generalizable to other populations, public health professionals, environmental risk managers, and other decision makers use ATSDR's EI results to determine if current conditions warrant additional sampling, if specific populations are at risk of adverse health effects, and if interventions are required to minimize or eliminate human exposure. ATSDR's primary goal for EIs is not to identify the source of the exposure, but to determine whether exposures may present a health risk. The EI design and preliminary results of the EI will be presented with a discussion of potential health implications for area residents.

Keywords: B-particulate matter, C-air

MO-SY-C3-135

Estimating Hydrogen Sulfide Emissions and Ambient Air Concentrations from Beef Cattle Feedlot Operations

C. V. Muianga; Agency for Toxic Substances and Disease Registry (ATSDR), Atlanta, GA

Abstract: Hydrogen sulfide (H_2S) is a colorless, highly toxic gas that has a characteristic odor of rotten eggs. H_2S is released through bacterial decomposition of organic matter containing sulfur, such as manure. H_2S emissions from concentrated animal feeding operations (CAFOs), such as beef cattle feedlots, can result in occupational and community exposures and possible health concerns. Accurate measurement of H_2S emissions and concentrations on or near large cattle feedlots can be challenging and sometimes infeasible because of the variety of emission sources and variations in daily and seasonal levels. Sophisticated sampling and analytical instruments are also needed, and sampling can be time-consuming, and costly. This makes dispersion modeling a desirable alternative method for exposure investigations associated with CAFOs. We used the American Meteorological Society/Environmental Protection Agency Regulatory Model (AERMOD) with published emission rates, closest national weather station data and feedlot capacity to estimate receptor ambient concentrations in varying downwind distances from the source. Published H_2S emission rates ranged from 0.025 g/s to 3.6 g/s. The respective maximum hourly measured concentrations onsite were 4.2 µg/m^3 (3 ppb) in dry conditions and 2.0 mg/m^3 (1,440 ppb) during worst case conditions, e.g., rainfall, agitation of liquid manure lagoon. The maximum hourly modeled ambient air concentration at the closest populated census block receptor (i.e. less than half kilometer) is 300 µg/m^3 (214.3 ppb) downwind of a typical 55,000 cattle head and approximately 85 hectares of feedlot pen surface. Among many factors, moisture content, temperature of air and manure, the daily sulfur intake of the animals, the size of feedlot, the housing type, and lot design influence air emissions and ambient air concentrations. An exposure assessment strategy that combines air dispersion modeling with air sampling and monitoring will provide a better understanding of H_2S emission rates, concentrations at the source, and potential exposures in communities near feedlots.

Keywords: Odorous gases; feedyards; open-lot beef cattle, dispersion modeling, AERMOD

MO-SY-C3-136

Measuring Particulate Matter, Hydrogen Sulfide, and Ammonia Levels in a Native American Community near Dairy Animal Feeding Operations (AFOs)

D. Gable, A. Young, A. Grober; Agency for Toxic Substances and Disease Registry, Atlanta, GA

Abstract: The Yakama Reservation, located in south-central Washington State, is home to many large dairy and beef animal feeding operations (AFOs). Residents in the city of Harrah, a small town on the Yakama Reservation, have expressed concerns about "bad odors" and possible exposures to air pollutants related to these AFOs. ATSDR collected ambient air measurements of pollutants over two 8-

week periods in 2014 and 2015. During both periods, ATSDR collected meteorological data and measurements of particulate matter ($PM_{2.5}$ and PM_{10}), ammonia (NH_3), and hydrogen sulfide (H_2S) from residential, commercial, and other locations close to AFOs. The measured concentrations were compared to chemical-specific, health-based comparison values (CVs) from ATSDR as well as other public health agencies. Preliminary findings indicate that the highest concentrations for H2S and NH_3, were below ATSDR health comparison values and were unlikely to pose a health concern. At each location, some of the $PM_{2.5}$ sample concentrations exceeded the 24- hour National Ambient Air Quality Standard (NAAQS) value for $PM_{2.5}$. The public health implications of $PM_{2.5}$ exposures to community members will be discussed.

Keywords: B-particulate matter, animal feeding operations, air pollutants

MO-SY-C3-137

Panel Discussion

P. J. Kowalski; Agency for Toxic Substance and Disease Registry, Atlanta, GA

Abstract: Panel Discussion

Keywords: A-exposure models, B-particulate matter, C-air

MO-SY-D3: Consumer Exposure Assessment: Tools and Information to Support a Fit-for-Purpose Approach to Exposure and Risk Assessment

MO-SY-D3-138

Advancing Models and Data for Characterizing Exposures to Chemicals in Consumer Products

K. Isaacs[1], K. Dionisio[1], K. Phillips[1], J. Wambaugh[2], P. Price[1], A. Guiseppi-Elie[1]; [1]US Environmental Protection Agency, Durham, NC, [2]United States Environmental Protection Agency, Research Triangle Park, NC

Abstract: EPA's Office of Research and Development (ORD) is leading several efforts to develop data and methods for estimating population chemical exposures related to the use of consumer products. New curated chemical, ingredient, and product use information are being collected from public sources including 1) chemical weight fractions (WFs) collected from Material Safety Data Sheets 2) WFs predicted from ingredient lists, 3) product use patterns developed from purchasing data and 4) chemical function data. The functional use data is also being used to develop chemical structure and/or property-based models for function and product weight fraction, which can aid in filling gaps in existing ingredient data. In addition, under the Exposure Forecasting (ExpoCast) project, high throughput (HT) analytical methods are being evaluated for characterizing the composition of different products, a potentially rich source of new data for modeling and evaluation. These new data streams can parameterize mechanistic models for use in both high-throughput chemical screening and prioritization and higher-tier applications such as targeted exposure assessments or chemical life-cycle impact evaluations. For chemical prioritization, ORD has developed the High Throughput Stochastic Human Exposure and Dose Simulation (SHEDS-HT) model, which predicts aggregate population exposures based on chemical WFs and use patterns in over 300 consumer product categories. A higher-tier model, the Human Exposure Model (HEM), is being developed to consider additional factors impacting variability in exposures and internal chemical concentrations, including physiological variability and demographic and longitudinal patterns in product use. HEM will incorporate agent-based models of product use derived from considering a "needs-based" (e.g., personal hygiene, pest control) framework. These data and models will improve the prediction of chemical exposures in support of risk-based decision making.

Keywords: C-consumer products, A-behavior, A-indoor environment, A - population exposure, A-exposure models

MO-SY-D3-139

Overview of OPPT Consumer Exposure Models, Use Descriptors, and Related Indoor Exposure Data

C. Fehrenbacher[1], C. Bevington[1], X. Liu[1], M. Lee[1], Z. Guo[2], H. Hubbard[2], A. Williams[2], C. Stevens[2], T. Hong[2]; [1]Environmental Protection Agency, Washington, DC, DC, [2]ICF, RTP, NC

Abstract: Consumer products and articles can be significant sources of indoor chemical exposures. Exposure estimates can be developed using measured exposure data, modeling approaches, or a combination of both. The U.S. EPA's Office of Pollution Prevention and Toxics has augmented their suite of exposure estimation tools with updates to the Consumer Exposure Model (CEM), the Indoor Environmental Concentrations in Buildings with Conditioned and Unconditioned zones (IECCU) model, and the development of protocols to experimentally determine exposure parameters. Updates to product use, article use, and functional use category definitions through the OECD can be used to inform development of consumer exposure scenarios. Updates to CEM have increased the scope and versatility of the tool, expanding the exposure pathways and product/article use scenarios. OPPT has also developed IECCU, a more complex indoor exposure model that estimates chemical emissions and resulting air concentrations from consumer products, articles, and building materials based on EPA's Office of Research and Development's existing Indoor Air Quality and Inhalation Exposure (IAQX) and Indoor-Semi-Volatile Organic Compound (I-SVOC) models. OPPT has also completed a second version of indoor exposure testing protocols. The updated protocols have information on experimentally determining the chemical content in products or articles; emission rates, including short- and long-term emission rates, material-air partition coefficients; solid phase diffusion coefficients; and generation of particles by abrasion. The protocols also address direct transfer of chemical to dust in contact with an article; chemical transformation due to photolysis; migration rates, including chemical migration from an article into water, saliva, and sweat; and dermal exposure to chemicals transferred from liquids or particles to the skin. This presentation will provide an overview of the interrelatedness of consumer exposure models, available empirical data, and use descriptors in for estimating consumer exposures. The views expressed in this abstract are those of the authors and do not represent Agency policy or endorsement.

Keywords: C-consumer products, TSCA, use descriptions, product testing, exposure assessment

MO-SY-D3-140

REACH Consumer Exposure and Risk Tools

H. Qian[1], T. Dudzina[1], C. Rodriguez[2], R. Zaleski[1]; [1]EMBSI, Annandale, NJ, [2]Procter & Gamble, Brussels, Belgium

Abstract: Under the EU Registration, Evaluation, and Authorization of Chemicals (REACH) regulation, the uses of classified substances (with production/import tonnage >10 Ton per year) require the development of exposure scenarios to describe safe use conditions of the substance and then communicate these conditions throughout the supply chain. To meet this requirement, several consumer exposure and risk assessment tools at screening or higher tier level have been used since 2010, including ECETOC TRA, EGRET, A.I.S.E. REACT, and RIVM ConsExpo tools and an European Chemical Agency's Chesar platform. Recently, several of these tools have been updated to reflect advances in exposure and risk assessment approaches. An overview on these consumer tools and their features will be presented. Further, HESI Risk21 lookup tables developed based on the screening level REACH consumer tools (i.e. TRA and EGRET) will also be discussed. These lookup tables present a novel perspective on how to use the existing REACH tools and methods in a high throughput manner.

Keywords: A-exposure models, C-consumer products, A-risk assessment, REACH

MO-SY-D3-141

Tiering Consumer Product Exposure Tools

C. Cowan-Ellsberry[1], R. Zaleski[2], B. Greggs[3]; [1]CE2 Consulting, Cincinnati, OH, [2]ExxonMobil Biomedical Sciences, Inc., Annandale, NJ, [3]Soleil Consulting, Sanibel, FL

Abstract: New research initiatives have resulted in increased ability to understand and predict how consumers are exposed to chemicals in products used in and around the home and office. Part of these efforts has focused on advancing the accuracy of consumer product exposure models that quantify daily exposures to ingredients in these consumer products. In fact, within the past year an updated version of EPA's Consumer Exposure Model (CEM) which is used within OPPT (Office of Pollution Prevention and Toxics) for new and existing chemical assessments was released. In addition in the USA, EPA's NERL (National Exposure Research Laboratory) developed the SHEDS-HT tool to provide exposure assessments for prioritization of large numbers of chemicals used in consumer products. These tools join several tools that have been developed for the REACH process to predict consumer exposures - ECETOC TRA, ECHA's CHESAR and ESIG EGRET. The models are the same in that they all include 1) Inhalation, 2) Ingestion and 3) Dermal exposure from use of consumer products and in many cases, also include indirect or secondary exposures for these routes, e.g., exposure to chemical on surfaces from product use. However, these tools represent varying levels of sophistication and have different objectives and goals within consumer exposure assessment. In this presentation, the tools and their predictions will be compared to assist in understanding how these tools complement each other and can be positioned within a tiered exposure framework.

Keywords: A-exposure models, C-consumer products, C-consumer/personal care products

MO-SY-D3-142

Panel Discussion for ISES Symposium Topic: Consumer Exposure Assessment: Tools and Information to Support a Fit-for-Purpose Approach to Exposure and Risk Assessment

R. Zaleski[1], A. Guiseppi-Elie[2]; [1]ExxonMobil Biomedical Sciences, Inc., Annandale, NJ, [2]Environmental Protection Agency, Research Triangle Park, NC

Abstract: The presentations in this symposium build upon each other by providing information on latest advances in available consumer exposure tools developed by multiple sources for a range of purposes, then integrating this information by providing examples of use and discussing fit-for-purpose application, leading into the panel discussion. The panel discussion can then consider areas such as future directions for consumer exposure science development - what are the focus areas and opportunities? 1. Overall -Do we gain information by considering models within a fit for purpose framework? -Are model roles or purposes clear? -How does model purpose influence development? 2. Modeling Capabilities -Are there opportunities to leverage approaches across models? -How to take advantage of the full scope of tools available internationally? -What are future areas for model development? 3. Model Input Data -What are key model inputs and how might this data be obtained? -Are there ways to better use emerging technologies that enable data collection? -Do we need greater interaction with other disciplines? Social sciences, computational science?

Keywords: C-consumer products, A-exposure models, C-consumer/personal care products

MO-SY-E3: Childhood Multimedia Lead Exposures: Innovative Modeling and Data Collection Efforts by the Federal Family to Guide Public Health Decision-Making

MO-SY-E3-143

Community Engagement Efforts in Flint, MI

D. Russell; U.S. EPA, Flushing, MI

Abstract: This presentation will describe some of the unique aspects to engaging the residents of the City of Flint during the drinking water crisis that became an emergency in 2016. The series of events that led to a State and Federal emergency declaration created a feeling of mistrust toward government. EPA and its Federal partners developed several strategies for reaching out to the community to ensure proper information was delivered and to begin the long process of reestablishing some level of trust. The presentation will also demonstrate how a community became empowered, through government support, to begin the long term recovery and healing process.

Keywords: D-community

MO-SY-E3-144

Multimedia Lead Exposure Modeling, and Water Monitoring Perspectives

V. G. Zartarian, J. Xue, R. Tornero-Velez, J. Brown, T. Speth, J. Garland; USEPA, Boston, MA

Abstract: Drinking water and other sources for lead are the subject of public health concern following the Flint, Michigan drinking water and East Chicago, Indiana lead in soil crises. In 2015, the U.S. EPA's National Drinking Water Advisory Council recommended establishing a "health-based, household action level" for children based on exposure to lead in drinking water. A modeling approach, coupling the EPA's SHEDS-Multimedia and IEUBK models was developed to help determine what drinking water lead concentrations keep children's blood lead levels (BLLs) below specified values, considering exposures from water, soil, dust, food, and air. Related objectives were to evaluate the model estimates using real-world blood lead data; quantify relative contributions by the various media; and identify key model inputs. This analysis for the U.S. population of young children probabilistically simulated multimedia exposures and estimated relative contributions of media to BLLs across all population percentiles for several age groups. Modeled BLLs compared well with nationally representative BLLs (0%-23% relative error). Analyses revealed the relative importance of soil and dust ingestion exposure pathways, and associated lead intake rates; water ingestion was also a main pathway, especially for infants. Given the spatial and temporal variability of household lead water concentrations, there are uncertainties in water lead concentration data collected under the current regulatory sampling schemes. Local-scale data for the various multimedia model inputs and BLLs would be beneficial for extending the coupled model approach to other applications and specific communities. This methodology advances scientific understanding of the relationship between lead concentrations in drinking water and BLLs in children. Disclaimer: The views expressed are those of the authors and do not necessarily reflect the views or policies of the U.S. EPA.

Keywords: C-multimedia, A-aggregate exposure, B-metals, D-children, A-exposure models

MO-SY-E3-145

Establishing a new Blood Lead Reference Value

P. Breysse, A. Ettinger; Centers for Disease Control and Prevention, Atlanta, GA

Abstract: The impact of lead exposure on young children is well known and well documented. No safe blood lead level (BLL) has been identified. At the same time, continuous analyses of the U.S. population-based data show BLLs decreasing over time. In 2012, the former CDC Advisory Committee on Childhood

Lead Poisoning Prevention (ACCLPP) recommended elimination of the term 'blood lead level of concern' and replacement with a reference value to define when a child is lead exposed. This value, equal to the 97.5th percentile of blood lead levels for U.S. children age 1-5 years, is updated every four years based on data from two successive cycles of the National Health and Nutrition Examination Survey (NHANES). Since May 2012, the reverence value of 5 micrograms per decileter (mcg/dL), calculated using NHANES data from the combined 2007-2008 and 2009-2010 cycles, has been used. In 2016, CDC reported results from the 2011-2012 and 2013-2014 NHANES data cycles indicating that the 97.5th percentile BLL for children aged 1-5 years was 3.5 mcg/dL. CDC's current advisory group, the NCEH/ATSDR Board of Scientific Counselors, recommended that CDC lower the blood lead reference value, accordingly, to 3.5 mcg/dL. Establishing a new blood lead reference value that is lower than 5 mcg/dL requires CDC's consideration of several key issues: 1) Analytical precision and limit of detection concerns associated with approved clinical laboratory methods used to measure BLLs; 2) Education and guidance to parents, pediatricians, and state and local public health officials regarding child-specific response actions associated with BLLs equal to or greater than the reference value; and 3) Communication and collaboration with key federal partners (the U.S. Department of Housing and Urban Development and U.S. Environmental Protection Agency) whose regulations and enforcement policies have traditionally been based on CDC's "action level."

Keywords: D-children, A - exposure measurement, A-biomonitoring, B-metals, A - population exposure

MO-SY-E3-146

The Distribution of Lead-Based Paint Hazards in U.S. Housing

P. J. Ashley, W. Friedman, E. A. Pinzer; U.S. Department of Housing and Urban Development, Washington, DC

Abstract: The U.S. Department of Housing and Urban Development (HUD) has conducted surveys of lead-based paint (LBP) and LBP hazards in representative samples of U.S. housing in 1999 - the National Survey of Lead and Allergens in housing (NSLAH) - and 2005 - the American Healthy Homes Survey (AHHS). Both surveys were conducted in stratified random samples of permanently occupied, non-institutional housing in which children could reside, a sample frame of about 96 million (M) housing units for NSLAH and 106 M for AHHS. In each home, paint was tested using a portable X-Ray Fluorescence analyzer, dust wipe samples were collected from floors and window sills, and paint was visually assessed for condition. Soil samples were collected from the residential yards, and information was collected from an adult resident by questionnaire. AHHS results include: 37.1 M homes (34.9%) had some LBP and 23.2 M (21.9%) had one or more LBP hazards per federal definitions of deteriorated LBP, or lead in dust or soil. Of homes with LBP, 34.4 M (93%) were built before 1978. Homes built before 1940 had the highest prevalence of LBP hazards (66.8%). LBP and LBP hazards had regional prevalence differences (highest in the Northeast and Midwest). 3.6 M homes with a child < age 6 had at least one LBP hazard, including 1.1 M low income households (< $30,000/yr). Low income households had a higher prevalence of LBP hazards (29%) than higher income households (18%). Among low-income households, those receiving Government housing assistance had a lower prevalence of LBP hazards (12%) than unsupported ones (22%). Significant differences between NSLAH and AHHS findings included fewer homes with both interior and exterior LBP (from 20.3 M to 16.2 M), and homes with the highest levels of LBP ($\geq$ 10 mg/cm^2) (from 13.4 M to 6.4 M). HUD is currently preparing to conduct another national survey with anticipated completion in 2018.

Keywords: A-built environment, A - population exposure, B-metals, A-environmental justice

MO-SY-E3-147

FDA Total Diet Study data on lead in food: current understanding, data gaps, and future efforts

J. H. Spungen[1], A. Gavelek[1], R. Pouillot[2], T. Councell[1], M. Wirtz[1]; [1]U.S. Food and Drug Administration, College Park, MD, [2]Consultant, Buenos Aires, Argentina

Abstract: FDA's Total Diet Study (TDS), initiated in 1961, continuously monitors concentrations of lead and other contaminants in about 280 foods, based on quarterly sampling and analysis of these foods in four regions of the U.S.. Over the past three years, FDA has focused attention on modernizing and revitalizing the TDS program by developing a data management system for trend analysis; updating analytical methods to improve data quality; improving the sample collection protocol; and updating the TDS website. TDS concentration data for lead and other contaminants include high proportions of values below the limit of detection (LOD), and this presents challenges for statistical calculations. FDA has developed a novel statistical method, clustered zero-inflated lognormal distribution analysis, to estimate central tendency concentrations and confidence intervals in food-analyte pairs.

Keywords: C-food, B-metals, A - exposure measurement

MO-SY-F3: Quantitative High-Throughput Exposure Methods for Chemical Alternatives and Comparative Risk Assessment

MO-SY-F3-148

The landscape of existing models for high-throughput exposure assessment

O. Jolliet[1], P. Fantke[2], L. Huang[1]; [1]University of Michigan, Ann Arbor, MI, [2]Technical University Denmark, Lyngby, Denmark

Abstract: Models are becoming increasingly available to model near-field fate and exposure, but not all are suited for high throughput. This presentation evaluates the available models for modeling exposure to chemicals in cosmetics, cleaning products, food contact and building materials. It assesses their applicability to quantitative high throughput exposure assessment in CAA and CRA, looking in particular at the following characteristics: validity of main assumptions; availability of analytical solutions and model parsimony; availability of methods to estimate key inputs for a large number of chemicals and ability to easily handle large datasets. For building materials a series of diffusion-based models have been developed to predict the chemicals emissions from building materials to indoor air, but existing models require complex analytical or numerical solutions, which are not suitable for LCA or HTS applications. Thus, existing model solutions needed to be simplified for application in LCA and HTS, and a parsimonious model has been developed by Huang et al. (2017) to address this need. For SVOCs, simplified solutions do exist, assuming constant SVOC concentrations in building materials and steady-state in indoor air (Little et al., 2012; Liu et al., 2013), but they do not well account for SVOC sorption into indoor surfaces and absorption into human skins (Huang et al., 2017). Thus a more comprehensive simplified solution is needed for SVOCs . For personal Care Products, a mass balance model that accounts for skin permeation and volatilization as competing processes and that requires a limited number of readily available physiochemical properties would be suitable for LCA and HTS purposes. Thus, the multi-pathway exposure model for chemicals in cosmetics developed by Ernstoff et al.constitutes a suitable basis and can be refined in the future. The review will also address models available for modeling chemicals in cleaning products and other indoor used chemicals.

Keywords: A-chemical prioritization, A-exposure models, A - population exposure, A-indoor environment, B-SVOCs

MO-SY-F3-149

Identifying Potential Alternatives using Chemical Functional Use

K. Phillips[1], J. Wambaugh[2], C. Grulke[2], K. Dionosio[1], K. Isaacs[1]; [1]U.S. Environmental Protection Agency, Research Triangle Park, NC, [2]U.S. Environmental Protection Agency, Research Triangle Park, NC

Abstract: The National Research Council noted in its 2014 report *A Framework to Guide Selection of Green Alternatives* that "focusing on function can provide opportunities for innovation in safer chemicals and materials." This presentation highlights efforts to explore such opportunities. Functional use data was collected for more than 14,000 chemicals from multiple public data sources. Harmonized functions were assigned to all chemicals via hierarchical clustering ensuring that a single functional use was assigned to a single chemical regardless of multiple, but similar, names assigned by the different data sources. These harmonized functions were then used to develop quantitative structure-use relationship models which use a chemical's structure and physicochemical properties to predict whether or not a chemical could serve any of 41 possible functional roles (e.g., surfactant, fragrance). Using these models, the Tox21 chemical library (consisting of ~8500 chemicals) was screened for chemicals that could serve a functional role in a product. Tox21 provides *in vitro*, high throughput screening data of chemical bioactivity. Tox21 bioactivity was used to compare chemicals with novel predicted uses with chemicals that were already known to be serving that functional role. Chemicals with lesser bioactivity that serve the same role become potential substitutes. These methods have identified more than 1600 candidate alternatives that could undergo higher-tier screening as safer, greener chemical alternatives. ***The views expressed here are those of the authors and do not necessarily reflect the view or policies of the U. S. EPA.***

Keywords: A-chemical alternatives, A-chemical prioritization, C-consumer products

MO-SY-F3-150

Integrating Toxicity, Toxicokinetic, and Exposure Data for Risk-based Chemical Alternatives Assessment

J. Wambaugh[1], B. A. Wetmore[2], K. Mansouri[3], B. Ingle[2], R. Tornero-Velez[2], R. Judson[1], K. Isaacs[2], K. Phillips[2], C. Nicolas[3], R. W. Setzer[1], R. S. Thomas[1]; [1]U.S. Environmental Protection Agency, Research Triangle Park, NC, [2]U.S. Environmental Protection Agency, Research Triangle Park, NC, [3]Scitovation, Research Triangle Park, NC

Abstract: In order to predict the margin between the dose needed for adverse chemical effects and actual human exposure rates, data on hazard, exposure, and toxicokinetics are needed. *In vitro* methods, biomonitoring, and mathematical modeling have provided initial estimates for many extant chemicals. Providing predictions for novel compounds, however, will need to rely on screening massive chemical libraries and drawing inference from chemical structure (*e.g.*, quantitative structure activity relationships). This presentation will review the challenges and opportunities for alternatives assessment based upon high throughput tools for exposure and hazard. In vitro high-throughput screening (HTS) assays, such as the U.S. Federal Tox21 consortium and the U.S. EPA Toxicity Forecaster (ToxCast) program, have generated bioactivity data for thousands of chemicals. For some endpoints (e.g., estrogen receptor and androgen receptor), models trained on these HTS data allow predictions for novel compounds. In tandem with ToxCast, the Exposure Forecaster (ExpoCast) program provides toxicokinetic (TK) data and exposure estimates to provide context for HTS data. Libraries of TK data, largely obtained from *in vitro* assays, have served as training sets for machine learning models capable of estimating TK for novel compounds. Biomonitoring data obtained by the US CDC National Health and Nutrition Examination Survey and consumer product formulation data have similarly been used to develop models for exposure prediction (mg/kg/day). Integration of these methods provides a timely, risk-based prioritization strategy that characterizes the dose relationships between *in vitro* bioactivities and predicted human exposure. *The views expressed here are those of the authors and do not necessarily reflect the views or policies of the U. S. EPA.*

Keywords: A-risk assessment, A-chemical alternatives, A-chemical prioritization, A-exposure models, A-green chemistry

MO-SY-F3-151

A First Case Study of a Life Cycle-Based Alternatives Assessment (LCAA)

P. Fantke[1], L. Huang[2], M. Overcash[3], E. Griffing[3], O. Jolliet[2]; [1]Technical University of Denmark, Kgs. Lyngby, Denmark, [2]University of Michigan, Ann Arbor, MI, [3]Environmental Clarity, Inc., Reston, VA

Abstract: Chemical alternatives assessment (AA) is an emerging screening-level method to replace hazardous chemicals with safer alternatives. Current AA frameworks, however, suffer from gaps in addressing exposure and life cycle impacts, which can leave trade-offs unidentified. Exposure needs to cover various population groups including workers, consumers and the general public, while life cycle impacts need to focus on categories relevant for a given AA chemical-product application. We systematically define the scope of AA and identify key elements for quantitatively considering exposure and life cycle impacts. Our approach is evaluated in a case study, through which we outline future research needs to fully operationalize a consistent and Life Cycle-based Alternatives Assessment (LCAA). We build on a flexible mass balance-based modeling system yielding cumulative multimedia transfer fractions and exposure pathway-specific Product Intake Fractions defined as chemical mass taken in by humans per unit mass of chemical in a product. When combined with chemical masses in products and further with toxicity information, this approach is a resourceful way to inform AA. Our case study reveals that replacing DEHP by DIHP as vinyl flooring plasticizer shifts from cancer to non-cancer disease burden and shows that plasticizers contribute between 55 and 85% to total toxicity burden from flooring. Comparing toxicity-related outcomes with outcomes from other life cycle impacts emphasizes the relevance of toxicity impacts for this chemical-product application. Our results demonstrate (a) how assumptions used in different assessment methods can be aligned in a manner that can avoid contradictory results, (b) how all relevant life cycle impacts can be consistently considered and compared, thereby avoiding burden shifting that could result from disregarding chemical and product life cycles, and (c) how the most relevant impacts across all life cycle stages are prioritized.

Keywords: A-chemical alternatives, A-chemical prioritization, A-exposure models, A-life cycle analysis, C-consumer products

MO-SY-F3-152

Suspect Screening and Non-Targeted Analysis of Coupled Soil and House Dust Samples

A. Marcotte[2], S. Laughlin-Toth[2], J. Grossman[3], J. R. Sobus[1], E. M. Ulrich[1]; [1]U.S. Environmental Protection Agency, Research Triangle Park, NC, [2]ORISE- U.S. Environmental Protection Agency, Research Triangle Park, NC, [3]ORAU - U.S. Environmental Protection Agency, Research Triangle Park, NC

Withdrawn

Withdrawn

MO-SY-G3: Exposure, cumulative risk, and epidemiology in communities near upstream energy development Part 1

MO-SY-G3-153

An industrial activity model approach to differentiating exposure categories for oil and gas epidemiological studies

W. Allshouse, J. Adgate, B. Blair, L. McKenzie; Colorado School of Public Health, Aurora, CO

Abstract: Oil and gas (O&G) production in the United States has increased in the last 15 years and these operations release pollutants to air. Health studies have relied on proximity to O&G wells as an exposure metric, typically using an inverse distance weighting (IDW) approach. Since O&G emissions are dependent on multiple factors, such as development phase, production volume, and scale, a dynamic model is needed to describe the variability in air pollution emissions over space and time. We used information on Colorado O&G activities, production volumes, and air pollution emission rates to create a spatiotemporal industrial activity model. The Spearman correlation coefficient between our modeled intensities and measured emissions was 0.56. We used our monthly model output to adjust the IDW well count metric for intensity and applied this metric to households in Greeley, Colorado, which is in the middle of the densely developed Denver-Julesburg basin. Our intensity-adjusted metric increased the dynamic range of "exposure" by 30-fold compared to the unadjusted IDW metric. Our approach distinguishes between high intensity events such as locations undergoing drilling, hydraulic fracturing/flowback, or large-scale production pads and lower intensity ones such as single well pads with small production volumes. This industrial activity model improves upon the proximity model approach and accounts for both inter- and intra-phase variability. As operators continue to move toward larger multi-well pads that reduce the overall geographic footprint of O&G development, individuals that live near those pads may be subject to much longer periods of high intensity and it will become more important to account not just for the density of wells, but the intensity of activity.

Keywords: A-exposure models, A-geospatial analysis/GIS, A-epidemiology

MO-SY-G3-154

Spatial Patterning of Hydraulic Fracturing Wastewater Injection Wells: Environmental Justice in Ohio

G. Silva, J. Warren, N. Deziel; Yale School of Public Health, New Haven, CT

Abstract: Background: Hydraulic fracturing of shale formations to extract natural gas produces 2-14 million liters of wastewater per gas well, which are commonly disposed of in Class II underground injection wells. We evaluated whether these disposal wells were disproportionately sited in vulnerable communities in Ohio. **Methods** Using data from the American Community Survey and Ohio Department of Natural Resources, we classified all 9238 Ohio census block groups with respect to median household income, percent population white, population density, percent high school education or higher, median age, median household value, presence of an injection well, number of hydraulically-fractured gas wells, and location on a shale formation. We modelled the probability of presence of an injection well with respect to these sociodemographic, gas well, and other spatially-varying covariates using a Sparse Spatial Generalized Linear Mixed Model for Areal Data, allowing control for spatial correlation/spatial confounding. All analyses were carried out in the Bayesian setting, providing a flexible framework to correctly characterize uncertainty associated with parameter estimation. **Results** In the multivariable logistic regression analysis, increased median income was associated with lower odds of a block group containing an injection well (odds ratio for a $10,000 increase in median income: 0.82 (posterior mean), 95% Credible Interval: 0.71-0.94). Block groups situated on shale formations were more likely to include an injection well, while an increased number of gas wells was associated with lower odds of an injection well. **Conclusion** Our findings suggest that injection wells in the state of Ohio are sited disproportionately in block groups with lower median income. Research is needed to understand whether these vulnerable populations are at risk of increased chemical exposures or adverse health effects due to these disposal sites.

Keywords: A-environmental justice, A-geospatial analysis/GIS

MO-SY-G3-155

Community Engaged Volatile Organic Compounds Sampling in Guernsey County, Ohio

E. Haynes[1], J. Quirolgico[1], R. Roberts[2], R. Shepler[1], R. Jandarov[1]; [1]University of Cincinnati, Cincinnati, OH, [2]Guernsey County Resident, Cambridge, OH

Abstract: Waste generated from natural gas extraction (NGE) contains toxicants with known reproductive and developmental risks. One leading mechanism for disposal of NGE waste is injection into Class II wells. Ohio has over 200 active class II wells. In 2014 Ohio injected 22 million gallons of NGE waste into these wells. Guernsey County, Ohio has six (6) wells experienced spills from the transport of these waste fluids and concern has been raised regarding a potential NGE solid waste disposal site in the county. We partnered with local emergency management and county officials to identify air sampling locations. The goal of the project was to determine air quality near active NGE waste injection wells and compressor stations and baseline air quality prior to the development of additional solid and liquid waste injection wells. We used batch-certified-clean evacuated stainless steel canisters and 24-hour flow controllers. Volatile organic compounds (VOCs) were analyzed using cryogenic preconcentration gas chromatography with mass spectrometry. We also collected a 24-hour formaldehyde concentration using an SKC passive sampling device. Formaldehyde concentration was analyzed using liquid extraction with acetonitrile using high performance liquid chromatography with ultraviolet detection. A total of four (4) samples were collected at each location (n=10) over a 2-week sampling period. Thus, 40 samples were obtained between October and November, 2016 by a trained community member. Of the 75 analyzed VOCs, 19 were detected in all samples. Analyses included distance from nearest roadway, NGE waste injection well or compressor station. The most common VOCs found in the samples were acetone, chloromethane and formaldehyde. Community engagement was critical for the success of the study.

Keywords: A - ambient monitoring, B-VOCs, D-community

MO-SY-G3-156

Techniques for Estimating Community Exposure from Hydraulic Fracturing Operations

M. McCawley, J. Barre, M. Nye, T. Knuckles; West Virginia University, Morgantown, WV

Abstract: The techniques of trace element analysis, satellite Aerosol Optical Depth, in-situ measurements, distance and depth of well are being used to assign dose in communities near hydraulic fracturing operations. Each method has its strength and weaknesses but they have not previously been compared. Being able to perform simultaneous assessments of all of these techniques has been possible for an experimental well drilled in West Virginia, using standard operating techniques by the drillers who were cooperating with the monitoring program. Results indicate that for the river valley topography of the region it was possible to assess some exposure from the well pad to the surrounding community at a distance of at least 7 km. This was done using trace quantities of airborne magnesium analyzed from high volume filter samples. Simultaneously, PM2.5 concentration data was collected using direct-reading photometric analyzers. These dust concentration values were shown to compare favorably to MODIS satellite data for analysis of aerosol optical depth (AOD) for a number of nearby locations. The satellite AOD measurements were then compared to previously proposed, indirect measures of community exposure derived from distance to the well and natural gas production values for surrounding sites with less success.. The conclusion was that trace magnesium concentration does appear to be a method for assessing environmental exposure for the time and location used in this study. This could provide a rationale for the findings of health effects outside the previous limits of detectable air emissions. The valley along which sampling occurred, however, may act to funnel the air contaminants and retard dispersion. Traffic from the well pad could not be ruled out from the current work as another source of wide spread air contaminant emissions. Further work is planned to use the tools assessed here to provide more information on that question.

Keywords: A - exposure measurement, A - ambient monitoring, A-geospatial analysis/GIS, B-particulate matter, C-air

MO-SY-G3-157

Truck and multivehicle truck accidents near Colorado oil and gas operations

B. Blair, J. Hughes, W. Allshouse, L. McKenzie, J. Adgate; University of Colorado, Aurora, CO

Abstract: Unconventional oil and gas (O&G) operations raise a number of public health concerns, such as the potential impacts from increased trucking activity in communities that host these operations. In this work, we use two approaches to evaluate truck accidents in relation to O&G activities in the State of Colorado. First, we calculate the empirical rate of truck accidents by computing the ratio of accident count and county population. When comparing counties with increased O&G operations to counties with less O&G activity, we find that counties with more activity have greater rates of truck traffic accidents per capita (Rate Ratio = 1.07, $p < 0.05$, 95% CI: 1.01 to 1.13). Second, we lay a square grid over the eleven counties of interest and count, for each cell of the grid, the number of truck accidents, the number of multivehicle accidents with injuries, the number of homes, and the number of wells. We then apply hurdle count models to the gridded data, using the accident counts as the outcomes and the number of homes and number of wells as independent variables. We find that both independent variables are significant predictors of truck accidents and multivehicle truck accidents. These accidents are of concern since they can have a direct impact on the people who live near O&G operations.

Keywords: A-workplace, D-community

MO-PL-A4: Novel Methods for Spatial Analysis

MO-PL-A4-158

Fine-scale spatio-temporal variation in particulate matter in a small wood-burning town revealed by a network of continuous low-cost sensors

I. Longley, G. Olivares, S. Edwards; NIWA Ltd, Auckland, New Zealand

Abstract: Following a co-location study, 15 monitoring nodes, based on the Plantower 3003 dust sensor, were deployed across a small New Zealand town for a minimum of 19 days in winter 2016. Rangiora has previously been identified as a town where wintertime concentrations of particulate matter breech the National Environmental Standard of 50 µg m^{-3} over 24 hours, and emissions are dominated by wood combustion for home heating. The location of the town on a coastal plain in the lee of a mountain range lead to complex air flows especially at dusk when nocturnal boundary layers form and emissions peak. The sensors were initially co-located with a TEOM-FDMS operated by the regulatory authority. $PM_{2.5}$ concentrations were then measured every minute at the 15 locations distributed across the town (area approx. 17 km^2, pop. 17,000) for 19 days (and at some of the sites for longer). This was supported by the deployment of three temporary automatic weather stations on the periphery of the town. A further co-location study was conducted at the end of the campaign. All 15 sensors reported valid data throughout the campaign. The co-location study revealed high correlation between the Plantower sensors and the TEOM-FDMS with very stable performance. No interferences were detected (e.g. related to air temperature or humidity). The sensitivity of each individual sensor varied but was sufficiently stable to be adjusted for with a simple linear function. On average the data revealed a large degree of spatial variation in concentrations across the town. The spatial variation was not consistent over time but highly complex. A tendency for concentrations to peak earlier in the east and later in the west was observable and broadly consistent with shifts in wind direction. Inspection of the time series indicated that some sites were subject to temporary elevations in concentrations (anomalies with respect to typical or larger-scale patterns) that imply localised near-source and plume-strike impacts.

Keywords: A-sensor technology, B-particulate matter, C-air, D-community

MO-PL-A4-159

A novel air quality model fusion method for improving spatial resolution (250m) of ambient air pollutant exposure

J. Bates[1], A. F. Pennington[2], X. Zhai[1], M. Friberg[1], F. Metcalf[1], M. Strickland[3], L. Darrow[3], J. Mulholland[1], A. Russell[1]; [1]Georgia Institute of Technology, Atlanta, GA, [2]Emory University, Atlanta, GA, [3]University of Nevada, Reno, Reno, NV

Abstract: Improving spatial resolution of air pollutant concentration estimates is critical for epidemiologic studies with spatially resolved health data. Air quality exposure assessment is often limited by spatially sparse monitoring data or model estimates that are either spatially coarse (chemical transport models) or lack chemistry (dispersion models). Therefore, two novel model fusion methods were developed to create spatially resolved, chemically comprehensive air pollutant concentration estimates, one for gaseous pollutants and one for particulate matter. The methods use computationally efficient linear combinations and mass conservative bilinear interpolation to fuse model results at two scales: the chemical transport model CMAQ at a 12km resolution, which accounts for numerous emission sources and complex chemistry, and the line dispersion model RLINE at a 250m resolution, which simulates the dispersion of inert, primary roadway pollutants. These methods are applied to Atlanta, GA to obtain daily 24-hr averaged $PM_{2.5}$ and 1-hour maximum CO and NOx from 2002-2011 at a 250m grid resolution, but these methods can be applied at different spatial scales and for other pollutants. Resulting estimates are spatially and temporally resolved air pollutant concentrations that include contributions from secondary processes and primary emissions from a comprehensive set of regional and local sources. Fused results show an improvement in both spatial R and temporal R^2 compared to CMAQ and RLINE, providing more accurate concentration estimates for improved exposure assessment. Results also capture steep spatial gradients in pollutant concentration near roadways, reducing exposure error in intraurban studies on the health effects of vehicle emissions. Capturing steep spatial gradients near roadways while also simulating effects of regional sources that can influence daily variation in pollutant concentrations can potentially improve exposure estimates in epidemiologic studies.

Keywords: C-air, B-particulate matter

MO-PL-A4-160

Improved Community Air Quality Monitoring Networks: Achieving Sensitivity to Pollutant Gradients Using Lower-Cost Sensors and Machine Learning

N. Zimmerman[1], A. Presto[1], S. Kumar[2], J. Gu[2], E. Robinson[1], A. L. Robinson[1], S. Ramachandran[1]; [1]Carnegie Mellon University, Pittsburgh, PA, [2]Sensevere LLC, Pittsburgh, PA

Abstract: Regulatory air quality monitoring typically relies on a small number of widely separated monitoring sites to assess population-scale exposure. However, air pollutant concentrations may exhibit significant spatial variability depending on local sources and features of the built environment, which may be poorly captured by the existing monitoring regime. Recently, low-cost sensing strategies have enabled denser air quality monitoring networks in urban areas. However, low-cost sensors can be sensitive to environmental conditions and pollutant cross-sensitivities, and these issues are poorly addressed by linear calibration models. In this study, machine-learning based calibration strategies (random forest) were used to calibrate the Sensevere Real-time Affordable Multi-Pollutant (RAMP) sensor package, which measures CO, NO_2, O_3, and CO_2. Calibrations were developed across 19 RAMP sensor packages using training and testing windows spanning August 2016 through February 2017. Following calibration, average mean absolute error on the testing dataset was 38 ppb, 10 ppm, 3.5 ppb and 3.4 ppb for CO, CO_2, NO_2 and O_3, respectively (~5-15% difference vs reference monitors). To test the sensitivity of the RAMPs to pollutant gradients, a network of 12 RAMPs was deployed for two month-long campaigns in Pittsburgh, PA. The stations were deployed at sites along the urban-rural transect and in downtown locations with a range of traffic, restaurant and tall building densities. Preliminary analysis shows that the sensitivity of the RAMPs was sufficient to detect pollutant differences between sites. For example, 3 sites each within 15m of a busy roadway showed elevated NO2 and suppressed O3 concentrations compared to an urban background site. The RAMP sensor network has been expanded to 30 nodes in Pittsburgh with plans to deploy an additional 20 nodes. This will help support better epidemiological models, aid in policy planning, and identify areas where more assessment is needed.

Keywords: A - ambient monitoring, A-sensor technology, A-environmental justice, A-statistical methods

MO-PL-A4-161

Predicting Daily $PM_{2.5}$ Concentrations in Texas Using High-Resolution Satellite Aerosol Optical Depth

X. Zhang[1], Y. Chu[2], K. Zhang[1]; [1]The University of Texas Health Science Center at Houston School of Public Health, Houston, TX, [2]The University of Texas Health Science Center at Houston School of Public Health, Houston, TX

Abstract: Background: The state of Texas contains many of most populous areas, prevalent diesel vehicles, and numerous industrial facilities that foreshadow a potential public health impact associated with particulate matter with aerodynamic diameter less than 2.5 μm ($PM_{2.5}$). Understanding the relevant public health impact requires a detailed knowledge of $PM_{2.5}$ exposure, but the current monitoring system is not capable of providing adequate spatial or temporal resolution for large cohort studies to conduct exposure assessments. **Objective**: The purpose of this study is to estimate ground-level $PM_{2.5}$ concentrations using satellite-retrieved Aerosol Optical Depth (AOD) in the state of Texas. **Methods**: We used gridded AOD values generated from the Moderate Resolution Imaging Spectroradiometer (MODIS) satellites data based on the Multi-angle Implementation of Atmospheric Correction (MAIAC) algorithm, which can provide 1km spatial and daily temporal resolution AOD data at a global coverage. In this study, we developed a mixed-effect regression model to estimate the ground-level $PM_{2.5}$ concentrations using MAIAC AOD, land use features, geographic characteristics, and weather conditions in the state of Texas for a study period of 2008-2013. **Results**: The results show high model fitting R^2 (0.83 - 0.88). The model performance was strongly influenced by dusts from unpaved roads and sand storms in the west of Texas, then yielded reasonable cross-validated R^2 (0.63 - 0.71). The multi-year predictions illustrated an unbalanced decrease of $PM_{2.5}$ concentrations between urban areas and rural areas from 2008 to 2013. Urban areas tended to have higher levels of reduction in air pollution. **Conclusions**: The MAIAC AOD was firstly used in estimating ground-level $PM_{2.5}$ concentrations in the state of Texas. The multi-year estimates enabled us to study the spatial variation of $PM_{2.5}$ levels at fine scale,, which can provide exposure metrics for epidemiological studies.

Keywords: C-air, A-exposure models, A-geospatial analysis/GIS,

MO-PL-A4-162

Air Quality Land Use Regression Model Robustness from Routine Mobile Monitoring Using Google Street View Cars

K. P. Messier[1], S. Gani[2], R. C. Vermeulen[3], C. Portier[4], R. Alvarez[4], J. Apte[2]; [1]Environmental Defense Fund/ UT-Austin, Austin, TX, [2]University of Texas at Austin, AUSTIN, TX, [3]Institute for Risk Assessment Science, Utrecht University, Utrecht, Netherlands, [4]Environmental Defense Fund, AUSTIN, TX

Abstract: Aim: Air pollution exposures are associated with deleterious human health effects and increased mortality. Exposure assessments developed from Land Use Regression (LUR) models based on mobile monitoring are typically dependent on purpose-oriented mobile monitoring campaigns. Here, we utilize air quality data from a large routine mobile monitoring campaign using specially equipped Google Street View cars. Goals of this analysis are to (1) develop a LUR model from 1 year of routine mobile monitoring, and (2) to assess how much mobile monitoring data is required to develop a robust LUR model. **Methods:** We collected an extensive dataset of mobile air quality NO measurements in Oakland, CA using two Google Street View cars with fast-response (1 Hz) chemiluminescence monitors. The dataset presented here incorporates ~1100 h and >28,000 km of on-road data collected over 150 days of monitoring. We reduce the temporally resolved dataset to spatial medians at every ~30 meter point within our study domain. First, we develop a spatial LUR model that accounts for spatial autocorrelation in the data for the entire study domain. Second, we utilize a Monte Carlo (MC) analysis to systematically shrink our dataset by unique drive days and road coverage. LURs developed from each MC run are evaluated by their predictive performance and selection of explanatory variables. **Results and Conclusions:** We successfully developed a spatial LUR model of NO in Oakland, CA based on routine mobile monitoring. Results indicate road type, land use, and the port as important explanatory variables. Preliminary results for the subsampling analysis indicate that a substantially less extensive sampling campaign would have been sufficient to develop a stable LUR. These results will be useful for air quality researchers and regulators interested in the ability of mobile monitoring to complement other exposure assessment approaches.

Keywords: A - exposure measurement, A-geospatial analysis/GIS, A-exposure models

MO-SY-B4: Toxic chemical water contaminants in low and middle income countries Part II: Analytics, risk analysis, and mitigation strategies

MO-SY-B4-163

Using a risk-based approach to rank toxic chemicals in drinking water to support prioritization of risk mitigation strategies in low resource settings

J. H. Redmon; RTI International, Research Triangle Park, NC

Abstract: Risk analysis is critically important in facilitating evidence-based, transparent decision making. Risk ranking is the systematic analysis and ordering of hazards in terms of the likelihood and severity of adverse impacts. The results can provide a scientific basis to characterize water quality problems, issue advisories, make informed regulatory decisions, and prioritize risk mitigation strategies to improve water quality. While probabilistic risk assessment methods are the gold standard for chemical risk characterization, it is often infeasible in low to middle income countries (LMICs) given a lack of available financial, technical, and informational resources. The objective of this presentation is to provide an example of a multi-staged risk-based approach that can be used to rank risks associated with toxic chemicals in water, and allow for the prioritization of effective risk mitigation approaches in resource-constrained settings. The first stage is to define the scope, which includes identifying the purpose, selecting what will be ranked (e.g., chemicals, water sources, and geographical region), and screening for relevance on the overall risk potential. This stage often includes a literature review and discussion with local experts to identify which chemical contaminants may be present in specific water sources and regions. The second stage is to develop the approach, which consists of selecting the risk ranking method, choosing metrics to characterize likelihood of exposure and severity, and collecting applicable data and information. The final stage is conducting the analysis and reporting results, assumptions, and limitations. Risk ranking results can be incorporated into risk prioritization efforts that consider other factors such as mitigation feasibility in a systematic manner. As new data and information become available, this approach can be iteratively refined to more effectively identify and mitigate toxic chemical exposures and risks.

Keywords: A-chemical prioritization, A-risk assessment, C-water, D-community, toxic chemicals

MO-SY-B4-164

Cost-Effective, Scalable Field and Laboratory Approaches for Quantitation of Established and Emerging Chemicals in the Environment

K. E. Levine, J. M. Harrington, P. Presler-Jur, A. Essader, J. Davis; RTI International, Research Triangle Park, NC

Abstract: Globalization has resulted in increased income and has provided economic growth opportunities across the developing world, stemming in large part from the relocation of chemical, pharmaceutical, waste disposal, and other manufacturing activities from established to emerging economies. Despite the economic opportunities that this manufacturing diversification has provided, it has also led to the unintended proliferation of chemical pollutants, including industrial effluents, electronic waste, agrochemicals, pharmaceuticals, and personal care products, in the environment. Unfortunately, low and middle income countries (LMIC) often lack the regulatory infrastructure and the needed resources to adequately assess the impact of these increasingly ubiquitous pollutants. Easy to use, cost-effective and scalable advances are desperately needed in laboratory and field analytics to enable LMIC's to monitor, mitigate, and ultimately prevent the release of toxic chemicals. This presentation will describe field and laboratory analytical capabilities presently applied in resource-constrained LMIC settings for the monitoring of environmentally relevant chemicals and highlight several emerging technological approaches that could extend analytical capacity. Advances in 'gold standard' laboratory-based approaches will be described, along with novel sampling approaches that could significantly reduce logistical costs and barriers. In addition, the potential utility of inexpensive paper and silicon-based sensors will be described for application in monitoring environmental contaminants.

Keywords: A-analytical methods, A - exposure measurement, D-susceptible/vulnerable, A - ambient monitoring, D-environmental justice

MO-SY-B4-165

Volatile organics in open dumps and their removal using municipal solid waste biochar

M. Vithanage; National Institute of Fundamental Studies, KANDY, Sri Lanka

Abstract: Open dumps are the method of solid waste disposal in most developing countries including Sri Lanka. Volatile organic compounds (VOCs) are carcinogenic priority pollutants that are released to the environment through the open dumping sites. This study intended to assess the VOCs in leachates of open dumps in Sri Lanka and evaluate the potential of municipal solid waste biochar (MSW-BC) for remediating VOCs. Pyrolyzed MSW-BC was characterized for its physicochemical properties. Furthermore, benzene, toluene and xylene (50 µg/L) were investigated for removal based on pH, kinetics and sorbent (1-10 g/L), sorbate (10-300 µg/L) concentrations. The analytical setup for experiment followed by EPA 524 and concentration of VOCs in aqueous media were measured by static head-space GC-MS. Results explore that, textural properties of the biochar, pore volume (0.013 Cm^3g^{-1}) and specific surface area (108 m^2/g) indicated its potential for higher adsorption capacity. Electrostatic interactions cannot be omitted, as can be seen from the dynamic adsorption behaviors in edge experiment at room temperature for benzene, toluene and xylene removal. Maximum VOCs removal were obtained at pH>7 whereas, highest adsorption at 24 hours reaction time on pH 9.0, 8.3 and 9.0 recorded as 85.4, 87.0 and

89% (42.7, 43.5 and 44.6 µg/g) respectively for benzene, toluene and xylene. Freundlich fitting could explain isotherm data with good accuracy for both benzene and toluene (R^2= 0.955, 0.988) while for xylene it was the Langmuir. Besides, recorded maximum adsorption capacity was about 218.2, 257.7 and 670 µg/g for benzene, toluene and xylene. Moreover, well fitted kinetics model of pseudo first order supposed non-dissociate molecular adsorption into adsorbent. Hence, a heterogeneous process involved with physisorption between adsorbate molecules and biochar surface can be suggested as the removal mechanism for benzene and toluene while chemisorptions and physisorption both are responsible for xylene removal. The results indicated the potential of MSW-BC as an efficient material for removing low concentrations of VOCs like benzene, toluene and xylene from landfill leachete or wastewater.

Keywords: B-VOCs, C-water, D-community, A-built environment, A-green chemistry

MO-SY-B4-166

A potential never-ending story of chemical water pollution in LMICs: Proliferation of legacy and replacement PFAS

J. DeWitt[1], C. Higgins[2]; [1]East Carolina University, Greenville, NC, [2]Colorado School of Mines, Golden, CO

Abstract: Per- and polyfluoroalkyl substances (PFAS) are organofluorine compounds with widespread applications in industrial and consumer products. They have been in production for decades and more than 3,000 are thought to be on the global market. Most of these compounds lack a comprehensive understanding of their environmental and human exposure routes due to a lack of information on their life cycles. PFAS enter source waters through industrial releases, wastewater treatment plant discharges, stormwater runoff, use of PFAS-containing firefighting foams, and land application of contaminated biosolids. They are highly persistent and can be widely transported through atmospheric deposition and water currents and as a result, have been detected in myriad waters throughout the world. Past and ongoing production and use of PFAS has led to, and will continue to lead to, global distribution of PFAS in the environment, wildlife, and humans. The very high persistence of PFAS also leads to poorly reversible human exposure. While production and use of legacy compounds, including perfluorooctanoic acid (PFOA) and perfluorooctane sulfonate (PFOS), has declined, production of fluorinated alternatives has increased, ensuring that continued presence of PFASs in the aquatic environment. Compounding the problem is the lack of toxicity data on these alternatives, an understanding of their environmental fate and transport, including distribution into humans, and inefficient treatment processes for contaminated waters. Therefore, PFAS represent a potentially never-ending chemicals management issue.<!--[if supportFields]><span style='font-size:11.0pt; font-family:"Times New Roman",serif;mso-fareast-font-family:"MS Mincho"; mso-fareast-theme-font:minor-fareast;mso-ansi-language:EN-US;mso-fareast-language: EN-US;mso-bidi-language:AR-SA'><span style='mso-element:field-end'></span></span><![endif]-->

Keywords: B-POPs, C-water

MO-SY-B4-167

Activation of Biochar Adsorbents with Base and Ash Leachates for the Removal of Organic Micropollutants in Low-Cost Water Treatment

M. J. Bentley[1], J. Kearns[2], R. Summers[1]; [1]University of Colorado Boulder, Boulder, CO, [2]North Carolina State University, Raleigh, NC

Abstract: Biochar produced by pyrolysis of waste biomass is a local alternative to activated carbon for the removal of organic micropollutants in low cost water treatment. In this study, biochars were prepared in the laboratory and field with a novel activation process using base and ash leachates to improve sorption capacity. *The objective of this study was to quantify the increase in sorption capacity gained through activation, and compare sorption by activated biochar with a commercial activated carbon (AC) reference sorbent.* The effects of feedstock and pyrolysis temperature (500 - 800 C) were also investigated. Field biochars were produced using a Top-Lit-Up-Draft (TLUD) Gasifier manufactured from

200 L surplus metal drums. Batch mode uptake of anionic human/veterinary antibiotic sulfamethoxazole (SMX) from DI water, lake water, wastewater effluent, stormwater, and combined sewer overflow water was quantified for biochars and AC. Base and ash activation showed an increase in SMX sorption by 2-5 times compared to raw biochar in all waters tested, and the performance of ash activated biochar met or exceeded the performance of AC in all waters tested. These results suggest that base and ash activation are effective in increasing sorption of organic micropollutants in low-cost water treatment scenarios. Ash and base activation of biochar has the potential to provide high quality treatment for waters impaired by organic micropollutants in low-cost settings such as developing communities and small systems.

Keywords: Biochar, Organic Micropollutants, A-global health, B-pharmaceuticals, C-water, C-streams

MO-SY-C4: ATSDR Exposure Assessments: Science, Collaboration, and Communication to Inform Public Health Decisions in diverse communities

MO-SY-C4-168

Characterizing exposure to dioxins/furans and other toxins from a biomass power plant near L'Anse Indian Reservation, MI.

M. P. Caudill; ATSDR, Chicago, IL

Abstract: ATSDR evaluated the potential health impacts of the L'Anse Warden Electric Company's (LWEC) operations on L'Anse, a village of 2,011 residents, in Michigan's Upper Peninsula. LWEC is adjacent to residential areas and the Keweenaw Bay Indian Community L'Anse Reservation is about 1 kilometer downwind to the north and east. LWEC originally burned fossil fuels, then in 2007 applied for conversion to biomass fuel, allowing it to burn railroad ties treated with pentachlorophenol (PCP) and creosote, chipped wood, and shredded tires. After the change, local residents complained of particles depositing on their property and also expressed concerns about health impacts. The US Environmental Protection Agency provided soil sample results for dioxins/furans, polycyclic aromatic hydrocarbons (PAHs), PCP, and toxic metals, as well as modeled estimates of particulate matter and toxins in air. PAH and dioxin concentrations in soil were higher near LWEC and central L'Anse than other areas further away. ATSDR performed a health risk screening on the full dataset and detailed cancer and non-cancer calculations on pollutants that exceeded screening levels. ATSDR found that dioxins and furans in soil could contribute to developmental effects in one area of L'Anse. However, the dioxin/furan chemical fingerprint did not match samples in areas that were downwind of LWEC and is likely due to another source. PAH profiles in soil were consistent with typical urban sources, including historic emissions from vehicles and power plants. The estimated increase to cancer risk due to PAH and dioxin exposures was not significant. Health risks from air emissions, based on modeling, were minimal. ATSDR recommended additional soil sampling in the areas that were not previously tested.

Keywords: B-SVOCs, A-risk assessment

MO-SY-C4-169

Spatial, temporal and meteorological trends in benzene air concentrations in Corpus Christi Refinery Row, Texas

J. Durant, D. Langmann; ATSDR, Atlanta, GA

Abstract: *Abstract:* Using the R package, *openair,* the Agency for Toxic Substances and Disease Registry (ATSDR) analyzed spatial, temporal, and meteorological effects on benzene air concentrations near petrochemical facilities in Corpus Christi, Texas. ATSDR used data collected by the Texas Commission on Environmental Quality (1994-2010), the Corpus Christi Air Quality Project (2005-2010), and industry sponsored monitoring (1996-2010). Temporal resolution of the monitoring data from these three networks ranged from 15-minute or 1-hour auto gas chromatograph data (Auto GC) to 24-hour

SUMMA™ canister samples. Cross-network concentrations were comparable at sites geographically closest to each other, with the relationships affected by wind direction. The data for all three networks showed strong seasonality, with higher concentrations occurring during the winter and autumn. The Auto GC monitors showed a strong diurnal pattern. Over time, concentrations of benzene decreased across all networks and monitoring locations. Conditional probability polar plots indicated higher levels of benzene associated with emissions from petrochemical facilities, with transportation sources also influencing some monitoring locations. To address how well these monitoring locations captured spatial variability from stationary sources of benzene, AERMOD version 15181 was used to model point and area source emissions from the 2011 National Air Toxics Assessment for Nueces County. Modeled benzene concentrations were less than monitored results by a factor of 2-6, depending on the monitor location. However, the model results captured the seasonality and spatial variability seen in the monitoring data and indicated that these monitoring locations were well-situated to capture high mean and 24-hour concentrations of benzene from the petrochemical plants.

Keywords: A-exposure models, C-air, B-VOCs, A-environmental justice

MO-SY-C4-170

Assessing Children's Lead Exposure in a Colorado Neighborhood near a Historic Smelter

L. Rosales-Guevara[1], B. Tierney[1], D. Dorian[1], B. A. Anderson[1], C. Nevin-Woods, D.O.[2]; [1]CDC/ATSDR, Atlanta, GA, [2]El Paso County, Colorado Springs, CO

Abstract: Background: The purpose of this investigation was to identify lead exposure in Pueblo, CO, where a historic smelter left lead contaminated soil and a slag pile in residential areas, and to convey the results to the community, local government, state and federal environmental health agencies, and area primary care providers. **Methods:** After obtaining consent/assent, we administered an in-person questionnaire and tested venous blood lead levels (BLLs) in 135 participants, including children <6 years, women who were pregnant, and women of childbearing age (without occupational lead exposure) residing within 0.5 miles of the historic smelter and slag pile. **Results:** Factors that increased risk for exposure in this area included population demographics (i.e., Mexican-American population) and a large percentage of older houses. The highest BLLs were detected in children < 6 years old with pica behavior. Among 33 children tested, 3 had BLLs exceeding the CDC reference level of 5 µg/dL and of those, one exceeded a BLL of 10µg/dL. In addition, a 7-year-old child had a BLL above 5µg/dL. The BLL 25th, 50th, and 75th percentiles for the youngest age groups (9 months to <6 years old, and 6 to < 12 years old) suggest higher exposure to lead compared to children's BLLs reported in the National Health and Nutritious Examination Survey. As a result of the investigation, Pueblo City County Health Department conducted medical follow-up, healthy home inspection, and health education, and the EPA conducted additional soil sampling. EPA used the results of this investigation to support soil remediation decisions. **Conclusions:** The increased BLLs can harm children's health. Sufficient scientific evidence indicates that BLLs of 5 µg/dL or less may cause neurological, cognitive, and attention related behavioral effects in children. Therefore, primary prevention should be a priority to prevent exposures.

Keywords: A-biomonitoring, D-children, C-soil, D-community, B-metals

MO-SY-C4-171

Assessing Children's Lead and Arsenic Exposure in Two Rural Arizona Communities near a Copper Smelter

B. Gerhardstein[1], B. Tierney[2], J. Rayman[1], B. A. Anderson[2]; [1]Agency for Toxic Substances and Disease Registry, San Francisco, CA, [2]Agency for Toxic Substances and Disease Registry, Atlanta, GA

Abstract: Background: ATSDR investigated lead and arsenic exposures in Hayden and Winkelman, Arizona, the site of historic and ongoing copper smelter operations. Lead and arsenic concentrations in air and non-residential soils had been found above background levels. Residents requested biological testing to address concerns about their exposure to lead and arsenic in the environment. **Methods:** In

2015, ATSDR administered a questionnaire and tested venous blood lead and spot urine arsenic concentrations in a convenience (i.e. volunteer) sample of 83 residents from 29 households. Participants included children, adolescents, and women of childbearing age (20-40 years). We evaluated local air monitoring data for 2015. **Results:** Of 54 children age 1-11 years, 2 had blood lead levels (BLLs) above the CDC reference level of 5 micrograms per deciliter (µg/dL). Another two children 1-5 years had BLLs between 4-5 µg/dL. The median BLLs for children and adolescents tested were about two times higher than median BLLs reported in the National Health and Nutritious Examination Survey (NHANES). Median total and inorganic arsenic levels (creatinine corrected) for participants were similar to NHANES median levels. The smelter was shut down for maintenance when we collected blood and urine samples. During the shutdown, levels of lead and arsenic in air were about seven and eight times lower. respectively than other times in 2015. **Conclusions:** Children and adolescent participants had higher BLLs than the U.S population, putting them at increased risk for neurological, cognitive, and attention related behavioral effects. Since arsenic has a short half-life in the body, the lower levels of arsenic in air during the testing may have affected participants' urinary arsenic levels. ATSDR is planning to retest urinary arsenic at a time when the smelter is operating normally. ATSDR recommended several steps government, industry, and community members can take to reduce lead and arsenic exposures.

Keywords: A-biomonitoring, D-community, B-metals, D-Southwest-specific, C-air

MO-SY-C4-172

Characterizing manganese inhalation exposure and the evaluation of neuropsychological health outcomes in two Ohio towns.

M. A. Colledge[1], J. Julian[2], R. Bowler[3], V. Gocheva[3], E. Kornblith[4], C. Beseler[5], D. Lobdell[6], G. Bollweg[7]; [1]ATSDR/CDC, Evanston, IL, [2]United States Environmental Protection Agency, Chicago, IL, [3]San Francisco State University, San Francisco, CA, [4]Alliant International University, San Francisco, CA, [5]Colorado State University, Fort Collins, CO, [6]United States Environmental Protection Agency, Research Triangle Park, NC, [7]United States Environmental Protection Agency, Chicago, IL

Abstract: Manganese (Mn), a nutrient required for normal metabolic function, is also a persistent air pollutant and a known neurotoxin at high concentrations. Elevated exposures can result in a number of motor and cognitive deficits. Quantifying chronic personal exposures in residential populations studied by environmental epidemiologists is time-consuming and expensive. We developed an approach for quantifying chronic exposures for two towns that have been identified as having elevated air manganese (air-Mn) due to industrial emissions (Marietta and East Liverpool, Ohio). This was accomplished through the use of measured and modeled data in the communities studied. The modeled chronic exposures were used as the exposure metric for a cross-sectional health outcome study of adult residents. The purpose of the study was to evaluate relationships between environmental ambient Mn exposure and distance from the source and motor and cognitive function in exposed residents. A neuropsychological screening test battery of cognitive and motor function was administered to study participants (East Liverpool=86, Marietta=100). Statistically significant deficits in cognitive and motor outcomes were noted with increasing air-manganese exposures.

Keywords: A - ambient monitoring, A - exposure measurement, B-metals, A-exposure models, C-air

MO-SY-D4: Characterizing Consumer Habits & Practices to Refine Consumer Product Exposure Assessment

MO-SY-D4-173

CURRENT EXPOSURE ASSESSMENT PRACTICES IN THE EU FOR CHEMICALS IN CONSUMER PRODUCTS

N. Corea; SC Johnson, Camberley, United Kingdom

Abstract: Within the EU, the exposure models to assess consumer products are primarily selected on the basis of their regulatory acceptance. The available exposure models under REACH Regulation (EC) No 1907/2006 and the Biocidal Products Regulation (EU) No. 528/2012 and their data sources are discussed. Conservative interpretation of exposure data coupled with the conservative selection of default values within regulatory accepted models can often lead to apparent unfavourable risk assessment outcomes. In such instances, the exposure assessor may be required to conduct bespoke exposure studies simulating consumer use of the product within the home in order to refine exposure. In such cases, understanding consumer habits and practices associated with a particular product can enhance the study design. An example is provided where product use directions and label claims influence the dose applied, frequency of use and the choice of residential room for a biocidal product. Such studies can provide a significant refinement in exposure compared to higher tier models, but at high financial cost. Future exposure assessment practices can be improved by further examining and adapting existing methods and data sources from other regions where consumer habits and practices overlap. Although this has been conducted to some extent already in the EU, particularly using US data, there is scope to gain insight from more recent models. For example, the transfer of biocidal substances to food from the use of cleaning products has been well defined in the US-EPA's higher tier IDREAM model. Aligning exposure assessment approaches between the EU and US could lead to globally acceptable risk assessment outcomes, thus increasing efficiency for exposure assessors and regulatory submissions. The availability of additional scientifically robust higher tier models in the EU will also serve to minimise the need for resource-intensive measurement studies.

Keywords: A - exposure measurement, A-activity patterns, C-consumer products

MO-SY-D4-174

Data for High Tier Consumer Product Exposure Assessment

S. A. Tozer; Procter & Gamble, Surrey, United Kingdom

Abstract: Exposure estimates are a critical part of the equation used to calculate acceptable "safe levels" and conduct quantitative risk assessments for ingredients in consumer products. Many ingredients are ubiquitous, so the consumer can be exposed from multiple products and by multiple exposure routes and so aggregate exposure estimates are necessary. Robust aggregate exposure tools are now available for some consumer product categories that contain large habits and practices datasets, which allow product usage to be modelled at the high tier. However, data on ingredient usage in consumer products is not widely available, and so conservative estimates regarding these parameters are often made, and this conservatism is compounded the more consumer products are included in the scenario. To this end, methods are needed to obtain data and to produce realistic exposure estimations. In this presentation examples will be given of how to approach more complex probabilistic aggregate assessments at the high tier. Datasets were modelled using the subject oriented Creme Care and Cosmetics aggregate exposure tool to refine the data, including co-use data and product composition data (ingredient concentration and occurrence), demonstrating how such data can be acquired and utilised in exposure assessment. The presentation will also discuss how the cosmetic Industry are now conducting a number of ingredient surveys to obtain such representative composition data, at the individual formulation level. The data from these surveys will quantify the real distribution and occurrence of key ingredients in cosmetics and personal care products and provide more refined data for use in future aggregate exposure modelling.

Keywords: C-consumer/personal care products, A-aggregate exposure, A-exposure models, A-exposure factors, A - population exposure

MO-SY-D4-175

The ECETOC Exposure Database

C. O' Mahony; Creme Global, Dublin, Ireland

Abstract: Currently, a large number of data sources and tools exist for exposure assessment, and it can be challenging for a risk assessor to determine what tools and data are appropriate for a given purpose. Equally, it is can be difficult to understand what data gaps exist within various domains with regards to human exposure assessment. To address this question, the European Centre for Ecotoxicology and Toxicology of Chemicals (ECETOC) initiated a task force activity, entitled Guidance for Effective Use of Human Exposure Data in Risk Assessment of Chemicals. One of the aims of the task force was to present and review the current state of consumer exposure science by examining data needs, sources (including novel data sources), tools, gaps, and the latest innovations. As part of this activity, the task force performed a landscaping exercise of available exposure data and tools, in order to gain insight and understanding into what data resources exist for exposure assessment, their level of detail, and their domains of applicability. A database was created where data sources and tools were categorised into various ways and described with a harmonised set of terms. The same categorisation was also applied to what tools were identified as part of the exercise. In order to make it available to the public and keep the database up to date, an online tool has been created to house it. Users can freely register to use the database and access it via a searchable interface. Also included is a user dialog page, where users can discuss and suggest improvements and updates to the database, facilitating interaction with and amongst the exposure science community. This presentation will provide an overview of the database, its contents, and the online tool.

Keywords: A-exposure factors, A-exposure models, A-risk assessment

MO-SY-D4-176

Development of Product Categories Linking Ingredient Data and Consumer Habits and Practices for Exposure Prediction

K. Isaacs, K. Dionosio, K. Phillips, P. Price; US Environmental Protection Agency, Durham, NC

Abstract: Consumer product categorizations in models of near-field (residential) exposures to must provide a bridge between habits and practices information (Why is the product used? Who uses the product? How much do they use? How often do they use it?) and product ingredients (What chemicals are in a product? What fraction of the products in a category contains a specific chemical?) Defining categories that are appropriate for characterizing both types of information are a challenge; the categories must be specific enough to resolve population variability in prevalence, frequency, and mass of use, while also capturing differences in likely chemical constituents. Furthermore, the categories must be sufficiently specific as to allow the assignment of unique exposure scenario assumptions, such as microenvironment of use (e.g., indoors or outdoors), release form (e.g. spray versus liquid), release to various media, removal processes (e.g., rinse-off or wipe-off), and route-specific exposure factors (e.g., dermal surface areas of application, fraction of release in respirable form). While challenging, developing harmonized product categories allows for rapid parameterization of route-specific exposure scenario algorithms for new products and efficient utilization of new data on use or composition. We present here a hierarchical set of Product Use Categories (PUCs) for consumer product formulations that address these requirements; the PUCs are easily cross-walked to other category harmonizations (e.g., those recommended by the Organisation for Economic Co-operation and Development). These PUCs are being used to link the consumer product ingredient information in EPA's Chemical and Products Database to algorithms for scenario-specific use, fate, and exposure in EPA's Human Exposure Model.

Keywords: A-exposure models, C-consumer products, A-exposure factors, A - population exposure, Human Exposure Model

MO-SY-D4-177

Calendar-based consumer product use survey data: application to exposure modelling

B. M. Young[1], J. H. Driver[2], J. Blattner[3]; [1]Bayer CropScience, RTP, NC, [2]risksciences.net, LLC, Manassas, VA, [3]SC Johnson, Racine, WI

Abstract: The REJV National Pesticide Use Survey (2012-2013) was implemented in response to an EPA Data Call-In (DCI). The 365-day (calendar-based) survey was intended to address gaps in understanding residential pesticide usage, particularly related to co-occurrence of chemical active ingredients and/or pesticide products across residential scenarios. The final survey, which has been reviewed and approved for use by EPA, consists of 4,573 unique U.S. households' 12 month diaries of residential pesticide use. Calendar-based consumer product use survey data provided by the REJV survey can be used for refined aggregate and cumulative residential exposure modelling for pesticides. Case studies are provided that demonstrate the utility of the survey data to temporally profile chemical-specific product use events and related exposure scenarios. The case studies include evaluation, within and across households, of the incidence of product use by categories, frequency of use, temporal co-occurrence of product use, and related demographic characteristics.

Keywords: A-exposure models, A-longitudinal metrics, B-pesticides, C-consumer products

MO-SY-E4: Pediatric Manganese Exposure and Neurodevelopment across Diverse Populations

MO-SY-E4-178

Maternal Manganese During Pregnancy And Neurodevelopment In Young Children From The Infants' Environmental Health (ISA) Study

A. Mora[1], L. Cordoba[1], M. Quiros-Lepiz[1], A. Fajardo-Soto[1], D. R. Smith[2], J. A. Menezes-Filho[3], B. Eskenazi[4], D. Mergler[5], B. van Wendel de Joode[1]; [1]Universidad Nacional, Costa Rica, Heredia, Costa Rica, [2]University of California at Santa Cruz, Santa Cruz, CA, [3]Federal University of Bahia, Salvador, Brazil, [4]University of California at Berkeley, Berkeley, CA, [5]University of Quebec in Montreal, Montreal, Canada

Abstract: Background: Numerous cross-sectional studies of preschool-age children have observed that exposure to manganese (Mn) adversely affects neurodevelopment. However, few prospective studies have looked at the behavioral effects of maternal Mn concentrations during pregnancy in young children. **Methods:** We conducted a birth cohort study (Infants' Environmental Health Study, ISA) among families living near banana plantations aerially sprayed with Mn-containing fungicide mancozeb in Costa Rica. We measured Mn concentrations in maternal blood and hair samples collected at different times during pregnancy. We assessed behavioral problems in 242 children at 5 years of age (range = 4.6-6.5) using the Child Behavior Checklist (CBCL). We tested for linear and nonlinear relationships and assessed effect modification by child sex. Models were adjusted for maternal education, parity, gestational age, child sex, and child age and HOME score at five years. **Results:** Median (P25-P75) maternal blood and hair Mn averaged during pregnancy were 23.9 µg/L (20.2-28.5) and 1.9 µg/g (1.0-4.2), respectively. We did not find linear associations of maternal blood and hair Mn with any of the CBCL behavioral outcomes. However, we observed nonlinear associations of maternal blood Mn and internalizing problems [β for the middle tertile = 2.9 (95% CI: -0.5, 6.2) and for the upper tertile = 2.0 (-1.5, 5.4), compared to the lower tertile] among girls. We also found nonlinear associations of maternal blood Mn with anxiety/depression symptoms [β for the middle tertile = 1.1 (0.0, 2.3), compared to the lower tertile] and somatic complaints [β for the middle tertile = -1.2 (-2.2, -0.1)] among boys. **Conclusions:** Further studies, which include maternal and cord blood concentrations, as well as other elements, such as iron, are needed to understand whether maternal excess Mn during pregnancy is associated with behavioral outcomes in young children.

Keywords: A-epidemiology, B-metals, D-children, D-prenatal

MO-SY-E4-179

Neurodevelopment and Manganese Exposure in Rural, Appalachian Ohio

E. Haynes[1], H. Sucharew[2], K. Dietrich[1]; [1]University of Cincinnati, Cincinnati, OH, [2]Cincinnati Children's Hospital Medical Center, Cincinnati, OH

Abstract: Manganese (Mn) is both an essential element and neurotoxicant. Two communities in Ohio have the highest Mn exposure sources in the nation. The goal of this study is to address the communities' primary research question: "Does Mn affect cognitive development of children?" Children ages 7-9 were enrolled in the Communities Actively Researching Exposure Study (CARES) from Marietta, Cambridge, and East Liverpool, Ohio. CARES has been conducted using community-based participatory research principles. We conducted neurodevelopmental measurements on children including neurocognitive, neuromotor, behavioral and social. Blood and hair were analyzed for Mn and lead (Pb) and serum was analyzed for cotinine. Both low and high Mn concentrations in blood and hair were negatively associated with child IQ scores. There was a negative association between hair Mn and child IQ in East Liverpool children. Serum cotinine was also negatively associated with child cognitive function. Community engagement has enhanced the research process and the translation of study findings for the communities.

Keywords: A - population exposure, A-biomarkers, B-metals, A-second-hand smoke

MO-SY-E4-180

Genetic variants in Mn transporters

K. E. Wahlberg; Lund University, Lund, Sweden

Abstract: Manganese (Mn) is an essential nutrient in humans and acts as a cofactor in many enzymes important for the brain. Mn is therefore crucial for normal brain development and function. However, at elevated levels, Mn is neurotoxic and there is a narrow range between essential and toxic doses. Toxic Mn levels in the body can be caused by Mn exposure from e.g. industrial contamination of the environment or naturally occurring in food and drinking water. There is now emerging evidence that Mn exposure, even at moderate levels, can interfere with children's neurodevelopment and cause cognitive, motor and behavioral deficits. In addition to Mn exposure, genetic factors can affect Mn levels in the body and cause neurological symptoms. Mn is tightly regulated by homeostatic mechanisms involving several metal transporters. Recent studies have shown that rare loss-of-function mutations in metal transporter genes *SLC30A10*, *SLC39A8* (ZIP8) and SLC38A14 (*ZIP14*) can considerably affect Mn concentrations in different parts of the body including the blood, brain and liver, and have been linked to neurodegenerative symptoms and impaired motor skills, which demonstrates the importance of these genes in Mn homeostasis. We have shown that also more common single nucleotide polymorphisms (SNPs) in these genes significantly contribute to variations in Mn concentrations between individuals and also influence neurological outcomes. Based on recent research on a cohort of Italian children with environmental Mn exposure, we will discuss the influence of SNPs in Mn transporter genes on Mn homeostasis in different early life stages and how they may cause differences in neurodevelopment between children. We will also discuss how theses SNPs may contribute to differences in sensitivity to Mn exposure between children by affecting how efficiently Mn concentrations in the body are regulated. Furthermore, we will discuss the importance in considering genetics in studies of human Mn exposure.

Keywords: B-metals, D-susceptible/vulnerable, Neutoxicity, genetic susceptibility, D-children, Neutoxicity, genetic susceptibility

MO-SY-E4-181

MANGANESE EXPOSURE AND NEURODEVELOPMENT AMONG ITALIAN ADOLESCENTS RESIDING NEAR FERRO-ALLOY INDUSTRY

B. Claus Henn[1], M. Arora[2], Y. Chiu[2], J. A. Bauer[1], C. Austin[2], S. Guazzetti[6], H. Hsu[2], S. Zoni[5], C. Fedrighi[5], M. Peli[5], C. Benedetti[5], B. A. Coull[3], R. O. Wright[2], D. R. Smith[4], R. G. Lucchini[2]; [1]Boston University, Boston, MA, [2]Icahn School of Medicine, NY, NY, [3]Harvard Chan School of Public Health, Boston, MA, [4]University of California, Santa Cruz, Santa Cruz, CA, [5]University of Brescia, Brescia, Italy, [6]Public Health Agency Reggio Emilia, Reggio Emilia, Italy

Abstract: Introduction: Growing epidemiologic evidence suggests that environmental exposure of children to manganese (Mn) is associated with decrements in neurobehavior. However, substantial gaps remain in our understanding of Mn health risks among adolescents and in relation to developmental timing of exposure, child sex, and varying biomarkers of exposure. Methods: As part of the Public Health Impact of Manganese Exposure (PHIME) study, we measured Mn concentrations in blood, hair, nails, saliva, and urine collected from 11-14 year old Italian children (N~720) living near ferromanganese industry. Mn levels were also measured in deciduous teeth to represent prenatal, postnatal, and cumulative early childhood exposure periods. Neurobehavioral assessments, conducted concurrent with biological sample collection, included the Wechsler Intelligence Scale for Children, Conners Behavior Rating Scale, Virtual Radial Arm Maze, and California Verbal Learning Test. Multivariable models were used to evaluate associations between Mn and neurobehavior. Results: We observed inverse associations between childhood Mn with cognition (e.g., saliva and verbal IQ, $\beta=-1.4$, $p=0.004$) and memory (e.g., saliva and short recall, $\beta=-0.3$, $p=0.001$). In contrast, early life Mn (teeth) was associated with improved scores of cognition (e.g., digit span backward, $\beta=0.4$, $p=0.05$), behavior (e.g., teacher-reported inattention, $\beta=-0.07$, $p=0.04$), and memory (e.g., sum of intrusions, $\beta=-1.8$, $p=0.03$). We also observed sex-specific associations: a U-shaped association was seen only among girls for prenatal Mn and visual learning ($p_{GAM}=0.009$-0.02). Conclusions: These results suggest a complex relationship between Mn and neurobehavior, whereby associations may differ by timing of exposure and sex.

Keywords: A-epidemiology, B-metals, D-susceptible/vulnerable

MO-SY-E4-182

Uncovering prenatal and early childhood critical windows of Mn exposure using tooth-matrix biomarkers

M. Arora, Y. Chiu, R. G. Lucchini, R. O. Wright; Icahn School of Medicine at Mount Sinai, New York, NY

Abstract: Fetal and early childhood are considered critical windows of suceptibility, when manganese excess or deficieny may cause neruodevelopmental deficits. Recently developed tooth matrix biomarkers of manganese exposure offer an opportunity to study fetal exposure retrospectively. We measured Mn levels in dentine of shed teeth, representing prenatal, early postnatal, and cumulative childhood exposure windows, from 195 children in Italy. Pursuit Aiming, Luria Nebraska Motor Battery, as well as Tremor and Sway system from Computerized Adaptive Testing System (CATSYS) were measured at 11-14 years old. We examined the relationships of tooth Mn (ln-transformed) with motor function using generalized linear models and generalized additive models, adjusting for age, sex, and socioeconomic status index. Effect modification by sex was also examined. We found that higher prenatal Mn was associated with better body stability in boys in a number of sway tests (including mean sway, trasnversal sway, sagittal sway, sway area, and sway intensity), while it was associated with poorer performance in girls on all these metrics (all p for Mn × sex interaction<0.05). Higher prenatal Mn was also suggested to be associated with better hand/finger and eye-hand coordination in boys compared to girls in sex-stratified analyses, albeit interaction models did not reach statistical significance. For tremor, interestingly, higher early postnatal Mn was associated with increased center frequency in girls (p for interaction<0.01), but increased Mn level at the later postnatal period was associated with increased center frequency in boys (p for interaction=0.01). Our study, which used a direct measure of prenatal and childhood Mn exposure supports sex-specific critical windows of early life Mn exposure in relation to neuromotor function in adolescents.

Keywords: A-biomarkers, D-neonatal

MO-SY-F4: A Collaborative Trial for Integrating Exposome, High-Throughput Screening, and Non-Targeted Analysis Research

MO-SY-F4-183

Setting the Stage for EPA's Non-Targeted Analysis Collaborative Trial (ENTACT)

J. R. Sobus[1], E. M. Ulrich[1], A. Williams[2], A. Richard[2], C. Grulke[2], A. McEachran[3], S. Newton[1], M. Stryar[1], K. Isaacs[1], J. Wambaugh[2]; [1]US EPA, Research Triangle Park, NC, [2]US EPA, Research Triangle Park, NC, [3]US EPA, Research Triangle Park, NC

Abstract: The last decade has witnessed tremendous innovation in the use of alternative methods for evaluating human health risks posed by man-made chemicals. The Federal Tox21 Consortium and US EPA ToxCast program have collectively screened thousands of chemicals for bioactivity across hundreds of assays. Additionally, EPA's ExpoCast program has developed methods for predicting exposures to many of these same compounds, allowing for efficient risk-based chemical prioritization. On a parallel research track, efforts continue towards advancing knowledge of the human exposome. Rather than focusing prospectively on chemical safety, these efforts aim to identify primary causes/markers of disease, which may include exogenous and endogenous stressors. Due to differences in approaches, a disconnect exists between top-down exposome research and bottom-up high-throughput screening (HTS) efforts. Non-targeted analysis (NTA) tools can bridge this gap, since they can efficiently generate exposure data for compounds never before monitored, thereby enriching the chemical universe examined in risk evaluations and human health studies. Despite the enormous promise of NTA tools, however, benchmark methods do not yet exist, and limited consideration has been given to how NTA methods can be integrated with traditional targeted workflows in a regulatory context. Noting these challenges, EPA has developed an NTA research framework and collaborative trial to actively engage the exposome and HTS research communities. The research trial makes use of chemical substances from EPA's ToxCast testing library, and aims to identify the most suitable instrumentation, software, databases, methods, and workflows for performing NTA in support of exposome and HTS research. This presentation will introduce the framework and research trial, and describe steps being taken by EPA to cement NTA methods as indispensable public health research tools. *This abstract does not represent EPA policy.*

Keywords: A - exposure measurement, A-analytical methods, A-chemical prioritization

MO-SY-F4-184

EPA's Non-Targeted Analysis Collaborative Research Trial (ENTACT): Evaluating the State-of-the-Science for Non-Targeted Analysis

E. M. Ulrich[1], J. R. Sobus[1], A. Richard[2], C. Grulke[2], A. Williams[2], S. Newton[1]; [1]US EPA, Durham, NC, [2]US EPA, Durham, NC

Abstract: While thousands of chemicals have been profiled for bioactivity using high throughput screening, many of these chemicals are lacking exposure data, which hinders risk-based evaluation. Suspect screening (SSA) and non-targeted analysis (NTA) methods using high-resolution mass spectrometry (HRMS) offer new approaches to efficiently generate exposure data for a growing number of chemicals in commerce in a variety of environmental and biological media. We are conducting EPA's Non-Targeted Analysis Collaborative Research Trial (ENTACT) to evaluate a range of state-of-the-science SSA and NTA approaches. Four categories of experiments are underway: 1) ten standard chemical mixtures from the EPA's ToxCast library, 2) extracts of standardized environmental matrices including house dust, human serum, and environmentally deployed silicone passive samplers, 3) extracts of standardized environmental matrices spiked with known chemical mixtures, and 4) approximately 4600 single chemicals from the ToxCast library to produce reference spectra. This presentation will describe the preparation of mixtures, standardized matrices, and single chemical multi-well plates. Approximately 25 laboratories worldwide from academia, government, and private (i.e., vendor) organizations are enrolled in ENTACT. Each laboratory will use their gas and/or liquid chromatography systems and methods to maximize chemical space coverage, provide variations for comparison, and produce both instrument specific and publically available spectral resources. Initial overview results from participating laboratories will be discussed including: methods, software, and databases used; correctly identified chemicals, false negatives, and false positives. The goal of this work is to produce benchmark methods for analytical, reporting, and data analysis to facilitate further analyses and identify areas for improvement.

Keywords: A-analytical methods, B-mixtures, C-multimedia

MO-SY-F4-185

Using the US EPA's CompTox Chemistry Dashboard to Advance Non-Targeted Analysis and Exposure Research

A. McEachran[1], J. Grossman[2], S. Newton[2], K. Isaacs[2], K. Phillips[2], N. Baker[3], C. Grulke[3], J. R. Sobus[2], A. Williams[3]; [1]US EPA/ORISE, Durham, NC, [2]US EPA, Durham, NC, [3]US EPA, Durham, NC

Abstract: The use of high resolution mass spectrometry (HRMS) and non-targeted analyses (NTA) is advancing exposure science by enabling researchers to more completely define the exposome. However, confident structure identification of unknowns in NTA continues to present challenges to analytical chemists. Identification requires the integration of complementary data types to generate confident consensus structures; these data include the use of reference databases and source ranking algorithms, fragmentation prediction tools, and retention time prediction. The aim of our research is to generate and implement an identification tool and workflow for NTA within the US EPA's CompTox Chemistry Dashboard (https://comptox.epa.gov), a chemistry resource and web application containing chemistry data on ~750,000 substances. Data for chemical identification were incorporated from a variety of sources, including: functional use prediction models, PubMed references, and environmental media occurrence models. Data were assembled and a scoring-based identification scheme was empirically developed such that true positives were identified at the top of candidate chemical lists. This scheme was evaluated using two test sets: a known test set of chemicals and a blinded, unknown mixture. This scoring method for tentative and probable identification of unknowns resulted in increased identification performance over previous workflows. We will discuss development of a visualization tool within the Chemistry Dashboard where users can visualize the relative contributions of identification-specific metrics on a list of candidate structures. The scoring-based method and visualization tools indicate the capability of NTA structure identification within the Dashboard and provide an open, accessible tool for exposure scientists and mass spectrometrists. This abstract does not necessarily represent the views or policies of the U.S. Environmental Protection Agency.

Keywords: A-analytical methods, Non-targeted analysis

MO-SY-F4-186

Using structural similarity to estimate concentrations of known unknowns in suspect screening analyses

J. Grossman[1], A. McEachran[2], A. Marcotte[1], A. Williams[2], J. R. Sobus[1]; [1]US EPA, Research Triangle Park, NC, [2]US EPA, Research Triangle Park, NC

Abstract: Suspect screening analysis (SSA), based on high-resolution mass spectrometry (HRMS), has recently been adopted in environmental chemistry applications to enable broad-scale chemical surveillance and exposure monitoring. SSA is inherently a qualitative technique, and generally unable to provide quantitative information for a given compound in the absence of a conventional reference standard. Here, we introduce novel methodology that enables quantitation of known unknowns (i.e., HRMS features for which a probable structure can be put forth via identification algorithms and a custom SSA workflow) in samples without the use of conventional standards. Three blinded synthetic mixtures, based on US EPA ToxCast chemicals, were analyzed using liquid chromatography quadrupole time-of-flight mass spectrometry. Calibration curves were generated for tentatively identified compounds (via SSA), in the first two mixtures. SSA was then performed on the third mixture at multiple blinded dilutions. Molecular formulae were assigned to features in the third mixture and mapped to candidate structures using the US EPA's CompTox Chemistry Dashboard. Ranking/classification tools, along with novel semi-Data Dependent MS/MS methods, were used to propose and corroborate chemical structures for each formula. Chemical similarity coefficients were used to link candidate compounds from the third mixture to structurally-similar compounds with existing calibration data from the first two mixtures. Quantitation for each candidate compound, at each dilution, was performed using surrogate calibration curves, with a weighting scheme based on chemical similarity score (i.e., structurally-similar compounds were given more weight). The techniques described here allow for semi-quantitation of unknowns detected by SSA, and thus provide a means to generate exposure and dosage information to be considered in support of avant-garde chemical screening programs.

Keywords: A-analytical methods, A-statistical methods, A-risk assessment, B-mixtures, A-chemical prioritization

MO-SY-F4-187

Evaluating the Chemical Space of House Dust, Human Serum, and Silicone Wristband Passive Samplers Using Suspect Screening Analysis

S. Laughlin-Toth[1], J. Grossman[2], A. Marcotte[1], S. Newton[3], A. McEachran[4], A. Williams[5], E. M. Ulrich[3], J. R. Sobus[3]; [1]ORISE Program Participant - US Environmental Protection Agency, Durham, NC, [2]ORAU Program Participant - US Environmental Protection Agency, Durham, NC, [3]US Environmental Protection Agency, Durham, NC, [4]ORISE Program Participant - US Environmental Protection Agency, Durham, NC, [5]US Environmental Protection Agency, Durham, NC

Abstract: Human exposure to synthetic chemicals is a continuing public health concern as many compounds have yet to be fully evaluated for safety. As the number of chemicals continues to increase, environmental health scientists are working to rapidly identify which chemicals are most likely to pose health risks. Current tools supporting rapid risk-based chemical prioritization include high-throughput (HT) exposure monitoring and modeling. The standard analytical tool for HT exposure monitoring is high resolution mass spectrometry. This tool facilitates suspect screening analysis (SSA), a method used to identify large numbers of known chemicals in environmental and biological samples. This presentation describes the use of liquid chromatography mass spectrometry (LC-MS) and state-of-the-art SSA techniques to evaluate chemicals within extracts of standardized human serum, house dust, and silicone wristband passive samplers as part of the US EPA's Non-Targeted Analysis Collaborative Trial (ENTACT). Molecular features in these extracts were first identified using full scan MS spectra and then assigned chemical formulae using the US EPA Distributed Structure-Searchable Toxicity (DSSTox) database. Formulae were then matched against known structures using the US EPA CompTox Chemistry Dashboard. Additional data (e.g., exposure and bioactivity) were used to prioritize candidates for MS/MS analyses. All probable structures, based on MS/MS results and other diagnostic criteria, were cross-referenced against chemical standards with existing calibration data. Using structural similarity scores, calibration curves for standards were assigned to probable compounds in the media and used to produce concentration estimates. These screening-level results were ultimately compared to those previously obtained using targeted methods. The methods and results presented here demonstrate SSA as a viable technique for rapid identification and quantification of novel compounds in high-interest media.

Keywords: Non-targeted analysis, Standard reference material, Mass spectrometry, High-throughput, Non-targeted analysis, Standard reference material, Mass spectrometry, High-throughput, Non-targeted analysis, Standard reference material, Mass spectrometry, High-throughput, Non-targeted analysis, Standard reference material, Mass spectrometry, High-throughput

MO-SY-G4: Exposure, cumulative risk, and epidemiology in communities near upstream energy development Part 2

MO-SY-G4-188

Estimating Cumulative Exposure and Risks from Oil and Gas Development in the Denver Julesburg Basin

J. Adgate, L. McKenzie, B. Blair, W. Allshouse; University of Colorado, CSPH, Aurora, CO

Abstract: As oil and gas production increasingly moves into populated areas in the US and multi-well production facilities becomes increasingly common, exposures for nearby residents are incompletely characterized and risks uncertain. Individuals that live near those multi-well pads are likely subject to longer periods of development and a mix of chemical and non-chemical stressors that vary temporally and spatially. Using measurement and modeling data from the Denver Julesburg Basin, we present preliminary results of our cumulative risk assessment for a range of chemical and non-chemical stressors present in and around these sites. This approach combines data on fires and explosions, noise, air pollution, and truck traffic to develop exposure profiles for the population at risk residing near these sites. We present both quantitative and qualitative risk results and discuss strengths and limitations of cumulative risk models for this set of stressors.

Keywords: A-cumulative exposure, A-risk assessment

MO-SY-G4-189

A Tale of Two Cities: Associations between subclinical biomarkers and active oil and gas development

L. McKenzie[1], J. Adgate[1], B. Blair[1], J. Peel[3], J. Crooks[2]; [1]University of Colorado Denver, Aurora, CO, [2]National Jewish Hospital, Denver, CO, [3]Colorado State University, Fort Collins, CO

Abstract: Technological advances in directional drilling and hydraulic fracturing have resulted in a boom of drilling and production of US petroleum reserves. It is increasingly common for oil and natural gas development to encroach on populated areas, exposing residents to a range of chemical, physical and psychosocial stressors that can potentially impact health and quality of life. We will present results from a cross sectional study of 80 volunteers living in two cities, Greeley and Fort Collins located in Northeast Colorado's Denver Julesburg basin, from 2015 to 2016. These two cities were selected as they are in close proximity yet have a notable different in oil and gas activities: Greeley has experienced extensive oil and gas development, while Fort Collins has fought these operations such as through a now overturned moratorium on drilling. We collected up to three repeated measures of hair cortisol, blood pressure, atrial augmentation index, TNF-α, IL-1β, and IL-6 for each volunteer. These subclinical biomarkers are in the biologic pathways that link psychosocial stress and air pollution to changes in cardiovascular function, systematic inflammation, and subsequent adverse health outcomes. We will compare the subclinical biomarkers between the two cities and the variation observed with varying inverse distance well counts as a metric of exposure to oil and gas development. Our systematic collection of subclinical biomarkers in these two populations provides data, tools and community networks needed to design and facilitate subsequent prospective studies that will offer insight on the health effects associated with this rapidly expanding worldwide industry.

Keywords: A-biomarkers, A-epidemiology, A - population exposure, Oil and Gas Development

MO-SY-G4-190

Residential proximity to unconventional oil and gas wells, drinking water contaminants, and health symptoms in the Ohio Utica shale formation

E. G. Elliott[1], X. Ma[3], D. L. Plata[2], B. P. Leaderer[1], L. McKay[1], C. J. Pedersen[1], C. Wang[1], C. Gerber[1], T. Wright[1], M. Brennan[2], N. Deziel[1]; [1]Yale University, New Haven, CT, [2]Yale University, New Haven, CT, [3]Yale University, New Haven, CT

Abstract: Background: More than 9 million Americans live ≤1 mile from an unconventional oil and gas (UO&G) well, potentially placing their drinking water at risk of contamination. We conducted a community-based, cross-sectional study in Ohio to assess relationships between residential proximity to UO&G wells, drinking water contaminants, and health symptoms. **Methods:** We collected drinking water samples and interviewed participants (n=66) about their water source and health symptoms. We analyzed water samples for 16 volatile organic compounds and gasoline-range organic compounds (GRO) using gas chromatography and flame ionization detection. We calculated residential proximity to UO&G wells using geocoded addresses and well locations from the Ohio Department of Natural Resources. We used parametric, non-parametric tests, and multivariable logistic regression analysis to assess relationships between UO&G well proximity, water contaminants, and health. **Results:** Detection rates of benzene, bromodichloromethane, bromoform, and dibromochloromethane were 2-5 times as high among homes <2 km versus ≥2 km from UO&G wells (p<0.1). Concentrations of these 4 contaminants and GRO were 2-6 times as high among homes <2 km versus ≥2 km from UO&G wells (p<0.05). Concentrations were below US Environmental Protection Agency Maximum Contaminant Levels. After adjusting for potential confounders, those <2km of UO&G wells were 3.9 (95% confidence interval [CI]: 1.1-14) times as likely to experience neurologic symptoms and 2.6 (95%CI: 0.92-7.4) times as likely to have new/increased respiratory symptoms versus those ≥2 km from UO&G wells. **Conclusions:** We observed correlations between residential proximity to UO&G wells and (1) toxic organic compounds in drinking water and (2) health symptoms. Though we cannot definitively link these compounds and symptoms to UO&G wells, our findings underscore the need for further investigation into exposures and health impacts of UO&G development.

Keywords: A - exposure measurement, A - population exposure, A-epidemiology, B-VOCs, C-water

MO-SY-G4-191

Community-driven approach to understanding exposures at urban oil drilling sites

J. Johnston[1], A. Collier[2], S. Navarro[3], M. Hannigan[2]; [1]University of Southern California, Los Angeles, CA, [2]University of Colorado Boulder, Boulder, CO, [3]Esperanza Community Housing, Los Angeles, CA

Abstract: Los Angeles, California is the largest urban oil field in the country and home to thousands of active oil wells in very close proximity to homes, schools and parks. Oil and gas drilling operations are associated with a complex mixture of air emissions, including volatile organic compounds (VOCs) which have been elevated near drill sites. Exposure to such VOCs is associated with a range of acute health symptoms, including eye, nose, and throat irritation, impaired lung function, dizziness, fatigue and asthma exacerbation. Locally community-based organizations approached researchers to better characterize air pollution related to drilling sites in urban, environmental justice neighborhoods. We deployed 12 low-cost active air quality monitors around two oil drilling sites equipped with electrolytic, metal oxides and infrared sensors to continuously measure (at one-minute intervals) volatile organic compounds (VOCs), ozone (O3), carbon monoxide (CO), and carbon dioxide (CO2). Using two different VOC sensors we hope to differentiate methane and total non-methane hydrocarbon signals, which could provide direct assessment of oil drilling operations, and the remaining pollutants help to characterize the relative contribution of nearby traffic as well as provide context for regional air pollution issues. We will describe this approach, calibration of the data with air district sensors, and preliminary results to identify intermittent plumes of VOCs from the drilling sites. We will then discuss what unique information we learned through linking quantitative and qualitative information and how this may enhance future studies. This detailed characterization of a neighborhood site in a community impacted by multiple sources of air the potential to empower communities to understand key exposures and develop potential interventions.

Keywords: A - ambient monitoring, B-VOCs, D-environmental justice, D-susceptible/vulnerable

MO-SY-G4-192

Panel Discussion on Exposure, Cumulative Risk, and Epidemiology in Communities near Upstream Energy Development

L. McKenzie[1], J. Adgate[1], N. Deziel[3], E. Haynes[4], M. McCawley[5], A. Miller[2]; [1]University of Colorado Denver, Aurora, CO, [2]National Institutes of Health, Bethesda, MD, [3]Yale School of Public Health, New Haven, CT, [4]University of Cincinnati, Cincinnati, OH, [5]West Virginia University, Morgantown, WV

Abstract: 21st Century extraction of O&G from unconventional reserves has been associated with a wide range of stressors. Potential stressors include exposure to hazardous air pollutant, water pollutants, noise, and light, as well as psychosocial impacts. Many individuals living near these operations have complained of physical health effects, noise, odors, sleep disturbance, stress, and disruption to quality of life. To best understand the full impact of O&G on communities, exposure science needs to integrate expertise from a diverse community of researchers. The panel will lead a discussion on current and future research needs that to address the chemical, physical, and psychosocial exposures that may be associated with 21st oil and gas development.

Keywords: A - exposure measurement, A-epidemiology, A-cumulative exposure, A-risk assessment, Upstream Energy Development

MO-PO: Monday Poster Session

MO-PO-193

Composition of particulate matter (PM) originated from desert dust storms in the Eastern Mediterranean region.

S. Achilleos[1], P. Koutrakis[1], M. Hadjicharalambous[2], E. G. Stephanou[3], M. Iakovides[4]; [1]Harvard University, Boston, MA, [2]Cyprus University of Technology, Limassol, Cyprus, [3]The Cyprus Institute, Nicosia, Cyprus, [4]University of Crete, Hearklion, Greece

Abstract: Dust storms are naturally occurred events that transport dust particles far from the desert of origin and can reach the Americas, Europe, Mediterranean, and Middle East. These events occur several times per year in the Mediterranean region and have been associated with increased cardiovascular and respiratory mortality and hospital admissions. However, it is still not clear which component(s) of dust particles are associated with human health effects, mainly due to lack of information on the organic and inorganic composition of the dust particles. Therefore, we collected PM samples, in presence and absence of dust episodes, from two cities in the Eastern Mediterranean region (fine and coarse PM from Limassol, Cyprus; fine PM from Heraklion, Greece), and analyzed them for mass, elemental and organic carbon, secondary inorganic ions, metals, and polycyclic aromatic hydrocarbons(PAHs). Heraklion samples were classified based on the origin of the blowing air mass in three categories: a) pure dust (PD), when air was blowing directly from Sahara/Middle East desert; b) mixed dust (MD), when air mas originating from Sahara/ME desert and passing from other European countries before reaching the sampling site; and c) non-dust (ND). Limassol samples were classified to dust (DD) and non-dust (ND) because of small sample size. During dust storms, both fine and coarse mass increased from two to five times higher than ND days; the highest mass concentrations were observed during PD days. On overall, dust storms had a high impact on mineral dust PM mass (defined as the sum of the oxide forms of crustal elements) and ΣPAH were in higher concentrations during ND than DD for both locations and PM size. However, ΣPAH were found higher when dust plume was passing from other European countries ($median_{MD}$ = 1.21 ng/m^3) as opposed to blowing directly from Sahara desert ($median_{PD}$ = 1.04 ng/m^3). This classification (PD *versus* MD *versus* ND) can be useful for future studies and risk assessments.

Keywords: B-particulate matter,

MO-PO-194

Personal Fine Particles ($PM_{2.5}$) Mass and Components Exposures in the General Population of Hong Kong: Variability and Contributing Factors

X. CHEN[1], T. J. Ward[2], J. Cao[4], S. Lee[3], J. C. Chow[7], S. Yim[5], G. Lau[1], K. Ho[6]; [1]The Chinese University of Hong Kong, Hong Kong, China, [2]University of Montana, Missoula, MT, [3]The Hong Kong Polytechnic University, Hong Kong, China, [4]Xi'an Jiaotong University, Xi'an, China, [5]The Chinese University of Hong Kong, Hong Kong, China, [6]The Chinese University of Hong Kong, Hong Kong, China, [7]Desert Research Institute, Reno, NV

Abstract: Personal fine particles ($PM_{2.5}$) exposure study was conducted in the general population (48 subjects, 18-65 years of age) of Hong Kong during the summer and winter of 2014-2015. Personal $PM_{2.5}$ exposures ranged from 3.5 to 110.9 µg/m^3 with an overall mean of 35.4 (±19.5) µg/m^3, which was comparable to the stationary ambient $PM_{2.5}$ levels (35.3 ± 19.4 µg/m^3). We applied mixed-effects models to further investigate the determinants of personal exposure to $PM_{2.5}$ mass and chemical components, as well as within- and between-individual variance components. Additional model inputs include information from individual's activity diaries. In the present study, the within-individual variance dominated the total variability for all investigated exposure data except for $PM_{2.5}$ and EC. Ambient $PM_{2.5}$ concentration was the dominated predictor of (R^2_β = 0.03-0.59, $p < 0.001$) and the largest contributor (14.2-99.0%) to personal exposures for $PM_{2.5}$ mass and components. Moreover, season and occupation were important factors (R^2_β: 0.03-0.06, $p < 0.05$) that affect personal $PM_{2.5}$ exposures and contribute to modifying personal-ambient correlations (Pearson's *r*, 0.46-0.73). Subsequently, adjusted mixed-effects models including activities as covariates, identified season (NH_4^+, NO_3^-, elements), occupation (OC, EC, Ca^{2+}, S, Zn, Pb), time spent at home (Ca^{2+}, SO_4^{2-}, oxalate), time spent in transit (EC, Ti, V, Fe), cooking (NH_4^+, NO_3^-), and cleaning (OC, V) as significant determinants for $PM_{2.5}$ components, contributing to 5.5-96.8% of the variability in personal exposures. Our results indicated that ambient measurements from fixed sites were insufficient as a proxy of personal exposures in Hong Kong, especially for adult subjects who were more active or less sedentary. We suggest it is of great importance to perform long-term monitoring to evaluate the complexity of factors that contribute to personal exposures.

Keywords: A - exposure measurement, A-activity patterns, A-exposure factors, B-particulate matter, A - population exposure

MO-PO-195

The Personal exposure of Traffic policemen in Ulaanbaatar, Mongolia

E. Erdenebayar; Mongolian National University of Medical Sciences, Ulaanbaatar, Mongolia

Abstract: Introduction. Air pollution in Ulaanbaatar /UB/ and other major towns has been increasing and greatly exceed international guidelines. UB has seen a large growth in car and truck traffic over the last decades and with daily traffic jams that can last several hours. Pollution from cars and trucks is a significant contributor to air pllution in UB. Traffic policemen are at increased risk for air pollution health effects due to their working in dense traffic areas. **Goal**. The purpose of this study was to characterize pollution exposure to traffic policemen and compared to policemen working in an office environment. **Methods**. The study was conducted in warm season (July and September) in Ulaanbaatar and included 55 police officers, which were 40 traffic policemen (TP) and 15 office policemen (OP). Full-shift personal samples were collected to assess several ambient pollutants, including $PM_{2.5}$, respirable dust, nitrogen dioxide, carbon monoxide, black carbon during the working hours for each of study subjects. The samplings were used following instruments: SidePak Aerosol Monitor AM 510 for $PM_{2.5}$, Triplex cyclone for respirable dust and Black carbon (BC), Lascar EL-USB CO for Carbon monoxide, Ogawa passive sampler for nitrogen dioxide, respectively. **Results.** The mean (SD) $PM_{2.5}$, or respirable dust or NO_2 or BC or CO concentration for traffic policemen and office policemen were 110 (114.62) µg/m^3 and 45.65 (40.51) µg/m3, or 300.0(249.67) µg/m^3 and 89.2(30.03) µg/m^3, or 41.0 (26.12) ppb and 11.03(7.36)ppb, or 422235.9(49578.75) ng/m^3 and 7107.53(8734.27) ng/m^3, or 2.31(2.61)ppm and 0.17 (0.31) ppm, respectively. Traffic policemen's exposure to air pollutants was significantly greater than of office policemen. **Conclusion**. We concluded that traffic policemen were exposed to air pollutants, especially traffic related pollutants, in warm season. Traffic policemen who work in the evening shift, were exposed greater than in the morning shift.

Keywords: A-exposure factors, B-particulate matter, C-air

MO-PO-196

Daily and diurnal trends in $PM_{10\,\&\,2.5}$ collected by a TEOM 1405-DF on the Hopi Reservation of Arizona.

M. A. Alshammari[1], S. Hadeed[1], G. S. Honanie[2], A. Mahkewa[2], M. Paukgana[1], M. K. O'Rourke[1]; [1]University of Arizona, Tucson, AZ, [2]Hopi Tribe, Kykotsmovi, AZ

Abstract: Air quality data from Native American reservations in Northern Arizona is limited. The Hopi Tribe is interested in air quality assessment to evaluate the impact of regional coal burning power plants and local household burning of wood and coal on ambient air quality. The combination of regional, local and indoor combustion may influence the self-reported asthma rate (25.5%) experienced by tribal members. As part of our larger study examining health disparities, we have installed a TEOM 1405-DF to monitor regional PM_{10} and $PM_{2.5}$ at the Hopi Mission School in Kykotsmovi, Arizona. The Hopi value knowledge although some have concerns about the impact of air quality information on traditional practices. Following approval of the tribal resolution granting permission for the study, construction of the site began with leveling the ground and pouring the concrete pad in August 2016. In September and October the container, electricity, back-up power and TEOM were placed at the site. Operation of the site began on October 28, 2016. Adjustments were made in November of 2016 and again in February of 2017. Preliminary winter data indicate that ambient concentrations of $PM_{2.5}$ range from 1.9 to 5.3 µg/m^3. PM_{10} ranges from 0 (during rainstorms) to 8.0 µg/m^3. Diurnal patterns of typical clear days show the highest concentrations of both PM_{10} and $PM_{2.5}$ during night time and lowest values during the late afternoon. As expected days with high relative humidity have low PM_{10} concentrations. $PM_{2.5}$ may remain unaffected during the winter since home heating with coal and wood may be more prevalent during cold conditions. These are only preliminary results.

Keywords: A - ambient monitoring, A - population exposure, D-Southwest-specific, D-First Nation, B-particulate matter

MO-PO-197

A New Approach for Deriving Numerical Relationships Between Different Elongate Mineral Particles (EMPs) Definitions

Y. Shao[1], G. Ramachandran[2]; [1]University of Minnesota, Minneapolis, MN, [2]Johns Hopkins University, Baltimore, MD

Abstract: Different dimensions of elongate mineral particles (EMPs) have been proposed as being relevant to respiratory health end-points but industriies still monitor the NIOSH-regulated EMP only.The goal of this study is to develop a mathematical approach for deriving numerical conversion factors between these exposure metrics for each mine-similar exposure group (SEG) combination of the Minnesota Taconite Worker Health Study. These factors could allow us to create job exposure matrices (JEMs) for any EMP metric from our existing NIOSH EMP data. There are a total of 162 mine-SEG combinations in our study, and in each combination, we assume that the overall EMP dimensions (lengths and widths) follow a bivariate lognormal distribution. A Bayesian approach was used to predict these distributions based on priors for EMP dimensions derived from a limited number (N=92) of area samples that were then updated using the SEG-specific personal EMP measurements (1267 personal samples). All EMP samples were analyzed using ISO-TEM methods. Once the EMP size distribution was ready, each of the selected EMP definitions ('NIOSH EMP', 'Chatfield asbestiform EMP', 'Chatfield non-asbestiform EMP' and 'Suzuki EMP') were then applied to the bivariate lognormal distribution, and the volume under the curve corresponding to the length and width for that definition is the percentage of the total number of EMPs for that definition. We then derived numerical conversion factors between any two EMP definitions, which would be the ratios of the volumes under the bivariate posterior EMP distribution for those two definitions. There are a total of 2,791 EMPs identified from the area samples, and 11,190 EMPs identified from the personal samples. For the final predicted conversion factors, the median [IQR] of the conversion factors of "Chatfield asbestiform EMP", "Suzuki EMP", and "Chatfield non-asbestiform EMP" to the "NIOSH EMP" were 0.38 [0.10-1.17], 8.59 [5.05-20.15], and 13.6 [8.3-25.4] respectively.

Keywords: A - exposure measurement, A-statistical methods, D-occupational, A-industrial hygiene, A-exposure models

MO-PO-198

Relationship between Reactive Oxygen Species (ROS) activity and Cytotoxicity of Ambient Particles

Y. Wang[1], M. Plewa[2], V. Verma[1]; [1]University of Illinois at Urbana-Champaign, Urbana, IL, [2]University of Illinois at Urbana-Champaign, Urbana, IL

Abstract: Associations of particulate matter (PM) concentration with cardiopulmonary disease and mortality have been reported in several studies. Threat of PM to public health is likely due to reactive oxygen species (ROS) generation by PM. However, only a handful of studies could demonstrate a consistent correlation between biological toxicity endpoints and ROS formation measured in acellular assays. In this study, we targeted to investigate these associations by measuring the acellular ROS generation and cytotoxicity of ambient particulate matter (PM) samples collected from an urban site. The cytotoxicity of water-soluble PM is assessed by exposing aerosol extracts to Chinese hamster ovary (CHO) cells. Two methods were used to assess the ROS generation ability - 1. dithiothreitol (DTT) loss (conventional DTT assay); and 2. hydroxyl radical generation through redox cycling in the DTT system. Inhibition of CHO cells by soluble PM was observed, and half maximal inhibitory concentrations (IC_{50}) were calculated for the PM samples, which range from 43.6% to 74.0% (percentage of soluble PM extract). Interestingly, we found strong and significant correlation between ROS generation in the DTT system and IC_{50} ($R^2 = 0.60$, $p<0.05$). To further understand the chemical components, which might drive the response of both DTT and cytotoxicity assays, we measured the concentrations of metals and water-soluble organic carbon (WSOC) in PM extracts. Fe, Cu, Zn and WSOC were significantly correlated with both IC_{50} and ROS generation in DTT assay. Further chemical analysis of the collected samples in underway and would be presented in this talk. However, Our preliminary results indicate that measuring the ROS generation from PM could be used as a biologically relevant metric to assess the toxicity and health impacts from ambient aerosol pollution.

Keywords: C-air, A - exposure measurement, B-metals

MO-PO-199

Citywide validation and improvement of the MAIAC aerosol product in Lima, Peru

J. Bi[1], B. Vu[1], A. Wildani[2], Y. Wang[3], A. Lyapustin[3], Y. Liu[1]; [1]Emory University, Atlanta, GA, [2]Emory University, Atlanta, GA, [3]National Aeronautics and Space Administration, Greenbelt, MD

Abstract: Background: Satellite-observed aerosol optical depth (AOD) have been widely applied to estimate the ground-level concentrations of fine particulate matters (PM2.5). One of the state-of-the-art AOD products is the Multi-Angle Implementation of Atmospheric Correction (MAIAC). While MAIAC AOD has been shown in North America and Europe to be a robust parameter of ground-level PM2.5 estimation, its quality in other parts of the world has not been extensively evaluated. **Objectives:** We conducted a validation analysis on Terra and Aqua MAIAC AOD in Peru and developed a gap-filling model to improve its coverage in Lima, Peru. **Methods:** We validated the accuracy of MAIAC AOD in Lima by comparing it with the AOD measurements from the aerosol robotic network (AERONET). A moving average (MA) model was then applied to impute the missing AOD observations in order to enhance the dataset's spatial coverage and quality. **Results:** The proportion of available MAIAC AOD data in Lima was ~20%. MAIAC AOD had a moderate correlation (r = 0.60) with the AERONET measurements. After applying an MA model with fixed window sizes (5 km for the spatial coverage and 7 days for the time span) on the time series of MAIAC AOD, the proportion of available data raised to 31% with a slightly decreased correlation (r = 0.53). **Conclusions:** MAIAC AOD has a reasonable but still limited quality in Lima in terms of its accuracy and spatial coverage. Improvements in the MA model by introducing flexible window sizes and new gap-filling strategies and the application of machine learning techniques are expected to further increase the quality of the dataset in Lima.

Keywords: A - exposure measurement, B-particulate matter, C-air

MO-PO-200

Distributions of real time black carbon concentration and its association to PM2.5 concentration at an urban hotspot in Seoul, South Korea : Preliminary study results

S. Kim[1], S. Lee[1], H. KIM[1], J. Lee[1], S. Park[1], J. Park[2], B. Park[3]; [1]Soonchunhyang University, Asan, Korea (the Republic of), [2]KOREA TECH, Cheonan, Korea (the Republic of), [3]Korea Environment Corporation, Incheon, Korea (the Republic of)

Abstract: Among the many airborne contaminants, the black carbon (BC) has been recently identified as a carcinogenic compounds, and new scientific evidence has led to a recognition of its representativeness as indicator of exposure to combustion PM, i.e., diesel traffic sources and the significant role of BC as one of the short-lived climate forcers. This study conducted to obtained distributions of BC concentrations and its association to PM.2.5 concentration at an urban hotspot of Seoul, South Korea. Real-time PM2.5 measurement was conducted using real-time portable monitors of Sidepak AM510 personal Aerosol Monitor (TSI Inc., St. Paul, MN) then results were compared with those obtained from national monitoring site. Comparative measurement of BC concentrations was conducted using single wave length (SWL) light absorption devices and multi channel wave length (MWL) light absorption devices. Descriptive analyses were conducted for continuous variables of hourly mean values of BC and PM2.5. Associations between BC and PM2.5 concentrations in different sampling time (morning, afternoon and evening) were evaluated with multivariate regression models after adjusting for temperature, relative humidity, wind direction and wind speed. The SAS statistical package (version 21.0) was used for statistical analyses. BC concentration increased 7.5 % by unit increase of PM2.5 concentration in morning and evening rush hour period of the weekend whereas we obtained 5.5% increase in morning rush hour period of weekdays ($p<0.05$). Degree of the increase rate was different depending on sampling time of weekdays or weekend. Our study provides compelling evidence supporting necessity of immediate monitoring of BC concentration at urban areas for emission source control and management as BC is known as a better indicator of harmful particulate substances from combustion sources (i.e., traffic or power plant) than undifferentiated PM mass.

Keywords: A - ambient monitoring, B-particulate matter, A - population exposure, C-air

MO-PO-201

Facilitating Collaboration Among South Asian Countries towards Better Understanding of Air Pollution

P. Pant[1], A. Gurung[2]; [1]University of Massachusetts, Amherst, MA, [2]University of Texas at Austin, Austin, TX

Abstract: Rapid urbanization and motorization has led to air pollution as one the most significant environmental hazards faced by countries in South Asia. Despite high air pollution in the region leading to large environmental, human health, and economical burden, scientific understanding of air pollution is very limited and can create bottlenecks for effective intervention strategies. Most of our current understanding of air pollution and its impacts are based on studies conducted in developed countries, and the patterns are likely to differ in developing countries due to differences in sources, environment, and population characteristics. Limited understanding in air pollution in South Asia relates to lack of data, limited access to resources, and competing policy, health, and research challenges. This is further enhanced by poor understanding of past and ongoing work, and repetition of work due to limited communication. There is also a large need to enhance research capacity with priority to training local researchers as well as enhancing technical capacity. These challenges are noted across South Asia showcasing an urgent need to foster collaboration both within as well as between countries. We are creating Air South Asia (ASA), an open platform for air quality in South Asia with the following objectives: ~ Creation of an open-access digital repository of past and present ongoing work (i.e. research projects and publications, non-governmental efforts, media stories etc.) in the region. ~ Creation of an open forum to facilitate communication among researchers in the region, and an entry point for global researchers interested in conducting research in the region. ~ Dissemination of information through an air quality newsletter and showcasing 'champions' and 'best practices'. Mentorship and professional development for young researchers. We seek ideas and feedback from members in the academic community to make the platform interactive and informative.

Keywords: A-global health, A - ambient monitoring, A-environmental regulation, C-air

MO-PO-202

Environmental Exposures to Particulate Matter among Rural, High Risk and Underserved Children in Eastern NC

S. Cho[1], G. Kearney[2], L. Wei[2], M. McCombs[1], D. Collier[2], K. Gowdy[2], L. Johnson[3], B. Kilburg-Basnyat[2], A. Kuranga[2], N. Siripanichgon[1], T. Blount[3]; [1]RTI International, Research Triangle Park, NC, [2]East Carolina University, Greenville, NC, [3]Vidant Medical Center, Greenville, NC

Abstract: Eastern N.C. has the highest prevalence of childhood asthma and asthma related emergency department visits in the state. However, little research has examined the association between asthma and environmental exposures in this rural community. The aim of this community-based collaborative project is to assess indoor air quality and environmental respiratory exposures, personal behavior, respiratory clinical symptoms and metabolic profiles among underserved children with poorly controlled asthma living in rural eastern N.C. Of the 25 participants in the study, 10 (40%) were female and 15 (60%) were male. 19 (76%) identified as Black Non-Hispanic, 4 (16%) as Hispanic, 1 as White Non-Hispanic, and 1 as Asian. 13 (52%) participants were ages 5 - 9, 8 (32%) were ages 10 - 14, and 4 (16%) were age 15 or older. 1 participant out of the 20 respondents reported a smoker who residing in the home. PM10 filter samples were collected for continuous 72 hours at personal and indoor levels for chemical speciation for PM mass concentrations, tobacco smoke, pesticides, metals, and endotoxin, while real-time PM exposures were measured at a personal level using RTI MicroPEM. Personal exposure to PM10 ranged between 1 and 166 ug/m3 with the average of 36 ug/m3. Indoor PM10 levels were between 8 and 106 ug/m3 with the average of 50 ug/m3. Real-time data indicated that 65 % of children wore the MicroPEM more than 50% of waking time. For a high compliant, high exposure group, they were exposed to 50 ug/m3 or greater level of PM10 for at least 10% of monitoring time. In our knowledge, this is the first study to assess and characterize environmental exposures to air pollutants for rural NC asthmatic children in detail and at accurate levels. Further analysis on biomarkers and linkage with exposure endpoints will determine areas to target for personal intervention to reduce asthma attacks and severity of asthma among children living in rural environments.

Keywords: A - exposure measurement, A-epidemiology, D-children, A-sensor technology, D-susceptible/vulnerable

MO-PO-203

Variability of bisphenol-A concentrations in first morning, bedtime, and 24-hour urine samples in 50 North Carolina adults over a six-week period

M. K. Morgan[1], M. Nash[1], D. B. Barr[2], J. R. Sobus[1]; [1]USEPA, Research Triangle Park, NC, [2]Emory University, Atlanta, GA

Abstract: Bisphenol-A (BPA) is a high-production volume chemical that is used to make a number of consumer products and packaged goods. Many cross-sectional studies have frequently reported detecting BPA in urine. However, limited data exist on the temporal variability of urinary BPA concentrations. The major objectives of this study were to: 1) quantify the levels of BPA in first-morning void (FMV), bedtime, and 24-hour samples as concentration, specific gravity (SG) corrected, creatinine (CR) corrected, and excretion rate values for 50 adults over a six-week monitoring period; and 2) determine if these correction approaches decreased the variability in urinary BPA levels. In 2009-2011, a convenience sample of 50 adults (19-50 years old) was recruited from residential settings in North Carolina. Participants collected urine samples (FMV, bedtime, and 24-hour) during weeks 1, 2, and 6 of the six-week monitoring period. Urine samples (n=2335) were analyzed for total BPA concentrations by high-performance liquid chromatography/tandem mass spectrometry. The preliminary data show that BPA was frequently detected (98%) in all of the urine samples. Median levels of BPA were consistently the highest in 24-hour samples (2.08 ng/mL, 2.36 ng/mL-SG, 2.53 ng/mg-CR, and 2.64 ng/min) and the lowest in FMV samples (1.73 ng/mL, 1.61 ng/mL-SG, 1.64 ng/mg-CR, and 1.38 ng/min) across all four methods. Intraclass correlation coefficient (ICC) estimates for BPA showed poor reproducibility ($\leq$ 0.40) for all urine sample types and methods over a day, week, and six weeks. The highest ICC value of 0.40 occurred for CR-corrected bedtime voids collected over a week. To obtain a reliable average biomarker estimate (ICC=0.80) for BPA, these results indicate that at least six bedtime urine measurements would be needed per adult over a week. In conclusion, these results suggest that bedtime voids may be the preferred sample type to collect in future studies to adequately assess BPA exposures in adults.

Keywords: B-BPA, A-biomarkers, A-biomonitoring, A-longitudinal metrics

MO-PO-204

Perfluoroalkyl substances in the Fernald Community Cohort: Exposure distributions over time and associations with thyroid and kidney function and obesity

B. E. Blake[1], S. M. Pinney[2], S. E. Fenton[1], K. Ferguson[3]; [1]National Institute of Environmental Health Sciences, Durham, NC, [2]University of Cincinnati, Cincinnati, OH, [3]National Institute of Environmental Health Sciences, Durham, NC

Abstract: Perfluoroalkyl substances (PFAS) are a diverse class of chemicals used in industrial settings and consumer products to make materials resistant to heat, water, and oil. PFAS exposure in the general population is common, and PFAS have been associated with altered kidney function, disrupted thyroid hormones, and are suspected obesogens. We investigated a panel of PFAS in participants of the Fernald Community Cohort (FCC), who were identified as at risk for high PFOA exposure based on living in zip codes bordering the Ohio River. Subjects had up to 3 repeated measures of serum PFAS (N=210 participants, N=517 total measurements). We compared changes in serum PFAS concentrations over time with those observed in the National Health and Nutrition Examination Survey (NHANES) at similar time points. We tested the association between PFAS measures and concurrent outcome measurements of obesity and kidney and thyroid function, including body mass index (BMI), kidney function (estimated glomerular filtration rate, eGFR), and thyroid stimulating hormone (TSH). We used linear mixed effects models to examine each association, adjusting for participant age, sex, and year of sample collection among other confounders. No associations were observed in models of PFAS and BMI or TSH. We observed significant positive associations between 2-(N-methyl-perfluorooctane sulfonamido) acetate (β=1.49, SE=0.70, p=0.04) and 2-(N-ethyl-perfluorooctane sulfonamido) acetate and eGFR (β=1.82, SE=0.71, p=0.01) and significant negative associations between perfluorononanoate (β=-2.96, SE=1.30, p=0.02), perfluorohexane sulfonate (β=-2.26, SE=0.86, p=0.01), and perfluorodecanoate and eGFR (β=-2.40, SE=1.00, p=0.02). Geometric means for serum perfluorooctanoic acid (PFOA) were threefold higher in the FCC compared to NHANES from 1999 to 2008 while perfluorooctanesulfonate (PFOS) levels did not differ between the groups. These results suggest PFAS may affect kidney function in a compound-specific manner.

Keywords: , A - exposure measurement, A-biomarkers

MO-PO-205

In vivo x-ray fluorescence measured toenail manganese as a biomarker of exposure among welders

A. J. Specht[2], X. Zhang[1], J. Weuve[3], L. H. Nie[1], M. Weisskopf[2]; [1]Purdue University, West Lafayette, IN, [2]Harvard T.H. Chan School of Public Health, Boston, MA, [3]Boston University, Boston, MA

Abstract: Overexposure to manganese (Mn) has been shown to cause Parkinson's-like symptoms and possibly have neurologic effects at even low levels of exposure. Mn exposure needs to be monitored to help identify threshold exposure levels and help guide public policy. Recently, toenail has been shown to be a reliable biomarker of Mn exposure reflecting an exposure period of about 6 months to a year. However, exposure assessment using toenail requires collection of clippings and analysis using inductively coupled plasma mass spectrometry (ICP-MS). The ICP-MS analysis process is costly and time consuming, and some people do not want to provide toenail clippings or cannot provide enough clippings for ICP-MS analyses leading to problematic missing data issues in epidemiologic studies. A novel measurement technique utilizing x-ray fluorescence can be applied to the measurement of toenails in vivo. This would eliminate the difficulties collecting nail clippings and drastically reduce the cost of analysis allowing for much larger studies. In this study, we calibrated a portable x-ray fluorescence (XRF) device for measurement of toenail Mn using Lucite soft tissue phantoms, plaster-of-Paris bone phantoms, and doped Mn epoxy resin toenail phantoms. Using these phantoms we identified a detection limit of 2.4 ppm. In order to validate this device, we then measured the toenails of a group of 16 welders. Welders have previously been shown to be a particularly vulnerable population for Mn overexposure. We found that using the XRF toenail measurements 40% of the workers had measureable Mn levels, and, consistent with other studies, Mn measurements correlated with motor function coordination testing. In future work, the technique will be evaluated for measurement of other metals in toenail and applied in epidemiological studies of potentially vulnerable or exposed populations.

Keywords: B-metals, A-biomarkers

MO-PO-206

German Environmental Specimen Bank (ESB): Time Trend of the Internal Exposure to Phthalates from 1988 - 2015

T. Weber[1], H. M. Koch[2], A. Conrad[1], M. Rüther[1], P. Apel[1], M. Kolossa-Gehring[1]; [1]German Environment Agency (UBA), Berlin, Germany, [2]Institute for Prevention and Occupational Medicine of the German Social Accident Insurance – Institute of the Ruhr-Universität Bochum (IPA), Bochum, Germany

Abstract: Background The wide use of phthalates as plasticizers leads to ubiquitous exposure of humans. Since 2015, the phthalates di-isobutyl phthalate (DiBP), di-n-butyl phthalate (DnBP), butyl benzyl phthalate (BBzP) and di(2-ethylhexyl) phthalate (DEHP), which are known endocrine disruptors and classified as toxic to the reproductive system, are listed in Annex XIV of the REACH regulation and require authorization before use. A restriction proposal was submitted to ECHA in 2016. **Methods** The German Environmental Specimen Bank (ESB) annually collects and cryo-archives human tissues in four different German locations. To evaluate the time trend of phthalate exposure in 24 h-urine samples of the ESB data of two existing time trend studies (1988 - 2003 and 2002 - 2008) were combined with new data (2007 - 2015). This dataset covers changes in human exposure over a time period of 27 years. Altogether, 24 h-urine samples from 1163 males and females living in the city of Münster were measured by LC-MS/MS and analyzed for this study. **Results** Analysis of the metabolites of the in Annex XIV listed phthalates show an overall decrease of median concentration in 2015 to about 10% of the starting concentration in 1988. A similar trend was shown for the metabolites of the not regulated di-methyl-phthalate and di-ethyl-phthalate (35% and 25% respectively in 2015 compared to 2007) which were measured for the first time. In contrary, metabolites of the not regulated di-iso-nonyl-phthalate show an increase over time. **Discussion and outlook** The results suggest a direct and fast impact of regulation on the exposure to phthalates. While plasticizers remain *an important group of chemicals in* today's society, research has to continue monitoring the exposure to not regulated alternatives, both phthalates and non-phthalates. **Acknowledgement** The ESB is funded by the German Ministry for the Environment, Nature Conservation, Building and Nuclear Safety (BMUB).

Keywords: A-biomonitoring, A-biomarkers, B-phthalates, A-chemical alternatives

MO-PO-207

Urinary measurements of bisphenol A in children, mothers and fathers from Slovenia: Overall results and determinants of exposure

J. Snoj Tratnik, T. Kosjek, E. Heath, D. Mazej, M. Horvat; Jozef Stefan Institute, Ljubljana, Slovenia

Abstract: In the present study, the first national data on urinary bisphenol A (BPA) levels was evaluated with regard to potential dietary and non-dietary exposure sources, and compared between three population groups, differing by age, gender and place of residence. First morning urine was collected from children (6-11 years), their mothers (≤ 50 years) and fathers/partners (≤ 50 years), living in urban and rural area of Slovenia. Along with the basic questionnaire providing general population characteristics, socio-economic status and dietary habits, BPA-specific data was collected, including consumption of food or beverages from plastic and canned containers, presence of white dental fillings, use of specific consumer products and hormonal treatments. The collected urine samples were analysed for free and conjugated BPA using GC-MS/MS. The urinary BPA levels in children, mothers and fathers were low, with geometric means of 1.51, 0.79, and 0.20 µg/g creatinine, respectively. The levels were comparable with the levels reported for other European countries and were below the current health-based guidance values in all participants. In line with large-scale surveys, our data showed age-dependant BPA levels, with the highest BPA urinary levels in the youngest age group. In mothers, hormonal interaction predominated association between BPA and dietary sources that was otherwise clearly observable in children (canned food/drink and food from plastic material) and fathers (canned food/drink). In children, white dental fillings were observed as a potential source of BPA exposure. The study clearly indicated physiological and life-style differences between the population groups studied and set priorities for the up-coming studies.

Keywords: A-biomonitoring, B-BPA, A - exposure measurement, C-consumer/personal care products, D-children

MO-PO-208

Biomonitoring Six Per- and Polyfluoroalkyl Substances Included in the U.S. EPA Third Unregulated Contaminant Monitoring Rule

L. Blum, S. Donovan, D. Schiller, B. Yue; NMS Labs, Willow Grove, PA

Abstract: Per- and polyfluoroalkyl substances (PFAS) are an assortment of substances used in a variety of industries and consumer products. As a result of their use, these persistent chemicals are found in the environment including drinking water. The U.S. EPA tests drinking water for six PFAS contaminants under the Third Unregulated Contaminant Monitoring Rule (UCMR3), which includes perfluorobutanesulfonic acid (PFBS), perfluoroheptanoic acid (PFHpA), perfluorohexanesulfonic acid (PFHxS), perfluorononanoic acid (PFNA), perfluorooctanoic acid (PFOA), and perfluorooctancsulfonic acid (PFOS). A quantitative assay for the linear components of these substances has been developed to test serum specimens by using HPLC separation with negative-ion electrospray tandem mass spectrometry (LC-MS/MS). Samples are prepared for analysis by adding analyte specific ^{13}Carbon isotopes as internal standards with subsequent protein precipitation. The analytical measurement range (AMR) is 0.05 to 10 ng/mL for PFBS,

PFHpA, PFHxS, and PFNA; and 0.5 to 100 ng/mL for PFOA and PFOS. The within and between day precision of 5 replicates measured for each of 3 days (n=15) are less than 10% CV for these analytes at three respective levels 0.05, 0.25 and 6.0 ng/mL for PFBS, PFHpA, PFHxS, and PFNA; and 0.50, 2.5, and 60 ng/mL for PFOA and PFOS, except for the within day precision values for PFHpA at 0.05 ng/mL (14.2%CV) and PFNA at 6.0 ng/mL (10.8%CV). A population of individuals (n=151) with unknown exposures to these substances showed the following respective median and 97.5th percentile as measured for PFBS (<0.05, 0.06 ng/mL); PFHpA (<0.05, 0.47 ng/mL); PFHxS (0.99, 5.75 ng/mL); PFNA (0.42, 1.44 ng/mL); PFOA (1.12, 4.13 ng/mL); and PFOS (1.77, 11.5 ng/mL). These values are comparable to other studies measuring these substances in the general population. This assay in serum can be used for biomonitoring those PFASs tested in drinking water under EPA UCMR3.

Keywords: A-biomonitoring, A-analytical methods, B-POPs

MO-PO-209

Laser Ablation Inductively Coupled Plasma Spectrometry calibration for the Quantification of Manganese in Human Hair

B. Reiss[2], N. Seixas[1], E. A. Riley[1], C. D. Simpson[1]; [1]University of Washington, Seattle, WA, [2]The University of Arizona, Tucson, AZ

Abstract: Mn-containing particles in welding fumes are difficult to measure. Laser ablation inductively coupled plasma mass spectrometry (LA-ICP-MS) seems to be suitable for analyzing Mn in hair. Hair calibration standards with known Mn concentrations do not exist. The objective of this study was to calibrate a LA-ICP-MS for measuring ^{55}Mn in individual hair strands ($MnH_{(La)}$). Two calibration standards were created: Gelatin mixed with Mn (MnG) and acid digested bulk hair ($MnH_{(B)}$). ^{55}Mn measurements were normalized with ^{34}S. LA-ICP-MS was used to analyze 53 hair strands from 22 bulk hair samples (Mn: 0.3 to 51.5 µg/g) and 10 gelatin-Mn mixtures (0-250 ppm). Bulk hair samples were collected from 47 welding school students, cut to 1 cm, washed, and analyzed for ^{55}Mn. A mixed model was used to assess the relationship between ln-transformed $MnH_{(La)}$ and MnG levels with ln-transformed Mn/S ratios measured in individual hair strands. Both ln-transformed $MnH_{(La)}$ and MnG levels were significantly associated with ln-transformed Mn/S ratios. An increase of 8.0% Mn/S [95% confidence interval (CI): 6.0-11.0%] was associated with a 10% increase in the standard concentrations. Both, the difference between the gelatin and bulk hair calibration Mn/S intercepts (p=0.101), and the interaction term for gelatin and bulk hair (p=0.968) were not significant. These results indicate that the calibration slopes are equivalent and that both MnG and $MnH_{(B)}$ are suitable calibration standards for measuring Mn levels in individual hair strands using LA-ICP-MS.

Keywords: A - exposure measurement, A-biomonitoring, B-metals, A-biomarkers, A-epidemiology

MO-PO-210

Impacts of Cold Weather on Emergency Hospital Admission in Texas, 2004-2013

T. Chen, K. Zhang; University of Texas Health Science Center at Houston School of Public Health, Houston, TX

Abstract: Background: Cold weather has been identified as a major cause of weather-related deaths in the U.S. While the effects of cold weather on mortality has been investigated extensively, cold-morbidity associations are less well studied. The purpose is to examine impacts of cold weather on emergency hospital admission (HA) in 12 major Texas Metropolitan Areas (MSAs) for the 10-year period, 2004-2013. Methods: Our study used a two-stage approach to examine the cold-HA association. We first applied distributed lag non-linear models (DLNM) to estimate cold effects for each MSA. A random effects meta-analysis was then used to estimate pooled effects. Percent increase in risk and corresponding 95% confidence intervals (CIs) were estimated as with a 1 °C decrease in temperature below a MSA-specific threshold. Age-stratified and cause-specific HA were modeled separately. Results: The majority of Texas MSAs were associated with an increased risk in HA ranging from 0.1% to 3.3% with a 1 °C decrease in temperature below cold thresholds. Pooled effect estimate was 1.4% (95% CI: 0.9%, 1.9%) increase in all-cause HA risk with 1 °C decrease in temperature. Cold wave effects in Texas were also examined and observed in most eastern and southern MSAs. Effects of cold on all-cause HA were highest among people over 75 years old (2.0%, 95% CI: 1.1%, 3.0%). Pooled estimates for cause-specific HA association were strongest in pneumonia (3.4%, 95% CI: 2.8%, 4.0%), followed by chronic obstructive pulmonary disease (3.2%, 95% CI: 2.1%, 4.5%) and respiratory diseases (2.4%, 95% CI: 2.0%, 2.9%). Conclusion: Cold weather generally increases hospital admission risk significantly in Texas, and cold effects were spatially heterogeneous across Texas. Our findings can provide insights to design better intervention strategies for targeted vulnerable populations towards reducing adverse health effects of cold weather.

Keywords: A-epidemiology, A - population exposure, A-climate change, D-susceptible/vulnerable

MO-PO-211

Evaluating effects of global change on patterns of aeroallergens and ground-level ozone across the contiguous US

T. Cai[1], Z. Mi[1], X. Ren[2], Y. Zhang[2], P. Georgopoulos[1]; [1]Rutgers University, Piscataway, NJ, [2]Rutgers University, Piscataway, NJ

Abstract: The prevalence of Allergic Airway Disease (AAD) is growing globally, resulting in increased numbers of emergency department visits and hospitalizations. Clinical studies show that AAD can be exacerbated by the synergistic action of aeroallergens such as pollen and fungi, and atmospheric pollutants such as ozone. Furthermore, climate change has critically affected atmospheric processes involved in the dynamics of air pollution systems and emissions of natural pollutants such as pollen and spores. Previous studies, involving data from nationwide observations of airborne pollen counts of selected plant species in conjunction with climatic factors, indicated that the start date and length of pollen season, the average peak value and annual total of daily counted airborne pollen have been affected substantially by the changing climate. The present study investigates co-occurrences of peak ozone concentrations and peak pollen counts across the contiguous United States (ConUS). Analysis of observed pollen counts and ozone concentrations at 58 pollen monitor stations were conducted. Concentrations of pollen in base years (2001-2004) and future years (2047-2050) are simulated with a customized version of CMAQ (the Community Multiscale Air Quality model) employing a grid with 36 km by 36 km horizontal resolution, while corresponding ozone concentrations at similar resolutions for the above timeframes were conducted by a multi-university/agency consortium including USEPA. The study employs spatiotemporal correlation analysis to examine patterns of co-occurring ozone and pollen concentrations. The outcomes of this study provide information that could support development of strategies for managing health-impacts of co-occurring photochemical pollutants and aeroallergens.

Keywords: A-climate change, A - ambient monitoring, B - ozone

MO-PO-212

Heat and Hydration Assessment of Migrant Grape-Workers in Sonora, Mexico

R. Wagoner; University of Arizona, Tucson, AZ

Abstract: The demanding physical labor that makes agricultural production possible in Northern Mexico is exacerbated by extreme temperatures, migration, and lack of health and safety oversight on a national level. This study addressed heat and hydration in a commercial grape farm in Northern Mexico where 95% of the land is classified as arid or semi-arid. Notably, one-third of Mexico's heat-related deaths reported from 2002 to 2010 occurred in agricultural workers in this region. Migration status, low socio-economic ranking, and poor literacy rates contribute to the vulnerability of workers in this region to heat exposure. Furthermore, there is growing concern about the direct and indirect health effects of climate change, especially in regions such as Sonora that have limited access to resources and occupational health control methods. A total of 38 participants were recruited for three sampling periods during spring and summer 2016. For each sampling period, an oral questionnaire was administered in Spanish; core body temperature was measured using ingestible thermometers; urine was collected and analyzed for specific gravity; and the effective Wet-Bulb Globe Temperature (WBGT) and metabolic rate were calculated. The majority of participants listed Chiapas as their home state, nearly half spoke an indigenous language as well as Spanish, most were between the ages of 18 and 24, and none had completed high school. The effective WBGT was higher during the summer. However, the core body temperature of workers in the spring were not significantly different than the core body temperatures of workers in June and August. As indicated by urine specific gravity, the majority of workers in all months were either mildly or clinically dehydrated. These results indicate the need for enhanced administrative and engineering controls, policies, and binational collaborations to reduce levels of heat stress and dehydration in agricultural workers in Northern Mexico.

Keywords: D-occupational, D-susceptible/vulnerable, D-environmental justice, A-climate change, A-workplace

MO-PO-213

Mercury Co-Benefits of Climate Policy in China

K. Mulvaney[1], N. E. Selin[2], A. Giang[3], S. Y. Kwon[3], M. Muntean[4], M. Li[5], C. Li[6], D. Zhang[7], V. J. Karplus[8]; [1]University of North Carolina at Chapel Hill, Chapel Hill, NC, [2]Massachusetts Institue of Technology, Cambridge, MA, [3]Massachusetts Institue of Technology, Cambridge, MA, [4]European Commission, Joint Research Centre, Institute for Environment and Sustainability, Ispra, Italy, [5]Massachusetts Institute of Technology, Cambridge, MA, [6]Massachusetts Institute of Technology, Cambridge, MA, [7]Massachusetts Institute of Technology and Tsinghua University, Cambridge, MA, [8]Massachusetts Institute of Technology, Cambridge, MA

Abstract: National commitments on the Minamata Convention on Mercury interact with other global environmental objectives, such as the Paris Agreement on climate change. Understanding the interactions between mercury and other policy interventions can help decision-makers identify more effective policies that can address multiple environmental issues simultaneously. Here, we examine how China's national climate policy focused on decarbonization affects the country's ability to meet its environmental goals related to mercury. To address environmental goals, we assess how mercury co-benefits (positive side effects that are peripheral to a policy's main goal) of a national climate policy in China could contribute to the country's commitments under the Minamata Convention. We examine climate policy scenarios in 2030 corresponding to various levels of carbon intensity reductions in addition to a business-as-usual scenario and a scenario that implements end-of-pipe controls aligned with the requirements of the Minamata Convention. Economic analysis from a computable general equilibrium model of China's economy (C-REM) provides information on changes in economic activity resulting from the climate policy scenarios. Using this economic data, we scale 2007 mercury emissions from the Emissions Database for Global Atmospheric Research (EDGAR) in a variety of sectors to 2030. We then use a global atmospheric transport model (GEOS-Chem) to project changes in mercury deposition at the regional scale in China for each policy scenario, and evaluate the resulting spatial distribution of mercury co-benefits. We find that mercury co-benefits of climate policy can lead to reductions in emissions comparable to end-of-pipe controls. We also find that regions exhibiting the greatest reductions in emissions across the policy scenarios also exhibit the greatest reductions in total mercury deposition.

Keywords: A - population exposure, A-climate change, A-environmental policy, A-geospatial analysis/GIS, B-metals

MO-PO-214

Quantifying recent associations between meteorology and multipollutant day types to inform future air quality projections

J. L. Pearce[1], J. Runkle[2], H. Chang[3], S. E. Sarnat[3], L. Waller[3]; [1]Medical University of South Carolina, Charleston, SC, [2]North Carolina State University, Raleigh, NC, [3]Emory University, Atlanta, GA

Abstract: Background: Changes to the climate system will impact air quality and related health effects through changes in exposure patterns. Objective: Establish influence of meteorology on multiple pollutants and estimate the potential changes in daily mixtures experienced due to expected changes in climate. Methods: We obtained four years (2011-2014) of daily average CO, NOy, SO2, O3, PM2.5 EC, PM2.5 OC, PM2.5 NO3, PM2.5 NH4, and SO4 and weather data for multiple cities in the Southeastern US. We fit generalized additive models (GAMs) to establish present day associations between daily pollution and weather. Future climate conditions were obtained for the years 2030-2040 from downscaled climate projections and fitted GAMs were then used to predict corresponding responses of daily pollution levels. Self-organizing maps (SOMs) were then used to identify categories of days based on multipollutant conditions (i.e., multipollutant day types (MDTs)) and class frequencies were used to establish differences in present day and future air quality. Results: We found MDTs identified days were conditions ranged from relatively clean days, high single pollutant days (e.g., SO2), to high combination days (e.g., CO, NOy, EC). Classifying days under our future climate scenario revealed that the largest increases in MDTs frequencies occurred on days characterized by moderate-to-high O3 pollution (21% increase), days dominated by relatively moderate-to-high CO, NOy, and EC (10.5% increase), and days with elevated SO2 and OC. The largest decreases occurred for relatively clean days (10.7% decrease); we show that these days transition to a similar profile with higher O3 in the future. Conclusion: We find combining multipollutant day typing (SOM), GAMs, and future climate predictions provides a complementary suite of tools for investigating potential air quality changes driven solely by future meteorological conditions.

Keywords: A-climate change, B-mixtures, A-statistical methods

MO-PO-215

Land-Use Regression Models of Parcel Level Intra-Urban Surface Temperature in Three Cities in Massachusetts.

K. J. Lane, J. I. Levy, Y. Zhang, P. Fabian; Boston University School of Public Health, Boston, MA

Abstract: Surface temperatures have been shown to vary substantially at the intra-urban scale. Models are needed that allow fine-scale spatial predictions of exposure to surface temperatures to conduct intra-urban exposure assessment and health analyses. The aims of this work were to (1) develop land use regression models of surface temperature at the parcel level in three cities in Massachusetts (Boston, Springfield and Worcester); (2) explore the relative importance of spatial and temporal factors on intra-urban surface temperature; and (3) compare models between cities. Thermal imaging data at a 30 m x 30 m resolution were obtained during flyovers from winter 2014 - summer 2016. In each city, image data were available for >30 days over the study period, and were used to calculate mean parcel surface temperature. Parcel scale regression models were developed for each city and for warm and cool seasons. Independent covariates in land use regression models included land use classification, year built, property value, impervious surface area ratio, normalized difference vegetation index (NDVI), wind direction, wind speed, temperature, relative humidity, population density and distance from major water bodies. Preliminary results indicate that regression models explained between 32 to 48% of the variability in mean parcel surface temperature. Proximity to major water bodies explained the greatest amount of the surface temperature variation in each city (adjusted-R^2=0.11-0.28), with differential effects by season. Land use and NDVI explained an additional ~10% of the variation, but had considerable differences in magnitude and strength of association between cities. Significant housing predictors included year built and building value. Our models can be used to predict fine scale surface temperature based on land use, housing, and meteorology, expanding our understanding of the drivers of urban-heat islands and associated health effects.

Keywords: A-exposure models, A-geospatial analysis/GIS

MO-PO-216

Report-Back for Location-based Personal Monitoring Data on Individual Experienced Temperature in Outdoor Workers

J. Runkle[1], L. Thompson[2], M. Sugg[2]; [1]NCSU, Asheville, NC, [2]Appalachian State University, Boone, NC

Abstract: Background: The report-back process involves developing data visualization methods for communicating results to participants with varying levels of scientific literacy. Limited research has examined the benefits of reporting back participatory monitoring results, particularly as it pertains to inspiring behavioral changes to reduce workplace heat exposure. **Objective**: Evaluate the individual data report-back of geo-located environmental data in grounds workers from three geographic regions in the Southeast. **Methods**: Data report-back sessions were held to provide participants with a packet containing a variety of visualization formats for interpreting personalized heat exposure in comparison to their aggregate worksite values. Surveys were administered to examine participants' response to their report-back information. Descriptive analysis were performed in SAS (version 9.4). **Results**: A total of 66 workers responded. Workers felt their results were personally useful (96%), helped educate about their workplace health risks (95%), and improved their understanding of heat as a health hazard (94%). At baseline, 26 percent of workers reported being very likely to change their work behavior to reduce heat exposure compared to 46 percent of workers post-monitoring study. A large proportion of participants with 10+ years of experience (88%, n=25) reported being more likely to make behavioral changes compared to only 23 percent of workers with less than 5 years of work experience (n=13). Workers agreed that each of the presentation formats (i.e., heat maps, box plots, summary tables, heart rate graph, and heat recommendations) in the report-back were helpful in understanding their results. **Discussion**: The report-back of location-based environmental monitoring results has the potential to be a powerful tool for workers to evaluate and respond to their personal exposure profiles. Our findings suggest that report-back is a valued part of the study for many participants.

Keywords: A - ambient monitoring, A - exposure measurement, A-activity patterns, A-climate change, A-workplace

MO-PO-217

Exposure assessment through modeling approach for urban air pollutants using climate indicators

F. Naseem, A. Rashid; PMAS Arid Agriculture University, Rawalpindi, Pakistan

Abstract: Weak environmental legislations and poor implementation of laws in developing countries are impending challenges for environmental health. Moreover, lack of evidence based data and research gaps restrict policy makers to link health and environment into practice. To address this issue we analyzed 12 months data that are considered essential to identify relationships between respiratory problems and variations in air pollution in an urban environment based on climate driven factors. Air pollutants susceptible to climate change were identified through time series model and proved relevance in causing adverse health affects among urban population of Rawalpindi city. We observed a strong correlation between respiratory responses (shortness of breadth, irritated cough and inflammation of submucosa) with that of air quality. Our results further demonstrate that life cycle of air pollutants which includes oxides of nitrogen, sulphur, ozone, particulate matter and pollens are affected by chemical as well as meteorological factors, such as temperature, humidity, wind, solar radiation and precipitation. Based on ARIMA model from "time series expert modeler" category, four strong climate predictors were identified in the final model (Ljung-Box statistics=81.78; stationary R^2=0.69; p<0.000). Since diesel burning in urban areas of Pakistan is unprecedented, therefore spatial variability of most of air pollutants and their emission are influenced by climate-change signals. Hence changing climatic conditions induced by anthropogenic emissions of greenhouse gases may be expected to have significant effects on air quality and subsequent environmental health outcomes especially at local to regional scales. In conclusion we strongly stress the need to create multidisciplinary research teams where skills of geographers, GIS experts, meteorologists and modeling experts could be integrated to estimate likely health impacts of global environmental change.

Keywords: A - population exposure, A-exposure models, A-climate change, B-particulate matter, A-statistical methods

MO-PO-218

The Astonishing role of soil lead in the exposure of children: Lessons from Hurricane Katrina.

H. W. Mielke; Tulane University School of Medicine, New Orleans, LA

Abstract: A critical characteristic for long-term resilience and sustainability of communities is the quality of the urban exposome. An impediment to adult members of a community is children's chronic lead exposure and its clinical consequences including disorders of health, learning, and behavior. There is no safe level of lead exposure and this revelation is confounded by the reality that there is no effective intervention after exposure has occurred. In August, 2005, Hurricane Katrina flooded 80% of New Orleans. Comparison of the before and ten years after soil lead condition of 172 communities and the matched blood lead of children was accomplished by using GIS methods to organize, describe, and map the data. Simultaneous decreases occurred in soil lead that were accompanied by exaggerated declines in children's blood lead outcomes. Health and welfare disparities between inner and outer communities continue to exist between soil lead and children's blood lead. The lead dust deposition in the soil reservoir are an astonishingly holistic source of both ingestible and inhalable lead dust. A major lesson from Katrina is that transporting low lead soil into urban communities is beneficial and a low cost method for primary lead prevention. **References**: -Mielke HW, Gonzales CR, Powell ET, Mielke PW Jr. Spatiotemporal exposome dynamics of soil lead and children's blood lead ten years after Hurricane Katrina: New grounds for primary prevention. *Environ Int* **2016**, DOI:10.1016/j.envint.2016.06.017. -Mielke HW, Gonzales CR, Powell ET. Soil lead and children's blood lead disparities in pre-and post-Hurricane New Orleans (USA). *Int J Environ Res Public Health* **2017**, DOI:103390/ijerph14040407. -Laidlaw M, Poropat A, Mielke HW, Ball A. Could past roadsie emissions of lead from petrol be associated with current incidence of Dementia? Accepted, 18 April **2017**, manuscript number: *The Lancet*-D-17-01430.

Keywords: A-built environment, D-children, A-geospatial analysis/GIS, A-environmental policy, B-metals

MO-PO-219

Associations Between First Trimester Blood Metal Levels and Childhood Obesity in the NEST Cohort

S. S. Park[1], Y. Luo[2], T. Darrah[3], R. L. Maguire[1], J. Tzeng[2], C. Hoyo[1]; [1]North Carolina State University, Raleigh, NC, [2]North Carolina State University, Raleigh, NC, [3]The Ohio State University, Columbus, OH

Abstract: The nephrotoxic and neurotoxic roles of metals such as cadmium, mercury, and lead are well known, now there is a growing recognition that metals may also be associated with cardio-metabolic outcomes. There is also a pressing need to investigate the health impacts of mixed metal exposures as these exposures are environmentally relevant. To this end, we measured first trimester blood levels of 24 metals via inductively coupled plasma mass spectroscopy (ICP-MS) in 98 mother-child dyads in the Newborn Epigenetics STudy (NEST) cohort. We used the Partial Least Squares (PLS) method to generate a mixed metals model which associates with an increased BMI-Z score at five years of age. We adjusted for maternal BMI, gestational weight gain, smoking, sex of the child, and breastfeeding for ≥ 3 months. Metals significantly contributing to the model included aluminum, nickel, and vanadium ($p<0.05$). We validated the significance of this model through a permutation analysis. In sum, we identified a group of metals that together may contribute to obesity in childhood. Further studies are needed to confirm these findings.

Keywords: A-epidemiology, B-metals, B-mixtures, D-children

MO-PO-220

A Literature Review and Data Mining Project to Identify Associations between Stressors and Health Outcomes for Children

S. Utile[2], N. S. Tulve[1], K. Thomas[1]; [1]Environmental Protection Agency, Durham, NC, [2]Student Services Contractor assigned to NERL/ORD, Research Triangle Park, NC

Abstract: Children are often more vulnerable than adults to the effects of environmental contaminants found in their everyday environments. Their dynamic growth and unique interactions with the built, natural, and social environments may result in a greater susceptibility to chemical and non-chemical stressors and potential adverse health consequences. Our objective was to examine approaches for evaluating interrelationships between chemical and non-chemical stressors and children's health outcomes. Using several search engines (e.g., Google Scholar, Web of Science) and websites (e.g., EPA and NIEHS websites), we compiled over 2400 peer reviewed articles published by the EPA/NIEHS Children's Environmental Health and Disease Prevention Research Centers (Children's Centers) for the period 1998-2016. We focused on Children's Centers publications because of the emphasis on chemical exposures, non-chemical stressors, and multiple child-specific lifestages. Full citations, keywords, and abstracts were entered into an EndNote database. Topic-specific libraries were created for specific chemical (e.g., metals, phthalates, pesticides) and non-chemical (e.g., built environment, food access, cultural practices) stressors and associated health outcomes (e.g., obesity, asthma). Prior to data extraction, manuscripts were reviewed to determine if they contained extractable data, information on chemical and/or non-chemical stressors and the health outcome of interest, and did not pertain to animal studies. Extracted data were entered into a database modeled after the Comparative Toxicogenomics Database. One topic-specific library focused on stressors and childhood obesity, with 156 Children's Center publications. Of these publications, 14 described chemical (e.g., phthalates, pesticides) stressors and 11 identified non-chemical (e.g., food types) stressors associated with childhood obesity. Preliminary analyses of the childhood obesity and stressors topic specific library will be presented.

Keywords: D-children

MO-PO-221

Household dust is a good predictor of children's lead exposure

S. Cao[1], X. Duan[2], X. Zhao[1], C. Sun[3], F. Wei[4]; [1]Chinese Research Academy of Environmental Sciences, Beijing, China, [2]University of Science and Technology Beijing, Beijing, China, [3]Chinese Center for Disease Control and Prevention, Beijing, China, [4]China National Environmental Monitoring Center, Beijing, China

Abstract: China is one of the main producer and consumer of lead in the world. As one of the typical lead-related activities, lead-bearing ore mining and refining influences the environmental quality and poses high risks to human health, due to large amounts of pollutants are emitted into the environment, particularly into the surrounding dust. Children's lead exposure depends heavily on multi-pathway and source-specific exposure, which have been rarely studied around typical ore mining and refining area. The objectives of this study are to 1) quantify the lead levels in the blood of children living around typical ore mining and refining area, China, as well as 2) lead levels in their household dust, and 3) estimate the contribution of household dust to the children's blood lead. Thus, lead contents and lead isotopic ratios ($^{207}Pb/^{206}Pb$ and $^{208}Pb/^{206}Pb$) in blood and household dust were studied. Lead contents in indoor and outdoor dust were 1358.08 and 2116.18 mg kg^{-1}, respectively, which showed a significant variance and were ~ 10 times higher than the thresholds of China (300 mg kg^{-1}).The children's blood lead level was 22.26 µg dL^{-1}, quite higher than the threshold in the criteria of China (10 µg dL^{-1}). Lead contents in indoor and outdoor were quite correlated ($p<0.01$, $r=0.77$), and the household dust lead showed a significant correlation to the children' blood lead ($p<0.01$, $r=0.60$). The isotopic ratios in the blood were 2.117±0.022 for $^{208}Pb/^{206}Pb$ and 0.851±0.021 for $^{207}Pb/^{206}Pb$, similar to those of the slag (2.121±0.018, 0.849±0.001 for $^{208}Pb/^{206}Pb$ and $^{207}Pb/^{206}Pb$) and dust (2.122±0.048, 0.852±0.027 for $^{208}Pb/^{206}Pb$ and $^{207}Pb/^{206}Pb$, respectively), suggesting the slag was the main pollution sources and household dust was the important exposure pathway of children's lead exposure. The study further indicated the slag in the mining and refining activities is the dominate pollution source of lead in children's blood, and highlight the household dust to children's lead exposure.

Keywords: A - exposure measurement, A-biomarkers, B-metals, D-children

MO-PO-222

Measuring Children's Personal Exposure to Household Air Pollution Using the Enhanced Children's MicroPEM

R. Chartier[1], M. Dherani[2]; [1]RTI International, Research Triangle Park, NC, [2]University of Liverpool, Liverpool, United Kingdom

Abstract: Three billion of the world's poorest people rely on solid fuels (wood, crop wastes, dung, charcoal and coal) and simple open fires and stoves for their everyday cooking needs. This results in high levels of exposure to smoke pollution among family members, including young children. These exposures to household air pollution (HAP) have been linked to numerous adverse health outcomes. The RTI MicroPEM (240g) has been used to measure personal exposure to particulate matter (PM) in numerous HAP exposure studies. The MicroPEM has been validated as a personal exposure monitor and deployed in field studies with adult and school-aged children. However, this device can be burdensome for small children (< 5 years). Therefore, RTI developed a smaller version of the MicroPEM, the Enhanced Children's MicroPEM (ECM). The ECM is 47% smaller and weighs 90g less than the MicroPEM. This decrease in size and weight reduces the physical burden on the child while the existing performance MicroPEM is maintained or improved. ECMs were deployed in a small pilot effort (N=50) in rural Malawi. ECMs were worn by children falling into one of 5 age groups (0-1, 1-2, 2-3, 3-4, and 4+ years of age). The youngest children were found to have the highest mean personal exposure levels (58.6 μg/m^3), and exposures decreased with age, reaching a minimum in the 3-4 year group (29.2 μg/m^3). Additionally, the ratio of the mother's $PM_{2.5}$ exposure, measured using the MicroPEM, to that of her child increased from a minimum at age 0-1 year (1.9) to a maximum at age 3+ (5.6). This small sample of child HAP exposure data indicate children likely spend more time away from their mother, and therefore the cookstove, as they become older. Positive participant feedback regarding the wearability of the ECM was received and 49 of the 50 mothers indicated they would be willing to have their children wear the ECM for future exposure measurements.

Keywords: A - exposure measurement, A-sensor technology, B-particulate matter, D-children

MO-PO-223

Association between body mass index and waist circumference with urinary paraben concentrations among adults, NHANES 2005-2014

L. Quiros-Alcala, M. D. Boyle; University of Maryland, Rockville, MD

Abstract: Parabens are a group of synthetic chemicals commonly used as antimicrobial preservatives in cosmetics, food, and pharmaceuticals. Although human exposure to select parabens is ubiquitous, their potential health effects have not been fully explored, and one recent *in vitro* study suggests that parabens may be linked to obesity while an *in vivo* study reported an inverse association between exposure and weight gain. This is the first study to examine the cross sectional association between exposure to four parabens (methyl, ethyl, butyl, and propyl) and measures of obesity (body mass index, BMI and waist circumference, WC) in 6,195 adults (>20 years) and 2,220 children (6-19 years) using a nationally representative sample of the U.S. population. We used multivariate regression analyses to test these associations, controlling for age, gender, race/ethnicity, urinary creatinine, smoking, diet, and physical activity. For adults, both BMI and WC were inversely associated with butyl (BP), ethyl (EP), methyl (MP), and propyl parabens (PP). BMI decreased by 1.29 (95% confidence interval (95% CI): -1.62, -0.96) kg/m^2 and WC decreased by 3.28 (95% CI: -4.07, -2.49) cm for every unit increase in log-10 MP concentrations, and BMI decreased by 0.95 (95% CI: -1.20, -0.71) kg/m^2 and WC decreased by 2.47 (95% CI: -3.10, -1.84) cm for every unit increase in log-10 for PP. Detection of BP was associated with a 1.43 (95% CI:

1.86, -1.00) kg/m^2 decrease in BMI and a 3.53 (95% CI: -4.62, -2.44) cm decrease in WC, while detection of EP was associated with a 1.83 (95% CI: -2.29, -1.38) kg/m^2 decrease in BMI and a 4.40 (95% CI: -5.45, -3.34) cm decrease in WC. Preliminary results indicate similar findings among children. These associations warrant further examination to confirm our findings and further studies are needed to elucidate the potential mechanisms by which these agents could impact metabolism.

Keywords: C-consumer/personal care products, A - population exposure, A-epidemiology, D-children

MO-PO-224

The Fernald Community Cohort: Lessons Learned from a Large Environmental Epidemiology Cohort with a 25 Year Heritage

S. M. Pinney, C. Xie, R. Leung, J. M. Buckholz; University of Cincinnati College of Medicine, Cincinnati, OH

Abstract: Background: Members of the Fernald Community Cohort (FCC, N=9782) received medical screenings from 1990-2008 (18 years). Although eligibility for the medical monitoring program was based on living within 5 miles of a uranium refinery, exposure dosimetry studies conducted by CDC demonstrated that much of the cohort never experienced uranium exposure beyond usual background levels, and thus there are both uranium-exposed and non-exposed persons in the cohort. Cohort follow-up continues to the present. **Methods:** Members of this multi-generational cohort completed yearly questionnaires (total 96,091) on a wide array of health risk factor and broad exposure information and received a comprehensive medical examination every 2 or 3 years (total 43,283) including clinical laboratory tests, mammograms, PFTs, EKGs, and chest x-rays. Over 160,000 aliquots of blood and urine samples are stored in -80 freezers. Cohort members gave broad consent for use of their data and biospecimens for future research, and data and biospecimen sharing was implemented in 2001, with multiple improvements to the policy over time with experience in resource sharing. **Results:** Data are coded and reside in a SAS database linked to the biospecimen repository. Learnings include planning early for extensive protocol and data documentation including a searchable data dictionary, a biospecimen inventory system and quality assurance procedures, actively encouraging and tracking publication of study findings, and budgeting for cohort maintenance and enhancement costs into the future. Seventy-six research projects have resulted in 38 published manuscripts, reporting the findings of studies of uranium and other environmental exposures, such as perfluoroalkyl compounds and bisphenol-A, and predictive biomarkers for breast, lung, prostate cancer, and abdominal aortic aneurysm. http://med.uc.edu/eh/research/projects/fcc. The FCC is registered with ClinicalTrials.gov. **Funding:** P30-ES006096, R834788, T32ES10957

Keywords: A - population exposure, A-biomarkers, D-susceptible/vulnerable, D-community, B-radiation

MO-PO-225

Integrating exposure knowledge and serum exposomics as a new approach to biomonitoring: An application in firefighters and office workers

R. Grashow[1], V. Bessonneau[1], R. R. Gerona[2], A. Wang[3], J. Trowbridge[4], T. Lin[2], R. A. Rudel[1], R. Morello-Frosch[4]; [1]Silent Spring Institute, Newton, MA, [2]University of California, San Francisco, San Francisco, CA, [3]University of California, San Francisco, San Francisco, CA, [4]University of California, Berkeley, Berkeley, CA

Abstract: The exposome offers investigators an opportunity to systematically and efficiently explore the full spectrum of environmental exposures in occupational and other cohorts. Known as general suspect screening, this approach uses liquid chromatography and high resolution mass spectrometry (LC-HRMS) to rapidly screen for a vast array of chemicals without the limitations of *a priori* hypotheses, allowing for the identification of previously unstudied but significant exposures. In this study, serum samples of 86 women firefighters from the San Francisco Fire Department and 84 office workers from San Francisco were analyzed using non-targeted LC-QTOF/MS. We matched all detected molecular weights of parent ions against a custom in-house library of 739 environmental chemicals of interest for exposures in these professions and chemicals relevant to breast cancer etiology. We detected 620 chemicals that matched 300 different molecular formulas, including phthalate metabolites, phosphate flame retardant (PFR) metabolites, phenols, pesticides, nitro- and nitroso- compounds, and per- and polyfluoroalkyl substances (PFASs) in both firefighters (FF) and office workers (OW). The average number of chemicals from our library that were detected in serum samples was 44 and 32 in FF and OW, respectively. We also observed that FF were more frequently exposed and/or exposed to higher levels of some phthalates and phenols compared to OW. We selected ten chemicals to validate and quantify in the serum samples based on the following criteria: 1) higher peak areas in FF samples, 2) higher detection frequency in FF samples, 3) concern with regard to breast cancer, 4) genotoxicity, and 5) not currently measured in large biomonitoring studies. This approach presents a novel and powerful method for using non-targeted HRMS analysis in occupational cohorts to reveal exposures to previously unstudied chemicals.

Keywords: A - exposure measurement, A-biomonitoring, D-occupational,

MO-PO-226

Top-Down Toxicology: an Experimental Application of the Pesticide Exposome in Honey Bee Queens

J. Milone, D. Tarpy; North Carolina State University, Raleigh, NC

Abstract: The exposome concept was developed for human exposure science and is often limited to observational research for exposure assessment. By adapting this framework for the investigation of pesticide exposure in honey bees (*Apis mellifera*) we are able to experimentally apply the exposome paradigm to a model species for social behavior and an essential agricultural pollinator. Studies of pesticide exposure in honey bees have historically focused on single pesticide dose-responses and have failed to accurately characterize exposure conditions faced by developing queens. Realistically, honey bees are exposed to numerous chemicals across multiple pathways during sensitive periods of development. In this study we reared queens in realistic exposure environments simulating elevated and reduced oral and contact exposures via pollen and wax. Pesticide treatments consisted of a combined mixture of insecticides, fungicides, and herbicides that have been commonly detected within commercial colonies and the toxicity of each mixture was based on field relevant Hazard Quotients. We sampled queens as newly emerged adults and at the onset of reproductive activity to quantify direct effects of developmental exposure, as well as sampling the offspring of mature queens to identify downstream effects on colony function. Our analysis includes measures of queen morphology, viral load intensity, mating number, insemination quality, and colony-level measures of health and disease prevalence. By understanding the effects of multi pesticide exposure across queen life history we have a more thorough understanding of honey bee exposome and multi-stressor interactions driving pollinator losses.

Keywords: B-pesticides, D-wildlife

MO-PO-227

Toward Capturing the Exposome: Biomarker Variability and Co-exposure Patterns in the Shared Environment

M. Chung[1], R. Sundaram[3], K. Kannan[2], G. Buck Louis[4], C. J. Patel[1]; [1]Harvard University, Boston, MA, [2]Wadsworth Center, New York State Department of Health, Albany, NY, [3]The National Institutes of Health, Rockville, MD, [4]The National Institutes of Health, Rockville, MD

Abstract: Background: Variation in exposure levels in the population is dependent on time, location, and sex. One specific factor that may influence exposure co-variations is shared household; specifically, co-variation of exposures may be influenced by both shared household (shared environment) and unique environment of the individuals (not attributable to shared household). Objective: We aimed to examine the influence of shared household in correlation of endocrine disrupting chemical (EDC) exposures systematically in couples living together. Methods: We leveraged a cohort with 501 couples and measured 128 serum and urinary EDCs from 13 different classes and estimated 1) sex-specific difference between genders; 2) variance explained by shared household; and 3) Spearman's rank correlation coefficient (r_s) for females, males, and couples. Results: After accounting for household, sex only influenced some biomarkers of the exposome, such as polyfluoroalkyl substances. On the other hand, shared household explained less than 20% of the total variance in 11 EDC classes. A "dense" correlational co-occurrence of the exposome was observed in males and females separately, especially among the persistent organic EDCs (e.g. polychlorinated biphenyls). For both females and males, within-class correlations of persistent organic EDCs had an absolute median r_s ranging between 0.29 to 0.45. In contrast, all classes had that below 0.21 in couples. Conclusions: Chemical co-exposure patterns, especially for persistent organic EDCs, can be dense, but similar in males versus females separately. However, although individuals spend a significant time at home, shared environment generally only accounted for less than 20% of the co-variation in levels of all markers of the exposome, and understanding the individual co-exposure patterns is a necessary step to capture the variations. In addition, the correlations between exposures have analytical implications for environment-wide association analyses.

Keywords: B-mixtures, A-biomarkers, B-phthalates, B-POPs, B-flame retardants

MO-PO-228

A Quantitative Health Risk Assessment for Stachybotrys chartarum- toxic mold

N. Sunger[1], B. Prasad[2], P. Morgan[1], E. Lennon[2], C. N. Haas[2]; [1]TE-West Chester University, Chester Springs, PA, [2]Drexel University, Philadelphia, PA

Abstract: The inhalation exposure to *Stachybotrys chartarum* (SC) has been playing a critical role in public health due to its ability to produce toxins that are implicated in cases of acute idiopathic pulmonary hemorrhages (AIPH) in infants. Currently, there is no regulatory standard for acceptable level of SC spores/toxins in indoor environment. The goals of this study were to determine the risk of death in infants using the maximum reported concentration of toxins and spores in water-damaged residences and further provide a benchmark dose level above which an adverse effect may occur in human infants. The best fit dose-response models were derived from published animal studies and by using 10% as the benchmark response level, permissible exposure limits were obtained as 413 SC-spores/m3 or 3.94E-05 mg-toxin/m3. The predicted risk of death in infants for acute 24-hour exposure to toxins ranged from 1E-4 to 3E-6. Uncertainty in risk estimates was characterized by conducting sensitivity analysis. The results indicated that while toxin exposure is inconclusive for AIPH in infants, a health hazard after inhalation of SC- toxins exists in human infants. Given the controversies surrounding the exposure and risk characterization of SC molds, these estimates should be interpreted with caution.

Keywords: C-indoor, A-risk assessment, C-air, D-children, A-exposure models

MO-PO-229

Data for Policy - Human biomonitoring's contribution to evidence-informed environmental health policy-making

A. Conrad, E. Rucic, C. Schulz, G. Schwedler, M. Rüther, P. Apel, M. Schmied-Tobies, T. Weber, M. Kolossa-Gehring; German Environment Agency (UBA), Berlin, Germany

Abstract: Human Biomonitoring (HBM) data reflect the total amount of a chemical transferred into the human body. Therefore, HBM provides key information for environmental-health policy-makers. Current results of the German HBM system - German Environmental Survey (GerES) and Environmental Specimen Bank (ESB) - highlight important aspects of further strengthening HBM as tool for evidence-informed policy-making: **a) Following up trends - HBM can yield early warnings or demonstrate the success of regulation:** German ESB data reveal an increase in glyphosate exposure (fraction of 24 h-urine levels ≥ LOQ increased from 10% in 2001 to 40% in 2015). In contrast, Pentachlorophenol - banned as pesticide in Germany in 1989 - decreased in blood plasma from 21 µg/L in 1987 to 0.5 µg/L in 2010 (geometric means). **b) Combining HBM with questionnaire and other monitoring data - elucidate routes of exposure:** GerES 2003-2006 revealed that cotinine in 3-14 years-old children's morning urine is associated with maternal and paternal self-reported smoking. In the same study, uranium in urine was correlated with uranium in drinking water. No significant associations were found for other metals like nickel or cadmium. **c) Evaluating the health-relevance of HBM results - assessment values as a vital tool:** HBM values for 19 compounds, derived by the German HBM Commission, are presently available. More than 90% of the ESB samples collected in 2010 and analyzed for PFOA exceeded the HBM-I value. This indicates that health effects cannot be ruled out with sufficient certainty and warrants increased attention in policy-making. The German Environment Agency (UBA) contributes its long-standing experience in transferring HBM data into policy to the **European Joint Programme HBM4EU**. As one main goal, this initiative makes the appropriate HBM data and complementary information available to policy-makers thus supporting targeted regulatory action in environmental health across Europe.

Keywords: A-biomonitoring, A-environmental policy, A-epidemiology, A-chemical prioritization, A-biomarkers

MO-PO-230

Complementing *in vitro* Hazard Assessment with Exposure and Pharmacokinetics Considerations for Chemical Prioritization

C. Tan[3], J. A. Leonard[1], C. Stevens[4], K. Mansouri[2], D. Chang[5]; [1]Oak Ridge Institute for Science and Education, Oak Ridge, TN, [2]ScitoVation, Durham, NC, [3]US EPA, Durham, NC, [4]US EPA, Athens, GA, [5]US EPA, Washington, DC

Abstract: Traditional toxicity testing involves a large investment in resources, often using low-throughput *in vivo* animal studies for limited numbers of chemicals. An alternative strategy is the emergence of high-throughput (HT) *in vitro* assays as a rapid, cost-efficient means to screen thousands of chemicals across hundreds of pathway-based toxicity endpoints and to aid in chemical prioritization for more extensive testing. Such HT *in vitro* methods, along with integration of HT *in silico* predictions of population exposure levels and pharmacokinetic (PK) characteristics, act as the foundation for HT risk assessment. Underlying uncertainties in predicted exposure concentrations or PK behaviors could significantly influence the prioritization of chemicals, though the impact of such influences is unclear. In the current study, a framework was developed to incorporate absorbed doses, clearance, and *in vitro* dose-response data into a PK/pharmacodynamic (PD) model to allow for placement of chemicals into discreet priority bins. In addition to measured values obtained from literature, values predicted for absorbed doses using a HT exposure model or for clearance using a quantitative structure activity relationship model, were entered into the PK/PD model to evaluate the impact of their uncertainties on the prioritization process. Scenarios using predicted absorbed doses resulted in a larger number of bin misassignments than scenarios using predicted clearance rates, when compared to placement of chemicals into bins using literature-reported measured values. Prioritization is more robust to uncertainties in clearance due to physiological constraints, whereas the large magnitude of differences between exposure predictions resulting from numerous possible exposure scenarios is the cause of increased errors in prioritization of chemicals into bins.

Keywords: A-chemical prioritization, B-pesticides, A-risk assessment

MO-PO-231

Defining the relationship between individuals' aggregate and maximum source-specific exposures

P. Price; US EPA, Research Triangle Park, NC

Abstract: The concepts of aggregate and source-specific exposures play an important role in chemical risk management. Aggregate exposure to a chemical refers to combined exposures from all sources and source-specific exposure refers to combined exposures from a specific source of a chemical. Both types of exposures can occur by multiple routes and exposure pathways and both can be determined on a population and individual level. The issue for aggregate and source-specific exposures is that while an individual's source-specific exposures may all be acceptable, an individual's aggregate exposure may not. Examples of regulatory issues that reflect this relationship include the "relative source contribution" used in setting ambient water quality standards, the regulation of sources under the Toxic Substances Control Act within a framework of aggregate exposure, and the assessment of impacts for life-cycle impact assessments. This talk presents a method (Maximum Aggregate Ratio or MAR) for evaluating the relationship between the two metrics. MAR is the ratio of the maximum source-specific dose of a chemical received by an individual divided by the individual's aggregate dose. The MAR can be viewed as a parallel concept to the Maximum Cumulative Ratio (MCR) which has been used in mixture risk assessments. The approach can be used to explore the relationship between interindividual variation in source-specific doses and interindividual variation in aggregate exposures and the impacts of controlling smaller sources on aggregate exposures. Case studies of the use of MAR to assess aggregate exposures to solvents used in consumer products are presented. The views expressed in this abstract are those of the author and do not necessarily reflect the views or policies of the U.S. EPA.

Keywords: A-aggregate exposure, C-consumer products, C-multimedia, A-risk assessment,

MO-PO-232

A Computational Framework for Modeling Multisystem Biological Effects of Multiroute Human Exposures to Ozone and Associated Air Pollutants

L. Chao, D. Mukherjee, C. Weschler, H. Kipen, P. Georgopoulos; Rutgers University, Piscataway, NJ

Abstract: Human exposures to ozone and associated photochemical air pollutants constitute a persisting and widespread problem around the globe. In addition to causing respiratory effects, photochemical pollution also impacts the cardiovascular, immune, integumentary and other physiological systems. Dermal contact can be a significant exposure route, with skin being both a barrier and target organ for pollutants. Reactive contaminants such as ozone often exert their detrimental effects via the generation of reactive oxygen species (ROS). For example, ROS result from reactions of ozone with skin lipids during dermal contact and from reactions with lung lining fluid components following inhalation. These secondary ROS initiate series of cascading events, such as release of pro-inflammatory mediators, infiltration of immune cells, activation of aryl hydrocarbon receptor (AhR) pathways, etc. Other physiological systems are subsequently affected: for example, the respiratory-originated pro-inflammation mediators can enter the circulatory system, initiate neuroendocrine-immune crosstalk and subsequently affect heart rate variability. The present work demonstrates new interconnected modules for the exposure biology of ozone and associated photochemical pollutants in the human integumentary, respiratory and cardiovascular systems. These modules are designed as components of the MENTOR (Modeling Environment for Total Risk) multiscale computational platform for whole-body human exposure, dosimetry, toxicokinetics and toxicodynamics. MENTOR has been under continuing development at the Computational Chemodynamics Laboratory (CCL) of EOHSI and employs a spectrum of systems dynamics modeling approaches, combining differential equation and agent-based methods to quantify overlapping Adverse Outcome Pathways (AOPs) involving multiple scales (biomolecular, cellular, histological, organ) and physiological systems and endpoints, resulting from multiple exposure routes.

Keywords: B - ozone, A-biomarkers, C-air

MO-PO-233

Re-evaluating the Inhalation Unit Risk for Chloroprene

S. Sax, K. Mundt, R. Gentry; Ramboll Environ, Amherst, MA

Abstract: Chloroprene is used in the production of polychloroprene (Neoprene), a synthetic rubber. In 2010, US EPA published an Integrated Risk Information System (IRIS) "Toxicological Review of Chloroprene," concluding that chloroprene was "likely to be carcinogenic to humans." This was based on findings from a 1998 National Toxicology Program (NTP) chronic inhalation bioassay showing multiple tumors within and across animal species; results from occupational epidemiological studies; a proposed mutagenic mode of action; and structural similarities with 1,3-butadiene and vinyl chloride. Using mouse data from the NTP study and a number of conservative assumptions, US EPA calculated an inhalation unit risk (IUR) factor for chloroprene of 5 x 10^{-4} per µg/m^3, the 5th highest IUR for chemicals classified as known or likely/probable human carcinogens (excluding metals or coke oven emissions) and orders of magnitude higher than IURs for vinyl chloride, benzene, and 1,3-butadiene - known human carcinogens. Studies were available at the time of the US EPA review demonstrating that the mouse is the species most sensitive to the effects of chloroprene exposure and that humans are far less sensitive due to differences in pharmacokinetics. These pharmacokinetic differences also are consistent with the lack of evidence observed in the most robust occupational epidemiological literature of exposed workers. The application of a validated PBPK model for chloroprene that allows species-specific estimation of internal exposure metrics (i.e., the amount of chloroprene metabolized per gram of lung tissue) suggests that the current IUR from EPA is too high by a factor of about 100. These findings highlight the need to update the chloroprene IUR to reflect the state of the science. An updated IUR will be presented applying the state of the science and approaches consistent with recommendations by the National Research Council seeking to improve toxicological reviews under the IRIS process.

Keywords: A-risk assessment, B-VOCs, C-air, D-occupational

MO-PO-234

Countering compounding conservatism - acute risk assessment trends in EU and JMPR

J. Stewart; BASF, Durham, NC

Abstract: Dietary risk assessment models used by authorities and international associations such as JMPR to protect consumers from the perils of pesticides in their food have the difficult challenge of safeguarding a diverse population while still representing realistic exposure scenarios. Many risk assessment models support a tiered approach which starts with a very conservative initial estimate of exposure, followed by more realistic refinement, if necessary. Recently Europe and JMPR are considering changes to their acute risk dietary model which will significantly increase the conservatism of the exposure estimate while precluding meaningful refinement. This poster will mathematically assess the conservatism of the current and proposed EU and JMPR acute risk assessment models and suggest modifications to the model which could provide an acceptable level of conservatism that is protective for consumers, but represents a more realistic exposure scenario.

Keywords: A-risk assessment, A-exposure models

MO-PO-235

Inhalation cancer risk estimation of source-specific personal exposure for particulate matter bound polycyclic aromatic hydrocarbons (PAHs) based on Positive Matrix Factorization (PMF)

B. Han[1], N. Zhang[1], J. Xu[2], W. Yang[1], X. Wang[1], Z. Bai[1]; [1]Chinese Research Academy of Environmental Sciences, Beijing, China, [2]University of Washington, Seattle, WA

Abstract: In previous studies, inhalation risk was estimated using conventional risk assessment method, which was normally based on compound specific analysis, and cannot provide substantial data for source-specific particulate matter concentrations and pollution control. In the present study, we applied a new risk analysis method, which was a synthetic combination of source apportionment receptor model and risk assessment method, to estimate cancer risks associated to individual PAHs coming from specific sources. Personal exposure PM samples referring to an elderly panel were collected in a community of Tianjin, Northern China in 2009, and 12 PAH compounds were measured using GC-MS. Positive Matrix Factorization (PMF) was used to extract the potential sources and quantify the source contributions to the PAHs mixture. Then, the cancer risk of each modeled source was estimated by summing up the cancer risks of all measured PAH species according to the extract source profile. We also used the bootstrap simulation inherent in EPA PMF5.0 for the uncertainty analysis. The final results indicated that the overall cancer risk was 2.11×10^{-5}, with the largest contribution from gasoline vehicle emission (42.4%). Unlike other risk estimation studies, this study was successful in combining risk analysis and source apportionment approaches, which allow to estimate the potential risk of all source types and provided suitable information to select prior control strategies and mitigate the main air pollution sources that contributing to health risks.

Keywords: B-particulate matter,

MO-PO-236

Insulation Usage and Asbestos Exposures: Historical Trends and Exposure Assessment

E. Hsu, D. Fung, C. barlow; Cardno ChemRisk, Boulder, CO

Abstract: It has been known since the 1960s that individuals who routinely handled or worked in the vicinity of asbestos-containing insulation are at increased risk of developing asbestos-related diseases. The trends of asbestos usage in insulation and the resulting airborne exposures from insulation work shifted over time as a response to advancements in industrial hygiene, knowledge of disease risk, and regulatory actions. Using the general trend of exposure intensity over time and comprehensive exposure analysis methods, we estimated the asbestos exposure for tradesmen who worked directly or indirectly with asbestos-containing insulation. A total of more than 3,800 industrial hygiene air sampling measurements were obtained from various database searches and industrial hygiene surveys representing four countries between 1965 and 1989. The measured airborne fiber concentrations were classified into comparable task/activity groups, aggregated, and then stratified by variables such as time period, sample type, and environment. To perform a quantitative exposure assessment, we constructed a model day for an insulator with descriptions of typical duties that an insulator encountered and then quantified exposures of tradesmen working with and around insulation over time. In general, the highest concentrations were generated during mixing and removal of thermal insulation. Exposure concentrations in the U.S. decreased over time with a distinct drop between 1969 and 1971, likely due to the passing of the 1970 OSH Act and subsequent actions taken by industry. Quantifying the airborne fiber concentrations for work with asbestos-containing insulation is an important step in characterizing the potential direct and indirect asbestos exposure not just for insulators, but workers in peripheral trades.

Keywords: A-cumulative exposure, D-occupational

MO-PO-237

Multimodal Physicochemical Characterization of Tire Crumbs Used at Synthetic Turf Fields

E. M. Heithmar, S. Gardner, K. Kovalcik, G. Momplaisir, A. Williams, M. Medina-Vera, T. Jones-Lepp; U.S. Environmental Protection Agency, Las Vegas, NV

Abstract: As part of a U.S. Federal Research Action Plan to study possible environmental and human health implications of tire crumbs used as infill for synthetic turf, the Environmental Protection Agency (EPA) is evaluating the potential of human exposure to various chemicals from tire crumbs. Physicochemical characterization of tire crumbs from playing fields and from manufacturing plants, including minor- and trace-element composition, as well as particle size and general morphology, is fundamental to this evaluation. EPA chose a multimodal approach to these characterizations: high resolution - inductively coupled plasma mass spectrometry (HR-ICPMS) for sensitive determination of trace-element concentrations of sample digests; X-ray fluorescence (XRF) for minor- and trace-element characterization of individual particle size fractions without digestion; scanning electron microscopy (SEM) for size distributions and morphology of fine particles; and energy dispersive X-Ray spectrometry (EDS) for information on elemental composition of selected particles imaged by SEM. Samples for elemental measurements by HR-ICPMS were digested using a modified version of EPA Method 3051A. Low detection limits and suppression of spectral interferences in complex matrices are benefits of HR-ICPMS. Particle Size Analysis (PSA) was performed on samples using sequential sieving. A flotation procedure was used to separate the sand from the crumb in samples with significant sand. A screening X-ray fluorescence analysis was performed on crumb size ranges separated during the PSA analysis, as well as on un-sieved samples. SEM imaging of fine particles was performed with a 24 keV electron beam and electron backscatter detection (BSD). Particles with significant non-carbon elemental composition were analyzed by EDS. This presentation describes the performance and demonstrates the complementary information provided by this multimodal physicochemical characterization of tire crumbs.

Keywords: B-particulate matter, B-metals, A-analytical methods

MO-PO-238

A multi-site recycled tire crumb rubber characterization study: recruitment strategy and field sampling approach

K. Benson[1], A. Grober[1], J. Truhe[1], K. Scruton[1], K. Thomas[2], M. K. Morgan[2], C. Croghan[2], E. A. Irvin-Barnwell[1]; [1]CDC/NCEH-ATSDR, Atlanta, GA, [2]U.S. Environmental Protection Agency, Office of Research and Development, National Exposure Research Laboratory, Research Triangle Park, NC

Abstract: Recently, concerns have been raised by the public about the safety of tire crumb rubber infill used in synthetic turf fields. In response, the 2016 *Federal Research Action Plan on Recycled Tire Crumb Used on Playing Fields and Playgrounds* (FRAP) was developed to examine key environmental and human health questions resulting from the use of tire crumb rubber. One specific objective of the FRAP was to characterize the chemicals, potential emissions, and toxicity of tire crumb rubber. The goal of the tire crumb characterization study was to analyze new and aged/used tire crumb rubber for a variety of chemicals and to characterize field use patterns and maintenance procedures using a structured questionnaire. The aim was to recruit and sample 40 synthetic turf fields with tire crumb rubber infill. Ten fields in each of the 4 U.S. census regions were targeted. We used a convenience sampling approach for recruitment and online search engines to locate contact information for potential fields. Study inclusion criteria included a maximum of two outdoor fields per facility with fields having different installation years or different installation companies. In addition, we recruited tire recycling/crumb rubber manufacturing facilities located across the U.S. A total of 306 community field owners/managers were contacted for potential participation. Sample collection was completed at 40 synthetic turf fields, including 21 community fields and 19 military installation fields. The final field count per census region was 9 Northeast fields (5 outdoor, 4 indoor), 13 South fields (11 outdoor, 2 indoor), 8 Midwest fields (2 outdoor, 6 indoor) and 10 West fields (7 outdoor, 3 indoor). Additionally, samples were collected at nine tire recycling/crumb rubber manufacturing plants across the country, including both ambient and cryogenic processes. Recruitment challenges will be discussed and field use and maintenance characteristics will be summarized.

Keywords: A-epidemiology

MO-PO-239

Athletes' Selected Micro-Activities on Turf Fields: Utilizing Extant Videography for Quantification of Events During Soccer, American Football, and Field Hockey Play

K. Hibbert[1], M. K. Morgan[1], G. Grissom[2], S. Utile[2]; [1]U.S. Environmental Protection Agency, Research Triangle Park, NC, [2]U.S. EPA, Research Triangle Park, NC

Abstract: Concerns have been recently raised about the potential exposures of athletes to chemicals when playing on synthetic turf fields. Previous research has shown that micro-activities (i.e., hand-to-mouth and skin-to-surface contacts) are important factors in people's exposures to chemicals in outdoor settings. However, no published data have been identified for exposure-related micro-activities of athletes engaged in various sports on synthetic turf fields needed for exposure modeling. The objective of this study was to quantify the frequency of micro-activity events of participants playing soccer, field hockey, and (American tackle) football on synthetic and natural turf fields from publicly-available videos. Using the social media website YouTube, extant videography was systematically mined for children and adults playing soccer/field hockey or football for a minimum of 15-minutes or 10-minutes, respectively. A total of

60 players were identified playing soccer (children: $n = 10$; adults: $n = 10$), field hockey (children: $n = 10$; adults: $n = 10$), and football (children: $n = 10$; adults: $n = 10$). Videos were downloaded as mp4 files and viewed on a computer using Windows Media Player software. Trained technicians tallied on paper the frequencies of hand-to-mouth, object-to-mouth, hand-to-turf, and object-to-turf events of each athlete. Frequency events for each type of micro-activity were normalized to one hour per athlete. Preliminary results showed no significant differences in frequencies of micro-activities depending on type of field. Also, there were no significant differences for individual micro-activities by age (children vs. adults). ANOVA analysis revealed significantly higher ($p < 0.001$) hand-to-mouth, object-to-mouth, hand-to-turf, and body-to-turf events for football players compared to soccer and field hockey players. This information suggests that type of sport played may have greater impact on potential exposures to chemicals than age or field type.

Keywords: A-activity patterns, A-behavior, A-exposure factors

MO-PO-240

Characterization of Semi-Volatile Organic Chemicals from Tire Crumb Rubber

M. S. Clifton[1], D. Mills[2], X. Liu[1], K. Thomas[1]; [1]U. S. Environmental Protection Agency, RTP, NC, [2]National Student Services Contractor administered by Oak Ridge Associated Universities to U.S. Environmental Protection Agency, RTP, NC

Abstract: Recycled tire crumb rubber (TCR) is often used as infill material in synthetic turf playing fields as well as some playgrounds. Concerns have been raised about the safety of this material and a multi-agency Federal Research Action Plan on Recycled Tire Crumb Used on Playing Fields and Playgrounds was developed to investigate key factors that could impact the environment and human health. Here we present work done to characterize semi-volatile organic compounds (SVOCs) from direct solvent extraction of TCR and airborne emissions experiments. A wide range of SVOCs (including PAHs, phthalates, and chemicals related to rubber manufacturing) were selected for targeted analysis. Solvent selection, extraction techniques, and instrument parameters were investigated in order to better understand the TCR material and to develop the methods and appropriate QAQC required for sample analysis. TCR samples were collected from nine tire recycling plants and 40 synthetic turf fields across the U.S. and were divided into subsamples for characterization experiments. TCR and emissions samples collected on PUF were extracted with 1:1 acetone:hexane. Emissions experiments were conducted at 25^{0}C, 46 % Relative Humidity (RH) and 1 h^{-1} air change (ACH) rate, and 60^{0}C, 6.6 % RH, 1 h^{-1} ACH in 53 L dynamic emission chambers. Data were acquired for all samples using GC/MS/MS in MRM mode with a calibration range of 0.1-500 pg/µL. MQLs, which were derived based on accuracy of standards compared to the calibration curves, ranged from 0.1-10 pg/µL. Non-targeted analysis was also performed by acquiring data by GCMS in scan mode (50-550 m/z) and then deconvoluting and library matching the spectra to tentatively identify components. Data obtained from the SVOC analyses will be used in conjunction with the other analyses that were conducted as part of the Federal Research Action Plan to identify key TCR chemical constituents, aid exposure assessment, and inform future studies related to TCR exposure.

Keywords: B-SVOCs, A-analytical methods, B-phthalates, ,

MO-PO-241

Characterization of Formaldehyde Emissions from Tire Crumb Rubber in Small Environmental Chambers

X. Liu[1], K. Krebs[1], M. Allen[3], K. Thomas[2], M. Stryar[2], D. Greenwell[1]; [1]U.S. Environmental Protection Agency, RTP, NC, [2]U.S. Environmental Protection Agency, RTP, NC, [3]Jacobs Technology Inc., Tullahoma, TN

Abstract: Concerns have been raised about the safety of recycled rubber tire crumbs used in synthetic turf fields and playgrounds in the United States. To support federal efforts to better characterize recycled tire-derived surface materials, dynamic small chamber tests were conducted at the US EPA National Risk Management Research Laboratory Small Chamber Facility to measure potential formaldehyde emissions from tire crumb rubber materials collected from nine tire recycling facilities and forty synthetic turf fields around the U.S. During tests, approximately 15 grams of tire crumb rubber materials were placed in the center of a 53 L dynamic emission chamber on an aluminum weighing pan for 24 hours before air samples were collected using 2,4-dinitrophenylhydrazine (DNPH) cartridges sampling at a rate of 200-400 mL/min for 90 minutes. The emission chambers were housed in temperature-controlled incubators. An OPTO 22 data acquisition system continuously recorded mass flow controller outputs, temperature, and relative humidity (RH) in the chamber and inlet air. Tests were conducted (N=82) under two chamber conditions, respectively. Formaldehyde concentrations were determined by solvent extraction and analysis by HPLC with Diode-Array Detector. Chamber background and field blank samples were collected for each test. DNPH-formaldehyde detection in selected samples was confirmed by LC/TOFMS. In addition, six duplicates and two time series tests were performed under each set of chamber conditions. The results show that measured formaldehyde concentrations in the chamber at 1 h^{-1} air change (ACH) rate, 25 °C, 46 % RH, were low and close to the chamber background level. Formaldehyde concentrations measured in the chamber at 1 h^{-1} ACH, 60 °C, 6.6 % RH, which may represent synthetic field surfaces under hot ambient conditions, were greater than the chamber background for most of the material samples. This research will provide important information for further human exposure study.

Keywords: C-air,

MO-PO-242

Arguments used to justify the ongoing use of asbestos: Myths or facts?

J. P. Ramos-Bonilla[1], E. Algranti[2], F. Cavariani[3], D. Marsili[4], A. Mazzeo[7], R. Pasetto[4], V. Santana[6], B. Terracini[8], A. Trotta[5], P. Comba[4]; [1]Universidad de los Andes, Bogotá, Colombia, [2]Fundacentro, Sao Paulo, Brazil, [3]Centro di Riferimento Regionale Amianto, Civita Castellana, Italy, [4]Istituto Superiore di Sanità, Rome, Italy, [5]Universidad Nacional de Lanus, Lanus, Argentina, [6]Universidade Federal da Bahia, Salvador, Brazil, [7]University of Bologna, Bologna, Italy, [8]University of Turin (Now Retired), Turin, Italy

Abstract: Approximately 2 million tons of asbestos are used each year at the global scale. Only 59 countries have banned asbestos. The WHO estimates that 125 million people are still exposed to asbestos in the work environment. Several studies conducted in a large number of countries have shown that the current use of asbestos frequently results in extremely high asbestos exposures in countries where the material is still legal. The aim of this presentation is to explain some of the most common arguments used by the asbestos industry (i.e., and in some cases legislators and decision makers), to justify the ongoing use of asbestos at the country level, and to block or delay regulatory interventions. These arguments have been identified by the authors during the work they have developed to reduce the severe health effects resulting from past or current use of asbestos in Brazil, Italy, Argentina, and Colombia. The main arguments used that will be discussed are: 1 - Argue that there is a potency difference between chrysotile and amphibole asbestos in terms of their carcinogenic effect; 2 - State that non-friable asbestos containing products are effective in preventing asbestos exposure; 3 - Argue that there are no safer substitutes for asbestos; 4 - State that local regulations should be delayed because of the lack of local evidence of adverse health effects or asbestos exposure among local communities; 5 - Argue that banning asbestos will have devastating consequences for families that depend on the asbestos industry for their livelihood; 7 - State that asbestos is only an occupational, not a public health problem; 8 - Argue, combining all the above-mentioned arguments, that a “safe/controlled” use of asbestos, instead of banning asbestos, is the right path to follow. The international peer-reviewed scientific literature published on the asbestos problem does not support these arguments. The authors have no financial conflict of interest to declare.

Keywords: A-global health, A - population exposure, A-environmental regulation, D-environmental justice,

MO-PO-243

Asbestos: Is the "safe use" concept supported by peer-reviewed scientific studies?

M. Valenzuela, J. P. Ramos-Bonilla, M. Giraldo, S. Gallo-Murcia, J. Pineda, L. Santos; Universidad de los Andes, Bogotá, Colombia

Abstract: All forms of asbestos (i.e., amphiboles and serpentines) are carcinogenic to humans. Exposure to asbestos causes mesothelioma and cancers of the lung, larynx, and ovary. More than half of the occupational cancer deaths are attributed to asbestos exposure. However, the asbestos industry argues that both asbestos and asbestos containing products can be used safely, which is summarized in the "safe/controlled use" concept. A literature review was conducted analyzing studies published between 2010 and 2015, to determine if the scientific literature supports the "safe/controlled use" concept. Two types of studies were included in the review: Asbestos exposure assessment studies, and asbestos adverse health consequences studies. An initial group of 155 studies were identified, and most of them (n=134; 87%) presented evidence of either high asbestos exposures or asbestos adverse health effects. These 155 studies included populations exposed to both amphibole and serpentine asbestos, and asbestos exposures reported could have occurred decades ago. To analyze more closely the evidence supporting the "safe/controlled use" concept, a subset of 44 studies that analyzed populations exclusively exposed to chrysotile asbestos or populations recently exposed to asbestos (i.e., after 2001) were identified. Most of these studies (n=36; 82%) present evidence that suggest that a "safe/controlled use" of asbestos is not being achieved. Furthermore, most of the studies that present evidence that support the "safe/controlled use" concept have a conflict of interest declared. A concerning low number of studies were identified in countries that are currently the major producers or consumers of asbestos. Thus, it is not clear which evidence is used to support the "safe/controlled use" concept. These results were recently published: Valenzuela M. et al., *Recent Scientific Evidence Regarding Asbestos Use and Health Consequences of Asbestos Exposure*, Curr Envir Health Rpt (2016) 3:335-347.

Keywords: A - population exposure, A-environmental justice, A-global health, A-workplace, D-occupational

MO-PO-244

Exposure Data For Toothpastes

M. Chiter[1], M. Gomez-Berrada[1], I. Boudieres-Laffont[1], A. Rielland[2], S. Guillou[2], D. De Javel[2], P. Ferret[1]; [1]Pierre Fabre Dermo-Cosmétique, Toulouse, France, [2]Eurosafe, Saint Grégoire, France

Abstract: Purpose: Toothpastes are intended to cleanse the oral cavity, freshen the breath etc. Depending on the principal effect claimed and the country of marketing, they are considered as cosmetics, Over-The-Counter drugs or quasi-drugs. No matter their status, these products must be safe. For this purpose, safety assessors need accurate exposure information. However, exposure data are limited and does not take into account specific consumer groups such as children. The main aim of this study was to determine the daily quantity and the frequency of use of toothpaste according to subjects' personal habits and to compare them between age and sex categories. Furthermore, the influence of toothpaste texture (gel/paste) and toothbrush kind (manual/electric) were also assessed. **Methods:** This study was performed with 104 families (206 adults, 169 children). They were selected among 1448 people who answered a preliminary web-based survey. Subjects used their own products (toothpaste/toothbrush) over a three-week period according to their personal habits. Products were weighed at the start and completion of the study in order to determine the total amount of product used. Toothpaste texture and toothbrush information (size/form/kind) were also recorded. **Results:** The mean, standard deviation and P90 value of the amounts used per day were as follows: Adults: 1.82±0.88g(3.04g);Women: 1.69±0.82g; Men:1.95±0.92g; Children: 1.10±0.62g(2.19g);Girls: 1.15±0.62g; Boys:1.05±0.62g. The mean, standard deviation and P90 value of frequency of use per day were as follows: Adults: 1.87±0.48(2.30); Women: 1.96±0.47; Men:1.79±0.49; Children: 1.74±0.46(2.10); Girls: 1.84±0.43; Boys:1.64±0.46. **Conclusion:** This study provides exposure data (amount/frequency) for toothpastes, which could be useful for safety assessors. It also reveals differences between category of age and sex. In order to understand these differences, factors that could affect the exposure were also studied (kind of toothbrush and toothpaste).

Keywords: C-consumer/personal care products, D-children

MO-PO-245

Identifying Novel Data Sources to Refine Consumer Product Exposure Assessment: Multi-stakeholder Collaboration Addressing the Behavioral Data Gap in Consumer Product Chemical Exposure

M. Embry[1], T. J. Buckley[2], R. Zaleski[3]; [1]HESI, Washington, DC, [2]USEPA, Research Triangle Park, NC, [3]ExxonMobil Biomedical Sciences, Inc., Annandale, NJ

Abstract: Even though it is known that consumer products provide an efficient means of chemical exposure and are a dominant source, our ability to estimate the nature or extent of exposure is stymied by the lack of data relating to consumer habits and practice patterns. The irony is that this data exists but for different purposes, i.e. marketing, and is generally not available for purposes of chemical exposure assessment. Product use surveys are generally carried out to better understand product markets, and public availability of these private, targeted surveys remains limited, as this information is costly to obtain. In 2017, the ILSI Health and Environmental Sciences Institute (HESI) convened a collaborative, multi-stakeholder group to explore novel data streams and data sources that could refine exposure estimates for consumer products via providing detailed information on consumer product habits and practices. The major data gaps that severely limit our ability to reliably approximate exposure and risk to consumer products include: 1) Information on what consumer products are being purchased and used and by whom; 2) How the consumer products are being used, and; 3) How environmental health factors into purchase/use decisions, e.g. what are the knowledge, attitudes, and beliefs about environmental health that underlie purchase and use? The data to answer these questions exists in various forms and places (e.g. retailers and manufacturers track this information to enhance sales). The overall goal of this collaborative effort is to identify where these data exist and to develop strategies to access information on consumer product habits and practices. This presentation will highlight the efforts of this HESI committee, whose initial aims are to identify potential data streams, with follow-up projects potentially focused on collecting, disseminating, and analyzing these data.

Keywords: A-activity patterns, A-behavior, C-consumer products, A-exposure factors

MO-PO-246

Characteristics of exposure factors for consumer products in Korean women and children

M. Lim[1], J. Park[2], K. Kim[3], K. Lee[1]; [1]Graduate School of Public Health, Seoul National University, Seoul, Korea (the Republic of), [2]Seoul National University, Seoul, Korea (the Republic of), [3]Seoul National University of Science and Technology, Seoul, Korea (the Republic of)

Abstract: Concerns about potential health risks of chemicals in consumer products like cosmetic, personal care products, food containers are growing. Especially infant and children are more vulnerable to chemical exposure compared to adults. Exposure to these chemical could be determined by consumer products usage pattern. An accurate measurement of the usage patterns of consumer products is important for realistic exposure assessment. The mother's usage pattern of consumer products and children's use of consumer products are interrelated. The aims of this study were to determine exposure factors of consumer products for child and mother and analyze the relationship between consumer exposures of caregiver and child. We determined the exposure factors of 12 kinds of cosmetics (3 basic cosmetics, 1 UV protection products, 3 hair products, 3 body products and 2 cleansing products) for adults and 10 kinds of consumer products (2 cosmetics, 3 oral supplies and 7 household products) for children and 11 kinds of food containers for household. Survey was conducted on 505 mother-infant pairs from Oct. to Dec. 2015 in Seoul and the metropolitan area by using structured questionnaires. The number of subjects were determined by proportionate quota sampling based on the population composition ratio in children's sex and age distribution from 0 to 4. All cosmetics investigated in this study were used on a daily basis and usage rates ranged from 52.1% to 98.0%, except 9.9% for hair styling product and 7.7% for deodorant. The frequency of food intake by food containers ranged 2.52 to 17.39 times a month. The use of children's oral supplies varies according to the age of the child. There were a significant difference in the mother's usage rates of lotion, hair products and vinyl package food by age of children. These exposure characteristics and factor data would be useful input data for exposure and risk assessment for chemical regulation.

Keywords: A-exposure factors, C-consumer/personal care products, D-children

MO-PO-247

Fluoride and arsenic uptake by bone char: quantifying uptake mechanisms for low-cost household water treatment systems

M. Fitzstevens[1], J. Kearns[2], A. Krupp[4], S. Mitchell[4], O. Duckworth[1], J. M. Harrington[3], D. Knappe[2], M. Polizzotto[1]; [1]North Carolina State University, Raleigh, NC, [2]North Carolina State University, Raleigh, NC, [3]Research Triangle Institute, Research Triangle Park, NC, [4]Caminos de Agua, San Miguel de Allende, Mexico

Abstract: Globally, 200 million people and 150 million people consume drinking water contaminated with fluoride and arsenic, respectively. Little is known about co-exposure to these contaminants, and recent studies show that they may have synergistic effects. In contaminated regions, many rely on well water and live in remote rural regions of low and middle-income countries. Bone char (BC) has been studied as an appropriate, low-cost sorbent for fluoride and arsenic in these settings. The efficacy of fluoride sorption to BC varies widely, depending on contaminant concentration and water chemistry (especially pH), bone type, char production conditions, and post-pyrolysis char treatments. The objectives of this research are to 1) quantify the reaction kinetics and uptake mechanisms of F and As for bone chars produced in the laboratory and in Guanajuato, Mexico, where groundwater is frequently co-contaminated with As and F, and to 2) develop bench-scale approaches to quantify fluoride uptake by BC. Reaction kinetics and uptake of As and F by BC are elucidated using batch kinetics tests and isotherm experiments for As and F separately and in combination, using laboratory simulated groundwater and natural groundwater from village wells in Guanajuato, Mexico. Bench-scale rapid small-scale column tests (RSSCTs) are conducted to simulate field-scale fluoride removal by bone char and are compared to field-scale column data. This work will inform the design of a household-scale treatment system for fluoride to be deployed in communities in Guanajuato, Mexico.

Keywords: C-water, B-mixtures, A-global health, A - population exposure, water engineering

TUESDAY, OCTOBER 17, 2017

TU-PL-A1: Air Pollution Measurement Error

TU-PL-A1-248

Implications of Exposure Measurement Error for Interpreting Epidemiological Results for Studies of Exposure to PM, NO_2, or SO_2

J. Richmond-Bryant, T. Long; U.S. EPA, Rtp, NC

Abstract: In epidemiologic studies of the health effects of ambient air pollutant exposure, measurements or models are used to estimate exposure concentration with error, which propagates to effect estimates. We critically evaluate potential biases and uncertainties resulting from exposure measurement error in air pollution epidemiological studies, using as examples particulate matter [PM, typically smaller than 2.5 μm in diameter ($PM_{2.5}$)], nitrogen dioxide (NO_2), and sulfur dioxide (SO_2), which have different spatial concentration distributions. We review the available literature on the impact of different types of exposure measurement error for community time-series and long-term average epidemiologic study designs on the health effects of PM, NO_2, and SO_2 exposure in the ambient air. The drivers of error are different for each pollutant. Differences in spatial distributions of PM, NO_2, and SO_2 and accuracy of their respective modeling techniques (e.g., satellite-CMAQ-fusion models for $PM_{2.5}$, land use regression or spatiotemporal models for NO_2, inverse distance weighting or dispersion models for SO_2) may influence exposure error. For epidemiological time-series studies of short-term exposure, exposure error tends to bias effect estimates towards the null and increase confidence intervals for PM, NO_2, and SO_2. For epidemiological studies of long-term average exposure, exposure error may bias the effect estimate in either direction and increase confidence intervals for all pollutants studied. Disclaimer: The views expressed in this abstract are those of the authors and do not necessarily reflect the views or policies of the U.S. Environmental Protection Agency.

Keywords: A-exposure factors, A-exposure models, A-epidemiology

TU-PL-A1-249

Do Differences in Exposures Explain the Observed Heterogeneity in $PM_{2.5}$ - Mortality Associations across U.S. cities?

L. K. Baxter, K. Dionisio, P. Pradeep, L. Neas; U.S. Environmental Protection Agency, Research Triangle Park, NC

Abstract: Multi-city population-based epidemiological studies of short-term $PM_{2.5}$ exposures and mortality have observed heterogeneity in risk estimates between cities. Factors affecting exposures (e.g. pollutant infiltration, time-activity patterns) not captured by central-site monitoring data can differ between communities potentially explaining some of this heterogeneity. This analysis evaluates exposure factors as potential determinants of the heterogeneity in city-specific associations between $PM_{2.5}$ and mortality. Exposure factor variables were created based on housing characteristics, commuting patterns, heating fuel usage, and climatic factors data from national surveys. When survey data was not available, air conditioning (AC) prevalence was predicted utilizing machine learning techniques. We examined the heterogeneity in 312 city-specific PM2.5-mortality health effect estimates using inverse variance weighted linear regression with inverse variance weights. Prevalence of central AC predicted via machine learning techniques showed a strong relationship (R^2 = 0.78) with the observed prevalence in surveyed cities. The national estimate (0.96% increase in total non-accidental mortality for a 10 μg/m3 increment in $PM_{2.5}$ at lag 1) decreased by 0.12% (95% confidence interval (CI) -0.27 to 0.03) for an interquartile increase in the predicted prevalence of central AC. The other determinants examined were also found to modify the PM2.5-mortality association to varying degrees: median house size (+0.09%, 95% CI 0.01 to 0.18, for a 1 room increase in the median number of rooms), percentage of heating fueled by oil (+ 0.15%, 95%CI 0.10 to 0.20, for a percentage point increase), and heating degree days (HDD) (+0.31%, 95% CI 0.12 to 0.50, for a 1 day increase). In our analysis, the health impact of $PM_{2.5}$ on mortality increases in cities with larger homes, more heating fueled by oil, more HDD, and less central AC.

Keywords: A-epidemiology, A-exposure factors

TU-PL-A1-250

How well do epidemiologic studies that use composite monitor values capture underlying variability in ambient ozone levels across larger urban areas in the U.S.?

Z. Pekar[1], D. Simon[1], T. Luben[2], B. Wells[1]; [1]US EPA, RTP, NC, [2]US EPA, RTP, NC

Abstract: Monitoring data are often used as a surrogate for population-level exposure in epidemiologic studies of ozone. Many studies use ozone measurements from individual monitors to create an average or "composite monitor" value which is then used as a surrogate for population exposure within a given study area. Given that ozone can vary across an urban area and that people may be exposed to different levels of ozone as they move within an urban area, the use of composite monitors may introduce additional exposure measurement error. Our research includes statistical analyses designed to explore the degree to which composite monitors capture patterns of ozone at individual monitors, including (a) degree of correlation between a daily time-series of composite monitors and individual monitors in urban areas, (b) degree to which composite monitors track individual monitors on days where individual monitor ozone values are above thresholds of 40, 50, 60, 70, and 80 ppb, (c) degree to which the correlation between composite monitors and individual monitors differs for different daily metrics used in risk modeling (e.g., 1hr max, 8hr max, 8hr mean, 24hr avg), and (d) degree to which use of population-weighted composite monitor values differs from traditional composite monitors in tracking variability at individual ozone monitors. In addition, we examine whether the distribution of composite monitor ozone values differs systematically from the distribution of ozone values at the individual monitors comprising the composite monitor. We find that composite and population-weighted composite monitor time series are highly correlated with individual monitor time series for most monitoring locations. However, this correlation is attenuated when looking across days with ozone concentrations higher than 50 ppb. Based on our research, we discuss potential implications for epidemiology studies that use the composite monitor approach to characterize population level exposure to ozone.

Keywords: A-epidemiology, A - ambient monitoring, A - population exposure, A - exposure measurement, B - ozone

TU-SY-B1: Disentangling disparities in exposures: body burdens, personal care products, and the indoor environment

TU-SY-B1-251

Racial/ethnic disparities in cumulative exposures to phthalates and parabens and implications for uterine fibroid size

A. R. Zota[1], J. Opoku-Anane[2], J. Robinson[3], G. Moawad[4]; [1]George Washington University Milken School of Public Health, Washington, DC, [2]UCSF, San Francisco, CA, [3]MedStar Washington Hospital Center, Washington, DC, [4]George Washington University School of Medicine and Health Sciences, Washington, DC

Abstract: In the U.S., black women are disproportionately impacted by uterine fibroids. They experience larger tumors and faster tumor growth, and consequently have a three-fold higher risk of hysterectomies than white women. However, the etiology of uterine fibroids and root causes of these racial/ethnic disparities remain poorly understood. The objective of this pilot study was to characterize racial/ethnic differences in exposures to phthalates and phenols and to examine their associations with uterine fibroid size, a marker of clinical severity. We recruited 29 premenopausal women with either small (<3 cm) or large fibroids (>=6cm) who were seeking surgical care for their fibroids in Washington DC. Spot urine samples were collected and measured for 17 phthalate metabolites, xx phenols, and xx phthalate alternatives. In addition to examining chemicals individually, we developed a cumulative phthalate score, cumulative phenol score, and cumulative EDC score. To calculate scores, participants' levels of each analyte were dichotomized as high (>25th percentile, assigned a "1") or low (<25th percentile, assigned a "0") and summed across chemicals. Our study sample was 59% black, highly educated, and privately insurance; 72% had large fibroids. Compared to white women, black women had higher cumulative chemical exposures and were more likely to have large fibroids ($p<0.05$). After adjusting from race and body mass index, cumulative phthalates exposure was associated with large fibroid size (adjusted odds ratio = 2.54 (95% CI: 1.00, 6.45)). These preliminary results suggest that racial/ethnic disparities in EDC exposures could have consequences for women's gynecological health.

Keywords: B-phthalates, A-cumulative exposure, A-environmental justice, A-epidemiology, women's health

TU-SY-B1-252

Endocrine disrupting chemicals in hair products marketed towards women of color

J. S. Helm[1], M. Nishioka[2], J. G. Brody[1], R. A. Rudel[1], R. Dodson[1]; [1]Silent Spring Institute, Newton, MA, [2]Battelle Memorial Institute, Columbus, OH

Abstract: Personal care products are a source of exposure to endocrine disrupting and asthma-associated chemicals. Because use of hair products differs by race/ethnicity, these products may contribute to observed exposure and health disparities. We tested 18 hair products used by Black and Hispanic women (hot oil treatment, anti-frizz/polish, leave-in conditioner, root stimulator, hair lotion, and relaxer) for 66 chemicals, including UV filters, cyclosiloxanes, glycol ethers, fragrances, alkylphenols, ethanolamines, antimicrobials, bisphenol A, phthalates, and parabens. We found cyclosiloxanes, parabens, and the fragrance marker diethyl phthalate (DEP) at the highest levels, and DEP most frequently. Root stimulators, hair lotions, and relaxers frequently contained nonylphenols, parabens, and fragrances; anti-frizz products contained cyclosiloxanes. Hair relaxers for children contained five chemicals regulated by California's Proposition 65 or prohibited by EU cosmetics regulation. Targeted chemicals were generally not listed on the product label. Hair products used by Black and Hispanic women and children contained multiple chemicals associated with endocrine disruption and asthma. The prevalence of parabens and DEP is consistent with higher levels of these compounds in biomonitoring samples from Black and Hispanic women compared with White women. These results highlight the potential contribution of hair products to exposure disparities and provide a focus for efforts to reduce exposures.

Keywords: C-consumer/personal care products, D-environmental justice, B-phthalates, A - exposure measurement

TU-SY-B1-253

Characterization of Seafood Consumption and Mercury Exposure in Chicago Asian Communities

M. Turyk[1], J. Zhang[2], Y. Liu[1], Z. Xiao[2], N. Chavez[3], H. Liu[4], S. Buchanan[2]; [1]University of Illinois at Chicago, Chicago, IL, [2]University of Illinois at Chicago, Chicago, IL, [3]University of Illinois at Chicago, Chicago, IL, [4]Midwest Asian Health Association, Chicago, IL

Abstract: Methyl mercury (MeHg) crosses the placenta and has been shown to affect neurodevelopment as well as cardiovascular risk in adults. The Asian population in the U.S. has elevated blood mercury compared with the non-Asian population due to higher frequency of fish consumption. East and Southeast Asians in Chicago, Illinois are a growing population that may have elevated health risks from fish consumption. We conducted a community assessment of 325 Chinese, Koreans and Vietnamese in Chicago consisting of a survey of fish consumption and cultural dietary practices, hair sampling for Hg, and analysis of fish purchased in Chicago markets for Hg and PCBs. Mean age was 44 years (range 18, 80), 84% were female, 49% had an annual income of <$20,000, 96% were born outside of the US and 42% spoke only non-English at home. Fish consumption rates differed by ethnic group, with 67% of Koreans consuming fish 3 or more days/week compared to 43% of Vietnamese and 37% of Chinese participants. Seafood consumption preferences varied among the ethnic groups and included species with a range of contaminant levels. Salmon, tilapia, bass, pompano and tuna were the most popular among Chinese; while salmon, catfish, pompano, mackerel and tilapia; and anchovies, tuna, mackerel, salmon and croaker were top choices among the Vietnamese and Korean, respectively. Elevated hair Hg levels (>1ug/g) were detected in 18% of Chinese, 13% of Vietnamese and 11% of Korean participants. Risk factors for elevated hair mercury included frequent fish consumption, recent immigration to the U.S. and traditional Asian dietary patterns. Species of fish and cultural fish consumption practices that are associated with elevated hair mercury and with average daily Hg and PCB intake will be presented. These findings will be used to develop messages to decrease exposure to contaminants while maintaining consumption of healthy nutrients in fish in these communities.

Keywords: B-metals, A-biomonitoring, A-epidemiology, C-food, D-community

TU-SY-C1: Integrating Community Engaged Research into the Exposure Science Paradigm

TU-SY-C1-254

Community assessment of exposure to traffic-related air pollution in near-highway neighborhoods

J. L. Durant[2], W. Zamore[3], E. Reisner[3], D. Brugge[1]; [1]Tufts University, Medford, MA, [2]Tufts University, Medford, MA, [3]Somerville Transportation Equity Partnership, Somerville, MA

Abstract: A community-engaged process was used to investigate exposure to ultrafine particles and other air pollutants near highways in and around Boston (MA, USA). Community partners assisted with study design, data collection, data interpretation, and reporting the results to the communities. The process led subsequent phases of the investigation in unexpected directions including indoor air quality interventions, public ordinances and additional monitoring of air pollution emission from traffic tunnels and airplanes.

Keywords: A - ambient monitoring, A-exposure models, B-particulate matter, C-air, D-community

TU-SY-C1-255

Citizen Science in Exposure Assessment: From Ecosystem to Biomarkers of Environmental Health

S. Croisant[1], C. Elferink[2]; [1]The University of Texas Medical Branch, Galveston, TX, [2]The University of Texas Medical Branch, Galveston, TX

Abstract: The UTMB Center for Environmental Toxicology (CET) supports translational environmental health research to improve early detection, prevention, and/or therapeutic strategies for environmentally-related disorders (Biomarkers and Analytical Diagnostics) and Community-Based clinical and public health research. We embrace involvement of impacted communities in our research projects and have employed or are employing Citizen Science in 1) a longitudinal study of Gulf of Mexico seafood which actively involved Gulf Fisherfolk in collection of samples, 2) a large-scale study of the coastal waters microbiome as a sentinel for environmental disturbances, and most recently 3) plans to integrate First Responders to capture environmental samples during Disaster Events. The CET utilized a Community-Based Participatory Research approach to establish a consortium and research proposal to understand the long-term health effects of exposure to seafood contaminated by the spill. The Gulf Coast Health Alliance: health Risks related to the Macondo Spill study utilized fishermen from across the Gulf States to develop a sampling schema and procure samples of a variety of seafood to characterize risk and develop and disseminate a seafood consumption calculator. The CET is building upon this work to further understand the impact of oil spills by establishing a baseline of diversity and species composition in microbial communities, microscopic populations of bacteria, fungi, algae, and other microorganisms, in near-shore Gulf of Mexico environments, and to monitor changes associated with oil pollutants. The CET

Disaster Research Response (DR2) program is a multi-faceted program to fast-track disaster research by developing partnerships with first responders, emergency management agencies, the community, and public health providers to enable rapid mobilization of research in the event of a disaster or an infectious disease outbreak. To meet DR2 objectives the CET is developing a roadmap for synergistically integrating its DR2 goals with first responder programs to explore opportunities to utilize existing Emergency Response framework and personnel and citizen scientists to capture samples and data as an event is unfolding.

Keywords: A-emergency response, D-community, A - population exposure, A - exposure measurement, C-food

TU-SY-C1-256

Community-engaged Assessment of Exposures to Particulate Matter in Imperial, CA Using a Network of 40 Community-operated Air Quality Monitors

E. Seto; University of Washington, Seattle, WA

Abstract: A community-engaged process was used to establish a network of 40 new air quality monitoring stations in Imperial, CA. Imperial Valley has the highest rates of hospitalizations and emergency visits for asthmatic school children in California. Poor air quality may be a contributing factor to this disproportionate disease burden, which has PM episodes that exceed PM2.5 and PM10 standards. There are many questions that residents have related to PM sources and whether hotspots exist where susceptible populations live. To address these questions, a community-engaged Research to Action process was implemented, in which three partners - The California Environmental Health Tracking Program, Comite Civico del Valle, and the University of Washington and other academics - collaborated on steps to improve community knowledge, participation, and action towards improving air quality. First, a Community Steering Committee (CSC) and Technical Advisory Group (TAG) were formed. Second, vulnerable areas were identified for data collection. Third, community input was gathered to identify specific siting locations for new air monitors. Fourth, 40 community air monitors were deployed. Fifth, data from the monitoring network was calibrated and validated in collaboration with the state air quality agency. Sixth, the IVAN Air website was developed to provide real-time air quality information to residents. The project is ongoing, and current community engagement activities are focused on working with the CSC, residents, and the TAG to collectively interpret the data collected in this project, and identify mitigations to reduce emissions and exposures. Furthermore, work is ongoing to sustain the monitoring network beyond the life the current grant. Already there has been considerable grassroots growth in interest in our community engagement process and monitoring approach from other community groups in California, which we hope to train and help establish similar projects in the future.

Keywords: A - ambient monitoring, A - population exposure, A-environmental justice, A-sensor technology

TU-SY-D1: The Role of Product Use Information in Quantitative Exposure Analyses

TU-SY-D1-257

Development and application of temporal consumer product use survey data to inform residential and aggregate exposure modeling

B. M. Young[1], M. Crowley[2], J. H. Driver[3], J. Blattner[4]; [1]Bayer CropScience, RTP, NC, [2]US EPA, Washington, DC, [3]risksciences.net, LLC, Manassas, VA, [4]SC Johnson, Racine, WI

Abstract: As part of the review of new pesticides or existing pesticides, EPA estimates exposure for consumer use pesticide products to determine whether they can be used safely according to the product label instructions. EPA has developed standard methodologies (SOPs), organized by exposure "scenarios" (e.g., lawns, pets) which aid in streamlining the risk assessment process and ensuring consistent regulation. The Residential Exposure Joint Venture (REJV) is a consortium of pesticide industry companies formed to address residential pesticide usage in the U.S. The REJV National Pesticide Use Survey (2012-2013) was implemented in response to an EPA Data Call-In (DCI). The survey was intended to address gaps in understanding residential pesticide usage, particularly related to co-occurrence of chemical active ingredients and/or pesticide products across residential scenarios. In its 2016 review, EPA concluded that the survey data are acceptable and reliable to support human health exposure and risk assessments related to pesticide registration decisions. The final survey consists of 4,573 unique U.S. households' 12 month diaries of residential pesticide use. Survey respondents entered product characteristics such as the EPA Registration Number and product name, the date and time of application, who made the application and whether other household members were present. Respondents also recorded various application characteristics, including site of application (e.g., on oneself, on a pet, to a garden); how the application was made (e.g., with an aerosol can, backpack sprayer, pet collar); and the type of application (e.g., broadcast, spot treatment, air/space spray). For each application record in REJV - across all households there are approximately 155,000 application records - the application characteristics can be "mapped" to the EPA residential SOP scenarios. Thus, the REJV survey provides a contemporary, temporal, i.e., calendar year, national pesticide use survey as a robust source of product use information to inform scenario-specific, aggregate, cumulative analyses for residential exposure assessments (deterministic and probabilistic).

Keywords: A-exposure models, A-longitudinal metrics, B-pesticides, C-consumer products

TU-SY-D1-258

Estimating concentrations of soil fumigants in ambient air using an atmospheric dispersion model and agricultural product use information to inform bystander exposure

I. Van Wesenbeeck, S. Cryer, O. deCirugeda Helle; Dow AgroSciences, Salem, OR

Abstract: Agricultural product use information is critical for parameterizing air dispersion models for simulating concentrations of volatile soil fumigants, such as 1,3-Dichloropropene (1,3-D), in air. 1,3-D is a multi-purpose fumigant for preplant treatment of soil to control plant parasitic nematodes, and to manage soil borne diseases in high-value crops in the USA and globally. The volatility of 1,3-D allows it to diffuse readily through soil pores for maximum pest control, however it also results in some evaporation from the soil into the atmosphere, where bystanders can potentially be exposed. Bystander exposure to atmospheric contaminants is dependent on their proximity to the source over time, their activity patterns, and the source 'strength' which is dependent on the volume of product applied, the application area, timing and the products volatility. Concentrations of 1,3-D in ambient air are modeled using SOFEA v3.0, an air dispersion modeling framework that simulates ambient air concentrations resulting from multiple 'real world' soil fumigations in a use region, and relies on AERMOD as the air dispersion model. Fumigant use parameters required by the model include application rate, date, depth and time as well as the spatial location and area of the treated field. The agricultural regions selected for modeling include areas in California, North Carolina, Washington, and Florida representing diverse agronomic use scenarios for 1,3-D soil fumigant in the USA. 1,3-D application data collected in each region is used to parameterize SOFEA which then simulates and generates outputs of annual average and 72h 1,3-D concentrations over a wide grid of receptors across each study region. The distribution of 1,3-D concentrations are used in conjunction with exposure factors, including residential mobility, time activity patterns, and age-specific inhalation rates and body weights, to assess potential long-term bystander exposure to 1,3-D in ambient air.

Keywords: A - ambient monitoring, C-air, B-pesticides, A - exposure measurement, A-risk assessment

TU-SY-D1-259

Residential and Population Generator (RPGen): Creating a detailed description of housing and occupants for predicting and describing chemical exposures

G. Glen[1], H. Hubbard[1], J. Levasseur[1], P. Price[2], K. Dionosio[2], D. A. Vallero[2], P. P. Egeghy[2]; [1]ICF, Durham, NC, [2]EPA, Research Triangle Park, NC

Abstract: The US EPA's Human Exposure Model (HEM) is a modular, stochastic model that simulates populations, assigns product use, and calculates direct and indirect exposure to chemicals in consumer products. The Residential and Population Generator (RPGen) module of HEM randomly samples empirical data from the Public Use Microdata Sample (PUMS) data, the American Housing Survey (AHS) and the Residential Energy Consumption Survey (RECS) to create a simulated population of primary individuals together with their household structures and housing characteristics. The PUMS, AHS, and RECS databases are linked by housing type, region of the country, local population density, household income, and family structure. Household characteristics focus on those that impact exposure to consumer products (e.g., types of appliances, number of cars, number of bathrooms, and size of house). Characteristics of family structure include the number of children, number of adults, and age and sex of each member. RPGen also assigns physiological variables (e.g. body weights and surface areas) to each primary individual using algorithms developed for EPA's "httk" R package. RPGen produces an output simulated population of primary individuals with associated geographic, demographic, housing, and physiological variables. This output can potentially be used in a variety of exposure models that require large, representative sample populations. In HEM, RPGen output drives an agent based model (ABM) which considers region, seasonality, and communal product use to develop the universe of consumer products to which the primary individual may be exposed. A series of rules based on the demographic and housing characteristics provided by RPGen further customize the consumer product use for each household. RPGen is flexible for use as a stand-alone module to produce detailed descriptions of synthetic populations, or for use with existing exposure models that require a large population for exposure calculations.

Keywords: A-indoor environment, A-exposure models, C-consumer/personal care products, A-exposure factors, Human Exposure Model

TU-SY-E1: Cheminformatics Tools to Support Exposure Analysis, Data Aggregation, and Modelling

TU-SY-E1-260

Exposure Science in The Comparative Toxicogenomics Database: Linking Chemical Stressors to Outcomes via an Exposure Ontology

C. Mattingly, C. Grondin, A. Davis, J. Wiegers, T. Wiegers; NC State University, Raleigh, NC

Abstract: The Comparative Toxicogenomics Database (CTD; http://ctdbase.org) is a freely available resource that provides manually curated information on chemical, gene, phenotype, and disease relationships to further our understanding of environmental exposures on human health. Four primary modules are independently curated into CTD, including chemical-gene interactions, chemical-disease and gene-disease associations, chemically-induced phenotype relationships, and environmental exposure data. The latter includes exposure measurements and effects on human populations (receptors). Approximately 50 characteristics of exposure studies are curated including nformation about exposure sources, population demographics, measured media and levels, locations and outcomes using standardized terms from the exposure ontology (ExO) and additional controlled vocabularies. Integration of these data into CTD provides a centralized, searchable repository of exposure data that facilitates meta-analyses and informs study design by allowing comparisons among experimental parameters. To date, over 800 unique chemical stressors and 500 disease/phenotype outcomes have been described (from over 1500 articles), and these data are linked to over 1.7 million chemical-gene-disease interactions and 80,000 chemical-phenotype interactions in CTD. Analysis tools in CTD reveal direct and inferred relationships among the data, and help generate, interpret and refine hypotheses relating to chemically-influenced diseases. CTD's centralization of exposure science data, integration with chemical-gene, disease and phenotype modules, and additional analysis tools provide a unique resource to advance our understanding of the molecular mechanisms of action of environmental exposures and their effects on human health.

Keywords: A-cumulative exposure, A-epidemiology, A-exposure factors, D-occupational

TU-SY-E1-261

Exploring environmental chemical space through HR/AM mass spectrometry and cheminformatics: The example of wastewater-derived organic micropollutants

L. Ferguson[1], G. J. Getzinger[2]; [1]Duke University, Durham, NC, [2]ETH Zurich, Zurich, Switzerland

Abstract: High-resolution accurate-mass mass spectrometry (HR/AM MS) provides a promising approach for comprehensive structural elucidation of numerous trace-level organic contaminants in environmental media. However, comprehensive sample characterization is made difficult by inefficiencies in the prioritization of detected features and postulated structures based on tandem MS analysis. We have developed a holistic, non-targeted screening workflow to address these challenges, which combines ultra-high resolution tandem mass spectrometry, computational mass spectrometry. We illustrate the utility of this approach to evaluate the occurrence of organic micropollutants in the effluent of a conventional domestic wastewater treatment facility. Composite effluent samples from the North Durham Wastewater Reclamation Facility (Durham, NC, USA) were analyzed using an Orbitrap Fusion tribrid tandem HR/AM mass spectrometer. Resulting data were aligned and componentized to yield 3,701 unique features, for which >91% had associated tandem mass spectral data. Custom data processing scripts were deployed to pipe relevant feature data to computational mass spectrometry tools for molecular formula assignment based on isotope pattern and fragment spectrum decomposition (SIRIUS, formulas assigned to 90.2% of features), structure assignment from the PubChem compound repository (postulated structures returned for >77% of features) and automated tandem mass spectral annotation and scoring (MetFrag CL, MAGMa, and CFM-ID). Finally, postulated structures were scored based on their structural similarity to possible environmental contaminants (i.e., compounds known to be produced in commerce) and to known wastewater pollutants. Structure descriptors were calculated using 42-molecular quantum numbers (MQN), a unique descriptor set that implements simple atom, bond and topology counts. After rigorous data filtering, 607 micropollutants were tentatively identified, representing classes of pharmaceuticals, transformation products, consumer chemicals, and industrial compounds. Overall, our results highlight the benefit of combining of state-of-the-art computational mass spectrometry tools with chemoinformatic approaches for increased efficiency and annotation rate in non-targeted analysis of built and natural environments.

Keywords: A-analytical methods

TU-SY-E1-262

New databases and cheminformatics tools for exposomics and toxicology

Y. Djoumbou Feunang; University Of Alberta, Edmonton, Canada

Abstract: The contribution to the human chemical exposome is determined by the exposure to exogenous chemicals (e.g. drugs, pollutants), and endogenous chemicals (e.g. lipids, hormones) that are produced or modified in response to external stressors. Nowadays, several freely accessible databases provide cheminformatics capabilities in addition to various types (e.g. toxicological, biochemical, spectral) of data for known chemicals. Unfortunately, the heterogeneity and variable incompleteness of the available data, even for known compounds, significantly hamper the assessment of the human exposome, as well as its effects on human health and the environment. Furthermore, only <2% of the peaks in high-resolution mass spectra collected from environmental or biological samples can be matched to any of the >100 millions molecules available electronically. This suggests that a significantly large portion of the human chemical exposome is populated by unknown structures. It is expected that most them arise from single or multi-step transformations of known molecules. Thus, understanding the metabolism of these chemicals could not only improve compound discovery and identification rates, but also provide a better assessment of the human exposome. We have developed a number of freely accessible cheminformatics tools that would help address these challenges and study the correlations along the alternate sides of the triangular structure-metabolism-spectra relationship. In this presentation, I will describe: 1) ClassyFire, a tool for automated structure-based taxonomic classification of chemicals; 2) BioTransformer, a software tool for predicting metabolic biotransformation products arising from human metabolism and environmental microbial degradation; 3) CFMID+, a software tool designed to accurately predict mass spectra for rapid compound identification; and 4) ContaminantDB, a comprehensive electronic database of nearly 100,000 known chemical contaminants.

Keywords: A - exposure measurement, B-pharmaceuticals, C-consumer/personal care products, Cheminformatics, Contaminants, Metabolism, Ontologies, MS Spectrometry, Computational metabolomics

TU-SY-E1-263

The US EPA CompTox Chemistry Dashboard - an integrated data hub for environmental chemistry

A. Williams[1], C. Grulke[1], A. McEachran[1], K. Dionosio[2], K. Phillips[2], K. Isaacs[2], J. Wambaugh[1], R. Judson[1], A. Richard[1]; [1]US Environmental Protection Agency, Durham, Research Triangle Park, NC, [2]US Environmental Protection Agency, Durham, Research Triangle Park, NC

Abstract: The U.S. Environmental Protection Agency (EPA) is integrating advances in biology, chemistry, and computer science to help prioritize chemicals based on potential human health risks. As an outcome of these efforts the National Center for Computational Toxicology (NCCT) has measured, assembled and delivered an enormous quantity and diversity of data for the environmental sciences including high-throughput in vitro screening data, in vivo and functional use data, exposure models and chemical databases with associated properties. Over the past decade, these data have been delivered as a series of software applications and databases. Recent work has focused on developing a new architecture that assembles the resources into a single platform. With a focus on access to Open Data streams, web service integration accessibility and a user-friendly web application, the CompTox Chemistry Dashboard provides access to data associated with ~750,000 chemical substances. These data include research data in the form of bioassay screening data associated with the ToxCast program, experimental and predicted physicochemical properties, product and functional use information and related data of value to environmental scientists. This presentation will provide an overview of the CompTox Chemistry Dashboard and its value to the community as an informational hub. *This abstract does not reflect U.S. EPA policy.*

Keywords: A-chemical prioritization, A-exposure models, A-risk assessment, C-consumer products, Cheminformatics, Online database

TU-SY-F1: Applying exposure science to increase the utility of *in vitro* data in efficacy and safety testing: research needs to support regulatory decision making

TU-SY-F1-264

Incorporating exposure driven approaches and in vitro data into regulatory decision making

N. Kleinstreuer; NTP Interagency Center for the Evaluation of Alternative Toxicological Methods (NICEATM), Durham, NC

Abstract: Toxicology in the 21st century (Tox21) has ushered in a scientific revolution focusing on novel alternative approaches to understanding chemical hazard and safety, with an emphasis on non-animal high-throughput screening (HTS) methods, adverse outcome pathways, and computational systems models to predict effects on human biology. Computational toxicology approaches are being combined with mechanistic frameworks to build biological pathway-based models integrating diverse streams of in vitro and in silico data. In vitro to in vivo extrapolation (IVIVE) is a critical tool for regulators to translate these in vitro bioactivity concentrations and in silico predictions into estimated in vivo doses, and compare those to relevant exposures, both measured and modelled. IVIVE techniques are being applied in a wide array of settings, from regulatory decision making in the Endocrine Disruptor Screening Program to researching the potential for life-stage specific effects of vascular disrupting compounds in the environment. The NTP Interagency Center for the Evaluation of Alternative Toxicological Methods (NICEATM) works with sixteen U.S. federal agencies with an interest in non-animal testing, via the Interagency Coordinating Committee on Validation of Alternative Methods (ICCVAM), and focuses on research efforts driven by agency priorities, both regulatory and scientific in nature. As a Tox21 partner, NICEATM applies computational toxicology techniques to build predictive models of chemical impacts on human health and disease pathways, and contextualize these predictions with integrated bioactivity exposure relationship estimates. Recent initiatives focus on the development of online tools in an "Integrated Chemical Environment" to make NICEATM analysis workflows (such as IVIVE) publicly available and provide datasets that are transparent, computable and searchable for model-building and evaluation.

Keywords: Tox21, high-throughput screening

TU-SY-F1-265

Physiologically-based kinetic modelling in risk assessment - reaching a whole new level in regulatory decision-making

C. Tan[1], A. Paini[2]; [1]US EPA, Durham, NC, [2]JRC, Ispra, Italy

Abstract: The European Union Reference Laboratory for Alternatives to Animal Testing (EURL ECVAM) hosted a two-day workshop with an aim to discuss the role and application of Physiologically Based Kinetic (PBK) models in regulatory decision making. The EURL ECVAM strategy document on Toxicokinetics (TK) outlines PBK models as a central feature for their potential use to integrate data generated by *in vitro* and *in silico* methods for absorption, distribution, metabolism, and excretion (ADME) in humans and to predict whole-body biokinetic behaviors. Experts were invited to identify, discuss and recommend challenges in application of PBK models to support regulatory risk assessment. The two primary challenges were i) constructing models that rely on *in vitro* and *in silico* methods for parameterization; (ii) assessing model credibility when *in vivo* kinetic data are not available for model evaluation how to validated or calibrated models with non-animal data. To address these challenges, the experts recommended a best practices workflow for guidance on the use of *in vitro* and *in silico* data in PBK models designed to support regulatory decision making. The outcomes of the workshop and recommendations of the experts will be summarized in this presentation. The presentation will also show results from an international survey on application of PBK models in science and regulatory submission. Analysis of the survey data will provide insights into key concerns in the PBK modeling community, so that recommendations can be made to promote the development and acceptance of PBK models in the safety assessment of chemicals. Disclaimer: This abstract has been cleared by the EPA but solely expresses the view of the author.

Keywords: A-risk assessment

TU-SY-F1-266

A road map for exposure-driven non-animal approaches

N. Gellatly, N. Burden, F. Sewell; NC3Rs, London, United Kingdom

Abstract: There are increasing numbers of non-animal approaches for generating data which can inform us about the biological effects of drugs and chemicals. This information has the potential to be useful in a wide range of settings from screening and prioritisation of candidate molecules to regulatory assessment.

Better consideration of exposure will be key to maximising the usefulness of this non-animal data in decision-making. This includes improved understanding of the likely exposures to both humans and the environment which can inform dose-setting and the interpretation of dose-response relationships. There is also an increasing appreciation of the need to better characterise the bioavailable rather than just nominal (or applied) concentration of a test material in an *in vitro* assay, and how this relates to the concentration that might be achieved at a target organ in real life. In these ways, exposure science will enable data generated using non-animal approaches to be more useful for decision-making purposes, supporting a reduction in the reliance on data from animal studies. In February 2017, together with Unilever, the NC3Rs hosted a scientific workshop which brought together a range of scientific disciplines in order to explore how exposure science can increase the utility of non-animal data in efficacy and safety testing and assessment. The workshop aimed to (1) increase awareness and build confidence in the application of exposure-driven approaches to support decision-making based on data from non-animal approaches across sectors; (2) build a community of scientists working in the area of exposure-driven safety assessment; (3) Identify gaps and challenge areas that need to be addressed to advance the application of exposure science capability; (4) Develop and publish guidance to facilitate the use and wider acceptance of exposure-driven non-animal approaches to inform and improve safety and efficacy decision-making.

Keywords: A-risk assessment, A-exposure models

TU-SY-G1: Responding to PFAS Contamination Across Multiple States: A Sticky Situation - Non-Stick Chemicals in Drinking Water, Biota, and Humans

TU-SY-G1-267

PFAS 101, and PFAS in a Prized Fishery

C. Bush; Mich Dept Health & Human Services, Lansing, MI

Abstract: Per- and polyfluorinated alkyl substances (PFAS) are manmade chemicals used in many consumer products and commercial and industrial applications. There are hundreds to thousands of individual chemicals within the PFAS “family.” PFAS releases occur during manufacturing processes and as a result of everyday use. Due to their chemical structure, PFAS do not easily degrade and are therefore persistent in the environment. Many of them are also bioaccumulative and have long-half-lives (years) in humans. In 2016, the U.S. Environmental Protection Agency issued Lifetime Health Advisory levels (LTHA) for two PFAS in drinking water - perfluorooctanoic acid (PFOA) and perfluorooctane sulfonate (PFOS). The LTHA, a non-regulatory number, is 70 nanograms per liter or parts per trillion, and is applied to the combined concentration of PFOA and PFOS. The LTHA protects against developmental effects, liver and kidney toxicity, and cancer. Some states have issued their own drinking water values, some of which are much lower than the LTHA. Some states have also set fish consumption guidelines, due to PFOS being highly bioaccumulative. In 2012, the Michigan health department issued a “do not eat” advisory for fish in several waterbodies near the former Wurtsmith Air Force Base (WAFB) in Oscoda, Michigan, due to very high PFOS levels detected in fish filets. The contamination is from use of PFAS-containing fire-fighting foam at WAFB. Since 2015, drinking water wells downgradient from WAFB have been shown to be impacted by the PFAS coming from the base. Oscoda’s economy relies heavily on tourists and anglers who use the abundant natural resources in the area. The PFAS contamination has stigmatized the town, potentially hurting local businesses and property values and worrying community members and WAFB veterans concerned about their health. The contamination investigation is ongoing, and the Michigan health department, along with other agencies, continues to provide outreach and public health support.

Keywords: C-water, A-environmental regulation, C-food, A-risk assessment

TU-SY-G1-268

PFAS Blood Testing in New Hampshire: An Evolving Response to Community Health Concerns

J. Sagona[1], K. Bush[1], C. Bean[1], A. Cosser[1], M. Levesque[1], B. Kernen[2], N. Whitley[1], J. Parrott[1], E. Daly[1], B. Chan[1]; [1]New Hampshire Department of Health and Human Services, Concord, NH, [2]New Hampshire Department of Environmental Services, Concord, NH

Abstract: Per- and polyfluoroalkinated substances (PFAS) are contaminants of emerging concern due to their widespread use in industry and in stain and water resistant commercial products. They are persistent in the environment and have a long half-life in humans. New Hampshire has identified multiple communities with PFAS-contaminated public and private drinking water supplies. Community demand and decisions by elected and public health officials have led to a biomonitoring program, coordinated by multiple state agencies, to provide individuals with PFAS blood levels and inform these communities about exposure levels. Biomonitoring began in April 2015 at the Pease Tradeport, a former U.S. Air Force base, and has expanded to include an area of southern NH (SNH) and, now, people served by the Merrimack Village District public water system. In 2015, 1,578 individuals exposed at Pease Tradeport had their blood tested for 11 PFAS compounds. Since July 2016, another 612 individuals from Pease and the two additional communities have had PFAS blood testing. Individuals exposed at Pease Tradeport had higher geometric mean (GM) and 95th percentile levels for perfluorooctanoic acid (PFOA), perfluorooctane sulfonic acid (PFOS), and perfluorohexane sulfonic acid (PFHxS) compared to U.S. background levels; individuals in SNH had comparatively higher GM and 95th percentile levels for PFOA only. Results so far have been consistent with identified PFAS drinking water contamination. Interpretation of test results, however, remains difficult and there is frustration with the uncertainty about how levels may relate to past, present or future health problems; thus, while biomonitoring provides useful information about levels of exposure, individual test results have limited ability to inform healthcare decisions. Biomonitoring should be conducted as a scientifically based community exposure assessment and should not be viewed as a way to inform individuals about adverse health risk.

Keywords: A-biomonitoring, A - exposure measurement, C-water, D-community

TU-SY-G1-269

Using a representative sampling approach to conduct biomonitoring in a large community

W. Hsu, Q. Gao, S. Forand, E. Lewis-Michl, S. Hwang; New York State Department of Health, Albany, NY

Abstract: The New York State Department of Health offered blood testing for perfluorooctanoic acid (PFOA) to residents in the Hoosick Falls area beginning in February 2016. A total of 2,908 participants were tested including people using Hoosick Falls Village public water, people using private wells, people who work in the area, and former residents. We identified 1,262 participants currently using Village water within 663 households. 2010 Census data show Hoosick Falls has 3,501 residents and 1,434 households. The sociodemographic distribution of our 1,262 participants was similar to that of the 2010 Census. We performed a single-stage cluster sampling where a random sample of households was drawn and every individual in the selected household was included. We repeated this process 100 times for 5% and 3% of the population, which is equivalent to 73 and 44 households (175 and 105 individuals) based on the 2010 Census. We also conducted simple random samples of 175 and 105 individuals. The geometric mean (GM) based on all 1,262 participants is 41.56 mg/L. The GMs and their 95% confidence intervals (CIs) from the single-stage cluster sampling are 41.31 mg/L (40.44, 42.20) and 40.57 mg/L (39.50, 41.66) at a sample size of 5% and 3% of the entire Village population. The random samples of 5% and 3% resulted in the GMs of 39.92 mg/L (39.42, 40.41) and 39.90 mg/L (39.26, 40.55), respectively. The results indicate that a statistically based sampling technique accurately generalizes our results from the overall population. Because the PFOA test is a very expensive and specialized test and few laboratories conduct PFOA blood testing, sampling approaches could save time, money and resources. In our case study, we could save approximately 95% in costs by using representative sampling. However, if the population is heterogeneous by certain factors, such as occupation or exposure levels stratification by these factors would be necessary for accurate estimates.

Keywords: C-water, D-community, A-biomonitoring, A-sampling methods

TU-PL-A2: Air Pollution and Noise

TU-PL-A2-270

A Combined Emission and Receptor-Based Approach to Modelling Environmental Noise in Toronto, Canada

T. H. Oiamo; Ryerson University, Toronto, Canada

Withdrawn

TU-PL-A2-271

Options and strategy to disentangle the cardiovascular health effects of noise and traffic related particulate matter

L. Dekoninck[1], L. Int Panis[2], D. Botteldooren[1]; [1]Ghent University, Ghent, Belgium, [2]VITO, Mol, Belgium

Abstract: Traffic emits noise and air pollution simultaneously. The physical link in the emission is a fundamental issue and mutual confounding of the health effects is expected. The epidemiological practice requires the identification of subpopulations with contrasting exposure to disentangle health effects. We summarize the options to identify the subpopulations and present a strategy to achieve this in practice. A first contrast is the physical difference in the vehicle noise and particulate matter emission. PM emission is related to the engine regime and peaks in dense and dynamic traffic while noise emission reaches its maximum at high speed on major roads but drops at low speed. The second contrast is based on the interpersonal variability of the mobility pattern. A third contrast is found in the differences in meteorological influences on the exposure to noise and PM. A fourth contrast is the difference in the diurnal exposure patterns. Sleep disturbance is a strong candidate for cardiovascular health effects and occurs off-peak during an activity with a low metabolic rate. An important uncertainty is the partially entangled outdoor to indoor correction for both burdens due to interactions between insulation, open windows and ventilation. The next layer is the internal exposure. Inhalation and size sensitive corrections for PM are required to assess the impact of exposure on in-body and organ specific mechanisms. For noise, the interacting biochemical and psychological components of annoyance, sleep disturbance, sensitivity and perception add a similar complexity in this layer. Important efforts are ongoing to identify noise parameters to include these aspects. While the listed aspects disentangle the exposure, the partially shared biomarkers of noise and PM entangle the assessment anew. Full integration of the multidisciplinary knowledge combined with simultaneous exposure assessments is a prerequisite to provide the required biomarker specific exposure indicators.

Keywords: A - population exposure, A-activity patterns, B-particulate matter, A-behavior, A-geospatial analysis/GIS

TU-PL-A2-272

Can exposure surfaces and GPS data predict personal exposures to air pollution and noise? Findings from a panel study

L. Minet[1], R. Liu[1], M. Valois[2], J. Xu[1], M. Shekarrizfard[1], S. Weichenthal[2], M. Hatzopoulou[1]; [1]University of Toronto, Toronto, Canada, [2]McGill University, Montreal, Canada

Abstract: Exposure surfaces developed from land-use regression (LUR) models are commonly used to quantify the risks of various air pollutants and noise on public health. Typically, exposures are associated with the home location of participants in a cohort. While a few studies have begun to incorporate data on the daily mobility to refine exposure estimates, often, the “mobility-based” exposures are not validated against personal data. During summer 2016, we conducted a panel study in the city of Toronto, Canada, which involved 46 participants who participated on two days. Participants were provided with instruments measuring levels of Ultrafine Particles (UFP), Black Carbon (BC) and noise, as well as a GPS. Over a 6-hours duration, they were free to pursue their daily activities with the constraint of walking for 2 hours outdoors. During the same summer a data collection campaign took place whereby the levels of UFP, BC and noise were measured in Toronto with the same devices via two different protocols. A mobile monitoring campaign using bicycles enabled the coverage of 3,895 unique road segments visited 5 to 6 times. In addition, 92 fixed points were also sampled 5 to 6 times. These simultaneous monitoring campaigns enabled the development of LUR models and associated exposure surfaces. The GPS coordinates of all participants were synchronised with the air pollution and noise data to generate estimates of personal exposures. These estimates were analyzed in terms of spatial variability throughout the city, and differences between indoor and outdoor levels. In addition, GPS data were intersected with the surfaces to derive “mobility-based” exposures. We are currently comparing the personal and “mobility-based” exposure estimates and identifying locations, and trajectories with strong and poor agreement between the two measures. We are also developing models that can correct “mobility-based” exposures using information on daily levels measured at a central location.

Keywords: A - exposure measurement, A-activity patterns, A - population exposure, A-exposure models, A-epidemiology

TU-PL-B2: Environmental Justice

TU-PL-B2-273

Social and environmental determinants of disaster preparedness among U.S. adults

C. Ekenga, Z. Lan; Washington University in St. Louis, St. Louis, MO

Abstract: Background: Epidemiologic studies have identified several sociodemographic risk factors for disaster-related morbidity and mortality, including low socioeconomic status and racial/ethnic minority status. Disaster preparedness can improve an individual’s ability to cope with the health effects of environmental disasters and public health emergencies. This study evaluated the relationship between sociodemographic, health, and environmental factors and disaster preparedness among U.S. adults. Methods: Data from the U.S. CDC’s 2006-2012 Behavioral Risk Factor Surveillance System (BRFSS) surveys were used to estimate the prevalence of four household preparedness items (3-day supply of water, 3-day supply of nonperishable food, a battery-operated radio with working batteries, a flashlight with working batteries). Multivariable logistic regression, accounting for the complex sampling design of the BRFSS, was used to evaluate associations between individual factors (age, gender, race/ethnicity, education, marital status, income, employment status, chronic disease status, perceived health, urban-rural residence) and household preparedness. Results: The study sample consisted of 96,137 adults aged 18 years and older who responded to questions about household preparedness. Overall, 42% (95% CI=40, 44%) of the sample reported all four preparedness items. Older participants and those with at least some college or technical school education were more likely to be prepared. Women participants (OR = 0.71, 95% CI=0.67, 0.75), participants with fair or poor perceived health (OR = 0.81, 95% CI=0.74, 0.88), and urban residents (OR = 0.91, 95% CI=0.85, 0.96) were less likely to report household preparedness. There were no differences between racial/ethnic groups. Conclusions: Our study identified subgroups with low levels of household preparedness. Further work is needed to identify and implement disaster preparedness interventions for these vulnerable populations.

Keywords: A-environmental justice, D-susceptible/vulnerable, A-emergency response, A-epidemiology

TU-PL-B2-274

Assessing Environmental Justice and Risk-based Coronary Heart Disease (CHD) Burden in Allegheny County, PA Using High-Resolution GIS-based Traffic-related Air Pollution Exposure Modeling.

J. P. Fabisiak[1], E. M. Jackson[2], A. Presto[3]; [1]University of Pittsburgh Graduate School of Public Health, Pittsburgh, PA, [2]University of Pittsburgh, Pittsburgh, PA, [3]Carnegie Mellon University, Pittsburgh, PA

Abstract: We examined if poor and minority residents in Allegheny County, PA were disproportionately exposed to black carbon (BC) and nitrogen dioxide (NO_2). Exposure estimates in census tracts were modeled via land use regression and then analyzed in relation to US Census data. Environmental Justice (EJ) tracts were defined as $\geq$ 20% of residents living in poverty and/or $\geq$ 30% of residents identifying as non-white. Exposure estimates for each pollutant were rank-ordered for all tracts in the county. Quartiles of increasing exposure (Q1 - Q4) were established and the location of EJ tracts noted along the continuum. The relative occurrence of EJ tracts in Q2-4 compared to Q1 progressively increased and reached a maximum in the Q4. For BC and NO_2, EJ tracts occurred 4 and 25 times, respectively, more often in Q4 relative to Q1. We applied a risk-based estimate to evaluate the distribution of CHD burden. First, we assigned the observed annual rate of CHD death in Allegheny County (144.7/100,000) to the tract with the median exposure value for each pollutant. That risk was then adjusted up and down in each ascending and descending tract according to the CHD mortality effect estimates determined by Gan *et al* (2010) (BC: RR of 1.06 / 0.752 μg/m^3 of BC, RR 1.04 /8.4 μg/m^3 NO_2). Pollution-attributable risk in excess of the lowest census tract was taken as the difference in risk between each tract and the lowest exposed tract and applied to the tract population to estimate CHD mortality burden. Countywide, an estimated 60 - 50 deaths were attributed to NO_2 and BC in excess of the lowest exposed tract. The population in all EJ tracts represents 28% of the county population, but they contain about 42% and 39% of the CHD mortality risk attributed to NO_2 and BC, respectively. EJ tract distribution is disproportionately skewed toward areas of high exposure to traffic-related pollutants and EJ residents may bear a greater risk for air pollution-related adverse health outcomes.

Keywords: A-environmental justice, A - population exposure, A-risk assessment, B-particulate matter, A-geospatial analysis/GIS

TU-PL-B2-275

School environment and links to student performance in an urban, mid-Atlantic Region

J. D. Berman[1], M. McCormack[2], K. Koehler[4], F. Connolly[3], M. Davis[4], F. C. Curriero[1]; [1]Johns Hopkins School of Public Health, Newington, CT, [2]Johns Hopkins School of Medicine, Baltimore, MD, [3]Johns Hopkins School of Education, Baltimore, MD, [4]Johns Hopkins School of Public Health, Baltimore, MD

Abstract: Conditions of school facilities, environment, and perceptions of safety and learning have individually been linked to academic performance. However, little research has considered the aggregate impacts of multiple school and community factors on student performance. We examined the association between school environment and community-level factors with academic achievement and attendance for grades 3-5 and 6-8 at 158 urban schools. Environmental characteristics included facility conditions, total length of nearby roads, and an EPA risk-screening environmental indicator (RSEI) based on emissions from the Toxic Release Inventory. Perceptions of school safety, learning, leadership, and institutional environment were acquired from a School Climate Survey. Neighborhood factors at the community statistical area, including demographics, crime, health, and SES, were acquired and assigned to schools based on location. Using results from the Maryland State Assessment (MSA), we considered outcomes of school-level performance in reading and math, plus attendance and chronic absence rates (ie. missing 20 or more days). Poisson regression adjusted for over-dispersion was used to model academic achievement and multiple linear models for attendance. Worsening facility and environmental conditions were associated with decreases in academic performance and attendance. Reading and math scores decreased by 1.0%-2.1% with worsening facilities, while attendance rates declined by 0.14%-0.26% and chronic absences increased by 0.70%-0.93% with each 1,000 unit increase in RSEI. Decreased perceptions of school safety resulted in a 1.1%-2.3% decrease in reading and math performance. All associations were robust to community measures of poverty and crime. These findings provide empirical evidence for the importance of the community and school environment, including building conditions and neighborhood toxic substance risk, on academic achievement and attendance measures.

Keywords: A-built environment, D-children, A-workplace, D-community

TU-SY-C2: Engaged Exposure Science and Epidemiology: Diverse Communities Driving Research

TU-SY-C2-276

40 Years and Three Generations: Science + Community Engagement + Public Health = Better Science and a Healthier Community

M. Marcus[6], J. Nyerges[1], M. Pearson[6], M. Cheatham[2], J. Keon[3], N. Keon[2], E. Lorenz[4], P. Muldoon[1], S. Schrader[5], J. Kesner[5], H. Barton[6], M. Terrell[6], A. Smith[6], K. Conneely[6], D. B. Barr[6], M. Marder[7], M. Jacobsen[6]; [1]PBB Citizens Advisory Board, Grand Rapids, MI, [2]Mid-Michigan District Health Department, Ithaca, MI, [3]Pine River Superfund Citizens Task Force, St. Louis, MI, [4]Alma College, Alma, MI, [5]NIOSH, Cincinnati, OH, [6]Emory University, Atlanta, GA, [7]California Environmental Protection Agency, Sacramento, CA

Abstract: The Michigan PBB Research Registry consists of ~6,000 people exposed to polybrominated biphenyls following their accidental introduction into Michigan's animal feed (1973). Multiple health problems found to be associated with this exposure include earlier puberty, more miscarriages, and breast cancer among women and more genitourinary problems among men. Research continued with the Registry for almost forty years with minimal community engagement, until recently, when the Registry was transferred to Emory University. The Emory researchers initiated community engagement activities with the goals of disseminating research findings to those affected, responding to community needs regarding education, incorporating community input into aspects of study design, and to facilitate long-term participation. These community engagement efforts have led to significant modifications of proposed outreach activities, research protocols, and the research priorities going forward. For example, the affected community's continued health concerns were added into study health questionnaires; the age range for study participation was expanded to include those exposed during childhood; additional exposed individuals were enrolled or re-enrolled into the Registry. Chemical workers and family members had been removed from the original Registry because of multiple exposures. In response to communtiy requests, the Emory team applied exposome methods (metabolomics) to evaluate their multiple exposures. Community concerns regarding multi-generational effects and continued high-levels of PBB body burden led to recently funded epigenetics study and clinical trial aimed at accelerating elimination of POPs. Community engagement efforts have evolved far beyond the original goals into a genuine community-based participatory research partnership, where the community has identified research questions, helped develop the research approach, and is an integral part of the research decision-making.

Keywords: B-flame retardants, Community Engaged Research, metabolomics, epigenetics, Community Engaged Research, metabolomics, epigenetics, Community Engaged Research, metabolomics, epigenetics, A-biomonitoring

TU-SY-C2-277

The Ramapough Nation: A Tribe in scientific crisis

J. Zelikoff[1], S. Cataldo[1], C. Mann[2]; [1]New York University School of Medicine, Tuxedo, NY, [2]Ramapough Lenape Tribal Nation, Ringwood, NJ

Abstract: Decades ago, the home to the Ramapough Lenape Turtle Clan was used as a dumping ground for tons of paint sludge and other toxic waste. Despite its Superfund status and extensive clean-ups, health problems including (but, not limited to) excessive incidences of cancer, cardiovascular disease, asthma and reproductive outcomes continue to run rampant in this Environmental Justice, Native American community. In response to community leader concerns, NYU COEC partnered with the Clan Chief and Tribal leaders to develop and implement a Concerns and Health Assessment Survey which queried Ramapough Native Americans and non-Native respondents concerning demographics, disease incidence, residence/workplace distance from contaminated sites and Indigenous sovereignty. Results revealed a high incidence of self-reported chronic illnesses in Native Americans compared to non-Natives. Center scientists and epidemiologists translated the findings to hypothesis-driven research to investigate exposure levels and the relationship between proximity to dumping with health outcomes. Chief Mann of the Turtle Clan will speak for the Clan regarding their "toxic legacy" and community-driven scientific research that has served to build capacity among the Tribe.

Keywords: D-community, D-environmental justice

TU-SY-C2-278

Defeating Environmental Exposures in Durham: A Community-Based Particpatory Research Approach

K. May[1], L. Smith[2]; [1]NC State University, Raleigh, NC, [2]Partnership Effort for the Advancement of Children's Health, Durham, NC

Abstract: Cadmium, arsenic and lead are ubiquitous toxic metals in the environment with a legacy of contamination in many urban landscapes. Environmental sources, including soils, house dust and water, act as stable external reservoirs from which these chemicals can be inhaled or ingested. As low-level exposure to these metals often elicits no acute symptoms, populations may be unaware of their exposure. From 2009 to 2011, over 2,500 women participated in the Newborn Epigenetic Study (NEST), a traditional cohort study that explored how environmental exposures affect health outcomes in Durham, North Carolina. Although NEST was not designed to look specifically at exposure to toxic metals, a follow-up study in 2015 revealed that women who had elevated levels of toxic metals in their blood/urine during NEST, also had soil and house dust contaminated with these metals. These women were geographically clustered in Northeast Central Durham (NECD), a neighborhood where 70% of the community members are African American or Hispanic, and the prevalence of cadiometabolic conditions, including type 2 diabetes and obesity is 15% and 72%, respectively, notably higher than the estimates for Durham County as a whole. After findings were reported to NECD community leaders and the state Health Department, the community requested additional data and involvement. Continued engagement with members of the NECD community has driven current research efforts to determine: the geographic boundaries of contaminated soils/house dust to define the exact population at risk, the sources and routes of toxic metal exposure, and develop culturally-relevant interventions to reduce exposure. What began as traditional epidemiology has transformed into community-based participatory research, utilizing best practices in engagement through community-university partnerships to assess exposures and implement interventions.

Keywords: A-environmental justice, A-epidemiology, B-metals, D-community

TU-SY-D2: Exposure to Personal Care Products and Cosmetics in Europe: Data, Models and Regulatory Challenges

TU-SY-D2-279

Exposure assessment within the Context of Cosmetic Ingredients in Europe

F. Almeida; Cosmetics Europe, Brussels, Belgium

Abstract: Safety Assessment is at the heart of the European Cosmetics Regulation. As such, together with the hazard assessment, exposure information is of key importance in ensuring that both ingredients and cosmetic products are safe for consumers and the environment. Historically, Cosmetics Europe has done a significant amount of work to guide and support the deterministic exposure assessment. This includes the studies done between 2005 and 2011 on habits and practices data for cosmetics in adults, which have been published in peer-reviewed publications and ultimately have been recognised by the Scientific Committee on Consumer Safety (SCCS) which has included the results of these studies in their Notes for Guidance as reference for all the cosmetics industry. Additionally, Cosmetics Europe has also published a tiered approach assessment for inhalation exposure which is being used by the industry in their ingredient and product safety assessment. Currently, new approaches have become available to refine exposure assessment for cosmetics, which includes a tiered approach for aggregated exposure and the development of probabilistic exposure models. These new tools can help in refining exposure which has now become even more prominent for the cosmetics industry due to the 2013 Animal Testing Ban. For some ingredients one new challenge is the total exposure assessment (from other sources outside cosmetics) which needs to be tackled by industry and regulators in the future. This presentation provides an overview of how exposure features in the work of Cosmetics Europe, the approaches taken, as well as regulatory significance and challenges.

Keywords: A-aggregate exposure, C-consumer/personal care products

TU-SY-D2-280

Exposure assessment within the Long Range Science Strategy at Cosmetics Europe

S. A. Tozer; Procter & Gamble, Surrey, United Kingdom

Abstract: The Cosmetics Europe (CE) Long Range Science Strategy (LRSS) Research Programme for 2016-20, seeks to develop new concepts for regulatory chemical safety assessment, and is focussed on exposure, eye irritation, genotoxicity, skin sensitization, bioavailability/ADME and systemic toxicity. The programme will also drive the process of scientific and regulatory acceptance for regulatory risk assessment purposes. Exposure is a key parameter in refining this safety assessment process and CE has a reputation for producing robust exposure data. As part of its LRSS programme, CE are undertaking a significant aggregate exposure project that explores scientifically robust methods for gathering representative cosmetic ingredient concentration data and assessing aggregate exposure to these cosmetic ingredients. In this project, data on the occurrence and concentration of key ingredients in the individual formulations on the market are obtained via large anonymous surveys and the data are modelled probabilistically, with subject-level habits and practices data from large populations, to generate robust aggregate exposure estimates that will be used for safety assessments. The cosmetic ingredients under study in this project will be common to those being examined elsewhere within the LRSS programme. This will ensure that assessments of chemical hazard via read across and physiologically based pharmacokinetic (PBPK) models developed can be considered in tandem with the aggregate exposure assessments, so that broader conclusions can be drawn on their overall safety and risk. This presentation will give an overview of the LRSS aggregate exposure study outlining how it fits into the wider systemic toxicity programme.

Keywords: A-aggregate exposure, A - population exposure, A-exposure models, A-risk assessment, C-consumer/personal care products

TU-SY-D2-281

Ingredient Concentration Surveys - A Key Input for Exposure Assessment

C. O' Mahony; Creme Global, Dublin, Ireland

Abstract: Exposure assessment modelling for the purposes of risk assessment typically follows a tiered approach. This involves starting with upper estimates of exposure based on deterministic models and using maximal values for model parameters, and proceeding to more refined estimates of exposure if needed. Refined exposure models seek to use more realistic model parameters rather than worst-cases, and the models themselves are often (though not necessarily) probabilistic in nature, allowing statistical distributions to be taken as input. For consumer product exposure assessment in particular, this requires developing estimates of ingredient concentrations in products rather than assuming chemicals occur at e.g. the maximum allowed concentration as authorised by a regulatory body. However, gathering this data can be challenging as it is not always clear how to characterise the statistical distribution of an ingredient in a product category on the market, as well issue of ingredient concentrations being proprietary information and therefore commercially sensitive. Ingredient usage has two distinct elements; occurrence, which is the proportion of products on the market within a specific category that contain a given ingredient, and concentration, i.e. the distribution of use-levels representing an ingredient in products on the market. This presentation will provide an overview of different approaches that can be taken to gather these key inputs for exposure assessment. In particular, emphasis will be given to addressing how to overcome the issue of gathering proprietary data, developing statistical distributions representative of use on the market, tonnage information, and how these inputs can be used in high-tier probabilistic exposure models.

Keywords: A-aggregate exposure, A-risk assessment, C-consumer/personal care products, A-exposure models

TU-SY-E2: The Role of PBPK Modelling in Holistic Exposure and Risk Assessment - Case Studies and Developments within the EU Project EuroMix

TU-SY-E2-284

PBPK Modelling of Bisphenol S and Bisphenol F

C. Karrer[1], T. Roiss[1], N. von Goetz[1], D. Gramec Skledar[2], L. Peterlin Mašič[2], K. Hungerbuehler[1]; [1]ETH Zurich, Zurich, Switzerland, [2]University of Ljubljana, Ljubljana, Slovenia

Abstract: With the high production volume chemical bisphenol A (BPA) being regulated more severely due to its known endocrine disrupting effects, replacement substances have conquered the market for different applications. The bisphenols S and F (BPS and BPF) are such replacements and can e.g. be found in foodstuffs, indoor dust, air, and thermal paper. Several studies showed that BPS and BPF just like BPA exert estrogenic and anti-androgenic effects. These effects have so far only been observed for the unconjugated substances. However, bisphenols are metabolized to glucuronides and sulfates in the liver and the gut. The velocity and extent of the detoxifying metabolism depends on substance and exposure route, respectively. Therefore, it is important to consider the corresponding specific pharmacokinetic behavior when assessing exposure and risk of bisphenol analogues. To address this issue, researchers have conducted physiologically based pharmacokinetic (PBPK) modelling for BPA, but not yet for BPS and BPF. Therefore, we have adjusted an existing PBPK model for BPA and complemented it with chemical-specific parameters for BPS and BPF. To characterize the metabolic behavior of the bisphenol analogues, we collected new data on their hepatic and intestinal glucuronidation. For estimating partitioning coefficients, we compared the most suitable quantitative structure-activity relationships (QSARs) available. For all model parameters, we derived distributions mirroring uncertainty and inter-individual variability. We then derived internal exposure estimates by coupling external exposure estimates with the PBPK models. For this, we compared internal exposures of BPA, BPS, and BPF after oral and dermal exposure to equal amounts and to at present realistic amounts obtained with higher tier external exposure modelling. The results show that replacing BPA with structural analogues does not necessarily lower the risk regarding endocrine disruption.

Keywords: B-BPA, A-aggregate exposure, internal exposure, PBPK modelling, endocrine disruptors

TU-SY-E2-285

Development of a toxicokinetic modeling approach to estimate the cumulative exposure to pyrethroids

P. QUINDROIT, C. BROCHOT; INERIS, Verneuil-en-Halatte, France

Abstract: Context and objective: Pyrethroids are ubiquitous insecticides used in many areas such as agriculture and human or veterinary medicine. The human biomarkers are usually the urinary concentrations of metabolites. The interpretation of these biomarkers to assess the environmental exposure of populations to a specific compound can be difficult since pyrethroids share metabolic pathways and common metabolites. In this work, we propose to develop a global model that links the external exposure to four pyrethroids (permethrin, cypermethrin, cyfluthrin and deltamethrin) to the urinary concentrations of their common metabolites (DCCA, 3-PBA, F-BPA and DBCA) and to apply this model to predict the exposure of the French population. Methods: A generic and gender-dependent physiologically based pharmacokinetic (PBPK) model was adapted to the toxicokinetics of the four pyrethroids, and one-compartment models were developed to describe the levels of metabolites in urine. The models for the parent compounds and the metabolites were connected at the level of the metabolic sites. In vivo, in vitro and in silico data were used for the model parametrization. Human toxicokinetic data were used to evaluate the model predictions. Several scenarios were tested, e.g. different pathways of exposure (oral, dermal and inhalation) or biological matrices (blood, urine). Realistic exposure scenarios were defined using food and air contamination data. The global model was then used to estimate the urinary metabolites concentration after a cumulative exposure to pyrethroids and results were compared with biomonitoring data of the French population (ENNS study). Results and conclusion: Results of the model evaluation showed a proper agreement between the model predictions and experimental data. The estimated cumulative exposure was used as input to the PBPK model. Exposure doses calculated for French population from the ingestion and inhalation of pyrethroids are similar to previous assessment. The median ratios between predictions of urinary concentrations after cumulative exposures and the data from the biomonitoring study were in a range between 0.58 and 1.5. Difficulties remain in estimating the inter-individual variability and extreme values. The global PBPK model can be used to estimate the biomarkers of internal exposure by taking into account cumulative exposure to pyrethroids.

Keywords: A-exposure models, A-cumulative exposure, B-pesticides, PBPK modelling

TU-SY-E2-286

A web-based toolbox linking kinetic models to risk assessment

W. D. Boer, H. Van der Voet, J. Kruisselbrink, M. Van Lenthe; Biometris, Wageningen, Netherlands

Abstract: In cumulative risk assessment for human health effects, an essential step is to link aggregate external exposure to multiple chemicals to effects at the level of the target organ, e.g. the liver. The aggregate exposure is modelled using data from mixtures of chemicals coming from dietary and non-dietary routes (dermal, oral, inhalation). In the Monte Carlo Risk Assessment (MCRA) toolbox, several cumulative models are available to estimate short or long term exposure. In the EU project EuroMix, new strategies are developed to link the external and internal exposures to mixtures of chemicals and to estimate health effects. Data are coming from in-vivo studies in rats or mice and in-vitro studies at the cellular level. PBPK modelling was identified as the critical link between exposure and effect: substance-specific kinetic models translate the external dose to the internal dose. A procedure for calibration of in-vitro and in-vivo dose response models is developed and being implemented in the MCRA toolbox. In this presentation the design of a web-based system that links data from multiple systems and links kinetic modelling to risk assessment will be demonstrated.

Keywords: A-aggregate exposure, A-exposure models, PBPK, calibration model, in-vitro in-vivo experiments

TU-SY-F2: Integrating Diverse Environmental Exposure Datasets to Study Human Health - from Satellite Sensing and Ground Monitoring to Personal Exposure Part (1): Resources, Technologies and Data Access

TU-SY-F2-287

Observing Air Pollution from Space: A Multi-Pollutant, Satellite-Based Health-Air Quality Index

B. Duncan[1], L. Lamsal[1], K. Cromar[2]; [1]NASA, Greenbelt, MD, [2]New York University, New York, NY

Abstract: As compared to surface networks of monitors, satellite data of air pollutants provide the unique advantage of spatial coverage, which increases statistical power that strengthens inference of the relation between pollutants and health outcomes. In the last five years, health professionals have used satellite data for health studies, such as exposure studies: stratospheric ozone for UV, nitrogen dioxide (NO_2), formaldehyde (HCHO), and aerosol properties for particulates. In this presentation, we will present the methodology and preliminary results of a multi-pollutant, satellite-based Health Air Quality Index (HAQI), a NASA Health & Air Quality Applied Sciences Team (HAQAST) initiative with UNICEF. HAQI coverage will include areas of the world with sparse or no surface monitors, including the world's polluted megacities. The pollutants included in the HAQI are NO_2, ozone (O_3), and fine particulate matter ($PM_{2.5}$). Since observing surface O_3 from space is currently not feasible, we will use O_3 from a NASA atmospheric model. Our effort also includes creating global air quality 5 and 10-day forecasts using the same NASA model.

Keywords: A-global health, satellite data

TU-SY-F2-288

Data Architecture and Infrastructure to Support Exposure Science

V. Kilaru; U.S. Environmental Protection Agency, Research Triangle Park, NC

Abstract: The modernization of exposure science called for by the NRC report requires a commitment and shift in the ways we treat data. Exposure science encompasses a wide variety of disciplines and hence constitutes a wide variety of data, from modeled and measured concentrations in physical media (air, water), to personal exposure assessments, to classic exposure models that incorporate time-activity patterns and microenvironments, to pharmacokinetic modeling and measures of internal dose. No longer can we see data a proprietary and to be horded. Organizations and researchers dealing with any part of this lifecycle need to adhere to the FAIR principles (Finable, Accessible, Interoperable, and Re-usable). The implications of FAIR involves the establishing data standards and conforming to ontologies but if real progress is to be made then data needs to move up the data value chain. This implies making adding further semantic content to data and making that content computable. The recent interest in exposure frameworks such as the Aggregate Exposure Pathways (AEP) goes to further underscore the need for a data architecture and infrastructure that can be the basis to fully realize the delineation of an exposome.

Keywords: A-exposure models, A - exposure measurement

TU-SY-F2-289

Emerging capabilities in personal exposure assessment

E. Seto; University of Washington, Seattle, WA

Abstract: Currently, there are multiple technologies and approaches to assess concentrations of environmental pollutants and personal and population exposures to them. Using air pollution as an example, sensing concentrations may leverage existing regulatory monitors, new densely sited environmental monitors, indoor vs outdoor monitors, personal monitors, and various forms of mobile monitoring, including vehicle, aircraft, drone, or satellite based sensing. An exciting prospect for understanding the complexity of exposures is the potential for merging data from all these approaches to better understand exposure. Three brief examples will show the diversity of exposure monitoring: First, an example illustrating the limitations of current regulatory air monitoring and the benefits of dense community monitoring will be presented from the Imperial Air study. Second, an example from a study of secondhand cigarette smoke exposure and asthma severity will show the value of personal indoor air monitoring. Third, an example of mobile data collection will illustrate the ability to efficiently collect fine spatially resolved data throughout a community for multiple exposure scenarios. Finally, a framework based on spatio-temporal modeling will be presented that may hold promise for integrating these diverse sensing modalities to estimate exposures.

Keywords: A - ambient monitoring, A - exposure measurement, A-sampling methods, A-exposure models, A-sensor technology

TU-SY-G2: Ensemble Learning for Air Pollution Exposure Assessment

TU-SY-G2-290

Assessing Daily Exposure to PM2.5 with Machine Learning and Remote Sensing

C. Brokamp[1], R. Jandarov[2], M. Hossain[1], P. Ryan[1]; [1]Cincinnati Children's Hospital Medical Center, Cincinnati, OH, [2]University of Cincinnati, Cincinnati, OH

Abstract: Aerosol optical depth measurements collected by Moderate Resolution Imaging Spectroradiometer instruments on satellites have been used to create PM2.5 exposure assessment models with high spatial and temporal resolution. Here, we compare the accuracy of several different modeling approaches. We have created daily PM2.5 predictions for 2002 - 2015 at a 1 km resolution throughout a seven-county area in Cincinnati, Ohio, USA. PM2.5 from the EPA's Air Quality System combined with ground-based measurements from our own PM2.5 sampling campaign were used to train models based on AOD measurements taken from the MODIS 3 km aerosol product. Spatiotemporal weather data, including planetary boundary height, temperature, humidity, precipitation, and wind speed from the North American Regional Reanalysis data as well as greenspace measures, the National Land Cover Database, and length of major roadways were also used to train the predictive models. Based on the previous performance of land use random forest models, we investigated the predictive accuracy of several different type of ensemble learners, including boosted linear models, boosted regularized linear models, bagged trees, and random forest, as well as commonly used land use regression models. Predictive error was assessed using 10-fold cross validation repeated ten times. Random forest models outperformed all other model types with a predictive R^2 of 0.79 and a predictive RMSE of 3.82 μg/m^3. Notably, boosted linear models had a predictive R^2 of 0.54 and a predictive RMSE of 5.76 μg/m^3. The random forest model was further improved by incorporating spatial relationships via grid coordinates. This model recreates both short- and long-term PM2.5 exposure histories useful for epidemiologic studies.

Using ensemble models can improve the predictive accuracy of spatiotemporal exposure assessment models using land use and remote sensing data.

Keywords: A - exposure measurement, A-geospatial analysis/GIS, A-statistical methods, B-particulate matter, C-air

TU-SY-G2-291

Ensemble Learning and Constrained Optimization Applications in Spatiotemporal Modeling of Long-Term Air Pollution Concentrations in Southern California

L. Li[1], F. W. Lurmann[2], J. Wu[3], R. Urman[1], C. Breton[1], F. Gilliland[1], R. Habre[1]; [1]University of Southern California, Los Angeles, CA, [2]Sonoma Technology Inc., Petaluma, CA, [3]University of California Irvine, Irvine, CA

Abstract: Models of air pollutants with high spatial and temporal resolution are often needed in epidemiological studies to estimate the health effects of long- and short-term exposures. Typical linear regression approaches are traditionally limited at capturing complex non-linear relationships and dependencies. We present our approach in modeling concentrations of nitrogen oxides and particulate matter in the southern California air basin at a high spatiotemporal resolution for extended periods of time and demonstrate improved prediction ability by using ensemble learning approaches including machine and deep learning. The main features of the models include combining short-term measurement data from several campaigns with routine, long-term ambient monitoring, remote sensing, meteorology and chemical transport model output data. Temporal basis functions are used to extract long-term and seasonal variation in pollution levels based on routine monitoring data. Spatiotemporal covariates are used to capture local scale spatial (eg, traffic density, line source dispersion model outputs) and temporal variation in emission sources and dispersion with non-linear, spline terms. Convolutional layers capture spatial autocorrelation. Ensemble learning approaches are used to optimize the performance of the model and minimize bias. Constrained optimization is used to adjust model predictions based on physical constraints. Using this modeling framework, we are able to predict concentrations of air pollutants in southern California at the local neighborhood and regional spatial scale and one- to two-week temporal resolution over a period of over 20 years.

Keywords: A-exposure models, A-geospatial analysis/GIS, B-particulate matter, C-air

TU-SY-G2-292

Approaches for Predicting Short-Term Traffic-Related Air Pollution with Irregular Monitoring Data

E. Noth[1], H. Minor[2], F. W. Lurmann[2], K. Hammond[1]; [1]University of California, Berkeley, Berkeley, CA, [2]Sonoma Technology, Inc., Petaluma, CA

Abstract: Our research on childhood asthma, immune dysregulation and birth outcomes requires estimation of short-term exposures with daily to weekly averaging times. We have compiled 16 years of airborne polycyclic aromatic hydrocarbon (PAH) sampling data collected in Fresno, a major metropolitan center in California's San Joaquin Valley. These data are of two basic types - continuous real-time monitoring data from the PAS2000, representing particle-bound PAHs of 3-rings are larger, and integrated 24-hour concentrations from filter sampling. The monitoring is quite irregular - the continuous monitors have many more observed days of PAHs but a smaller spatial extent. The integrated data have two periods of spatial saturation sampling (2002-2003, 2014-2015). Additionally, over the 16 years, there is a downward secular trend in the PAH concentrations. Our research compares the performance of regression-based air quality models to the spatiotemporal framework approach implemented through the R package SpatioTemporal. The regression model includes land use, meteorology, and estimates from a source dispersion model (Caline4) as predictive variables. The fixed site measurements are used as a temporal anchor in the regression and the integrated measurements provide the spatial-temporal distribution. Using a regression model with random effects of sampling location and date, we were able to capture most of the spatial variability (78%) and slightly over half the temporal variability (53%) observed in the data. Our implementation of the SpatioTemporal model uses a 24-hour averaging period for PAH prediction in Fresno, CA. We applied the model with 2 basis functions and 4,276 time points. The algorithm was provided with the same set of spatial covariates as the regression model. The two approaches produce comparable models, but greater inquiry is necessary to understand the nuances in the resulting exposure estimations.

Keywords: A-exposure models, A-geospatial analysis/GIS, C-air, D-community

TU-PL-A3: Occupational Exposure to Pesticides

TU-PL-A3-293

An exposure assessment of occupational exposures among small quantity pesticides users in amenity horticulture.

A. Connolly[1], K. Jones[2], K. Galea[3], I. Basinas[3], L. kenny[2], P. McGowan[4], M. Coggins[1]; [1]National University of Ireland Galway, Galway, Ireland, [2]Health and Safety Laboratory (HSL), Buxton, United Kingdom, [3]Institute of Occupational Medicine (IOM), Edinburgh, United Kingdom, [4]Irish Commissioners for Public Works, Co. Meath, Ireland

Abstract: An occupational biomonitoring study to assess gardeners and amenity horticulturalist exposure to pesticides was completed over the summer period of 2015. Spot urine samples were collected pre and post work tasks involving small volumes of pesticides products (ranging from 100 ml to 2 L of concentrated product used), samples were analysed for pesticide content. The active ingredients of interest in this study included glyphosate and fluroxypyr. Study results (GM (GSD) 0.66(1.11) $\mu g\ L^{-1}$ for glyphosate and GM (GSD) 0.29 (1.69) $\mu g\ L^{-1}$ for fluroxypyr) showed an exposure potential during amenity horticulture work, which warrants further investigation. Results also suggested that factors such as the re-use of PPE and contact with contaminated objects used during spraying could result in increased exposure concentrations. A larger study involving 24-hour biological monitoring, dermal exposure assessment and an assessment for the potential for inadvertent ingestion is now planned for the gardening season of 2017. Workers will be grouped into four similar exposure groups based on the pesticide application method used and whether they were working with glyphosate or fluroxypyr. Dermal exposure assessment will involve collecting wipe samples from the worker hands, gloves and potentially contaminated objects such as mobile phones, product containers and vehicle steering wheels. Wipe samples from the worker perioral region will also be collected before and after completion of work tasks alongside detailed contextual information regarding the worker, task and environmental conditions. Data collection and analysis will be completed by August 2017. Study results will be used to estimate total body burden and the contribution of dermal and inadvertent ingestion as exposure routes.

Keywords: A-biomonitoring, A - exposure measurement, A-industrial hygiene, B-pesticides, D-occupational

TU-PL-A3-294

Human exposure to spray drift: investigations into modelling the spray exposure of residents and bystanders and its variability

C. Butler-Ellis[2], M. Kennedy[3], R. Alanis[2], C. Tuck[2], C. J. Kuster[1]; [1]Bayer, Monheim, Germany, [2]Silsoe Spray Application Unit, Silsoe, United Kingdom, [3]Fera, York, United Kingdom

Abstract: The model for bystander and resident exposure to spray drift (BREAM) which is incorporated into current EFSA guidance, uses a mechanistic model to predict airborne spray and ground deposits, and an empirical model that relates airborne spray to deposits on the human body, from which dermal exposure can be calculated. The EFSA guidance requires the 75th and 95th percentiles of exposure distributions to be used to represent long-term and acute exposures respectively, and there has therefore been some scrutiny of the nature of these distributions, and whether the relatively high values for the 95th percentile can be reduced by addressing some of the uncertainties and the causes of variability. The empirical data used to relate bystander exposure and airborne spray has been reviewed and extended slightly. In addition, the contextual data relating to the application conditions for each pair of data were obtained from the original records and included in the spreadsheet to enable further analysis. This contextual data included nozzle and pressure, sprayer speed, distance downwind, wind speed measured at 2.0 m height above the ground, boom width, crop height, approximate bystander height, approximate bystander cross-sectional area. In the original model, the only factor that was considered in estimating the relationship between airborne spray and bystander exposure was the bystander height. An analysis of both the empirical data and model output distributions was undertaken to investigate the sources of variability. Some wind tunnel studies were undertaken to explore the collection efficiency of spray droplets under conditions similar to a bystander exposed to spray drift and it is shown that it is possible to develop a mechanistic model of spray collection based on an impact parameter, similar to previous work undertaken by May and Clifford (1967).

Keywords: A-exposure models, A-risk assessment, A-exposure factors, A-statistical methods, D-children

TU-PL-A3-295

Real-Time Particle Monitoring of Pesticide Drift from Two Different Orchard Sprayers

M. N. Blanco, R. A. Fenske, E. J. Kasner, E. Seto, E. Austin, M. G. Yost; University of Washington, Portland, OR

Abstract: Pesticide drift from agricultural spraying is a significant public health concern, affecting workers and surrounding communities. In Washington State, a majority of pesticide-related illnesses and application-related complaints involve drift. This study employed real-time particle monitors to characterize off-target spray drift during a series of orchard spray trials. The study was nested within a larger study that used micronutrients as tracers, and both active and passive sampling methods. Sections of an orchard block were randomly sprayed by alternating two orchard spray technologies - axial fan airblast (AFA) and multi-head fan tower (MFT) -- while ten Dylos DC1100 Pro real-time particle counters sampled aerosols generated by the sprayers from various locations in a neighboring block, ranging from 0-152 m (0-500 ft) downwind). Two meteorological stations collected wind speed, wind direction, temperature and relative humidity throughout the study period. Measurable aerosol drift levels were found at all downwind sampling locations for both sprayers. Significantly greater drift was associated with the AFA than the MFT sprayer below the canopy and at closer distances. Controlling for wind speed and height, the 75th drift quantiles were 126.3 and 47.9 pg/cm^3/min for the AFA and MFT sprayers respectively. Controlling for sprayer type and wind speed, the 75th drift quantiles were 38.5 and 20.3 pg/cm^3/min above and below the canopy respectively. In a restricted analysis looking at spray periods and controlling for distance from the sprayer, sprayer type, wind speed and height, every additional foot (0.305 m) away from the sprayer was associated with 0.1 pg/cm^3/min less drift. These results were consistent with results determined by passive sampling methods. Our findings indicate that real-time particle monitoring for pesticide aerosols can serve as an accurate and relatively inexpensive approach to characterizing pesticide drift.

Keywords: D-occupational, B-pesticides, C-air, A - exposure measurement, A-industrial hygiene

TU-PL-A3-296

Exposure To Thiabendazole In Mothers And Children From The Infant's Environmental Health Study In Costa Rica Title has to be Title Case

B. van Wendel de Joode[1], M. Mora[1], L. Cordoba[1], A. Mora[1], M. Öberg[2], J. Hoppin[3], C. Lindh[4]; [1]Universidad Nacional, Heredia, Costa Rica, [2]Karolinska Institute, Stockholm, Sweden, [3]NCSU, North Carolina, NC, [4]Lund University, Lund, Sweden

Abstract: Background: Thiabendazole (TBZ) is broadly used as a post-harvest fungicide in different crops, including banana where it is applied in packing plants. Teratogenic and endocrine disrupting effects of TBZ have been reported in mammals; however, studies on TBZ exposure are limited. Methods:

As part of the ISA study, a birth cohort study of mother-child pairs living near banana plantations in Matina County, we measured 5-hydroxythiabendazole (5-OH-TBZ), main metabolite of TBZ, in urine samples collected from mothers during pregnancy (445 mothers, 909 samples) and when their children were 1 year of age (326 mothers). Urinary 5-OH-TBZ concentrations were also measured in samples collected from 68 1-year-old children. Results: Urinary 5-OH-TBZ concentrations measured in samples collected repeatedly from the same mother were moderately correlated (ICC=0.48). Child 5-OH-TBZ concentrations were associated with maternal 5-OH-TBZ concentrations ($R^2=0.59$, $p<0.01$). Women working in banana plantations' packing plants had urinary 5-OH-TBZ concentrations about ten times higher than women who did not (during pregnancy: GM=0.84 vs. 0.08 µg/L; and at child age 1 year: GM=0.86 vs. 0.07 µg/L, respectively). Similarly, infants whose mothers worked in banana packing plants had higher 5-OH-TBZ concentrations compared to infants whose mothers did not have this job (GM=1.77 vs. 0.14 µg/L, respectively), suggesting that mothers may transfer TBZ residues to their children. Correlations were similar for women who breastfed their children compared to women who did not. Other factors such as maternal age, distance between residence and plantation, and general pesticide use at home were not related to 5-OH-TBZ concentrations. Conclusions: Mothers working in banana plantations' packing plants are exposed to TBZ and may transfer the pesticide to their home environment. Following the precautionary principle, measures are needed to reduce exposure at work and avoid take-home exposure.

Keywords: A - exposure measurement, A - population exposure, D-occupational, D-children, A-biomonitoring

TU-PL-A3-297

Chlorpyrifos and the health of agricultural workers and their families

R. A. Fenske; University of Washington, Seattle, WA

Withdrawn

TU-PL-B3: Air Pollution – Epidemiology

TU-PL-B3-298

Individual's Exposure and Emotional Response to Visual Impact of Particulate Matter Pollution: A Psychophysiological Study

J. Yang[1], X. Liu[1], J. Chen[2], M. Liu[1]; [1]Nanjing University, Nanjing, China, [2]Nanjing Foreign Language School, Nanjing, China

Abstract: As a leading risk factor for respiratory and coronary disease in China, air pollution has been studied extensively from physical health perspective. However, it is gradually recognized that exposure to air pollution also exert massive adverse impacts on individual's mental health. Particulate matter pollution (e.g. $PM_{2.5}$) is perceived as one of the most harmful air pollutants by Chinese lay public. High-concentration and large-scale PM pollution can damage visibility seriously, thus evoke individual's negative emotions. In this study, we administered a psychophysiological study to quantify association between individual exposure and emotional response to PM pollution. Participants were instructed to view a series of photos taken at the same site and occasion but with different $PM_{2.5}$ concentrations. The images were mixed with pictures selected from IAPS (International Affective Picture System) in case of fatigue. Three self-report scales on basic dimension of emotion (i.e. valence, arousal and motivation) were rated by participants after viewing each photo. Moreover, their psychophysiological signals, including skin conductance responses, electrocardiogram, electromyogram, were also recorded during the whole experiment. Obvious but diverse trends of different emotion indicators were observed. A 10% increase of $PM_{2.5}$ concentrations was associated with a 0.07 decrease of valence scores. Arousal scores, however, dropped down sharply when $PM_{2.5}$ concentration increased from 11 to 50μg/m^3, and tended to be quite stable when the number exceeded 50μg/m^3. Physiological responses indicating negative emotion status were significantly stronger if $PM_{2.5}$ concentrations were higher than 50μg/m^3. We conclude the policy implications revealed from our study, and discuss the feasibility for further applying this novel method to measure individual's emotional reactions towards various environmental risks.

Keywords: B-particulate matter, A-exposure models, A-environmental policy

TU-PL-B3-299

Amyotrophic lateral sclerosis and exposure to diesel exhaust in a Danish cohort

A. S. Dickerson[1], J. Hansen[2], O. Gredal[2], M. Weisskopf[1]; [1]Harvard T.H. Chan School of Public Health, Boston, MA, [2]Danish Cancer Society Research Center, Copenhagen, Denmark

Abstract: Background: Previous studies have suggested relationships between occupational exposures and risk of ALS. Specifically, some manuscripts have reported increased risk of ALS and other motor neuron diseases in occupations commonly exposed to diesel exhaust, such as truck and bus drivers. Additionally, studies have implied that residential exposure to air pollutants may increase risk of ALS. **Objective:** To investigate the association between occupational exposures to diesel exhaust and odds of ALS. **Methods:** 4600 ALS cases were identified from the Danish National Patient Registry with 459,906 sex- and birth year-matched controls from 1982 to 2013. We acquired occupational history since 1964 from the Danish Pension Fund. Diesel exhaust exposures and probability were estimated for each job and incorporated into a job exposure matrix, which was used to calculate cumulative and average exposures for jobs held at least 3 years before diagnosis/index dates. Adjusted odds ratios (aOR) and 95% confidence intervals (CI) for quartiles of cumulative and average exposure were obtained using conditional logistic regression analyses and stratified by sex. **Results:** 1090 (23.70%) cases and 110,988 (24.13%) controls were ever exposed to diesel exhaust at work. No significant results were seen in our investigation of cumulative exposures. However, we did observe a significant increase in ALS for those with the highest quartile for average daily exposure compared to no exposure (aOR = 1.17; 95% CI: 1.02, 1.34). Stratified analyses revealed that this association was only in males (aOR = 1.19; 95% CI: 1.03, 1.38). **Conclusion:** Our study suggests an association between consistently higher exposures to diesel exhaust. These findings support those of previously reported associations between ALS and occupations commonly exposed to diesel exhaust.

Keywords: D-occupational, A-epidemiology, A-cumulative exposure, B-particulate matter, A-workplace

TU-PL-B3-300

Estimating exposure to oxidative potential of ambient particulate matter using emission source impacts

J. Bates[1], R. Weber[2], J. Abrams[3], V. Verma[4], T. Fang[2], C. Ivey[5], M. Klein[3], M. Strickland[5], S. E. Sarnat[3], H. Chang[3], J. Mulholland[1], P. Tolbert[3], A. Russell[1]; [1]Georgia Institute of Technology, Atlanta, GA, [2]Georgia Institute of Technology, Atlanta, GA, [3]Emory University, Atlanta, GA, [4]University of Illinois Urbana-Champaign, Urbana-Champaign, IL, [5]University of Nevada, Reno, Reno, NV

Abstract: Oxidative potential of fine particulate matter ($PM_{2.5}$) may be a biologically relevant exposure metric for epidemiologic studies linking air pollution and health as particles may be capable of inducing oxidative stress. Oxidative potential of ambient water-soluble $PM_{2.5}$ was measured using an accellular dithiothreiol (DTT) assay at six locations (four urban and two rural) across the Southeastern United States from June 2012-July 2013. Two advanced source apportionment techniques, including the receptor model Chemical Mass Balance method (CMB) and the chemical transport model CMAQ-DDM, were used to estimate emission source impacts on ambient $PM_{2.5}$ during the study period. These sources were related to ambient oxidative potential using ordinary least squares regression with backward selection. The resulting models were applied to simulate spatial trends in concentrations of ambient oxidative potential across the Eastern United States and to develop a 10-year historical time series of estimated daily ambient oxidative potential in Atlanta, GA for an epidemiologic study. Regression analyses show that vehicles and fires significantly contribute to oxidative potential. Higher oxidative potential is generally seen in urban areas compared to rural areas due to higher vehicle activity and in the winter than the summer due to prevalent prescribed burning across the Southeastern United States. Furthermore, epidemiologic analyses find a significant association between estimated oxidative potential and asthma/wheezing and congestive heart failure emergency department visits in the Atlanta, GA region, even when controlling for $PM_{2.5}$ mass, supporting the hypothesis that oxidative potential, as measured using a DTT assay, may be a useful exposure metric for health studies. Using emissions source impacts can be useful for accurately assessing the spatial and temporal trends in exposure to the novel biologically relevant metric oxidative potential for future epidemiologic studies.

Keywords: B-particulate matter, C-air

TU-PL-B3-301

Relationships of Ambient Fine Particle Air Pollution, Endothelial Function, and Blood DNA Methylation Age: Implications for Renal Health

J. Nwanaji-Enwerem[1], M. Weisskopf[2], A. Baccarelli[3], J. Schwartz[2]; [1]Harvard Medical School & Harvard School of Public Health, Boston, MA, [2]Harvard T.H. Chan School of Public Health, Boston, MA, [3]Columbia Mailman School of Public Health, New York, NY

Abstract: Background: Research has demonstrated associations of long-term ambient fine particulate matter ($PM_{2.5}$) with renal function and DNA methylation (DNAm) age, an epigenome-wide DNA methylation-based predictor of chronological age. It has also been shown that the association of $PM_{2.5}$ component species with DNAm-age can be modified by normal variation of genes involved in endothelial physiology. **Objective**: Given that renal and endothelial physiology are closely related, we determined the relationship of DNAm-age with measures of renal function and examined the modifying role of DNAm-age on the relationship of $PM_{2.5}$ with renal function. **Methods**: This analysis included 559 participants from the Normative Aging Study with multiple visits between 2000 and 2011 (n=943 visits). We estimated one-year $PM_{2.5}$ and component species (ammonium and sulfate) levels at participants' addresses using the GEOS-chem model. Participants' blood, collected at each study visit, was used to derive DNAm-age and four renal function markers (serum creatinine, serum BUN, eGFR, and BUN-to-creatinine ratio). Blood DNAm-age was calculated using CpG sites on the Illumina HumanMethylation450 BeadChip. We estimated associations of DNAm-age with renal function using linear mixed-effects models adjusted for chronological age, lifestyle/environmental factors, and aging-related diseases. **Results**: A one-year increase in DNAm-age was positively associated with serum creatinine (β = 0.006; 95%CI 0.001, 0.010). DNAm-age was negatively associated with both eGFR (β = -0.34 mL/min/1.73 m^2; 95%CI -0.51, -0.18) and BUN-to-creatinine ratio (β = -0.07; 95%CI -0.13, -0.02). The data also suggested a modifying effect of DNAm-age on the association of one-year $PM_{2.5}$ and Ammonium levels with eGFR. **Conclusions**: DNAm-age is significantly associated with measures of renal function. Additional work is necessary to understand these relationships particularly in the contexts of $PM_{2.5}$ exposure, effect modification, and mediation.

Keywords: B-particulate matter, A-biomarkers, C-air, A - population exposure, A-epidemiology

TU-PL-B3-302

Impact of Fine Particulate Matter on Lipids/Lipoproteins in a Cohort of Midlife Women

X. Wu[1], R. Broadwin[1], R. Basu[1], B. Malig[1], K. Ebisu[1], S. Green[1], E. B. Gold[2], L. Qi[2], C. Derby[3], S. Park[4]; [1]CalEPA\OEHHA, Oakland, CA, [2]University of California Davis, Davis, CA, [3]Albert Einstein College of Medicine, Bronx, NY, [4]University of Michigan, Ann Arbor, MI

Withdrawn

TU-SY-C3: Inspiring the next generation: Bringing exposure science into the STEM community

TU-SY-C3-303

Making Chemical Exposures Relevant to Students: Strategies for integrating the concept of the Exposome into STEM Education

D. B. Haine; UNC-Chapel Hill, Chapel Hill, NC

Abstract: First introduced in 2005, the concept of the exposome has fueled a body of scientific research that has tasked scientists with being able to distinguish between endogenous and exogenous chemical exposures when it comes to assessing impacts to human health. This session will showcase a data interpretation and graphing activity based on the research of one scientist who investigates the mechanisms of carcinogenesis in response to exogenous exposure to cancer causing chemicals such as vinyl chloride. Specifically, this activity introduces students to the emerging science of the exposome as it invites students to compare the rate of formation and repair of three different DNA biomarkers that form in response to vinyl chloride exposure. Exposure science research projects provide students with relevant learning experiences that promote integration of science and engineering practices as well as demonstrate links among engineering, technology, science, and society. Partnering with scientists to develop activities that promote student engagement with real data, cultivates increased awareness of how chemicals in the environment impact DNA structure and function and ultimately human health. This session will highlight strategies for effectively engaging scientists in developing scientifically rigorous, STEM-focused educational activities that are aligned to state and national science standards and also address the realities of the science classroom.

Keywords: A-biomarkers, A-cumulative exposure, B-VOCs

TU-SY-C3-304

Understanding Air: A hands-on STEM package for learning about climate change and air pollution

K. M. Vandiver; Massachusetts Institute of Technology, Cambridge, MA

Abstract: We have designed hands-on LEGO™ representations of the molecules found in air. We can also model how airborne pollutants are produced and how they contribute to everyday health hazards and unseen exposures. We employ standard LEGO bricks that match the chemists' CPK colors for the elements such as oxygen, nitrogen, carbon, etc. Then we teach the environmental health content in the context of STEM chemistry. The bricks can be snapped together into specific shapes to model elements, compounds, and mixtures. In a chemical reaction such as combustion, the burning of fuel with oxygen molecules, atoms are not lost or gained. Instead, the atoms are rearranged into different groups-- clusters of atoms that we recognize as water and carbon dioxide molecules. The incomplete combustion reaction will create different chemical products. Over the past ten years, teachers using bricks as atoms have field-tested our biology, chemistry, and earth science lessons. The "Understanding Air" lessons provide practical examples for key chemistry concepts that teachers need to teach. Additionally, the bricks can be used to illustrate the photosynthesis reaction. When the carbon dioxide molecules are reacted to create glucose, starch, and cellulose molecules, the concept of carbon cycling is no longer an abstraction. Students can follow the carbon brick. The conference presenter will share best practices for the STEM classroom and for setting up learning stations at science festivals and community environmental health fairs. Visitors at these public events are guided by instructional pictures or "placemats" at the different learning stations. Families are drawn to our tables by the sight of the tempting LEGO bricks, while adults are drawn by curiosity.... How do LEGO bricks relate to atmospheric chemistry?

Keywords: C-air, D-children, A-climate change, D-community, STEM, education

TU-SY-C3-305

AQ Treks: Where Students and Other Citizen Scientists Explore Air Pollution in their Neighborhoods

J. W. Birks, C. Williford, D. Meyers, J. Hatz, J. Ellenburg; 2B Technologies, Inc. and GO3 Foundation, Boulder, CO

Abstract: AQTreks is a STEM project that provides a hands-on air pollution measurement experience to middle and high school students who use a Personal Air Monitor (PAM) to measure air pollutants along mobile "treks" of their own design. The PAM wirelessly connects to smart phones of every participant, and treks are displayed in real-time on a mobile app. The app also contacts AirNOW to display the nearest air pollution measurements at government monitoring stations as a quality check and for comparison purposes. Each trek is automatically uploaded via WiFi or 3G to a database for online graphing and display within a blog on the AQTreks website where students discuss their results with other students, teachers and scientists. The program includes an online curriculum where students learn about the sources, transformations, and sinks of air pollutants along with associated health effects. In its first incarnation, the air pollutants ozone and black carbon were measured along more than 500 treks by more than 100 schools. Highlights include comparisons of rural vs. urban exposures, discovery of increased pollution levels during pick-up/drop-off traffic at schools, comparisons of pollutant levels along busy and residential streets, analysis of exposures during commutes to school, a trek at a hydraulic fracturing site, treks from urban areas into the mountains, and investigation of emissions from different types of sources such as lawnmowers and buses. One school explored an area that is known to have an underground coal mine fire and even launched the ozone monitor on a balloon to 30 km in the stratosphere. The newly developed PAM makes use of much less expensive electrochemical sensors to measure the air pollutants CO, CO_2, NO, PM_{10}, $PM_{2.5}$, and $PM_{1.0}$ along with temperature, pressure and relative humidity. The PAMs are rented or loaned to schools for periods of 2-3 weeks, with functionality and calibration checks performed between each deployment for quality assurance purposes.

Keywords: A-sensor technology, D-children, D-environmental justice, B-particulate matter, air pollutants

TU-SY-C3-306

Water Quality Testing and Citizen Science for Rural Middle and High School Students

E. Haynes[1], R. Shepler[1], R. Roberts[2]; [1]University of Cincinnati, Cincinnati, OH, [2]Cambridge Schools, Cambridge, OH

Abstract: Water quality is an environmental health issue in rural Ohio, particularly for residents living near natural gas extraction waste disposal sites. In order to 1) provide outreach and education to local science educators and 2) equip students with data collection tools, we created Water Quality Test Kits. Each water quality test kit includes a student and teacher manual, devices for quantifying conductivity, pH, and temperature. Data is uploaded into an online platform, Google Fusion Tables, to enable real-time data sharing. During pilot testing with teachers, the manual and methodology was aligned with the state curriculum. Our NIEHS Center's Community Engagement Core and community partners have enabled partnership with over 30 classroom teachers and environmental learning programs, including the Guernsey County Soil and Water Conservation District and local 4-H programs. These kits will be used by over 2,200 students within the next year.

Keywords: A - ambient monitoring, A-geospatial analysis/GIS, D-community, C-streams, C-water

TU-SY-C3-307

Environmental Health "Science Take Out" Kits: Tools for teaching exposure science in the classroom - and beyond

K. Korfmacher; University of Rochester, Rochester, NY

Abstract: In 2014, the COEC partnered with Science Take-Out on an NIEHS-funded Phase I STTR grant, "Science Take-Out Kits for Environmental Health Education" (1 R41 ES023706-01). This project developed a series of hands-on environmental health science kits for use in formal (middle/high school) and informal educational settings. Science Take-Out is a University of Rochester start-up company that was created in 2008 to disseminate educational materials developed by the UR's Life Sciences Learning Center. Science Take-Out kits are fully assembled "labs in bags" that require no teacher prep and no additional lab equipment. They provide teachers with a convenient and cost-effective way to infuse engaging hands-on wet lab and manipulative modeling activities into their existing curriculums. The COEC and Science Take-Out staff first generated 21 different environmental health kit ideas, and then used survey responses from 432 teachers to narrow these to a list of nine prototype kits based on important environmental health concepts, connections to teachers' existing curriculums, and relevance to students' lives. Each of these prototype kits was reviewed by a focus group of area secondary school teachers, whose feedback was used to refine the kits and select the top eight kits for pilot testing. Kit topics include: antimicrobials, pesticides, breast cancer, lead, lung disease, healthy homes, skin cancer, and water quality. Almost 700 teachers from throughout the US volunteered to pilot test the kits; of these, 32 were selected to conduct pilot testing starting in February, 2015. The pilot testing results were incorporated in the final editing of the new environmental health kits, which are now in production. At each stage of review, the COEC provided feedback on accessibility and community relevance of materials. Over the past year, we have worked with outreach programs around the country to pilot several of these kits in informal community education settings. STO and the COEC plan to continue to adapt these kits to produce versions appropriate for diverse community settings and audiences.

Keywords: D-community

TU-SY-D3: Challenges and Opportunities: Assessing Exposures to Chemical Substances under Amended TSCA- Methods, Models, and Data - Part 1

TU-SY-D3-308

Overview of Amended TSCA: Pipeline of New and Existing Chemical Assessments

C. Fehrenbacher; U.S. EPA, Washington DC, DC

Abstract: On June 22, 2016, the Frank R. Lautenberg Chemical Safety for the 21st Century Act was signed into law thereby amending the Toxic Substances Control Act (TSCA), the Nation's primary chemicals management law. This new legislation is the first update to TSCA in 40 years. The new law includes several important enhancements to evaluating exposures for new and existing chemicals. These changes include an evaluation of both new and existing chemicals against a new risk-based safety standard that includes explicit consideration for vulnerable populations, evaluating hazards and exposures under the conditions of use, and evaluating the likely duration, intensity, frequency and number of exposures under the conditions of use. Implementation of statutory requirements for both new and existing chemcials is well underway within the U.S. EPA's Office of Pollution Prevention and Toxics. This presentation will focus on the key exposure assessment components of the new legislation, and will provide an overview of the process by which chemicals will be evaluated, upcoming deadlines related to implementation, implementation steps completed to date, and next steps in fully implementing the new legislation. This symposium will provide an important forum for the discussion and exchange of ideas related to implementing the most significant update to an environmental law in 20 years. The views expressed in this abstract are those of the authors and do not represent Agency policy of endorsement.

Keywords: A-environmental regulation

TU-SY-D3-309

Assessing Environmental Exposure under Amended TSCA: Models, Data, and Methods

E. M. Wong, J. Gallagher, N. N. Mottl, J. Todd, N. Orentas, H. Bethel; U.S. EPA, Washington, DC, DC

Abstract: On June 22, 2016, the Frank R. Lautenberg Chemical Safety for the 21st Century Act was signed into law thereby amending the Toxic Substances Control Act (TSCA), the Nation's primary chemicals management law. Implementation of the amended legislation is carried out by U.S. EPA's Office of Pollution Prevention and Toxics (OPPT) and includes assessments of both existing and new chemical substances. The amended TSCA legislation has important implications for exposure assessments, such as the mandate to determine whether the chemical substance presents an unreasonable risk of injury to human health or the environment under the conditions of use (including manufacturing, processing, distribution in commerce, use, or disposal). There are a variety of possible methods, models, and data sources that could be used to quantify exposures across a chemical substance's lifecycle. Measured data, such as enivonrmental releases, media concentrations, physical-chemical and environmental fate properties all help inform exposure scenarios and can provide important context for understanding environmental exposure. Mathematical modeling approaches can also be used either in conjunction with measured data and/or in the absence of robust data sets to estimate environmental concentrations. It is likely that both measured data and modeling approaches will be needed. The types of data and the media that need to be assessed in order to quantify exposures will vary by chemical and type of environmental assessment. This presentation will describe how the evaluation, integration and intersection of environemtnal exposure assessment methods, models, and data are primary factors during the risk evaluation process in the context of exposure assessment. We will provide an overview of environmental exposure assessment approaches pulling from examples from the first ten chemical scoping documents. The views expressed in this abstract are those of the authors and do not represent Agency policy or endorsement.

Keywords: A-ecological exposure

TU-SY-D3-310

Assessing Human Exposure under Amended TSCA

C. Bevington, C. Baier-Anderson, S. Sarraino, N. N. Mottl, H. Bethel, F. Branch; U.S. EPA, Washington, D.C., DC

Abstract: On June 22, 2016, the Frank R. Lautenberg Chemical Safety for the 21st Century Act was signed into law amending the Toxic Substances Control Act (TSCA), the Nation's primary chemicals management law. The TSCA legislation has important implications for human exposure assessment. Implementation of the legislation is an on-going process, which results in changes to how exposure assessments are conducted by the Office of Pollution Prevention and Toxics (OPPT). To meet the needs of the amended TSCA, OPPT will consider approaches to quantify aggregate and sentinel exposures, including for potentially exposed and susceptible subpopulations for all conditions of use across a chemical's lifecycle. Exposure scenarios are developed using physical chemical properties, environmental fate, likely exposure pathways, and chemical use information. Empirical and/or modeling approaches are used to quantify chemical concentrations in indoor or outdoor environments. The timing of exposure frequency and duration, associated averaging times for life stages of interest, and related exposure metrics (*e.g.*, hourly, one day, one year, sub-chronic, or lifetime average doses or concentrations) are also important considerations. Sentinel exposures consider high-end exposure(s) within and across exposure scenarios. OPPT will evaluate available biomonitoring data, which aggregates exposures across sources. Aggregate exposures consider the addition of multiple exposure estimates across multiple pathways, routes and scenarios. Assessing exposures for potentially exposed and susceptible subpopulations may require consideration of how identified exposure scenarios may contribute to greater exposure distinct to a subpopulation. Addressing aggregate and sentinel exposures and accounting for differentially exposed subpopulations across all conditions of use for a chemical's lifecycle will likely entail defining and evaluating all relevant potential exposure pathways and scenarios using a variety of approaches. The views expressed in this abstract are those of the authors and do not represent Agency policy or endorsement.

Keywords: A-aggregate exposure

TU-SY-D3-311

Data Gathering Methods for Existing Chemicals Risk Evaluations under the Amended TSCA: Proposal for Using Natural Language Processing to Prioritize the Literature Search and Review

M. Cawley[3], A. Varghese[1], H. Hubbard[3], C. Henning[3], F. Branch[2], A. Benson[2], I. Camacho[2]; [1]ICF International, Durham, NC, [2]U.S EPA, D.C., DC, [3]ICF, Durham, NC

Abstract: The amended Toxic Substances Control Act (TSCA) requires EPA to assess risks for existing chemicals. The risk evaluation must "integrate and assess available information on hazards and exposure for the conditions of use of the chemical substance, including information on specific risks of injury to health or the environment" and "describe the weight of scientific evidence for the identified hazards and exposure." To meet these requirements, EPA searched for all relevant data for the major components of an evaluation (engineering, exposure, fate, ecological hazard, and human health hazard), which produced thousands of results per chemical requiring manual review. In this presentation, we explore natural language processing (NLP), to automate discovery of relevant data from the peer reviewed literature. We describe the initial data gathering method for the first 10 chemicals undergoing risk evaluation focusing on the strategy for the exposure, engineering, and fate disciplines. We then explain how NLP can sort references into clusters using text patterns to prioritize extensive search results. We used a supervised form of clustering by adding a seed set (i.e., references identified *a priori* as relevant to the chemical) to the corpus to be clustered. Using NLP techniques, we compared the clustered results from the peer reviewed literature to the results from manually screening the same peer-reviewed references. Specifically, we calculated recall and elimination rate using clustering predictions compared to the entirely manual approach of reviewing all results. This approach requires a seed set which might not be easily obtained for future chemicals that have fewer known relevant references available. To address this, we also present the results using a generic, chemical-agnostic seed set that could be applied to future chemical assessments. *The views expressed in this abstract are those of the authors and do not represent Agency policy or endorsement.*

Keywords: A-environmental regulation, A-risk assessment, TSCA Systematic Review, A - population exposure, C-multimedia

TU-SY-D3-312

Characterization of Uses of Chemical Substances for Exposure Assessment

N. N. Mottl, C. Bevington, C. Baier-Anderson, N. Orentas, S. Sarraino, A. Corado; U.S. EPA, Washington D.C., DC

Abstract: On June 22, 2016 the Frank R. Lautenberg Chemical Safety for the 21st Century Act was signed into law thereby amending the Toxic Substances Control Act (TSCA), the Nation's primary chemicals management law. Characterization of use informs the exposure assessor on how to develop exposure scenarios for occupational, general population, consumer, and environmental exposure assessments. Existing use description categories include industrial sector, industrial use type/processing codes, functional use, commercial/consumer product, and article. Mutliple use categories can be considered holistically to define how a chemical is used in a given application across its lifecycle (manufacturing, processing, use, dispoal). Valuable sources of cross-walking use information for exposure scenario development include the EPA Chemical Data Reporting (CDR) database, Toxics Release Inventory (TRI), Pre-Manufacture Notice (PMN) submissions, and other use-categorization and reporting schemes from authoritative sources. Once a chemical's use descriptors are defined across its lifecycle, additional use-related information can be used to further develop exposure scenarios by informing who is using the chemical, how they are using it, and where it is beign used. The goal is to effectively mine use information to set up exposure scenarios that are descriptive and distinct from each other to effectively characterize how chemicals can result in exposure from different applications across their lifecycle. A systematic appraoch to applying use description categories has the added benefit of promoting transparency when communicating with stakeholders. Examples of applicable use information that could further inform exposure assesement include use patterns (amount, frequency, duration), method of application (spray, brush, roll-on, dip), physical form, and setting of use. The views expressed in this abstract are those of the authors and do not represent Agency policy or endorsement.

Keywords: TSCA, use categories, exposure scenario development

TU-SY-E3: Exposure, Hazard, and Risk Assessment: Putting Exposure Back in the Process

TU-SY-E3-313

Hazard versus Risk: Blurring of the Lines.

D. Kaden; Ramboll Environ, Boston, MA

Abstract: Risk assessment involves the estimation of the probability, impact, and other attributes of an adverse effect, through multiple components including hazard identification, dose-response and exposure assessment. Omitting key steps can distort results. The risk assessment paradigm put forward over 30 years ago (NAS, 1983) set out a framework for organizing risk science information based on four fields of analysis, which describe the use and flow of scientific information in the risk assessment process. (1) *Hazard Identification* determines whether the available scientific data describe a causal relationship between an environmental agent and demonstrated injury to human health or the environment. It does not examine how much of the agent is necessary to cause harm-just whether the agent is associated with harm at some exposure situation. (2) *Dose-Response Relationships* establish the quantitative relationship between exposure (or dose) and response in existing studies in which adverse health or environmental effects have been observed. (3) *Exposure Analysis* moves the assessment from the study of known populations (laboratory or epidemiology) in which dose (exposure) and response occur together, to the task of identifying and characterizing exposure in other potentially exposed populations. (4) *Risk Characterization* then puts it all together to characterize risk. However, under the guise of the precautionary principle, many risk assessments have skipped steps 2 and 3 altogether to arrive at a (inaccurate) risk characterization. This is a troubling trend, as it results in public fear and, sometimes, regulatory decisions being made without a true understanding of relative risks. In order to make informed decisions in risk management, all steps of the risk assessment process are necessary. Exposure pathways must be understood, and exposure magnitude estimated. The dose-response must also be understood. This session will present different examples and perspectives on the use of hazard identification as a surrogate for risk.

Keywords: A-risk assessment, A-chemical prioritization, A-environmental policy, A-environmental regulation

TU-SY-E3-314

Global Trends in Risk Assessment in Pesticide Regulation

I. D. Kelly; Bayer Crop Science, Research Triangle Park, NC

Abstract: For several decades the Environmental Protection Agency has been a key authority in the development and implementation of risk assessment and risk characterization as it relates to pesticides. The Agency has been a proponent of assessments that are risk rather than hazard based due to the clear anomalies that a hazard based system can introduce. The Delaney Clause was one such example. First introduced in 1958, and covering the use of food additives found to cause cancer in animals or humans, its application in the registration of crop protection chemicals was clearly limiting the introduction of new safe chemistries. The Food Quality Protection Act was passed in 1996 and put the assessment of human risk on a strong scientific basis. In recent years, however the European Union has become more active in pesticide regulatory processes. Cut-off criteria for “CMR” chemicals (carcinogen, mutagen or reproductive toxicant) have been introduced and criteria are also under discussion for endocrine disrupters. The divergence in the role hazard based cut-offs play in the regulatory processes of North America and Europe and its implication as it relates to both the global movement of food and to the development of innovative technologies will be presented.

Keywords: A-environmental regulation, A-risk assessment, B-pesticides, C-food

TU-SY-E3-315

Risk, Hazard and Precaution in the United States and Europe

J. B. Wiener; Duke University, Durham, NC

Abstract: The Precautionary Principle espouses regulation in anticipation of uncertain risks, which may entail regulation based on hazard alone. Some claim that the EU has become more precautionary than the US, by adopting more anticipatory regulation of risks -- such as hormones in beef, genetically modified foods, toxic chemicals, and climate change. But we find -- based on an array of case studies, and a large-N quantitative study -- that the reality is general parity plus selective application of precaution to particular risks on both sides of the Atlantic. Examples of more precautionary US regulation include stratospheric ozone depletion, fine particle air pollution, mad cow disease in beef and in blood, and choking hazards. This more complex pattern has implications for regulatory policy, international trade, and learning from policy variation. Further, improved versions of precaution should go beyond hazard to incorporate more comprehensive assessment of risks (including exposure), benefits, costs, and tradeoffs, as well as to incorporate 'provisionality' mechanisms that foster learning and adaptive updating over time.

Keywords: A-environmental regulation, C-air, A-climate change, C-food, Precaution

TU-SY-E3-316

Threats to the Use of Exposure Science and Public Health for the Regulatory Process

B. D. Goldstein; Univ Pittsburgh, Pittsburgh, PA

Abstract: I will build on Jonathon Wiener's presentation on the precautionary principle (PP). First I will relate the PP to Public Health theory, including the linkage with exposure analysis. Second, I will bring in the context of EPA's current regulatory issues. Conceptually, there are three levels of prevention in public health: primary which precludes the risk occurring; secondary which identifies those already at risk before the unwanted adverse event occurs; and tertiary which lessens the impact of an adverse effect. The broad range of definitions of the PP straddle both primary and secondary prevention. Risk assessment and management generally is a form of secondary prevention in that the risk is already present in which exposure science participates when it assists in the quantitative description of risk. Conceptually, identifying and interdicting exposure is central to environmental public health. When exposure science is used as a means to understand the potential interface of a new technology with receptors, it can be a very powerful tool for primary prevention or the stronger definitions of precaution. Exposure science has been a driving force in improving analytical chemistry of human body fluids and excreta so that ever more minute levels of exogenous agents can be measured. But the "vanishing zero" of analytical chemistry has greatly outstripped the ability of animal toxicology or of exposure science linked to epidemiology to provide the information necessary for risk-based rather than hazard-based regulation. A strong precautionary approach to eliminating hazard without waiting for the data about exposure needed for quantifying risk therefore decreases the need for quantitative exposure measurement of ambient media. Also of importance to the future of exposure science are bills working their way through congress that could greatly decrease the role of science at EPA, such as the HONEST Act and amendment of EPA's advisory processes.

Keywords: A-risk assessment, A-environmental policy, A-environmental regulation

TU-SY-E3-317

Using 21st Century Exposure Science to Improve Risk Assessment

A. Guiseppi-Elie; United States Environmental Protection Agency, Research Triangle Park, NC

Abstract: Exposure science is used across the Federal government to inform many important actions. Federal agencies rely on exposure science or exposure assessments to allow them to develop risk mitigation strategies for public health and environmental concerns. The more precise these exposure assessments can be, the more targeted and in many cases, more limited the risk mitigation approaches can be. For example, EPA uses exposure science to more fully inform risk reduction strategies under the Toxics Substances Control Act (TSCA), the Clean Air Act (CAA) and other environmental statutes. To ensure state of the art exposure science, it is critical that exposure science approaches, tools and methodologies keep pace with emerging technologies, regulatory program requirements and societal needs. Major scientific advances in the field are needed to accomplish that goal. In 2010, the United States Environmental Protection Agency (EPA) and the National Institute of Environmental Health Sciences (NIEHS) commissioned the National Research Council (NRC) to develop a report with the goal of advancing exposure science, its use, and its impact similar to the NRC's *Toxicity Testing in the 21st Century: A Vision and a Strategy*. This presentation will focus on the role of exposure science in decision-making, drawing on scientific advances, and how they enhance the risk assessment process. The presentation will highlight how recommendations from recent National Academies of Science reports, *Exposure Science in the 21st Century: A Vision and a Strategy* (2012) and *Using 21st Century Science to Improve Risk-Related Evaluations* (2016) bolster the premise that exposure science is critical in risk assessments.

Keywords: exposure science, exposure assessments

TU-SY-F3: Integrating Diverse Environmental Exposure Datasets to Study Human Health - from Satellite Sensing and Ground Monitoring to Personal Exposure Part (2): Exposure Modeling, Data Integration, and Applications in Epidemiology

TU-SY-F3-318

Observations from the 2017 ATS Workshop on Air Pollution Monitoring for Health Research and Patient Care

K. Cromar; New York University, New York, NY

Abstract: The American Thoracic Society (ATS), with financial support from the US EPA, NASA, and NIEHS, sponsored a 2017 workshop on "Air Pollution Monitoring for Health Research and Patient Care" which brought together international experts to discuss the current capabilities and near-term opportunities for improvements in air quality monitoring using various technological platforms (i.e., satellite, low-cost portable monitors, sub-regulatory stationary monitors, and modeling). The motivation for this workshop stemmed in part from the recognition that integration of these technologies is essential in order to accomplish the shared objective of improving air pollution monitoring for health research and patient care. Themes and observations from this workshop will be addressed with a specific emphasis on near-term opportunities and challenges.

Keywords: A - ambient monitoring, C-air

TU-SY-F3-319

OpenAQ: An Air Quality Open Data Resource and Global, Grassroots Community

C. Hasenkopf; OpenAQ, Washington, DC

Abstract: Thousands of government ground-based air quality monitors around the world publicly publish real-time air quality data; however, researchers and the public often do not have access to this information in the ways most useful to them. Even when the data are available for a given country, across countries these data are not available in a consistent, universal data format, which stifles the development of universal tools and colllaborations across sectors and geographies. To increase accessibility and to begin standardizing this disparate dataset, our community have built OpenAQ, an open-source platform. The source code for the platform is viewable at github.com/openaq. Currently, we are aggregating, storing, and making publicly available real-time air quality data ($PM_{2.5}$, PM_{10}, SO_2, NO_2, BC, CO, and O_3) for more than 5000 stations in 47 countries via an Application Program Interface (API) and web interface (openaq.org) where users can download customizable csv files. We will present the OpenAQ platform, and also share the ways in which this platform has been built and used by a global, grassroots community, and how this community has impacted the ways that governments around the world share their air quality data.

Keywords: A - ambient monitoring, A - population exposure, A-environmental justice, A-environmental policy, D-community

TU-SY-F3-320

Remote sensing air pollution, exposure modeling, and health effects

Y. Liu[1], X. Hu[1], Z. Ma[2], M. Strickland[3]; [1]Emory University, Atlanta, GA, [2]Nanjing University, Nanjing, China, [3]University of Nevada, Reno, Reno, NV

Abstract: In recent years, satellite remote sensing data has emerged as an important source information in estimating air pollution levels at global to local scales. Well-designed statistical models driven by satellite aerosol remote sensing data, together with meteorological and land use information are able to provide accurate daily PM2.5 concentration estimates at a fine spatial scale. In this presentation, three modeling examples are presented to demonstrate the latest developments in this research area. First, a two-stage spatiotemporal model incorporating MAIAC AOD data, meteorological fields, and land use variables was used to estimate PM2.5 concentrations at 1 km resolution and investigate the 10-year spatial and temporal trends of PM2.5 levels in the southeastern US. Our time-series analysis results indicate that the PM2.5 levels decreased ~20% in the study region and ~23% in the Atlanta metro area during the period between 2001 and 2010, and the largest drop occurred between 2007 and 2008. Second, a two-stage spatial statistical model was developed using the Moderate Resolution Imaging Spectroradiometer (MODIS) AOD and assimilated meteorology, land use data to estimate PM2.5 concentrations in China from 2004 to 2014. Analysis of predicted PM2.5 levels showed a mean annual increase of 1.97 µg/m3 between 2004 and 2007 and a decrease of 0.46 µg/m3 between 2008 and 2013. Finally, we estimated associations between satellite-estimated daily PM2.5 concentrations and ED visits for six pediatric conditions in the U.S. state of Georgia by urbanicity classification. We found that a 10-µg/m3 increase in same-day PM2.5 concentrations was associated with ED visits for asthma or wheeze (OR = 1.013; 95% CI: 1.003, 1.023) and upper respiratory infections (OR = 1.015; 95% CI: 1.008, 1.022); associations with previous-day PM2.5 concentrations were lower. Differences in the association estimates across levels of urbanicity were not statistically significant.

Keywords: A-exposure models, A-geospatial analysis/GIS, A-statistical methods, B-particulate matter

TU-SY-F3-321

Geographic Information Sciences, Sensor Systems and Data Fusion for Better Exposure Estimation

M. Jerrett; University of California, Los Angeles, Los Angeles, CA

Abstract: Recent advances in computing resources, data analytics, smart phone technologies and Geographic Information Science (GISc) are combining to enhance opportunities to refine human and environmental exposure assessment. In this critical review, I will summarize the current state of GISc, with particularly emphasis on how to fuse data from smart phones with other exposure science data on ambient environmental conditions to refine estimates of exposure to environmental stressors. I will begin with defining opportunistic, participatory and ubiquitous sensing systems. While nuances in definition exist, opportunistic sensing generally refers to sensors that run in the background on smart phones, which require little or no interaction with the user to acquire data. Participatory sensors, alternatively, obtain information from the phone or from other miniature sensors that do place burden on the participant because they must carry an additional device or supply direct information to the phone. Ubiquitous sensing refers to sensors embedded into landscape or onto regularly parts of the infrastructure such a public transit vehicles. I will survey recent developments in all types of sensing with the primary objective of understanding how these sensors can be integrated with other geographic information, remote sensing data, and modeled ambient data to improve exposure estimation. Although highly promising, to date sensor systems have not been widely used in epidemiological investigations or well integrated into Geographic Information Systems. I will conclude with explanations for the minimal use, conjectures on the future of the field, and how sensors may contribute more than they have to date to understanding exposures to environmental stressors.

Keywords: A - ambient monitoring, A - exposure measurement, A - population exposure, A-activity patterns, A-geospatial analysis/GIS

TU-SY-F3-322

Estimating exposure to nature and its relation to human health within the Nurses' Health Study Cohorts

P. James[1], J. E. Hart[2], F. Laden[2]; [1]Harvard Medical School & Harvard Pilgrim Health Care Institute, Boston, MA, [2]Harvard Medical School & Brigham and Women's Hospital, Boston, MA

Abstract: Exposure to nature may influence health by buffering exposure to air pollution, noise, and extreme temperatures; providing opportunities for physical activity; creating a setting for social interactions; or directly improving mood and cognitive restoration. Using data from the nationwide Nurses' Health Study prospective cohorts (N~237,000), we have linked measures of exposure to nature to health outcomes, including physical activity, mental health, cancer, and mortality. We have spatially joined nationwide satellite data on greenspace to the geocoded home addresses of participants for up to 30 years of follow-up to create time-varying metrics of long term exposure to natural environments. We are also working to incorporate higher resolution measures of nature derived from aerial photos (green space, tree cover, water) at 1 meter resolution to our cohorts. Finally, we are working to incorporate mobile health technology into our cohorts to better understand high spatio-temporal resolution exposure to green spaces at the minute-level using smartphones and accelerometry-based devices.

Keywords: A-ecological exposure, A - exposure measurement, A-geospatial analysis/GIS, A-epidemiology, A-sensor technology

TU-SY-G3: Perinatal Exposures in the Home Environment: Sources, Measurements, and Associated Health Outcomes - Part 1

TU-SY-G3-323

Investigating Toddlers' Exposure to Organophosphate Flame Retardants in the Home Environment

A. Phillips[1], S. C. Hammel[1], K. Hoffman[1], A. Lorenzo[1], T. F. Webster[2], H. M. Stapleton[1]; [1]Duke University, Durham, NC, [2]Boston University School of Public Health, Boston, MA

Abstract: Following the phase out of polybrominated diphenyl ethers, organophosphate flame retardants (PFRs) have been increasingly used to meet commercial flammability standards. Additionally, some PFRs, such as triphenyl phosphate (TPHP), can have alternate uses as plasticizers. Many recent studies have indicated ubiquitous human exposure to various PFRs and have highlighted their potential health hazards as neuro- and developmental toxicants. The current study utilized paired samples of air, dust, handwipes, and urine collected from a cohort of 180 toddlers in central North Carolina to examine the relative importance of inhalation, ingestion, and dermal absorption exposure pathways for common PFRs. TDCPP, TCPP, and TPHP were frequently detected in air and handwipes (33-100%); EHDP, 2IPPDPP, 4IPPDPP; 4tBPDPP were commonly detected on handwipes (73-100%); BDCPP, BCPP, BCIPHIPP, DPHP, ip-PPP, and tb-DPP were commonly detected in urine (80-100%). TCPP levels on handwipes, but not in air, were correlated with urinary metabolites, BCPP and BCIPHIPP (r_s=0.23, 0.15, $p<0.05$), suggesting dermal absorption plays a vital role in TCPP exposure. TDCPP levels on handwipes and in air were correlated to urinary BDCPP levels (r_s=0.37, 0.37, $p<0.01$), indicating dermal and inhalation pathways are both important pathways for this compound. TPHP levels in air, as well as EHDP, TPHP, 2IPPDPP, 4IPPDPP, and 4tBPDPP levels on handwipes were all correlated with urinary DPHP (r_s=0.18-0.36, $p<0.05$), signifying multiple sources for this urinary biomarker. TPHP levels in air were also correlated with ip-PPP and tb-DPP levels in urine (r_s=0.29, 0.40, $p<0.05$), likely due to exposure to commercial flame retardant and plasticizer mixtures like ITPs and TBPP. Results also suggest that personal characteristics and behavior are important predictors of exposure. Taken together, these findings provide important exposure data for potentially toxic chemicals during early childhood—a critical developmental window.

Keywords: A - exposure measurement, A-biomarkers, A-indoor environment, A-exposure factors, B-flame retardants

TU-SY-G3-324

Young Children's Exposures to Molds and Consumer Product Ingredients in their Homes

N. S. Tulve[1], M. S. Clifton[1], S. Vesper[1], P. P. Egeghy[1], K. Thomas[1], E. M. Ulrich[1], D. Stout[1], K. Isaacs[1], D. Werthmann[2], F. Rabito[2]; [1]United States Environmental Protection Agency, Research Triangle Park, NC, [2]Tulane University, New Orleans, LA

Abstract: Young children's physiology and unique interactions with the environment have been shown to influence their exposures to common chemical and biological agents. Many types of consumer products are present in homes, resulting in the potential for direct and indirect exposures to multiple agents. Biological agents present in homes can combine with chemical agents to potentially impact health. Our objective was to examine how lifestage, product use, and building factors affect children's exposures to selected chemical and biological agents in their home. Multimedia samples (air, dust, surface and hand wipe, sock) and supporting information were collected from 12 children (4-11 years old) living in 8 low-income homes in New Orleans, LA (2016). Surface wipe samples (wipe area=929 cm^2) were collected from counters and floors in kitchens and bathrooms. Preliminary consumer product chemical concentrations (pg/cm^2) included linalool: 1-8,500; limonene: <MDL-1,250; methyl paraben: <MDL-20,000; propyl paraben: <MDL-13,000; butyl paraben: <MDL-290; triclosan: <MDL-2,000; piperonyl butoxide: <MDL-18,000; permethrin: 1-36,000; fipronil: <MDL-340. Settled dust samples were collected from door jambs and analyzed for mold contamination using a DNA-based analysis. The ERMI (Environmental Relative Moldiness Index) scale, used to measure water damage and predict potential for unhealthy mold conditions, was calculated. The average ERMI value was 5.8 (scale: -10 to 20); however, in 2 homes the ERMI values exceeded 14, indicating mold contamination and potential water damage. All surface wipe samples contained measureable concentrations of the selected chemical agents. These homes showed evidence of water damage and mold growth, with 4 homes in the highest quartile for mold contamination. These preliminary results provide additional evidence that the young children in this pilot study were exposed to a combination of multiple consumer product chemicals and biological agents in their homes.

Keywords: C-consumer products, D-children

TU-SY-G3-325

Organophosphate flame retardants and phthalates in dust from children's bedrooms compared with whole-house measurements

M. Frederiksen[1], V. Andersen[2], L. Knudsen[3], L. Mikkelsen[4]; [1]Danish Building Research Institute, Aalborg University, Copenhagen, Denmark, [2]Green Lab Brugerinnovation, Ballerup, Denmark, [3]University of Copenhagen, Copenhagen, Denmark, [4]The Ecological Council, Copenhagen, Denmark

Abstract: Children are exposed to a variety of chemicals indoors. Their bedrooms can be potential hotspots for chemicals as large quantities of plastics, other synthetic products as well as electronic equipment are often present. In the current study, we investigated the levels of 10 organophosphate esters (OPEs) and 7 phthalates in dust sampled in bedrooms of 17 children (age: 2-6 y) and compared with whole-house dust samples. Dust samples from the children's bedrooms were sampled on a filter using a non-scraping nozzle connected to a vacuum cleaner. The whole-house dust samples were taken from the content of the domestic vacuum cleaner bags sieved to <75um. Furthermore, the parents were interviewed about e.g. the perceived amount of toys in the room, number of electronic devices and ventilation habits. Most OPEs (8 of 10) were found in all homes, including the three chlorinated OPEs: TCEP, TCIPP and TDCIPP. Furthermore, TCIPP and TDCIPP were found at the highest concentrations only exceeded by TBOEP (median 1470 ng/g). Significant correlation was obtained between the two sample types for the majority of OPEs (8 of 10). However, absolute values varied slightly with generally higher levels in bedrooms except for TCEP, TMPP and TBOEP but most differences were non-significant. The phthalates were dominated by DEHP (median 66 µg/g). It was approx. one order of magnitude higher than the second most abundant, DiBP. The mean on filters and in bags were not significantly different; however, there was not a significant correlation between the two sample types either. When comparing with previously published phthalate levels from Denmark the levels have decreased markedly. However, some variation may be caused by differences in study design etc. In conclusion, good correlation between OPE-levels was found between children's bedrooms and the whole house, though the absolute values varied slightly. However, for phthalates no correlation between sample types was observed.

Keywords: A - exposure measurement, B-flame retardants, C-indoor, D-children, B-phthalates

TU-SY-G3-326

Self-Reported Consumption of School Lunches and Urinary Phthalate Metabolite Levels in US Children

K. Ferguson[1], R. C. Lewis[2], J. Meeker[2], J. Colacino[2]; [1]NIEHS, Durham, NC, [2]University of Michigan, Ann Arbor, MI

Abstract: Certain phthalate diesters are common components of food packaging materials and can migrate to foods with contact, potentially making diet a major source of exposure. Accordingly, school lunches, which are regularly consumed by children in the US, may be an important contributor to overall dietary exposure to phthalates in this part of the population. In this study, we investigated the association between eating school-prepared lunch and urinary phthalate metabolites measured in school age children (6-19 years) in the National Health and Nutrition Examination Survey. Study participants were restricted to those who attended school and who reported whether or not they consumed school lunches. Analyses were performed separately in kids (ages 12 and below) and teens (ages 13 and above) because of differences in school lunch consumption, other potential exposure sources, and metabolism. From 2003-2014, 4510 participants in NHANES reported attending school and provided information on school lunch consumption. Of these, 20.3% reported never eating school lunch (0 days per week); 19.3% reported eating school lunch sometimes (1-4 days per week); and 60.4% reported always eating school lunch (5 days per week). In fully adjusted models, we observed significantly higher urinary phthalate metabolite levels in kids who always ate school lunch compared to those who never ate school lunch, particularly for the metabolites of di-2-ethylhexyl phthalate (DEHP). Kids who always ate school lunch had 26-31% higher urinary DEHP metabolites compared to kids who never ate school lunch ($p \leq 0.01$). Levels of other high molecular weight phthalates, including mono-carboxyoctyl phthalate and mono-3-carboxypropyl phthalate, were also significantly elevated in kids who always ate school lunch (25-43% higher, $p<0.01$). No significant differences in phthalate metabolites were noted in teens by these groups. School lunch may be an important source of exposure to phthalates in kids under the age of 12.

Keywords: A-biomonitoring, A-biomarkers, B-phthalates, D-children, D-environmental justice

TU-SY-G3-327

Exploring Children's Chemical and Non-Chemical Stressors by Non-Targeted Analysis of House Dust

D. Mills[2], A. Marcotte[3], J. R. Sobus[1], N. S. Tulve[1], E. M. Ulrich[1]; [1]US Environmental Protection Agency, Research Triangle Park, NC, [2]US Environmental Protection Agency, Research Triangle Park, NC, [3]US Environmental Protection Agency, Research Triangle Park, NC

Abstract: Children are at higher risk for adverse health outcomes because of their developing physiology and activities and behaviors, as compared to adults. Additionally, they have increased exposure to chemicals found in house dust due to the amount of time spent indoors, specifically on floors, and child-specific behaviors (e.g., mouthing activities). A pilot study was undertaken to understand whether chemicals found in house dust can indicate exposure to non-chemical stressors (e.g., poor nutrition). A non-targeted analysis method was used to examine dust samples collected from two distinct areas of the same home (upstairs and downstairs locations). Samples were sieved into two particle size fractions (<150 µm and 150 µm-1 mm), extracted in methanol and analyzed using LC/Q-TOF/MS. Over 10,000 unique aligned features were identified and 1,782 total features from both ionization modes (769 negative and 1,013 positive) matched to the U.S. EPA's DSSTox database and were further investigated using the U.S. EPA's CompTox Chemistry Dashboard. Spearman correlation coefficients showed near perfect agreement (r=0.98) between features across the two size fractions (26 vs. 55 unique features for small and large), and strong agreement (r=0.86) between features in upstairs vs. downstairs samples (150 vs. 139 unique features). Together, these results indicate high similarity for chemicals adsorbed to dust particles throughout the house. Based on tentative chemical assignments, the chemical stressors in highest abundance are classified as flame retardants, plasticizers, pesticides and personal care products. Further investigations will include the analysis of additional dust samples to focus on: 1) which size fraction and sample location provides more information regarding children's exposure; 2) whether or not tentatively-identified chemicals can be used as surrogates for non-chemical stressors; and 3) the likely sources of tracer compounds that relate to non-chemical stressors.

Keywords: A-analytical methods, D-susceptible/vulnerable, D-children, Non-targeted analysis, QTOF, Non-targeted analysis, QTOF

TU-PL-A4: Pesticide Tools & Processes

TU-PL-A4-329

Assessing Global Endocrine Disrupting Activity in Drinking Water with Bioassays: A Possible Tool for Population-level Exposure Assessment

R. R. Jones[1], M. Inoue-Choi[1], D. Stavreva[3], L. Varticovski[3], P. Weyer[4], M. D. Wichman[5], T. Cain[5], N. Chavis[2], B. Graubard[1], L. Beane Freeman[1], G. Hager[3], M. H. Ward[1]; [1]National Cancer Institute, Bethesda, MD, [2]George Washington University, Washington, DC, [3]National Cancer Institute, Bethesda, MD, [4]University of Iowa, Iowa City, IA, [5]University of Iowa, Coralville, IA

Abstract: Background: Some natural and anthropogenic substances in drinking water sources are known or suspected endocrine disrupting chemicals (EDC); few are regulated or routinely measured in U.S. public water supplies (PWS). Compared to measuring single analytes, bioassays that assess activity levels may better quantify general population exposure to a variety of EDC. **Methods:** We conducted a pilot study of 10 PWS in Iowa, where agriculture-associated water contaminants (e.g., pesticides, nitrate) are suspected EDC. Source (untreated) water samples and finished (treated) drinking water samples were collected in spring and fall and concentrated 200x via a U.S. Geological Survey protocol. We assessed global EDC activity (aryl hydrocarbon [Ah], androgenic [AR], thyroid hormones [Ty], estrogenic [ER], and corticosteroid [CS]) with novel assays that express nuclear steroid receptor constructs in mammalian cell lines. We quantified statistically significant activity relative to negative controls and used bivariate and multivariate methods to compare activity overall, by season, and by utility/sample characteristics. **Results:** Among 62 samples (31 per season), 69% had significant Ah activity, 52% AR, 3% Ty, 2% ER and 0% CS. Ah activity was greater in the spring (84% vs. 55% of samples; χ^2p=0.01), and in utilities with ≥median (≥3.19 mg/L) nitrate levels (p=0.04). AR activity was more common in the spring (p<0.001) and in utilities served by surface versus groundwater sources (p=0.02). We found non-significant within-season contrasts in activity across strata of treatment and regulated contaminants. Multivariable analyses were imprecise, but suggested that surface waters and untreated samples had greater levels of Ah and AR activity after adjusting for other factors. **Conclusions:** Specific types of EDC activities in PWS varied by season and utility characteristics. This holistic approach to measuring EDC may hold promise for characterizing EDC exposures through drinking water.

Keywords: C-water, A-epidemiology, A-exposure models, A-geospatial analysis/GIS, B-mixtures

TU-PL-A4-330

Efficacy and exposure risks of total release foggers

Z. DeVries[1], R. Santangelo[1], J. Crissman[2], R. Mick[1], A. Suazo[3], C. Schal[1]; [1]North Carolina State University, Raleigh, NC, [2]Columbia University, New York, NY, [3]NC Department of Agriculture and Consumer Services, Raleigh, NC

Abstract: Differences in pest prevalence in homes result in disparities in pesticide use patterns, the efficacy of pest control products, and health risks associated with exposure to pests and pesticides. Total-release foggers (TRFs) are popular over-the-counter insecticidal products commonly used to control cockroaches. However, little is known about the efficacy and health risks associated with these products. Four TRF products were evaluated for their efficacy at controlling German cockroach (*Blattella germanica*) infestations. Total release foggers failed to reduce cockroach population, while the application of gel-baits resulted in significant reductions in cockroach populations (70-93%) within one month after treatment. Resistance of apartment collected cockroaches to pyrethroids, the active ingredients in TRFs, was extensive (59-347x relative to a pyrethroid-susceptible population), while resistance to fipronil, a common active ingredient in insecticidal baits, was considerably lower (6-23x relative to a fipronil-susceptible population). Knock-down resistance (*kdr*), a mechanism of target site insensitivity that confers resistance to pyrethroids, was widespread, with 96% of apartment-collected populations containing at least one allele of the L993F *kdr* mutation. Swabs of surfaces in the kitchen revealed large quantities of pesticide residues throughout the kitchen, with dermal exposure rates estimated as high as 1.67 mg d^{-1} for cypermethrin and 4.35 mg d^{-1} for piperonyl butoxide. The ineffectiveness of TRFs at reducing pest populations, their similar monetary cost compared to highly effective bait products, and the high human pesticide exposure risks they carry, call into question their utility in the marketplace. The disproportionate use of TRFs by people of lower socio-economic status suggests that TRFs are contributing substantially to indoor environmental social injustice.

Keywords: A-environmental justice, A-indoor environment, B-pesticides, C-consumer products

TU-PL-A4-331

Improving Spatial Resolution of Pesticide Applications from Arizona's Pesticide Use Reporting System

M. Furlong, P. I. Beamer; University of Arizona, Tucson, AZ

Abstract: Estimating exposures to nearby agricultural pesticide applications from Arizona's Pesticide Use Reporting system may be subject to substantial exposure misclassification. Pesticide use is reported at the section level of the Public Land Survey System (PLSS), which is typically one square mile. However, pesticide application is field-specific, and there are on average 16 fields per section. We sought to improve the spatial resolution of reported pesticide applications and the accuracy of pesticide exposure assignments by linking pesticide use reports to their crop-specific fields within the reported section. To determine whether exposure classification differed between methods, we compared percent of residences exposed using the section level of resolution against the field-crop level of resolution for a specific crop (alfalfa), pesticide class (organophosphate pesticides), county (Yuma), and year (2011). We used ArcGIS to import pesticide use reports for OPs applied to alfalfa in Yuma County in 2011, the 2011 CropScape Cropland Data Layer for Arizona from the USDA, and a shapefile of geocoded residences in Yuma County, AZ from the county assessor's office. We identified all sections with reported OPs applied to alfalfa in 2011, and all alfalfa-specific fields with reported OP applications. Finally, we calculated the number and percent of residences within 0.5 km and 1 km of alfalfa-specific OP applications at the section level and at the alfalfa-field level. For 2011, 9% of 21,494 addresses in Yuma County were within 0.5 km of the edge of an OP alfalfa section, and 5% were within 1 km. With the field-crop method, 6% were within 1 km of an OP field application and 3% were within 1 km (section vs field $p<0.01$). In sum, 33-40% of those classified as living near an OP application are misclassified when using the section-level method. Incorporating crop information may improve spatial resolution and exposure misclassification when using Pesticide Use Reporting systems.

Keywords: A - exposure measurement, B-pesticides, A - ambient monitoring, A-geospatial analysis/GIS

TU-PL-A4-332

Pesticide residues dietary exposure in Europe

L. Villamar-Bouza; European Food Safety Authority (EFSA), Parma, Italy

Abstract: In the safety assessments of maximum residues levels (MRLs) for pesticides in the European Union (EU), or proposals for MRLs, the chronic and acute dietary consumer exposure to pesticide residues is estimated by using a calculation model developed by EFSA called PRIMo (Pesticide Residue Intake Model). National food consumption figures considering different diets and unit weights provided by EU Member States and the internationally agreed risk assessment methodologies are the basis of the model to assess the pesticide exposure for consumer through the diet. Based on these calculations and a comprehensive assessment of the available data, EFSA gives a reasoned opinion related to the risk to consumer. EFSA is the keystone of risk assessment regarding food and feed safety. In close collaboration with national authorities and in open consultation with its stakeholders, EFSA provides independent scientific advice and clear communication on existing and emerging risks to risk managers, which are the EU Member States and the European Commission. The current PRIMo version 2 is publicly available. EFSA is currently working in a third version of the model that it would include the EU comprehensive food consumption database and the most recent implementations in terms of risk assessment methodology. As part of its on-going mission to advance risk assessment, EFSA has started using an innovative approach to dietary exposure known as cumulative risk assessment in order to consider risks arising from the exposure to more than one pesticide. The outcome of the EU pesticide residues in food monitoring program states in its latest report that this is a common scenario. Science-based approaches and integrated risk assessments by using experimental data, models for pesticide residues intake estimations and monitoring data to from real exposure are EFSA's working tools to contribute to EFSA's mission on protecting European consumers' health and the environment.

Keywords: A-risk assessment, A-cumulative exposure, B-pesticides, A - population exposure, A - exposure measurement

TU-PL-B4: Respiratory Health Effects of Air Contaminants

TU-PL-B4-333

Survey for dust exposure and the respiratory effect in toner-handling workers. -10-year cohort-

A. Ogami[1], K. Ikegami[1], S. Michii[1], H. Ando[1], R. Sugano[1], N. Yanagi[1], M. Uehara[2], T. Higashi[3]; [1]Institute of Industrial Ecological Sciences, Univ. of Occupational and Environmental Health, Japan, Kitakyushu, Japan, [2]Institute of Industrial Ecological Sciences, University of Occupational and Environmental Health, Japan, Kitakyushu, Japan, [3]University of Occupational and Environmental Health, Japan, Kitakyushu, Japan

Abstract: AIM: 10-year cohort study was carried out in the toner-handling workers to evaluate the relationship between the dust exposure and the respiratory effect. METHODS: Subjects that were included in the analysis of this study were 82 employees of a Japanese business equipment manufacturer. Onset of pneumoconiosis, pulmonary fibrosis, granulomatous pneumonia and lung cancer were assumed as endpoints of the investigation, and blood markers (KL6, SPD, etc.), respiratory function index and the chest CT reading were used as substitute endpoints for before the onset of these diseases. Disease onset was determined via a self-administered questionnaire, blood tests and respiratory function tests were conducted once a year, while Chest CT examinations were conducted in the 1st, 5th and 10th year of the study. The work environment dust monitoring was measured following the Japanese standard protocol. The personal exposure of the toner-handling workers was measured using digital dust measuring apparatuses (PDS-2, LD-6N SIBATA Scientific Technology Ltd.). Total exposure index (TEI) was estimated by the result of personal dust exposure and the questionnaire. RESULTS: The average dust levels in the environment of the toner handling work decreased well below the ACGIH allowable concentrations through the period of the study. None of the endpoint diseases developed in toner-handling workers. Annual percent change for blood marker and respiratory function levels, and chest CT parameters were compared according to the TEI but no statistical significance was observed. CONCLUSION: The crucial respiratory diseases were not observed in the 10-year cohort study of the toner-handling workers. From various results, we concluded that harmful effects on health by toner exposure were not observed, and the risk of developing respiratory disease in the toner-handling workers with the controlled dust exposure levels was extremely low.

Keywords: A-workplace, A-biomarkers, A-epidemiology, A-industrial hygiene, A-risk assessment

TU-PL-B4-334

Prenatal Exposure to Environmental Mixtures and Asthma Risk Among Children

R. Khalili[2], S. Korrick[3], V. Vieira[1]; [1]University of California, Irvine, Irvine, CA, [2]University of California, Irvine, Irvine, CA, [3]Brigham and Women's Hospital, Boston, MA

Abstract: Prenatal exposures have been shown to alter fetal immune development and increase subsequent risk of childhood atopic disorders, including asthma. Children exposed to multiple environmental stressors may be particularly susceptible to asthma. We investigated the association between asthma risk and mixtures of environmental stressors among children born 2000-2005 to mothers living near a PCB-contaminated Superfund Site in New Bedford Harbor, Massachusetts. We identified 10,517 births from four towns surrounding the Superfund Site who were followed until 2010. There were 418 asthma cases identified from emergency department and hospital discharge records, an indication of persistent and poorly controlled asthma. We estimated prenatal exposure to multiple chemicals using biomarker measurements from the New Bedford Cohort study and building regression models as a function of covariates available in birth records. We examined the joint effects of DDE, distance to road, and maternal age at birth using a multi-dimensional smooth within a generalized additive model (GAM), adjusting for confounders including maternal education and prenatal smoking. Results suggest that children of younger mothers who were also exposed to the highest levels of DDE and lived closest to major roads were at greatest risk of developing asthma with an odds ratio of 1.8 and p-value for the significance of the mixture smooth term of 0.002. This distance to road and DDE pattern was not apparent for older mothers. PCBs, which were highly correlated with DDE, produced similar results in separate models. This example illustrates the utility of using GAMs for visualizing mixtures; the importance of both chemical and non-chemical stressors in determining asthma risk would not likely have been discovered using more traditional epidemiologic models.

Keywords: B-mixtures, A-epidemiology, D-children

TU-PL-B4-335

Particulate Exposure in Asthmatic Kids (PEAK); Fine and ultrafine particle exposures among inner-city children

E. Majd[1], K. Koehler[1], M. McCormack[2], N. Hansel[2]; [1]Johns Hopkins University, Baltimore, MD, [2]School of Medicine, Johns Hopkins University, Baltimore, MD

Abstract: The adverse effects of particulate matter (PM) air pollution, particularly fine particles ($PM_{2.5}$, particles with diameter less than 2.5 μm), on respiratory health are well known. However, recent toxicological studies suggest the potential for stronger association of the ultrafine (UFP, particles with diameter less than ~0.1 μm) portion of PM exposure with respiratory health. Despite the known challenges of assessing their exposure, understanding the health effects of UFP is a critical step to developing interventions aimed at reducing the harmful effects of PM exposure. The PEAK study aims to determine the association of particle size, microenvironment and peak exposure with respiratory effects in inner-city children with asthma. We examined personal UFP and $PM_{2.5}$ exposures in a cohort of 15 asthmatic children aged 10-17 at high temporal and spatial resolution. Each child was followed for 7 days with 4 days of continuous monitoring of UFP (expressed as LDSA, lung-deposited surface area), $PM_{2.5}$ and geographic location via GPS using a backpack. On days 2-5, spirometry and exhaled NO measurements were done and biomarker (urinary LTE4) was collected. Daytime and nighttime asthma symptom were obtained via text message from the child or parent. Highly variable exposure to UFP (1-hr mean: 49 $\mu m^2/cm^3$; range: 1-1413 $\mu m^2/cm^3$) and $PM_{2.5}$ (1-hr mean: 23 $\mu g/m^3$; range: 1-377 $\mu g/m^3$) with weak correlation ($R^2 < 0.3$) were observed. Day-to-day and within-day variability were large for both UFP and $PM_{2.5}$ with mean coefficients of variation of 213% and 286%, respectively. Asthma symptoms, mostly mild and moderate, were observed in 22% of the nights and 27% of the days. Both UFP and $PM_{2.5}$ exposures showed high peaks with large day to day and between-day variabilities. The weak correlation between the two exposure metrics demonstrates the value of measuring both. By the end of this study we will be able to investigate the association between acute exposure to PM and asthma symptoms.

Keywords: A - exposure measurement, B-particulate matter, C-air, A-exposure factors

TU-PL-B4-336

Short-Term Effects of Airport-Associated Ultrafine Particle Exposure on Lung Function and Inflammation in Asthmatics

R. Habre[1], S. Eckel[2], H. Zhou[1], T. Enebish[1], S. Fruin[1], E. Rappaport[1], T. Bastain[1], F. Gilliland[1]; [1]University of Southern California, Los Angeles, CA, [2]University of Southern California, Los Angeles, CA

Abstract: Background: Exposure to ultrafine particles (UFP, particles with aerodynamic diameter <100nm) is associated with airway inflammation in asthmatics. UFP number concentrations (PN) are elevated downwind of Los Angeles International Airport (LAX) due to aviation activity. **Methods:** We conducted a randomized crossover study of 22 non-smoking adults with mild to moderate asthma (Nov-Dec 2014, May-Jul 2015) to investigate short-term effects of exposure to LAX UFPs in adult asthmatics. Participants conducted scripted, mild walking activity on two occasions in public parks inside and outside of the high UFP zone. Spirometry, exhaled nitric oxide, and circulating inflammatory cytokines were measured before and after exposure. Personal UFP PN and lung deposited surface area (LDSA) and stationary UFP PN, black carbon (BC), particle-bound PAHs (PB-PAH), ozone (O_3), carbon dioxide (CO_2) and particulate matter ($PM_{2.5}$) were measured. Source apportionment analysis was conducted. Health models investigated within-subject changes in outcomes as a function of pollutants and source factors. **Results:** A high contrast of ~ 34,000 #/cm^3 was achieved with mean (std) PN at the exposure and control site of 53,342 (25,529) and 19,557 (11,131) #/cm^3, and LDSA of 54.2 (17.4) and 26.5 (11.9) cm^2, respectively. Mean particle size was 31.5 nm (std 9.0) at the LAX site. Principal components analysis differentiated aviation (PN), traffic (BC, PB-PAH), PM mass ($PM_{2.5}$, PM_{10}), and photochemistry (O_3) sources. The aviation UFP factor was elevated at the LAX site compared to the control (p=0.001) and was significantly associated with IL-6, a circulating marker of inflammation, in single (β=0.28, 95% CI=0.11 - 0.45, p=0.003) and multipollutant models (0.24, 0.05 - 0.42, p=0.017). **Conclusions:** Our study demonstrates acute inflammation effects of airport-related UFPs and the importance of multipollutant source apportionment models to disentangle the health effects of specific UFP sources in a dense urban area.

Keywords: A - exposure measurement, B-nanoparticles, C-air, A-epidemiology, B-mixtures

TU-PL-B4-337

Reconstructing Historical Exposures to Respirable Dust and Respirable Silica in the Taconite Mining Industry for 1956-2010

Y. Shao[1], G. Ramachandran[2]; [1]University of Minnesota, Minneapolis, MN, [2]Johns Hopkins University, Baltimore, MD

Abstract: As part of ongoing epidemiological studies for assessing the association between exposure to taconite dusts and the development of respiratory diseases, the goal of our study is to reconstruct the historical silica (RS) and respirable dust (RD) exposures of workers in Minnesota taconite industry from 1955-2010 by developing a job exposure matrix (JEM) that uses 9,128 RS and 19,408 RD industrial hygiene monitoring data. Historical RS and RD data were obtained for seven taconite mines from three sources:(1) the Mine Safety and Health Administration (MSHA) has online database records for all inspection results since 1978 for 13 MSHA Mine IDs with 4,303 RD monitoring records;(2) the mining companies' internal monitoring reports provided 14,417 RD records, most of which date from the late 1970s.(3) University of Minnesota in 2010 conducted 688 pairs of RS and RD measurements covering six active mines. Unlike RD data which can be directly read from IH reports, many historical RS data have to be calculated using RD and silica percentage information. 8,840 RS data were calculated using available silica percentage information. After data treatment, all these data were grouped into 7 mines and then into 8 departments. Within each department, we applied a two-level random-intercept model which assumes that the natural log of Y (RD or RS) changes over time at a constant rate. Among the 56 mine-department combinations, 13 combinations show a significant (p-value<0.05) decreasing RS and RD trend with the maximum rate of decrease of 3.3% per year;12 combinations show a significant increasing trend with the maximum annual increase rate of 8.7%. The estimated percent silica varies by department, with the maximum value usually occurring in the crushing department (8.5%-27.6%) and the minimum value occurring in the pelletizing department (2.6%-8.5%).The result of this study is a JEM by mine, department, and year for RD and RS that will form the basis for future epidemiological studies.

Keywords: A-exposure models, A-cumulative exposure, D-occupational, B-particulate matter, A - population exposure

TU-SY-C4: Engaging Citizens and Embracing Diversity in Air Pollution Exposure Research

TU-SY-C4-338

Shared Air/Shared Action: A cross-community coalition to understand air quality

W. Griswold; University of Memphis, Memphis, TN

Abstract: This presentation is an overview of a project led by four Chicago communities, with a support team of university researchers and nonprofit organizations, to develop a shared understanding of their air quality. Its focus is the process used to address each community's differing concerns and goals, while helping each other develop the capacity to manage and implement community air monitoring programs. Using a Participatory Action Research model, stakeholders have collaboratively developed air monitoring plans tailored to individual communities' contaminants of concern and cultural landscapes. Peer-to-peer mentoring is utilized to develop local capacity for air monitoring, to exchange information and skills between communities, and to create intercommunity networks for understanding shared and disparate air quality issues.

Keywords: A-environmental justice, A-sustainability, B-particulate matter, D-community

TU-SY-C4-339

"Low-cost" Sensors for Measuring Gaseous and Particle Air Pollutants: Results from Three Years of Field and Laboratory Testing, and Real-Life Sensor Applications

A. Polidori, V. Papapostolou, B. Feenstra, H. Zhang; South Coast Air Quality Management District, Diamond Bar, CA

Abstract: In an effort to inform the general public about the actual performance of commercially available "low-cost" air quality sensors, in June 2014 the South Coast Air Quality Management District (SCAQMD) has established the Air Quality Sensor Performance Evaluation Center (AQ-SPEC).This program aims at performing a thorough characterization of currently available "low-cost" sensors under ambient (field) and controlled (laboratory) conditions. In the field air quality sensors are operated side-by-side with Federal Reference Methods and Federal Equivalent Methods (FRM and FEM, respectively), which are routinely used to measure the ambient concentration of gaseous or particle pollutants for regulatory purposes. Field testing is conducted at one of SCAQMD's existing air monitoring stations in Rubidoux, California. Sensors that demonstrate an acceptable performance in the field are brought back to the lab where a "characterization chamber" is used to challenge these devices with known concentrations of different particle and gaseous pollutants under different temperature and relative humidity levels. Testing results for each sensor are then summarized in a technical report and, along with other relevant information, posted online on a dedicated website (www.aqmd.gov/aq-spec) to educate the public about the capabilities of commercially available sensors and their potential applications. During this presentation, the results from three years of field and laboratory testing will be discussed in detail along with our experience developing sensor networks for specific applications, including: fenceline monitoring of fugitive PM emissions from a waste disposal facility, community monitoring of VOC emissions downwind of refineries and other industrial facilities, real-time monitoring of ambient PM concentrations to validate satellite data, and others.

Keywords: A-sensor technology, B-particulate matter, B-VOCs

TU-SY-C4-340

Engaging Citizens in Air Quality and Exposure Research - Field Study Implementation

S. Cho[1], L. Cicutto[2], K. Crews[2], M. McCullough[2], W. Hawthorne[3], J. Harris[3], J. Thornburg[1], G. Thomas[4], M. Ogletree[4], G. Harshfield[5]; [1]RTI International, Research Triangle Park, NC, [2]National Jewish Health, Denver, CO, [3]Groundwork-Denver, Denver, CO, [4]Denver Dept. of Environmental Health, Denver, CO, [5]Colorado Dept. of Public Health & Environment, Denver, CO

Abstract: Background: As low cost air sensors become readily available, communities have opportunities to collect data to understand their air quality and exposures and contribute to needed datasets for environment and public health research, policy decisions, and urban planning. **Purpose:** This study aims to set-up an air quality sensor network of sites important to community and identify best approaches to translate the sensor-generated data so that it is useable by communities. **Methods**: Our project is being conducted in the communities of Globeville, Elyria and Swansea (GES), an environmental justice community north of downtown Denver where a major expansion of highway corridors is planned. Through a pilot study, the study team including the GES community, a community-based non-profit organization, state and city public health departments, and research institutes are developing and testing guidelines and protocols for field study design and setup and ambient air and personal air quality exposure data collection. The Community Advisory Committee (CAC) consists of various community stakeholders. Wearable real-time sensors including RTI MicroPEM will be used for measurements of $PM_{2.5}$ and NO_2. **Results:** The CAC identified site selection criteria. Next, community partners and stakeholders submitted sites for monitoring and their rationale. The CAC reviewed submitted sites and made final selections. Several barriers have been encountered that include inability to use utility poles and gaining permission to access property. Data collection will occur from June-Sept 2017. Technical research efforts will evaluate the usability of field protocols and monitoring platforms for both stationary and personal monitoring applications, preference of data output representations, data quality of low cost sensors, and spatial variability of air quality. **Conclusion:** This pilot study will provide useful data and insights for future projects and community groups interested in air quality monitoring for action.

Keywords: A - ambient monitoring, A - exposure measurement, A-sensor technology, A-environmental justice, D-community

TU-SY-C4-341

Addressing wood smoke with student citizen scientists in a multi-cultural rural setting

E. Austin[1], E. Seto[1], J. L. Black[2], E. Spalt[1], M. Tchong-French[1], K. Hartin[1], E. Min[1], O. Stampfer[1], C. Karr[1]; [1]University of Washington, Seattle, WA, [2]Heritage University, Toppenish, WA

Abstract: The lower Yakima valley of Washington state is a highly impacted wood smoke region with a large population of Yakama Nation and immigrant Latino families. The partnership between a local university (Heritage U) and University of Washington builds on an established science education program that links undergraduate mentors with area high school students. Over this three year project, University of Washington scientists are developing a tailored air pollution curriculum, providing low-cost air pollution sensor technology and research grade monitoring equipment to students and mentors. The mentored student projects will lead to a better understanding of the exposure and health impacts of wood smoke in a rural, low-income community. A detailed pollution literacy questionnaire and community needs assessment allow for curriculum effectiveness and community experience to be evaluated at the beginning and end of each school year. The experience in deploying low-cost sensors will inform future community based air pollution projects.

Keywords: A - exposure measurement, A-environmental justice, A-sensor technology, D-First Nation, D-community

TU-SY-C4-342

Low-Cost Air Quality Monitoring in Environmental Justice Communities in Allegheny County

S. Ramachandran[1], N. Zimmerman[1], S. Kumar[1], R. Tanzer[1], J. Gu[3], J. Downs[2], A. Presto[1]; [1]Carnegie Mellon University, Pittsburgh, PA, [2]Carnegie Mellon University, Pittsburgh, PA, [3]SenSevere, Pittsburgh, PA

Abstract: Allegheny County has a history of exceeding National Ambient Air Quality Standards for different pollutants, with major local sources near environmental justice (EJ) communities. Local hot spots and transient spikes may not be captured with traditional reference monitoring networks which tend to be sparsely distributed. Recently, low-cost sensors have enabled air quality monitoring with high spatial and temporal resolution. We have developed a new low-cost air quality monitor, the Real-time Affordable Multi-Pollutant (RAMP) sensor package, which measures up to four gaseous pollutants, carbon dioxide, and fine particulate mass ($PM_{2.5}$) with a Met-One neighborhood PM monitor (NPM). The options for gases include carbon monoxide, nitrogen dioxide, nitric oxide, sulfur dioxide, and ozone using electrochemical sensors, and volatile organic compounds (VOCs) using a PID detector. We use machine learning-based calibration strategies to convert gas sensor response into measurements comparable to conventional, high-cost monitoring networks. The Met-One NPM showed the most consistent performance for ambient PM, usually agreeing to within ±5% of each other and correlation coefficients (r^2) of 0.9 or higher. Citizen volunteers and schools were recruited through word-of-mouth to host the RAMPs, with 25 RAMPs already deployed across Pittsburgh and nearby suburbs. Preliminary data from EJ communities show $PM_{2.5}$ spikes as high as 120 $\mu g/m^3$ in Lincoln and around 50 $\mu g/m^3$ in Braddock and Clairton. We are working with local community groups to further expand the network in these communities. The RAMPs transmit data to a central web server over a GSM network, and a web interface (designed based on input from a Community Advisory Board) will provide real-time information from these monitors to community members, some of whom will be recruited for in-person surveys to inform risk communication strategies.

Keywords: A - ambient monitoring, risk communication, A-environmental justice, A-sensor technology, B-particulate matter

TU-SY-D4: Challenges and Opportunities: Assessing Exposures to Chemical Substances under Amended TSCA- Methods, Models, and Data - Part 2

TU-SY-D4-343

Overview and Examples of EPA/ OPPT Occupational Exposure and Releases Assessment and Tools

S. Prothero, G. Macek; US EPA, Washington, DC

Abstract: On June 22, 2016, the Frank R. Lautenberg Chemical Safety for the 21st Century Act was signed into law amending the Toxic Substances Control Act (TSCA), the Nation's primary chemicals management law. The Office of Pollution Prevention and Toxics (OPPT's) risk evaluation addresses occupational settings as part of the exposure assessment. Releases to the environment and occupational exposures may be considered simultaneously to gain assessment efficiencies. Key exposure parameters include duration, intensity, frequency and number of exposures under the conditions of use of a chemical, and key release parameters may be expressed similarly. Typically, measured data are preferred, but when data are not available, empirical or modeling approaches or assumptions are used. This presentation provides an overview of OPPT's approaches and tools used for assessing occupational exposures and releases and will cover two examples. The first example will describe how monitoring data and two-zone exposure modeling was used to estimate worker vapor inhalation exposure to the existing chemical 1-Bromopropane in uses such as spray adhesive application and vapor degreasing. The second example will describe how an EPA/OPPT-developed generic scenario on the use of metalworking fluids was used to estimate potential worker exposures and environmental releases in a new chemical review. The views expressed in this abstract are those of the authors and do not represent Agency policy or endorsement.

Keywords: A-exposure models, D-occupational, A-workplace, environmental release

TU-SY-D4-344

Exposure Modeling Tools and Databases for Consideration for Relevance to the Amended TSCA

K. Isaacs[1], J. Wambaugh[2], P. Price[1], J. Bare[3], K. Dionisio[1], K. Phillips[1], R. W. Setzer[2], C. Barber[1], B. A. Wetmore[1], R. Tornero-Velez[1], J. Orme-Zavaleta[1], R. S. Thomas[2], E. M. Wong[4], C. Fehrenbacher[4]; [1]US Environmental Protection Agency, Durham, NC, [2]US EPA, Research Triangle Park, NC, [3]US EPA, Research Triangle Park, NC, [4]US EPA, Washington, DC

Abstract: The Agency's Office of Research and Development (ORD) has a number of ongoing exposure modeling tools and databases. These efforts are anticipated to be useful in supporting ongoing implementation of the amended Toxic Substances Control Act (TSCA). Under ORD's Chemical Safety for Sustainability Research Program, new databases of how chemicals are used in commerce (e.g., in products or industrial processes) have been collated from dozens of public resources. These use data have enabled the development and application of high-throughput (HT) pathway-specific exposure models (e.g. the HT Stochastic Human Exposure and Dose model) and calibrated consensus exposure predictions for thousands of chemicals evaluated with biomonitoring data. These tools allow identification of the potential route of chemical exposure for certain conditions of use. A newer effort, the Human Exposure Model, incorporates new information on population variability (e.g., demographics, product use, geography, housing) that impacts exposures, and thus may be applicable to the identification of potentially exposed or susceptible subpopulations. To provide context to these predictions, new HT methods are being developed for converting external exposures to internal tissue doses, including methods that incorporate physiologic and lifestage variability. Finally, human-oriented exposure models also provide chemical release input to new ecological exposure models, which are being evaluated with public water monitoring data. While the amended TSCA provides an opportunity for ORD exposure modeling tools and databases to support amended TSCA, prior to any implementation the fitness-for-purpose of these projects (e.g., for prioritization, scoping, or risk evaluation) must be evaluated in the context of TSCA requirements. *The views expressed here are those of the authors and do not necessarily reflect the views or policies of the U. S. EPA.*

Keywords: A - population exposure, A-exposure models, A-chemical prioritization, A-environmental regulation, A-aggregate exposure

TU-SY-D4-345

Generating Exposure-Relevant Measurement Data for Potential Use in Support of TSCA Requirements

J. Wambaugh[1], J. R. Sobus[2], A. Williams[1], X. Liu[3], M. Stryar[2], C. Grulke[1], A. Richard[1], E. M. Ulrich[2], S. Newton[2], K. Phillips[2], K. Isaacs[2], C. Sonich-Mullin[3], J. Orme-Zavaleta[2], R. S. Thomas[1], C. Bevington[4], C. Fehrenbacher[4]; [1]U.S. Environmental Protection Agency, Research Triangle Park, NC, [2]U.S. Environmental Protection Agency, Research Triangle Park, NC, [3]U.S. Environmental Protection Agency, Research Triangle Park, NC, [4]U.S. Environmental Protection Agency, Washington, DC

Abstract: The EPA Office of Research and Development (ORD) has a number of ongoing projects which generate exposure measurements. These data may be useful in supporting ongoing implementation of the amended Toxic Substances Control Act (TSCA). Various types of exposure measurements include physical-chemical property information, monitoring (environmental, biological, and consumer product), and emission testing data. These data may allow identification of exposure levels or exposure potential for chemical substances and mixtures. ORD has recently developed the Computational Toxicology (CompTox) Chemistry Dashboard for dissemination of high-quality curated chemical structures, physicochemical properties, and exposure-related data. The CompTox Chemistry dashboard includes measured and predicted properties for over 700,000 chemicals. ORD is augmenting the data with new measured values for potentially challenging chemical classes. Further, data for many chemicals in consumer products, human serum, house dust, drinking water, and other environmental media are being generated by both ORD and contract laboratories to help refine exposure pathways and reduce uncertainty in exposure predictions. Suspect screening and non-targeted analyses (SSA/NTA) using high resolution mass spectrometry methods are being used with targeted confirmation for high-priority chemicals. Finally, both ORD and contract laboratories are generating data describing the emission and migration characteristics of chemicals from consumer products using testing chamber methods. While the amended TSCA provides an opportunity for ORD generated exposure data to support amended TSCA, prior to any implementation the fitness-for-purpose of these data (e.g., for prioritization, scoping, or risk evaluation) must be evaluated in the context of TSCA requirements. *The views expressed here are those of the authors and do not necessarily reflect the views or policies of the U. S. EPA.*

Keywords: A - exposure measurement, A-chemical prioritization, A-risk assessment, A-sensor technology, A - population exposure

TU-SY-D4-346

Chemical exposure assessment under the new Toxic Substances Control Act: An NGO's perspective

J. McPartland, L. McCormick; Environmental Defense Fund, Washington, DC, DC

Abstract: Under the Frank R. Lautenberg Chemical Safety for the 21st Century Act, which amends the Toxic Substances Control Act of 1976 (TSCA), the U.S. Environmental Protection Agency (EPA) is charged with establishing a robust system to identify, assess and regulate chemicals presenting risks to humans or the environment. Environmental Defense Fund (EDF) is a science-based advocacy organization that works to advance health-protective chemicals policies. This presentation will provide EDF's perspective on new statutory mandates particularly relevant to exposure assessment, such as identification of conditions of use, prioritization based on exposure potential, evaluation of aggregate and sentinel exposure, and consideration of potentially exposed and sensitive subpopulations. Observations from the first ten chemical scoping documents and opportunities for improvement will be discussed, including consideration of the chemical lifecycle, potentially exposed subpopulations including children and workers, reasonably forseen uses, issues associated with workplace exposure control methods (e.g., engineering controls and personal protective equipment), and the use of measured versus estimated data. Additionally, advances in exposure detection technology will be highlighted, specifically wearable chemical monitors, and how these emerging tools may increasingly inform chemical assessment activities such as those pursued under TSCA. Finally, opportunities for engagement by the academic and public interest community in TSCA implementation will be discussed.

Keywords: A-aggregate exposure, A-chemical prioritization, A-environmental policy, A-risk assessment, D-susceptible/vulnerable

TU-SY-D4-347

Challenges and Opportunities: Assessing Exposure to Chemical Substances under Amended TSCA

R. Zaleski; ExxonMobil Biomedical Sciences, Inc., Annandale, NJ

Abstract: Under the 2016 Lautenberg Chemical Safety Act amendments to the U.S. Toxic Substances Control Act (TSCA), data and information about both hazard and exposure potential will be central to EPA's risk based decision-making about TSCA chemicals. Under the amended law, EPA will evaluate chemicals under their conditions of use. As a result, exposure considerations, in particular, will be critical both to EPA's designation of chemicals as high or low priority for risk evaluations and in subsequent EPA risk evaluations of high priority chemicals. In recent years, exposure science measurement and predictive capabilities have advanced significantly. New understandings have evolved in 21st century hazard and exposure science. These developments include: - REACH tools on exposure assessment - Updated and new EPA exposure assessment models and tools - OECD approaches to harmonizing use categories and exposure scenarios - High Throughput (HTP) approaches to exposure that complement HTP toxicity testing All of these advances are useful to consider for integrating into transparent, consistent approaches to regulatory risk-based decision-making. This will enable both efficiency and credibility in regulatory decision-making. Learnings from experience to date with these new methodologies and tools will be discussed, focusing on relevance to TSCA.

Keywords: A-exposure models, A-chemical prioritization, A-risk assessment

TU-SY-E4: Exposure Assessment and Epidemiology for Regulatory Decision Making- Challenges and Opportunities

TU-SY-E4-348

Challenges for the Agrochemical Industry

J. E. Collins; CropLife America, Washington, DC

Abstract: Under the Federal Insecticide, Fungicide, and Rodenticide Act (FIFRA), US registration of a crop protection product for sale and use requires a data-rich registration package with specific Good Laboratory Practice toxicological studies to establish safety. The Federal Food, Drug and Cosmetic Act further establishes data requirements and information needed to establish safety of pesticide chemical residues in food. Required studies are the basis of the risk assessment for environmental and human health to establish no "unreasonable adverse effects on the environment." In December 2016, EPA posted its *Framework for Incorporating Human Epidemiologic & Incident Data in Risk Assessments for Pesticides.* Significant concerns with use of epidemiologic data in risk assessment exist, and no guidance has been developed for its use. Nevertheless, EPA Office of Pesticide Programs has proposed to integrate epidemiologic study outcomes in its risk assessments, with no guidance for design or use of epidemiological studies, or how EPA intends to integrate such outcomes when primary data are not available. No criteria exist for study or data quality. Such disparity in data quality requirements between primary *in vivo* and *in vitro* studies in the registration package, and the literature outcomes reported from epidemiologic studies begs the question as to how studies are assessed and how is a weight-of-evidence approach taken. Significant scientific questions arise as to the validity of some studies used by EPA when no hypothesis or mechanism of action is proposed. Attempts to integrate data or literature reports is confounded by concerns of scientific veracity when such studies cannot be replicated or verified. Use of such studies has resulted in undue conservative application of safety factors to pesticide registrations under the Food Quality Protection Act, and thus has the potential to inappropriately limit of use of value tools for agriculture.

Keywords: A-risk assessment, A-environmental policy, A-epidemiology

TU-SY-E4-349

Estimating pesticide exposures: areas of strength and weakness in epidemiologic studies

C. J. Burns; Burns Epidemiology Consulting, Sanford, MI

Abstract: Comparing the merits of toxicology and epidemiology studies can become like parents extolling their child's prowess at a school function or sporting event. As an epidemiologist, I consider the real-life exposure experience in human populations to be highly relevant and informative for public health. However, just as I know (deep down) that my daughter will never play soccer in the Olympics, the practice of exposure assessment by epidemiologists is not at a world class level. We, epidemiologists, can learn from the robust nature of toxicology studies. The US EPA Office of Pesticide Programs (OPP) released its Framework for Incorporating Human Epidemiologic & Incident Data in Risk Assessments of Pesticides in 2016. Therein, the OPP presented quality considerations (low, medium and high) for exposure assessment using questionnaires and for biomonitoring data. These score sheets are a starting point for discussion. Knowing that we must rely upon observational methods and that the human dose levels are much lower than for laboratory animals, the demand for excellence is high. Epidemiologists need to ensure that biologically relevant exposure data are collected, that the data are sufficiently transparent, that measurement error has been quantified and that multiple exposures have been assessed.

Keywords: A-epidemiology, A-exposure factors, B-pesticides

TU-SY-E4-350

Transparent and Systematic Reviews of Exposure Data in Environmental Epidemiology: Approaches and Case Studies

J. LaKind; LaKind Associates, LLC, Catonsville, MD

Abstract: There is a growing interest in systematic reviews of observational epidemiology research for weight of evidence assessments. Various checklist approaches to assessing studies have been published. However, they generally include vague approaches to evaluating the quality of studies under review; this limits the opportunity for transparent, consistent and reproducible approaches. The quality of exposure assessments in environmental epidemiology is highly variable and a method for describing quality across studies and across reviews should be used. A published instrument (the BEES-C; Biomonitoring, Environmental Epidemiology, and Short-lived Chemicals instrument) has been used to examine various bodies of literature, and the approach and results from these studies are described. Results of reviews for certain chemicals (e.g., BPA, 2,4-D), matrices (breast milk) and publication type (reviews) will be described.

Keywords: A-epidemiology, A-environmental policy, C-consumer/personal care products, A-biomonitoring, systematic review

TU-SY-E4-351

How Epidemiologic Studies Can Be Used Qualitatively and Quantitatively in the Regulatory Decision Making Process

T. Luben[1], E. Kirrane[1], J. Lee[1], A. Davis[1], S. jenkins[2], J. Sacks[1]; [1]U.S. EPA, RTP, NC, [2]U.S. EPA, RTP, NC

Abstract: Together, epidemiologic, animal toxicological and controlled human exposure studies provide valuable evidence to inform policy makers as they make science-based regulatory decisions to protect human health from the adverse effects of air pollution. Observational epidemiologic studies are unique among these three study types, as they are able to include susceptible or vulnerable individuals who might be at greatest risk, provide information about health effects occurring at real-world or ambient concentrations, and can include various relevant health outcomes (e.g., mortality, emergency department visits) that cannot be or are seldom evaluated in animal toxicological or controlled human exposure studies. The results from epidemiologic studies can be used in two ways to support regulatory decision making: (1) qualitatively, to inform the hazard identification aspect of risk assessment; and (2) quantitatively, to inform hazard identification, the dose-response relationship and/or the health impact (burden) component of risk assessment. This presentation will review the characteristics of epidemiologic studies that make them appropriate and very useful for informing each of these risk assessment components, and ultimately regulatory decision making. We will present examples from epidemiologic studies of lead (Pb), arsenic (As) and ambient particulate matter (PM) that were used qualitatively and/or quantitatively in recent EPA science assessments to demonstrate these points. **Disclaimer:** The views expressed here are those of the authors and do not necessarily represent the views or policies of the U.S. Environmental Protection Agency.

Keywords: A-environmental regulation, A-epidemiology, B-particulate matter, C-air, A-environmental policy

TU-SY-E4-352

Regulatory Human Health Risk Assessment Calls for Good Epidemiology Practice

H. Lynch[2], J. Goodman[2], Z. Yan[1]; [1]Dow AgroSciences, Indianapolis, IN, [2]Gradient, Cambridge, MA

Abstract: Toxicology and epidemiology both inform human hazard and risk assessment in regulatory settings. While toxicology studies are often conducted using standardized protocols (*e.g.*, Organisation for Economic Co-operation and Development [OECD] test guidelines) and in compliance with Good Laboratory Practice (GLP) guidelines, there are no such established standards for conducting epidemiology studies. Although the existing publications and frameworks provide guidance on how to evaluate epidemiologic study quality, they are most often high-level, generic and reliant upon expert judgement, rather than definitive and quantifiable instructions. Further, if epidemiology studies are to provide value to human health risk assessments, there is a need to improve the quality of these studies. Building on the available frameworks and GLP guidelines for toxicology studies, we have begun developing guidelines for conducting and reporting results from epidemiology studies. These "Good Epidemiology Practice" (GEP) guidelines are organized into several categories, each of which includes multiple requirements: objectives and study plans, study design, exposure characterization, outcome assessment methods, study results, and discussion. Establishing GLP-equivalent guidelines for epidemiology studies will not only aid data quality evaluation for existing studies, but also provide detailed instructions and standardization of future studies. The ultimate goal of this initiative is the best use of epidemiology studies for regulatory decision making and public health protection.

Keywords: A-epidemiology, A-risk assessment

TU-SY-F4: An Infrastructure for Generating Exposomes: Initial Lessons from the Utah PRISMS Platform

TU-SY-F4-353

Introduction to Utah PRISMS Platform

K. A. Sward; University of Utah, Salt Lake City, UT

Abstract: The Pediatric Research using Integrated Sensor Monitoring Systems (PRISMS) program was launched by the NIH/NIBIB in 2015 to develop a sensor-based, integrated system for measuring and monitoring environmental, physiological, and behavioral factors in studies of pediatric asthma and other conditions. The PRISMS program has 3 inter-related arms: development of wearable and stationary sensors relevant to epidemiologic studies of asthma; a data and software coordination center; and an informatics platform. We describe key informatics challenges and how the overall PRISMS infrastructure is designed to address those challenges. The Utah PRISMS informatics platform has 3 key components: a data acquisition pipeline that supports interactions with study participants; a data federation platform to integrate heterogeneous data sources, and a research support platform. The components are interconnected with each other and with the overall PRISMS program, forming a coherent whole. The Utah PRISMS informatics center is a collaboration by a team of diverse expertise including electrical, computer, chemical and industrial engineers, atmospheric and computer scientists, informaticists, and pediatric researchers in developing an infrastructure for generating exposomes.

Keywords: A - ambient monitoring, A-epidemiology, A-exposure models, D-children

TU-SY-F4-354

Utah PRISMS Data Acquisition System for Generating Exposomes

N. Patwari[1], P. Lundrigan[2], K. Min[1], K. Kelly[3], M. Meyer[2], K. A. Sward[4]; [1]University of Utah, Salt Lake City, UT, [2]University of Utah, Salt Lake City, UT, [3]University of Utah, Salt Lake City, UT, [4]University of Utah, Salt Lake City, UT

Abstract: We have architected, deployed, and tested a data acquisition pipeline for the Utah PRISMS platform, designed to simplify deployment by exposure scientists of networks of wireless sensors into a subject's home. Our primary goals are: 1) ease of deployment, 2) high data integrity, 3) open source, 4) low cost, and 5) low latency. We describe several design decisions tied to these goals and the types of deployments enabled by our pipeline. Our contributions are from sensor through server. We have built a sensing platform on a TI Beaglebone to read, buffer, and transfer data from sensors (including a Dylos DC1100) over a WiFi channel. We have developed software for a gateway, operating on a Raspberry Pi located in the home. The gateway provides: buffering capabilities to increase data integrity in case of outage; allows simplified sensor configuration and authentication; pulls data from sensors in a time-division protocol to avoid collisions; allows connection to a variety of "internet-of-things" (IoT) sensors and actuators via the open-source Home Assistant project; and allows for local data reduction, decisions, or actuation. We have deployed our pipeline in five types of deployments: 1) We deploy a variety of commercial IoT devices (motion sensors, proximity sensors, video cameras, smart plugs) operating on different wireless protocols (WiFi, Z-wave) to demonstrate their integration. 2) We deploy several (8-11) air quality sensors in a single house to ensure scalability and to collect data for spatiotemporal modeling studies. 3) We deploy in homes with children with asthma to study the airborne particulate matter in a home and how data feedback to its residents impacts them. 4) We have deployments in which a furnace fan is controlled to automatically turn on and filter the air whenever the particulate matter reading is high. 5) We deploy systems that use LoRa to enable wider area deployments for which WiFi has insufficient range.

Keywords: A - ambient monitoring, A-sensor technology, wireless networking

TU-SY-F4-355

Utah PRISMS Data Integration Platform for Generating Exposomes

R. Gouripeddi[1], M. Cummins[2], N. Burnett[1], R. Madsen[3], P. Mo[3], P. Warner[3], R. Butcher[3], K. A. Sward[2], J. Facelli[1]; [1]University of Utah, Salt Lake CIty, UT, [2]University of Utah, Salt Lake City, UT, [3]University of Utah, Salt Lake City, UT

Abstract: Performing exposure research requires streamlined integration of heterogeneous and disparate data sources for supporting a diverse set of use-cases. A platform supporting such integration would need to empower researchers by reducing complexities of working with multi-dimensional exposure Big Data and at the same time limiting semantic information loss associated with data transformations that might occur when integrating or converting to different formats. In presentation we discuss several exposomic research use-case architypes elicited from researchers and literature. Based on our findings, we developed a conceptual representation of an exposome for translational research that characterizes uncertainties in using exposomic data. Using these results, we are developing the Utah PRISMS Federated Integration Architecture which is a comprehensive, standards-based, open-source informatics platform that allows federation and integration of high-resolution clinically relevant exposome information from environmental, physiological, and behavioral sensors, computational models and biomedical data in a meaningful manner. Towards harmonizing different sensor data, we developed a data model that harmonizes metadata and measurements from different types of sensors. This model consists of Instrument, Deployment and Measurement Output domains that describe characteristics of a sensor device, its deployment in research settings, and its measurements respectively. The core data integration platform consists of an ontology server (OS) that stores and harmonizes semantic descriptions within different data resources (DR); a graphical metadata repository (MDR) that stores metadata artifacts of DR, data adapters (DA) to facilitate interoperability with DR; administrative and security components; an Event Document Store (EDS) - a Big Data store of project-specific data integrations that supports different traditional and data science analytical methods, as well as generation of analytic data models; and a federated query engine that orchestrates the integration process between the MDR, OS, DA and DR to populate EDS and analytical data models. We will share results from our pilot exposome generation efforts. We envision this infrastructure to support different types of environmental biomedical studies.

Keywords: A-aggregate exposure, Exposome generation, Big Data Integration, Data Modeling

TU-SY-F4-356

Mathematical Modeling and Uncertainty Quantification for Exposomic Studies

J. Facelli, A. Lund; University of Utah, Salt Lake City, UT

Abstract: It is difficult to measure the impact of air quality on human health at the individual level because populations are mobile and air quality data is reported at low geographic resolutions (>1 km2), which makes it difficult to characterize acute local variations in air quality and obtain a precise estimate of actual exposure. This presentation discusses a mathematical modeling framework that uses sparse environmental experimental data from mobile sensors, environmental monitoring and point sources and agent-based human activity estimates to provide a high-resolution spatial-temporal grid of individual exposure. Here we report on the development of an agent-based model to help simulate human activities and locations throughout an arbitrary day. Simulated households are constructed from census tables on a census block level. Each individual is randomly assigned an employment and school status based on aggregate data for the state of Utah. Activity profiles are generated from the American Time Use Survey produced by the BLS. The activity profiles are combined with the simulated households to build individual trajectories of activity and location over the desired region of study. This activity model is coupled with a high resolution PM 2.5 model built using multiple sources including EPA and Mesonet measurements, point source and automobile emission data and household emissions. The sensitivity of the combined model to the assumptions used is tested using Monte Carlo ensemble techniques. The methods proposed here provide a strategy for generating better exposome estimates for epidemiological studies, without using costly individual measurements of human movement and activity data.

Keywords: A-exposure models, A-activity patterns, A-aggregate exposure

TU-SY-F4-357

Presentation and Visualization of Exposomic Data for Research Use-cases

K. A. Sward; University of Utah, Salt Lake City, UT

Abstract: The interdisciplinary and multi-scale nature of exposure science, and the multiple different communities engaged in the research process, leads to a set of diverse research use-cases. The PRISMS platform is aiming to create a replicable, scalable infrastructure for designing and conducting exposomic studies; with pediatric asthma as the initial exemplar. Understanding the effects of the modern environment on pediatric asthma requires generating a complete picture of the contributing environmental exposures and socio-economic factors. Such an exposome requires integration of data from wearable and stationary sensors, regional environmental monitors, self-reported symptoms and other person-generated data, along with physiology data, medication use, and other clinical data. The platform needs to support rigorous, reproducible science. We leverage user-centered design approaches and human-computer interaction principles to design meaningful data presentations/visualizations for three primary stakeholder groups: (1) researchers designing and conducting studies, (2) families participating in studies, and (3) sensor developers and technology support personnel.

Keywords: A - ambient monitoring, A-epidemiology, A-exposure models, D-children

TU-SY-G4: Perinatal Exposures in the Home Environment: Sources, Measurements, and Associated Health Outcomes - Part 2

TU-SY-G4-358

Prenatal Exposure to Perfluoroalkyl Substances (PFAS) in the Home Environment and Associations with Birth Weight

J. A. Craig[1], T. F. Webster[1], K. Hoffman[2], S. C. Hammel[2], A. Phillips[2], A. Lorenzo[2], H. M. Stapleton[2]; [1]Boston University, Weymouth, MA, [2]Duke University, Durham, NC

Abstract: Per- and polyfluoroalkyl substances (PFASs) are chemicals with water and stain-resistant properties used in numerous consumer products. Because of their persistence and widespread use, many PFAS are nearly ubiquitous in the indoor environment and human serum. Previous studies have suggested that increased exposure to PFAS may be associated with decreased birth weight. We examined exposure to PFASs and associations with birth-weight and other outcomes in a cohort of 200 mother-child pairs from North Carolina. We used 1st trimester serum samples to determine prenatal PFAS exposure and obtained birth weight from medical records. In follow-up home visits with offspring at aged 3 to 6 years (median 54 months), we obtained additional serum samples and indoor environmental samples. Serum samples were analyzed for PFASs, including PFHxS, n-PFOS (linear), Sm-PFOS (monomethyl branched), n-PFOA (linear), Sb-PFOA (branched), PFNA, PFDeA, PFOSA, Et-PFOSA-AcOH and Me-PFOSA-AcOH. Preliminary data are available for 77 mother-child pairs. Detection frequency for serum from both mothers and children were >80% for PFHxS, n-PFOS, Sm-PFOS, n-PFOA, Sb-PFOA and PFNA. Although the magnitude of association was small, maternal and offspring serum concentrations were significantly correlated ($p<0.05$) for n-PFOA, Sm-PFOS and Me-PFOSA-AcOH suggesting similar sources of exposure for mothers and children. We did not find significant correlations between maternal and offspring serum concentrations for other PFASs. Mothers had significantly higher ($p<0.05$) serum concentrations than children of n-PFOS (median 4.5 vs 2.0 ng/mL), Sm-PFOS (median 1.4 vs 0.5 ng/mL) and PFNA (median 0.7 vs 0.4 ng/mL). Mothers and children had similar serum concentrations of PFHxS (median 0.9 vs 0.7 ng/mL) and n-PFOA (median 1.8 vs 1.6 ng/mL). Univariate analyses suggest that some prenatal PFAS exposures are inversely associated with birthweight and additional analyses with fetal growth are underway.

Keywords: D-prenatal, A-biomonitoring, D-children, PFAS

TU-SY-G4-359

Children's Exposure to Organophosphate Esters: Socioeconomic Factors and Associations with BMI

S. C. Hammel[1], K. Hoffman[1], A. Lorenzo[1], A. L. Phillips[1], M. Frenchmeyer[1], B. Flaherty[1], T. F. Webster[2], H. M. Stapleton[1]; [1]Duke University, Durham, NC, [2]Boston University, Boston, MA

Abstract: Organophosphate esters (OPE) are often used as flame retardants and plasticizers and are frequently detected in indoor environments. OPE exposure is widespread, which is concerning for children who are thought to have higher exposures and may be at risk for long-term adverse health effects. Here, we assessed how children's OPE urinary metabolite levels varied with socioeconomic factors such as race and mother's education. We also examined associations between OPE metabolites and child's body mass index (BMI) percentile. Children ages 3-6 and their families were recruited 2014-2016 as part of a study examining children's exposure to semi-volatile organic compounds. Three urine samples were collected from each child over 48 hours and pooled (n=181). Diphenyl phosphate (DPHP), bis(1,3-dichloroisoproyl)phosphate (BDCIPP), isopropylphenyl phenyl phosphate (ip-PPP), bis(1-chloro-2-isopropyl) 1-hydroxy-2-isopropyl phosphate (BCIPHIPP), and tertbutylphenyl phenyl phosphate (tb-PPP) were detected in >95% of samples. Compared to non-Hispanic white children, white Hispanic children had half the levels of BDCIPP ($10^{\beta}=0.5$, $p<.01$). Both white Hispanic and non-Hispanic black children had BCIPHIPP and ip-PPP levels that were significantly higher than white non-Hispanic children ($10^{\beta}=1.5$-1.9, $p<.05$). Higher levels of BCIPHIPP, ip-PPP, and tb-PPP ($10^{\beta}=1.5$-2.0, $p<.01$) were observed among children of mothers with lower educational attainment. When adjusted for child's sex and age, and mother's BMI, age at birth, race, and education, children with the highest tertile of urinary ip-PPP had BMI percentiles that were 15 percentile points higher than those with low ip-PPP ($p<.05$). Higher levels of tb-PPP were moderately associated with higher BMI percentile among males, but inverse associations were observed for females ($p=0.1$). Our results suggest that OPE metabolite levels vary by children's demographics and are associated with children's BMI, which could have long-term health impacts.

Keywords: A-exposure factors, A-indoor environment, B-flame retardants, A-epidemiology

TU-SY-G4-360

Prenatal EDC exposures and biomarkers of the chronic stress response

A. R. Zota[1], L. Romano[1], K. Coleman-Phox[2], N. E. Adler[3], B. Laraia[4], E. Epel[3]; [1]George Washington University Milken School of Public Health, Washington, DC, [2]University of California San Francisco, San Francisco, CA, [3]University of California San Francisco, San Francisco, CA, [4]University of California Berkeley School of Public Health, Berkeley, CA

Abstract: Background: Humans are widely exposed to endocrine disrupting chemicals (EDCs) through consumer products, foods, and cosmetics. EDCs may interact with the glucocorticoid receptor; however associations between EDCs and the chronic stress response have not been well studied in humans. **Objective**: We investigated associations between prenatal exposures to persistent EDCs and biomarkers of inflammation (interleukins (IL-6 and IL-10), tumor necrosis factor alpha (TNFα)) and cellular aging (leukocyte telomere lenth (LTL) in women during pregnancy and postpartum. **Methods**: Participants were low-income, obese/overweight pregnant women (n=106) in the San Francisco Bay area of California. Blood was collected at early second trimester and 3 and 9 months postpartum. Concentrations of polybrominated diphenyl ethers (PBDEs), hydroxylated PBDE metabolites (OH-PBDEs), polychlorinated biphenyl ethers (PCBs), and perfluoroalkyl acids (PFASs) were measured during the prenatal period. IL-6, IL-10, TNFα, and LTL were measured at all three time points. **Results**: In mixed models adjusted for age, race, BMI, parity, and education, some PFASs and PBDEs/OH-PBDEs were associated with inflammatory biomarkers. For example, a doubling in exposure to PFOS was associated with a 15.67% (95% CI: 2.10%, 30.13%) and 4.97% (0.21%, 9.43%) increase in IL-6 and TNFα, respectively. Similarly, a doubling in exposure to PBDE-99 was associated with a 13.29% (95% CI: -0.69%, 29.24%) and 4.25%

(95% CI: 0.05%, 7.92%) increase in IL-6 and TNFα, respectively. No consistent associations were observed between chemical exposures and IL-10 or LTL. **Conclusions**: These initial findings suggest that exposure to specific EDCs is associated with increased inflammation during pregnancy and postpartum. This study is one of the first to investigate the relationship between exposure to EDCs and biomarkers of chronic stress. In future work, we will examine whether perceived chronic stress modifies the observed associations.

Keywords: D-prenatal, A-epidemiology, C-consumer products, A-biomarkers, D-susceptible/vulnerable

TU-SY-G4-361

Prenatal Exposure to Organophosphate Flame Retardants and Children's Gestational Duration and Growth

K. Hoffman[1], H. M. Stapleton[1], A. Lorenzo[1], J. L. Daniels[2]; [1]Duke University, Durham, NC, [2]UNC Gilling School of Global Public Health, Chapel Hill, NC

Abstract: Aims: Organophosphate esters (PFRs) are often used as flame retardants and plasticizers. Animal data suggest exposure PFRs could impact children's growth and development. Our work aims to evaluate impacts of prenatal PFR exposure on the timing of delivery and children's early growth. **Methods:** The Pregnancy Infection and Nutrition Study enrolled North Carolina women in early pregnancy and conducted follow-up through delivery. Analyses were limited to mothers recruited 2002-2005, whose children participated in follow-up (n=349). Mothers collected urine samples in which PFR metabolites were assessed. Anthropometric information was collected at birth and through age 3 years (>2800 measures). **Results:** PFR metabolites were detected frequently, with BDCIPP, DHPH, ip-PPP and BCIPHIPP detected in >80% of samples. The median infant birthweight was 3353 g and infants were born at a median of 39 weeks gestation. Relationships between prenatal PFRs exposures and birth outcomes differed by infant sex. On average, women with the highest ip-PPP concentrations delivered girls 1 week earlier than women with the lowest levels (95% Confidence Interval: -1.9, -0.2 weeks). Similarly, higher ip-PPP levels were associated with lower birthweight, but not after standardizing for gestational age. Among males, maternal DPHP and ip-PPP were associated with increased gestation duration and decreased odds of preterm birth. Data also suggest that ip-PPP levels in prenatal urine samples may be associated with increased weight gain in toddlers. **Conclusions:** Cumulatively, our results indicate that prenatal PFR exposure may impact timing of birth. Prematurity is a leading cause of perinatal morbidity and mortality. Although most preterm babies survive, they are at increased risk of many chronic conditions throughout their lifetimes. Our results also indicate prenatal PFR exposure could impact later weight gain. Given the importance of early growth in long-term health, further investigation is warranted.

Keywords: A-epidemiology, B-flame retardants, D-children, D-prenatal

TU-SY-G4-362

Trajectories of Early Life PBDE Exposure in Relation to Neurocognitive Development in Children

W. Cowell[1], A. Margolis[1], V. Rauh[1], A. Sjodin[2], R. Jones[2], Y. Wang[1], S. Wang[1], J. Herbstman[1]; [1]Columbia Center for Children's Environmental Health, New York, NY, [2]Centers for Disease Control and Prevention, Atlanta, GA

Abstract: Polybrominated diphenyl ethers (PBDEs) are persistent environmental chemicals that were used extensively as flame retardants in furniture and furnishings. Prenatal and childhood exposure to PBDEs is associated with reduced neurocognitive performance, however, longitudinal patterns of early life exposure have not been previously examined in relation to neurodevelopment. We measured PBDEs in umbilical cord and/or venous blood collected from 334 African American and Dominican children enrolled in a New York City-based birth cohort between 1998 and 2006. At follow-up visits, research workers administered the Wechsler Intelligence Scales for Children (WISC-IV) (mean±sd age: 8±1 years) and the Children's Memory Scales (CMS) (11±1 years); scores are standardized to a mean±sd of

100±15. We conducted latent class growth curve analysis to estimate trajectories of BDE-47 (ng/g lipid) exposure over time. We then used multiple regression to examine exposure trajectories, as a categorical variable, in relation to continuous scores on the WISC-IV and CMS. All children had at least one measure of PBDEs at birth, 2, 3, 5, 7 or 9 years. We identified 4 trajectories of BDE-47 exposure: persistent low (33%), prenatal high (21%), toddler peak (36%) and sustained childhood high (11%). Compared to girls in the persistent low trajectory, girls in the sustained childhood high group performed significantly worse on the WISC-IV Working Memory Index (β [95% CI] = -6.48 [-11.65, -1.30]) and girls and boys in this group performed significantly worse on the CMS Immediate Recall Visual Memory Index (Girls: β [95% CI] = -6.83 [-13.02, -0.64]; Boys: β [95% CI] = -8.35 [-16.64, -0.07]). We did not observe significant effects on WISC-IV or CMS verbal memory domains. Children with sustained high exposure to BDE-47 throughout childhood (2-9 years) may be at risk for auditory working memory (girls) and visual memory (girls and boys) deficits compared to children with low exposure throughout childhood.

Keywords: B-flame retardants, A-longitudinal metrics, C-indoor, D-children, D-prenatal

TU-PO: Tuesday Poster Session

TU-PO-363

Assessing Indoor $PM_{2.5}$ Concentrations in Households on the Hopi Reservation

S. Hadeed[1], M. K. O'Rourke[1], G. Honanie[2], A. Mahkewa[2], M. Paukgana[2], C. Kelley[1], M. A. Alshammari[1], R. Canales[1], R. Harris[1]; [1]The University of Arizona, Tucson, AZ, [2]The Hopi Tribe, Kykotsmovi, AZ

Abstract: Introduction: A significant proportion of the Hopi Tribe burn coal, wood and other organic combustible materials for heating, lighting and cooking in homes. An estimated 37% and 33% of Hopi household report burning coal and wood for heating, respectively. The Hopi Tribe has expressed their concerns about indoor air quality from the combustion of these fuels. The purpose of this pilot study is to measure indoor $PM_{2.5}$ concentrations in homes and assess the effects of fuel type and housing structure on indoor levels. Method: Pilot data was collected during the 2017 heating season in four households on the Hopi Reservation. Indoor concentration of $PM_{2.5}$ was measured over a 24-hour period using real time monitors (pDR-1500) set at 1-minute logging intervals. Subsequent collection of primary fuel type and housing characteristics was recorded by field team members at the time monitors were placed indoors. Average $PM_{2.5}$ concentration was calculated by fuel and housing type. Results: In the four households sampled to date, the indoor mean (SD) $PM_{2.5}$ concentration was 18.7 (37.3) μg/m^3. Two modern households that used a combination of coal, wood and electricity for fuel had an indoor mean $PM_{2.5}$ concentration of 17.9 (36.0) μg/m^3 and 1-minute peak concentration of 332.8 μg/m^3. Average $PM_{2.5}$ concentrations for a hybrid modern-traditional home with coal and wood fuel was 33.0 (50.6) μg/m^3 and 6.1 (6.8) μg/m^3 for an electrically powered mobile home. Discussion: From this pilot study, households using coal and wood as primary fuel sources were found to have elevated indoor $PM_{2.5}$ concentrations compared to electrically powered homes. These initial results are similar to previous studies that found increased exposures and indoor $PM_{2.5}$ concentrations in households using combustible fuels as a primary fuel source. This project will continue to assess indoor air quality of households on the Hopi Reservation by various fuel sources and housing types during heating and non-heating seasons.

Keywords: A-indoor environment, B-particulate matter

TU-PO-364

Carbon Dioxide Accumulation Inside Vehicles: The Effect of Ventilation and Driving Conditions

N. Hudda[1], S. Fruin[2]; [1]Tufts University, Medford, MA, [2]University of Southern California, Los Angeles, CA

Abstract: Limiting the air exchange of passenger vehicles by closing windows and recirculating cabin air (RC) restricts the influx of roadway particulate pollutants and reduces both particle mass and particle number concentration. However, under these conditions the carbon dioxide (CO_2) and bioeffluents exhaled by the occupants accumulates. We modeled in-vehicle CO_2 accumulation by linking typical CO_2

generation rates with our previously characterized associations between AER, vehicle characteristics and driving conditions. We further characterized the factors (ventilation setting, vehicle age, speed, cabin volume, trip duration, and number of occupants) that allow CO_2 accumulation to reach concentration thresholds found in other studies to produce cognitive effects of concern such as fatigue or difficulty concentrating (e.g.,1200 and 2500 ppm). Ventilation setting was the primary determinant of CO_2 accumulation, with only the RC setting (and not outside-air intake) ever allowing CO_2 accumulation that exceed either threshold, particularly during longer trips with multiple occupants. However even under RC setting, the 2500 ppm threshold was generally not exceeded for one- or two-occupant, average length commutes (twenty-six minutes), and the 1200 ppm threshold was only exceeded in 10-30% of the fleet (e.g., newer cars at slower speeds) for one-occupant average commutes. Further examination of this fraction of the fleet in which these thresholds are expected to be exceeded indicated that they are most likely to be exceeded in below median age (8 year), compact to mid-size vehicles during rush hour or congested driving condition (30 miles/h) average- or long-duration (sixty-one minutes) commutes. For multiple passenger commutes and/or longer trips, RC ventilation should be periodically interrupted with outside air or restricted to keep CO_2 concentrations below thresholds of concern.

Keywords: A-indoor environment, A-exposure models, C-air, C-indoor,

TU-PO-365

Analysis and Identification of Ozone-Squalene Particulate Phase By-Products

B. Coffaro, C. Weisel; Rutgers, Highland Park, NJ

Abstract: Ambient ozone penetrates into homes exposing residents to ozone and its by-products. Recently, squalene, a natural component of skin oil, has been shown to be highly reactive with ozone resulting in both volatile compounds and particulate phase by-products. The latter is likely to include a series of carbonyls and organic acids that may be respiratory and ocular irritants. While a number of the volatile ozone-squalene by-products have been identified, the composition of the particles formed has not been well characterized. This work identifies and quantifies the particulate products formed from the reactions of ozone and squalene in a controlled setting that is representative of indoor conditions. Squalene was loaded onto a surface at a concentration of 0.252 mg/cm^2 to simulate a human forehead and 100 ppb ozone was passed over the surface and allowed to react within the vessel. Produced particles were extracted from Teflon filters and followed by derivatization of the compounds with BSTFA. Subsequent analysis and identification by GC-MS was conducted where succinic acid and levulinic acid were identified as the likely unique squalene-ozone reaction products, as well as nine other alcohols and carboxylic acids. The effect of varying concentrations of ozone in the air, duration of reactions, humidity levels and whether ozone continues to react with the by-products produced on the filter are being determined. This research is meant to identify compounds that people may be exposed to inside their home from this reaction taking place on the surface of their skin.

Keywords: A-indoor environment, C-air

TU-PO-366

Evaluating air pollutant exposure with air quality index (AQI)

Y. Huang[1], Y. Lin[2]; [1]National Kaohsiung First University of Science and Technology, Kaohsiung, Taiwan, [2]Fu-Jen Catholic University, New Taipei City, Taiwan

Abstract: Exposures to ambient air pollutants are often considered a continuous process with varying exposure concentrations. In some cases, exposure to air pollutants may be considered a potential hazard when pollutant concentrations exceed certain levels. Nonetheless, exposure assessment to ambient air pollutants often rely on estimates of long-term pollutant levels. Intermittent exposures to air pollutants at moderate levels have not been examined closely, especially with respect to the occurrence frequency, severity and persistence of air pollution events. In this study, air quality monitoring data were examined to describe the occurrence and severity of moderate air pollution events. Hourly monitoring records between

2005 and 2016 at a monitoring station in southern Taiwan were obtained from Taiwan EPA. After removing invalid records, AQI were calculated for $PM_{2.5}$, PM_{10}, and 1-hour and 8-hour ozone concentraions. Pollutant-specific AQI values were calculated by identifying the maximum value for each day, and the daily AQI values were determined as the maximum value from the pollutant-specific AQI. An air pollution day was defined as having daily AQI value above 100. With each air pollution day, pollutant-specific AQI values above 100 were also recorded. The results were examined with respect to pollutant concentration, frequency of occurrence, and any potential seasonal or long-term trends. In all, 4,006 daily air quality monitoring records were analyzed. With definition of AQI above 100 as an air pollution day, 2,712 days (67.5%) were identified within the examined period. On air pollution days, 2,605 days (96.0%) showed elevated $PM_{2.5}$ levels, 750 days (27.6%) of high PM_{10} levels, 932 days (23.3%) with 8-hour ozone levels, and 1,303 days (48.0%) with multiple pollutants. Implication from simultaneous exposure to multiple air pollutants warrants further exploration.

Keywords: A - ambient monitoring, A - population exposure, A-statistical methods, B-particulate matter

TU-PO-367

Real-Time Personal Ozone Monitoring Versus Ambient Ozone Monitoring in the Ironbound District of Newark, NJ

J. Therkorn[2], Y. Son[1], R. Laumbach[1], Q. Meng[1]; [1]Rutgers University, Piscataway, NJ, [2]Johns Hopkins University Applied Physics Laboratory, Laurel, MD

Abstract: Ozone (O_3) exposure can cause serious respiratory problems. Much of our current knowledge of the effects of O_3 is based on studies using data from central monitoring sites or passive O_3 sampling. Here, we present the first use of real-time, continuous (1-min) personal O_3 monitoring for children (n=15) using a highly sensitive monitor (Personal Ozone Monitor, 2B Technologies) in comparison with ambient federal monitoring data. Average personal measurement periods lasted ~2 days, and all measurements were taken between 2/2013 to 2/2016 in Newark, NJ. Comparison of personal O_3 exposure across subjects to time-matched 1-hr ambient O_3 concentration showed the average personal O_3 concentrations were about 1/3 of the ambient (7.4 ± 7.7 ppb for personal vs. 20 ± 13 ppb for ambient O_3). However, personal O_3 monitoring revealed higher max peaks (131.9 ppb for max personal O_3 vs. 56 ppb for max ambient O_3). A generalized linear mixed model approach with personal O_3 as the response variable suggested ambient O_3 was a significant predictor (p=0.005) indicating a slight increase in personal O_3 per unit change in ambient O_3 (Personal O_3 = 6.942 + 0.0394(Ambient O_3), R^2 = 24.6%). Ambient O_3 regression coefficient varied by season: 0.0119, 0.0335, 0.1149, and -0.0025, for winter, spring, summer and fall, respectively. The results indicate greater within than between subject variations in personal O_3 exposures, which is likely due to personal activities with a seasonal influence. This work indicates that personal O_3 exposures are much lower than levels predicted by ambient monitoring, suggesting health effects may occur at lower levels of O_3 than expected. Further analyses will explore additional predictors by subjects' activities and meteorological factors. Overall, this knowledge informs efforts to assess risks from O_3 and investigates interconnected elements of O_3 exposure assessment, including personal and ambient monitoring and time-activity patterns.

Keywords: B - ozone, A - exposure measurement, A-activity patterns, A-sensor technology, C-air

TU-PO-368

STATISTICAL FUSION OF PRESENT-DAY OBSERVED GLOBAL OZONE CONCENTRATIONS AND CCMI-1 MULTI-MODEL SURFACE OZONE ESTIMATES

M. Buechlein, M. Serre, J. West, Y. Zhang; UNC, Chapel Hill, NC

Abstract: With over 9000 surface monitoring stations globally, the Tropospheric Ozone Assessment Report (TOAR) has compiled the largest database of global ozone observations using metrics relevant for human health impacts, vegetation impacts, and climate impacts. Despite the effort to assemble a comprehensive global ozone observation database, the TOAR database still contains large regions with sparse data. Chemistry-climate models (CCMs) and chemical-transport models (CTMs) can provide estimates of air pollution concentration in regions where observed concentrations are sparse. Phase 1 of the Chemistry-Climate Model Initiative (CCMI-1) includes 20 CCMs and CTMs that use simulated meteorological conditions and chemistry to estimate air pollutant concentrations globally. We conducted the first global Bayesian Maximum Entropy (BME) data fusion of ozone observations and CCMI-1 model predictions. We used a Constant Air Quality Model Performance (CAMP) evaluation of the CCMI-1 model outputs to account for nonlinear and heteroscedastic performance of global models and to correct for consistent bias within the models. The observations and CAMP corrected global model outputs were then integrated in the BME framework to provide a comprehensive and robust global estimate of "present-day" surface ozone that can be used to estimate the global health burden of ambient ozone.

Keywords: B - ozone, A-geospatial analysis/GIS, A-statistical methods

TU-PO-369

Spectral Identification of Localized Roadway, Aircraft and Rail Transportation Noise Sources and Correlation with Ultrafine Particulates

D. Leaffer[1], J. L. Durant[1], D. Brugge[1], N. Hudda[1], B. Tracey[2], M. C. Simon[1], A. Hastings[3]; [1]Tufts University, Medford, MA, [2]Tufts University, Medford, MA, [3]US Department of Transportation, Cambridge, MA

Abstract: Accurately characterizing particulate matter (PM) and noise in exposure science has broad implications in public and community health surveillance and exposure mitigation. Noise and PM frequently occur from common sources, including roadway traffic, rail and aircraft, with localized wind speed and wind direction as important factors mitigating or elevating pollutant levels. Recent epidemiological studies have focused on understanding the role of ultrafine particulates (UFP; <100 nm diameter) on health effects as a public health priority. Due to simultaneous exposure of both particulates and noise from transportation sources it is difficult to characterize these exposures as individual or synergistic contributors to adverse health effects. Our methodology aims to better define and potentially identify UFP sources and exposures by utilizing spectral analysis of transportation noise and co-analyzing particle number count (PNC, a proxy of UFP) at near-roadway, near-railway line and near-airport urban sites. Spectral characterizations of selected samples of recorded transportation source noise measurements were analyzed for frequency by Fast Fourier Transform (FTT) and correlated to PNC measured concurrently; spectral fingerprints were identified for transportation noise sources and potential UFP sources by transportation class, such as truck, train or aircraft. Aircraft and rail spectra exhibited a distinct Doppler effect which was identified and linked to rail and flight schedules, then correlated to PNC time-series under varying meteorological conditions. With wind speed and direction preferential to PM transport, PNC varied by an order of magnitude or more while noise levels were not significantly affected. Spectral identification of transportation noise may provide a better predictive link between noise exposure and UFP in health studies.

Keywords: B-particulate matter, A-exposure factors, A - ambient monitoring, C-air, A-epidemiology

TU-PO-370

Characterization of Particulate Matter in Urban Traffic Roads and Health Effects on the Pulmonary System in Mice with Lung Inflammation

V. G. Venkatareddy[1], A. K. Fauzie[1], R. K. Thimulappa[2], A. A. Nori-Sarma[3], M. L. Bell[3]; [1]Mysore University, Mysore, India, [2]JSS Medical College, Mysore, India, [3]Yale University, New Haven, CT

Abstract: *Introduction:* Traffic is a major contributor to air pollution in Indian cities, and traffic-related exposure has been shown to induce acute inflammation in humans, both in chamber studies using diesel exhaust and in real-life environments. In urban centers, diesel exhaust particles are considered to be the hazardous pollutants released from automotive engines due to their aerodynamic and chemical characteristics. *Methods:* Emissions of $PM_{2.5}$ and PM_{10} were estimated using a mathematical model incorporating number of vehicles and their emission factors, and correlated significantly with experimental ambient PM concentrations. Elemental composition of PM was analyzed using Energy Dispersive X-Ray (EDX), also referred to as Energy Dispersive Spectroscopy (EDS) method. Preparation of BAL was done following Harrod's protocol with some modifications. Within 2h following the end of particle exposure (5-90 days) mice were anesthetized using chloroform; total cell count and viability determined by Nebur's hemocytometer. *Results:* EDX analysis found that particulate matter was dominated by black carbon (C) about 56% affected mostly by tail end pipe emissions. The samples were also rich in elements like silicon (Si), iron (Fe), calcium (Ca), aluminium (Al), sodium (Na) and potassium (K) either in single elements or as chemical compounds. There was a significant reduction of the bronchoalveolar lavage lymphocyte cell numbers (08.75) in mice after 30 and 90 days exposure. Percent viability of leucocytes has decreased and differential cell of macrophages and neutophils were increased in BAL fluid following mice exposure to $PM_{2.5}$ in different intervals. *Conclusion:* Exposure to the mixture of suspended particulate matter particles induces pathological changes, differential cell counts and inflammatory response in the lungs of mice in a dose and duration dependent pattern. The responses observed from the present study are associated with the bioavailable of inhaled particulate mixtures.

Keywords: A - ambient monitoring, B-particulate matter,

TU-PO-371

Smoke Sense - a crowd sourced study of health impacts of wildland fire smoke exposures

A. G. Rappold[1], K. rappazzo[1], S. Stone[2], D. Diaz-Sanchez[1], B. Hubbell[3]; [1]US EPA, Durham, NC, [2]OAR/OAQPS, Durham, NC, [3]US EPA, Durham, NC

Abstract: Exposures to wildland fire smoke are linked to a range of health outcomes, affecting wellbeing and productivity in the communities. We present the protocol and the results of the Smoke Sense study - the first crowd sourced study designed to quantify health impacts attributable to wildland fire smoke and to examine the ability of health risk communication strategies to improve health outcomes during wildland fire smoke events. The Smoke Sense study leverages smartphone app to deliver real time air quality and recommended health risk messages to participants and to facilitate input about the health symptoms and smoke experienced in the affected communities as well as the actions they take to reduce their exposure. The study adopts principles of gamification to engage participants to learn about how air quality impacts their health and actions they can take to reduce exposures. We will use regression techniques to estimate associations between smoke exposure and health outcomes and examine modifications through behavioural changes, demographic characteristics, knowledge of air quality and baseline health. Smoke Sense study will pilot during the summer months of 2017. We will present the protocol and the results from the pilot data collection including the magnitude and severity of cardiovascular, respiratory, eyes and ears, and other symptoms, medication usage and doctor's visits among the participants; actions, including behavioral changes, taken to reduce exposure; and the impact of those actions on the frequency of health outcomes. We will also present our efforts to engage communities and encourage participation. This study is expected to improve our understanding about the range of health outcomes experienced in the communities and assess whether health risk communication via mobile devices can improve public health outcomes. Disclaimer: The views expressed in this abstract do not necessarily reflect the views or policies of the U.S. Environmental Protection Agency

Keywords: D-wildlife,

TU-PO-372

Potential inhaled dose: a better metric of personal exposure to particulate air pollution

S. N. Chillrud[1], P. Kinney[2], C. M. Smith[1], J. Thornburg[3], Q. Yang[1], D. W. Jack[1]; [1]Columbia University, New York, NY, [2]Boston University, Boston, MA, [3]RTI International, Triangle Research Park, NC

Abstract: Personal exposure studies are seen as the gold standard in air pollution epidemiology, however, the majority of previous studies only measure pollution concentrations, ignoring variations in volumetric respiration rates. Potential inhaled dose, the product of concentration and minute ventilation, takes both into consideration and should theoretically be a better exposure metric. We carried out a pilot study in NYC with 43 healthy commuting bikers for testing feasibility and validating the different monitors. We recruited participants via the local public radio station and selected 43 participants who typically commute to and from work for approximately 45 minutes each way. Each subject was trained to self deploy and maintain the equipment for five 24 hr sessions over a two week period. Here, we report on results from the microAeth AE51 by AethLabs (black carbon), the RTI microPEM ($PM_{2.5}$), and the Hexoskin biometric shirt by Carré Technologies Inc. (minute ventilation). BC concentrations showed an average value of 3.4 μg/m^3 (median 1.8) during biking, three times of the average of 1.1 μg/m^3(median 0.55) during non-biking periods. In comparison, $PM_{2.5}$ averaged 16 μg/m^3 (median 11) during biking, less than twice of the averaged 14 μg/m^3 (median 6.2) during non-biking periods. Minute ventilation during biking averaged at 51 L/min (median 45) was five times of that averaged 10 L/min (median 8) during non-biking periods. The unit time of potential inhaled dose during biking was approximately 10 times the non-biking periods, accounting for 40-65% and 25-40% of total BC and $PM^{2.5}$ dose, respectively. Biking time only made up 6-8% of the monitored 24-hour period. This better exposure metric has important implications in air pollution epidemiology and public health risk assessment.

Keywords: A - exposure measurement

TU-PO-373

Characterizing Aggregated Exposure to Fine Particulate Matter from Indoor and Outdoor Sources

P. Fantke[1], O. Jolliet[2], J. Apte[3], N. Hodas[4], K. Stylianou[2], J. Evans[5], C. Weschler[6], T. McKone[7]; [1]Technical University of Denmark, Kgs. Lyngby, Denmark, [2]University of Michigan, Ann Arbor, MI, [3]University of Texas at Austin, Austin, TX, [4]California Institute of Technology, Pasadena, CA, [5]Harvard School of Public Health, Boston, MA, [6]Rutgers University, Piscataway, NJ, [7]University of California, Berkeley, Berkeley, CA

Withdrawn

TU-PO-374

Linking Green Space Attributes and Mental Health: Understanding the Connection between Perceived Safety and Perceived Restorativeness Using Immersive Virtual Environments

P. K. Baran, P. Tabrizian; NC State University, Raleigh, NC

Abstract: Exposure to green space provides restoration from depleted emotional and cognitive resources, but characterizing what green space exposure is difficult. Using immersive virtual environment (IVE) stimuli, we are able to simulate green spaces in urban environments, to better understand factors influencing green space benefits. Little is known about the attributes of green spaces that support mental restoration and the mechanisms through which green spaces promote restoration. We examine the effects of arrangement and density of vegetation on (1) perceived restorativeness and (2) whether perception of safety mediates the effects of these attributes on perceived restorativeness. 360° panoramas taken from two urban settings (plaza and park) were manipulated for vegetation arrangement (the number of vegetated segments along a boundary of a square shaped area) and density (the number of trees in each segment) to generate 18 IVE stimuli. The experiment had a 3 x 3 x 2 mixed factorial design with density (one, three and six trees per segment) as between-subjects treatment, and arrangement (1-, 2- and 4-sided) and setting type (park and plaza) as within-subjects treatments. Using a head-mounted display, participants (N=87) rated IVEs on perceived restorativeness and perceived safety. Vegetation arrangement affected perceived restorativeness across both setting types, but in opposite direction. More enclosed arrangements (4-sided) in plaza setting were perceived as more restorative, whereas in park setting the same arrangements significantly reduced perceived restorativeness. Vegetation density and arrangement significantly interacted with each other and with setting type to affect perceived restorativeness and perceived safety. Perceived safety mediated the effects of vegetation density and arrangement on restorativeness. IVEs can be used to explore specific aspects of broad exposures to better characterize impacts of the environment on human health.

Keywords: A-built environment, A - exposure measurement,

TU-PO-375

A Review of Stressors from the Built and Natural Environments Impacting American Indian Children's Health and Well-Being

N. Barros[1], N. S. Tulve[2], D. Heggem[3], K. Bailey[4]; [1]ORISE, United States Environmental Protection Agency, Research Triangle Park, NC, [2]United States Environmental Protection Agency, Research Triangle Park, NC, [3]United States Environmental Protection Agency, Las Vegas, NV, [4]United States Environmental Protection Agency, Cincinnati, OH

Abstract: Exposures to chemical and non-chemical stressors from children's environments during early development may contribute to differences in health and well-being outcomes. Children from American Indian tribes may have a disproportionate burden of these stressors from their man-made (i.e., built environment) and natural surroundings. Our objective was to identify stressors from the built and natural environments that may affect the health and well-being of American Indian children. Databases (ProQuest, PubMed, Web of Science) were searched with key words and search strings (e.g., Alaska Native child) to identify scholarly literature focused on stressors, exposures, and health and well-being for American Indian children. A total of 2,535 references were identified. References were then excluded if they did not discuss American Indian children or were not the primary cohort of interest; discussed Tribes outside the U.S.; described interventions; or did not provide information about stressors from the built or natural environments. Of the 23 remaining references 10 discussed stressors from the built environment and 13 from the natural environment. For the built environment (home), the main stressors were use of a wood stove for cooking and heating, presence of mold, presence of dust, lack of plumbing, presence of dirt floors, and overcrowding. For the natural environment, however, only chemical stressors were identified, focusing on maternal and adolescent exposures to persistent organic pollutants and metals around hazardous waste sites (through contaminated waterways) and abandoned mines. Preliminary findings identified a limited number of studies (<30), demonstrating a major information gap. Analyses provide preliminary information about the nature of chemical and non-chemical stressors from built and natural environments that may influence American Indian children's health and well-being, which may be distinct from other communities.

Keywords: A-built environment, A-exposure factors, C-water, D-children, D-susceptible/vulnerable

TU-PO-376

Study on personal exposure to fine particles for retired adults in Beijing

Q. Wang, N. Li, C. Xu, D. Xu; National Institute of Environmental Health, Chinese Center for Disease Control and Prevention, Beijing, China

Abstract: Aim The objective of this study was to assessment the personal exposure to fine particles ($PM_{2.5}$) and to explore the relationships between personal, outdoor and indoor $PM_{2.5}$ concentrations for retired adults in Beijing city. **Methods** The indoor, outdoor and personal PM2.5 samples were collected simultaneously by Teflon filters for five consecutive days during both winter (7 January to 14 March, 2016) and summer (22 June to 7 August, 2016) for 30 retired adults in Beijing city. The mass concentrations of PM2.5 were measured by gravimetric method. **Results** The $PM_{2.5}$ personal exposure concentrations were 54.8±25.8 (23.0-118.0) μg/m^3 and 55.6±18.0 (28.0-97.0) μg/m^3 during winter and summer sampling periods respectively. In both seasons, the personal exposure levels were higher than indoor concentrations (P=0.003 in winter and P=0.004 in summer by paired t test) and lower than outdoors ($P<0.001$ in winter and P=0.067 in summer by paired t test). The relationships between personal exposure and indoor concentrations (Person r=0.873 in winter and r=0.952 in summer) was stronger than outdoors (Person r=0.71 in winter and r=0.82 in summer) for both seasons. **Conclusions** The retired adults in Beijing were exposed to relatively high level of $PM_{2.5}$. The personal exposure levels were between indoor and outdoor concentrations, but more related to indoors.

Keywords: A - exposure measurement, A - population exposure, B-particulate matter, C-air, A-indoor environment

TU-PO-378

Very high levels of individual exposure biomarkers: A void in guidance for study investigators

S. M. Pinney; University of Cincinnati College of Medicine, Cincinnati, OH

Abstract: Background: Ever improving technology allows researchers to detect ever smaller amounts environmental toxins in biomedia. This information can have far reaching implications for participants, even if the health effects associated with the exposures are currently not well understood. Further questions arise when participant samples contain biomarker levels for exposures well above the expected limit, such as two times above the NHANES 95th percentile. What are the implications? Some study investigators have been reporting back the personal exposure biomarker measurements to study participants, and the potential benefits and detriments of this practice have been discussed in national workshops and in published papers. An issue that has not been addressed, and for which there is no established guidance, is an unanticipated finding of a very high level of a chemical or metal in a single study participant. **Objective: To** identify the medical, ethical, and legal issues presented with the finding of a very high level of an exposure biomarker in a study participant and to develop guidance for study investigators. **Methods:** At a workshop in September, 2017, funded by NIEHS, we will discuss the issues noted above. Should very high environmental biomarker findings be reported to the research participant's personal physician or public health officials? Are physicians equipped to answer the participant's questions? Do they even want to be involved? Do community physicians want to receive that information? Are they prepared to counsel their patients? A summary of the workshop discussion and recommendations will be presented. **Conclusions:** Through further research and discussion, we hope that guidance for researchers and primary care providers will emerge. **Funding:** P30ES006096, R13 ES027347

Keywords: A-biomonitoring, D-environmental justice, A-epidemiology

TU-PO-379

Community Driven University Partnerships to Assess Exposures and Risk Perceptions of Diné Communities following the Gold King Mine Spill

P. I. Beamer[1], J. C. Ingram[3], K. Chief[2]; [1]University of Arizona, Tucson, AZ, [2]University of Arizona, Tucson, AZ, [3]Northern Arizona University, Flagstaff, AZ

Abstract: On August 5, 2015, 3 million gallons of acid mine drainage was released from the Gold King Mine, eventually reaching the San Juan River - the lifeblood of the Navajo Nation. This talk will share the experiences of building community and university partnerships to quickly develop and implement a community-based risk assessment in the wake of this environmental disaster. Central to this effort has been the development of a network of Diné community partners from the affected chapters that have guided the university researchers in designing and implementing a culturally appropriate study that addresses the community's concerns. A key focus has been on building capacity for assessing environmental exposures through training of Diné tribal college students, environmental interns, and community health representatives. To date more than 40 students (20 were Diné students) and 25 community members have collaborated and participated in data collection, interpretation, and dissemination. Given the potential for future catastrophic mine spills in the Western United States, findings will be used to develop a model of community capacity-building aimed at empowering affected communities to collect samples, minimize impacts, and engage in informed-decision making.

Keywords: A-environmental justice, D-environmental justice, D-Southwest-specific

TU-PO-380

Use of satellite imagery to identify a target population for recruitment of households within a large rural tribal area.

C. Kelley[1], R. Harris[1], J. Tabor[1], M. K. O'Rourke[1], G. S. Honanie[2], L. Joshweseoma[2], S. Hadeed[1], M. A. Alshammari[1]; [1]University of Arizona, Tucson, AZ, [2]Hopi Tribe, Kykotsmovi, AZ

Abstract: Introduction: Study recruitment is increasingly challenging. In rural and tribal communities, which may lack traditional street addresses. Strategies are needed to identify and randomly select participants. The Hopi Environmental Health Project is seeking recruitment of households representative of housing type, fuel source and village. Hopi lands are >1.5 million acres, with 12 villages. **Methods:** Census block vectors for tribal lands were overlaid onto satellite imagery in Google Earth Pro 7.1. Potential household structures were marked with a geotag to record coordinates and an identifier describing block number, housing type, and region. Metadata were copied to an associated .KLM file and included block number, housing type, region, latitude, and longitude. Recruitment goals were set to be proportional to population size, based on the number of houses identified within village vectors. 'Pins' were randomized by village/region; latitude and longitude were used to generate web-addresses which were used by Hopi to locate dwellings and print maps. **Results:** The process identified 2512 potential 'houses' within 400 block vectors. This number is similar to US Census 2010. Initial recruitment of the first 30 target households identified several issues: some 'houses' were multi-family dwellings, or vacant, outbuildings of an occupied house, public building or a kiva, all of which were ineligible. Recruiters are tracking all households for eligibility; reasons for ineligibility are type of structure, probable fuel type, and response. **Discussion:** Some researchers suggest that a random recruitment process will not be effective on native lands. Our approach provides a baseline list that is "field-truthed" prior to recruitment. Recruitment statistics generated using this approach will provide insight to response rates by region and housing type. The Hopi project provides an opportunity to evaluate the described approach.

Keywords: A-epidemiology, D-Southwest-specific, D-First Nation, C-multimedia, A-geospatial analysis/GIS

TU-PO-381

Plasma Trans-Fatty Acid Levels in Fasting Adults from the National Health and Nutrition Examination Survey

H. C. Kuiper, S. P. Caudill, H. Vesper; CDC, Atlanta, GA

Abstract: Trans-fatty acids (TFA) are geometric isomers of naturally occurring cis-fatty acids, formed industrially via partial hydrogenation of vegetable oils or naturally in ruminant animals. High dietary TFA intake results in increased low density lipoprotein levels and has been associated with other risk factors for cardiovascular disease. Due to these adverse health effects, reduction of TFA consumption is a major public health objective. Food intake studies have provided estimates for TFA levels in the U.S. population; however, little is known about the TFA levels in humans. We developed and validated a new isotope dilution-gas chromatography-negative chemical ionization-mass spectrometry method for the quantitation of 27 FA including four major TFA in 100 µl of human plasma. FA were derivatized with pentafluorobenzyl-bromide, resolved on a 200 m Select-FAME column with hydrogen as the carrier gas, and analyzed in selected ion monitoring mode using negative chemical ionization. The limits of detection were 0.07 µM for palmitelaidic acid, 0.28 µM for elaidic acid, 0.43 µM for trans-vaccenic acid, and 0.02 µM for linoelaidic acid. The intraday and inter-day percent coefficients of variation ranged from 1-11%CV and 6-15%CV respectively. The mean accuracy for all four TFA was 102% (95%CI: 98%-107%). We used this method to quantitate four major TFA, palmitelaidic acid, elaidic acid, trans-vaccenic acid, and linoelaidic acid, in a nationally representative group of fasted adults in the U.S. population using National Health and Nutrition Examination Survey samples. The TFA were measured in 1613 participants from NHANES 1999-2000 and 2462 participants from NHANES 2009-2010, and were detected in all samples. Geometric means for the sum of the 4 TFAs were 80.9 µmol/l (95% CI: 75.7, 86.5 µmol/L) and 37.4 µmol/l (95% CI: 36.1, 38.8 µmol/L) in the NHANES 1999-2000 and 2009-2010, respectively. Overall TFA concentrations were 54% lower in NHANES 2009-2010 compared to 1999-2000.

Keywords: A-biomonitoring, A-analytical methods

TU-PO-382

Urinary Arsenic and Dietary Components in the General U.S. Population

R. Birch, J. Rogers, D. A. Marker, S. M. Viet; Westat, Rockville, MD

Abstract: Arsenic is a naturally occurring metalloid element. Inorganic forms are highly toxic while organic forms are less toxic. This study investigates urinary arsenic concentrations in the U.S. population aged 6 years and older, utilizing data from the National Health and Nutrition Examination Survey (NHANES) 2003-2010. The goal of this study is to evaluate the relationship between urinary arsenic and diet. Survival analysis was used to predict log-transformed arsenic concentrations as a function of demographic characteristics, year, alcohol intake, serum cotinine, urinary creatinine, body weight, intake of nine foods, and water intake. The foods were represented in modeling by two variables 1) whether a respondent consumed that food the previous day (Y/N) and 2) log-transformed amount consumed if the respondent consumed the food. Participants' consumption of a food was set to zero when it was <-1 on a log-transformed scale (0.368 grams).Statistical significance was assessed at $p<0.01$. Intake of rice, shellfish, or finfish the previous day is associated with increased urinary TAs, arsenobetaine, and DMA, and rice intake the previous day is associated with increased urinary MMA. Intake of all three resulted in urinary TAs 5.5 times higher, urinary DMA 2.9 times higher, and arsenobetaine 17.1 times higher compared to persons without intake of those foods the previous day. Intake of rice resulted in urinary MMA 1.2 times higher compared to no rice consumption. Each gram consumed resulted in additional increases in urinary arsenic. The three dietary components that contribute most to urinary arsenic concentrations are finfish, shellfish, and rice. Urinary concentration of arsenobetaine, DMA, and TAs were statistically associated with consumption of these foods. Urinary concentration of MMA was only associated with rice consumption. There were no statistically significant relationships between any foods of interest and urinary concentrations of the other arsenicals.

Keywords: C-food, B-metals

TU-PO-383

The OH-PAHs in urine can make a dyslipidemia, but not by a oxidative stress pathway

T. Wang[1], J. Xu[2], L. Zhang[1], B. Han[2], Z. Bai[2]; [1]Tianjin Medical University, Tianjin, China, [2]Chinese Research Academy of Environmental Sciences, Beijing, China

Abstract: Background: The hydroxylated polycyclic aromatic hydrocarbons (OH-PAHs) have been associated with oxidative stress and dyslipidemia, but the mechanism and extent of the impact have not been well evaluated in epidemiological studies. **Objectives:** This study focused on the relationship between OH-PAHs, 8-hydroxy-2'-deoxyguanosine (8-OHdG) and lipid profiles. **Methods:** In this cross-sectional study, 112 elderly were volunteered and finally 109 objects have complete data. Blood samples and morning urine samples were collected in 2011. OH-PAHs, creatinine and 8-OHdG were determined by GC-MS, spectrophotometry and an ELISA kit. Blood lipid profiles were analysed by an automatic biochemical analyzer. The relation of lipid profiles and 8-OHdG was detected by a 2-independent samples nonparametric test in different gender, smoking and alcohol status. After normalizing the concentration value, a linear regression model was used to detect the correlation between OH-PAHs, 8-OHdG and lipid profiles. **Results:** We found the median concentrations of OH-PAHs in urine were about 1-10 µmol/mol Cr. Lipid profiles and 8-OHdG had significant differences in different gender, but no significant differences in different smoking and alcohol status. The results of linear regression showed 2-OHNap, 9-OHFlu, 3-OHFlu, 2-OHFlu, 1-OHPyr and 6-OHChr separately a unit logarithmic concentration increasing to make an increasing of 0.305, 0.420, 0.456, 0.467, 0.496 and 0.324 lg(8-OHdG/Cr) ng/mg by adjusting BMI and age. 1-OHNap and 1-OHPyr showed a negative correlation to Apo(A) I. 1-OHPyr for LDL-C and 2-OHBcPhe for Apo(B) showed positive relationships. 8-OHdG hasn't showed significant correlation with lipid profiles. **Conclusions:** Our study showed OH-PAHs may cause level of 8-OHdG rise and dyslipidemia, but 8-OHdG didn't cause dyslipidemia. It indicated OH-PAHs may cause dyslipidemia through other routes, rather than oxidative stress.

Keywords: A-biomarkers, B-POPs, A-biomonitoring

TU-PO-384

Metallothioneins polymorphisms and blood/plasma trace elements in Slovenian mother-child pairs (CROME-LIFE+ study)

A. Stajnko[1], J. Snoj Tratnik[1], A. Trdin[1], M. Jagodic[1], A. France Štiglic[2], M. Horvat[1], I. Falnoga[1]; [1]Jozef Stefan Insitute, Ljubjana, Slovenia, [2]University Medical Centre of Ljubljana, Ljubljana, Slovenia

Abstract: Methallothioneins (MT) are metal(loid)s binding proteins characterised by low molecular weight (6-7 kDa), high cysteine content (~30%) and high metal affinity. They have several roles, including Cu and Zn, homeostasis, metal(loid)s detoxification, redox regulation and protection against oxidative stress. In humans various MT (sub)isoforms have been identified and are coded by at least 11 functionally active genes located on chromosome 16. Single nucleotide polymorphisms (SNPs) of these genes could modify proteins functions. The aim of present study was estimation of possible associations of selected MTs SNPs or their combination with levels of various trace elements determined in blood and/or plasma of participants. The study was focused on 179 non-occupationally exposed Slovenian mother-child pairs (mothers: mean age = 38.6 years; child: 7-8 years, 90 females, 89 males), which participated in CROME-LIFE+ study (2013-2016). DNA extracted from venous blood or saliva was used for genotypisation of 12 SNPs in MT genes (*MT1a, MT1b, MT1e, MT1f, MT1g, MT1x; MT2a; MT3; MT4*) and trace elements (Hg, As, Pb, Cd, Cr Zn, Cu, Mn, Se) determined in blood and plasma of participants. Obtained significant associations between genotypes of individual SNPs or their combinations and trace elements were further tested by multiple linear regression models for possible confounders such as age, gender, body mass index, education, smoking, sea food consumption, etc. Different associations were observed for blood samples of mothers (mostly with non-essential metal(loid)s, Pb, Hg and Cr) than that of children (mostly with essential metal(loid)s Zn, Cu, Se and Mn). Each kind of element (or SNP) had its own specific kind of associations.

Keywords: A-biomarkers, A-exposure factors

TU-PO-385

Apolipoprotein E polymorphisms and mercury concentrations in pregnant women from Croatian coastal region

A. Trdin[1], I. Falnoga[1], J. Snoj Tratnik[1], A. Stajnko[1], D. Mazej[1], J. Marc[2], M. Krsnik[3], J. Osredkar[3], I. Prpic[4], O. Petrovic[4], Z. Spiric[5], M. Horvat[1]; [1]Jozef Stefan Institute, Ljubljana, Slovenia, [2]Faculty of Pharmacy, University of Ljubljana, Ljubljana, Slovenia, [3]University Medical Centre Ljubljana, Ljubljana, Slovenia, [4]University Hospital Centre Rijeka, Rijeka, Croatia, [5]Green infrastructure ltd, Zagreb, Croatia

Abstract: Apolipoprotein E (apo E, gene *APOE*) is a plasma glycoprotein with central roles in lipid and neuronal metabolism. Some researchers point on its metal-binding and antioxidative properties. It has three major isoforms apo E2, apo E3, and apo E4 encoded by alleles ε2, ε3 and ε4 respectively. According to some studies, individuals with ε4 variant are supposed to be more susceptible to metal toxicity, including mercury (Hg). On the other hand, it is believed that apo E is associated with higher levels of cholesterol, vitamin D and calcium, which can be beneficial early in life. The purpose of present study was to estimate relations between *APOE* polymorphisms and concentrations of total Hg (tHg) and methyl Hg (MeHg) in pregnant females exposed to Hg through seafood consumption. Study database consisted of tHg levels in maternal hair, urine, milk, peripheral and cord blood, and questionnaires with personal and life style data of 221 Croatian pregnant women. MeHg was additionaly measured by CVAFS (Tekran 2700). Archived DNA extracts from maternal leukocytes and cord tissues were genotyped using TaqMan® SNP assay for *APOE* rs429358 and rs7412 (Applied Biosystems). Statistics: Wilcoxon rank-sum test and multiple linear regression (STATA). EU projects: PHIME, HEALS. Mothers were divided into ε4 carriers (genotypes ε3/ε4 and ε4/ε4) and ε4 non-carriers (genotypes ε3/ε3, ε3/ε2 and ε2/ε2). We identified 17% ε4 carriers; they had higher geometrical means of tHg in maternal blood (ε4 carriers 2.6 ng/g, ε4 non-carriers 2.0 ng/g; p=0.0165) and cord blood (ε4 carriers 4.0 ng/g, ε4 non-carriers 2.7 ng/g; p=0.0128). MeHg represented about 80% of tHg. After taking into account the influence of seafood intake, parity, age, body mass index and smoking the observed higher concentrations of blood tHg or MeHg in *APOE* ε4 carriers were no longer significant. In conclusion, our data do not support the association of *APOE* genotypes with tHg neither with MeHg levels at low to moderate Hg exposure.

Keywords: D-susceptible/vulnerable, A-biomarkers, B-metals, D-prenatal

TU-PO-386

Results of the first National Human Biomonitoring in Slovenia: levels of trace metal(loid)s in lactating women and their partners and sources of exposure

D. Mazej[1], J. Snoj Tratnik[1], Z. Slejkovec[1], M. Jagodic[1], V. Fajon[1], A. Stajnko[1], M. Krsnik[2], A. Sešek Briški[2], M. Skitek[2], D. Kocman[1], A. Kobal[1], L. kononenko[3], M. Horvat[1]; [1]Jozef Stefan Institute, Ljubljana, Slovenia, [2]University Medical Centre, Ljubljana, Slovenia, [3]Ministry of Health, Ljubljana, Slovenia

Abstract: The main objectives of the human biomonitoring programme (HBM) are to provide data on exposure of the inhabitants of Slovenia to environmental pollutants, to look at spatial differences in exposure, and to identify main sources of exposure. The study population included lactating *primiparous* women (n=535) and men (n=561) aged 20-40 years living in twelve areas across Slovenia covering urban and rural areas and areas potentially contaminated due to recent or past industrial activities. Concentration of potentially toxic (cadmium, lead, mercury, arsenic) and essential elements (selenium, copper and zinc) was determined in blood, breast milk, urine and concentration of mercury also in hair using ICP-MS and CVAAS/CVAFS. Exposure of the study population to toxic elements was comparable with other European populations, so was the status of essential elements. In general no risk for the study participants was identified, and there was no noticeable lack or excess of essential elements, except in some individuals. Monitored levels of potentially toxic elements were found to be associated to: 1) life-style determinants, particularly, for Cd (smoking and diet) and Hg (fish consumption and amalgam fillings), and 2) to environmental exposure, which is of most concern in the area of Mezica Valley (Pb) and the former mercury mining town of Idrija (Hg). Linking HBM and environmental databases confirmed geo-dependant environmental nature of Pb and Hg exposure in the two cases. Apart of the geographically-dependant Pb exposure, public water supplies seem to be an important source of Pb. As a possible source of Pb exposure in the urban environment, emission of particulate matter from residential sources and roads was revealed. The data obtained will allow a health risk assessment of Slovenian population as well as further development and implementation of risk reduction measures.

Keywords: A-biomonitoring

TU-PO-387

Estimated Absorption of Permethrin and Pentachlorophenol and Body Mass Index of NC and OH Preschoolers

S. A. Gaston[1], M. K. Morgan[2], N. S. Tulve[2], K. A. Gaetz[3], K. Lichtveld[4]; [1]ORISE/ US EPA, Research Triangle Park, NC, [2]US EPA, Research Triangle Park, NC, [3]North Carolina Department of Health and Human Services, Raleigh, NC, [4]University of Findlay, Findlay, OH

Withdrawn

TU-PO-388

Thought-Starter: The concept of human exposure-based toxicity testing for agrochemicals

C. Terry[1], A. Whyte[2], D. Juberg[1]; [1]Dow AgroSciences, Oxon, United Kingdom, [2]Dow AgroSciences, Abingdon, United Kingdom

Abstract: Agrochemicals are thoroughly evaluated to ensure human safety. Despite efforts to introduce *in vitro* methods, animal studies identify toxicological hazards and set reference doses for human health risk assessments. Very high dose levels are employed in animals, with no relevance to realistic human exposure scenarios. A relevant, exposure-based and 3Rs-focused approach would be to limit high dose levels to a fixed multiple of human exposure; this assessment assumed a limit of 1000-times the predicted human exposure level. A dataset was established to evaluate how this approach would have impacted the dose setting for toxicity tests of existing agrochemicals. Chronic and acute dietary exposure scenarios were considered for 161 agrochemical reviews published by EFSA (of these 140 had ADIs and 100 ARfDs). Values were obtained for ADI, ARfD, critical NOAEL and predicted human exposure levels. The results showed that for 67% of ADIs and 59% of ARfDs there was no toxicity at 1000-times the predicted maximum human exposure level. In the remaining cases, 1000X human exposure exceeded the critical NOAEL, but the majority of the exposure levels were still lower than the critical LOAEL. For the limited number of chemicals where exposure levels were higher than the critical LOAEL, a case-by-case analysis investigating the expected nature of effects was conducted. In conclusion, the majority of the agrochemicals reviewed in this study were tested in animals at high doses not relevant to human exposure. These high doses often result in unnecessary, non-specific toxicity, and understanding the scientific relevance of these effects at doses with no relationship to actual human exposures leads to additional animal testing. These data highlight the potential refinement that human exposure-based testing could provide to the safety assessment of agrochemicals from a 3Rs and scientific perspective.

Keywords: B-pesticides, A-risk assessment

TU-PO-389

Comparative Evaluation of Orchard Sprayer Technology Based on Pesticide Drift Exposure Potential

E. J. Kasner[1], R. A. Fenske[1], G. A. Hoheisel[2], K. Galvin[1], M. G. Yost[1]; [1]University of Washington, Seattle, WA, [2]Washington State University, Prosser, WA

Abstract: The airblast sprayer has been a standard tool for tree fruit pesticide application technology since its rapid and wide-scale adoption in the 1950s. Pesticide drift potential has increased because traditional airblast sprayer output no longer matches the reduced height and canopy volume of modern orchard architecture. To strike a balance between spray coverage and drift prevention, orchardists are adopting new sprayer technologies. We conducted 18 field-based trials to compare three spray technologies—one traditional sprayer and two tower sprayers—based on minimizing downwind drift exposure. We hypothesized that newer tower sprayers would have lower drift measurements due to shorter nozzle-to-tree distances and a smaller risk of canopy escape. Three micronutrient tracers (Zn, Mo, and Cu) were applied separately with each sprayer to the same one-acre orchard block. Tracer aerosols were collected downwind on 6 m passive samplers (continuous pipe cleaners) suspended from masts in 15 gridded locations and then analyzed by inductively coupled plasma mass spectrometry (ICP-MS). Results were expressed as tank mix volume equivalents deposited on each 2 m section (n=348) of the passive sampler to illustrate the vertical drift profile at each location. We used a mixed-effects model to describe potential exposure with sprayer and wind parameters as fixed effects and location as a random effect. Geometric means and standard deviations of deposition at locations greater than 25 m downwind were 33 (2.5), 29 (2.6), and 21 (2.8) μL for the traditional airblast, directed air tower, and multi-headed fan tower, respectively. Vertical profiles demonstrated greater deposition at the highest sampling level with increasing distance. Preliminary results indicate that the multi-headed fan tower sprayer reduced drift when compared to the other two sprayers. Systematic evaluation of orchard sprayers is essential for developing recommendations about pesticide drift reduction.

Keywords: A-exposure models, B-pesticides, C-air, D-occupational

TU-PO-390

Health Risk Assessment of Endocrine Disruptor Organophosphorus Pesticides Exposure through Dietary Intake of Fresh Vegetables for children in Tehran, Iran

M. Zare Jeddi[1], Y. Bruinen de Bruin[2], V. Mahdavi[3]; [1]Institute for Environmental Research (IER), Tehran University of Medical Sciences, Tehran, Iran (the Islamic Republic of), [2]Joint Research Centre of the European Commission, Ispra, Italy, [3]Laboratory of Pesticide Research,, Tehran, Iran (the Islamic Republic of)

Withdrawn

TU-PO-391

Assessment of Inorganic Bromide Concentrations in Crops following Soil Fumigation with Methyl Bromide

C. Curl[1], J. Marshall[2], M. Moll[2], M. Morra[2], J. Porter[1], C. Rogers[2], G. Shewmaker[2]; [1]Boise State University, Boise, ID, [2]University of Idaho, Idaho Falls, ID

Abstract: Background: In 2006, discovery of the pale cyst nematode in eastern Idaho created an immediate need for control measures to protect the Idaho potato industry from this high-consequence pest. Federal and state agencies developed a comprehensive eradication plan, which included repeated soil fumigations with methyl bromide (MeBr), all conducted according to label. As an unexpected consequence, high concentrations (>7,000 ppm) of bromide (Br^-) were found in some forage/hay crops grown on treated fields, and adverse health effects were observed in cattle consuming these crops. Objective: To assess Br^- uptake in crops grown in MeBr-treated fields, determine the effect of soil amendments on Br^- uptake, and establish background Br^- concentrations in untreated fields in the region. Methods: In a field fumigated with MeBr in 2013 and 2014, we established 4 replicate plots of each of four crops common to the region: corn, wheat, barley, and alfalfa. Potatoes were in an adjoining field fumigated once in 2014. We measured Br^- concentrations in 503 soil and plant tissue samples. Additional alfalfa plots were treated with either dry manure or potassium chloride alongside an unamended control. These were replicated 3 times and 393 alfalfa and soil samples were analyzed for Br^- concentrations. We also analyzed Br^- concentrations in 144 crop and soil samples from nearby fields untreated with MeBr. Results: Baled alfalfa and straw from wheat and barley contained significantly higher Br^- concentrations than corn ears and stalks, wheat and barley grain, and potatoes. Use of soil amendments significantly reduced Br^- uptake in alfalfa. With the exception of potatoes, crops grown in previously treated fields contained significantly higher concentrations of Br^- than those grown in untreated fields. Conclusion: Most concern around MeBr use has focused on stratospheric ozone depletion and acute toxicity following inhalation. Our results suggest that plant uptake of Br^- should also be considered.

Keywords: B-pesticides, A - exposure measurement, C-food, C-soil

TU-PO-392

Method development for the detection of pyrethroid metabolites in saliva

M. Wren, I. Yang, B. Buckley; Rutgers University, Piscataway, NJ

Abstract: Saliva may be used as a less-invasive alternative to urine or blood in pesticide biomonitoring. Salivary concentrations of lipophilic pesticides have been correlated to the unbound fraction level present in blood, making it a useful biomarker for occupational and environmental exposures. No method has previously been developed to quantify biomarkers of pyrethroid insecticides in saliva. The purpose of this study was to develop a method to detect three pyrethroid metabolites in saliva: *cis*-3-(2,2-dibromovinyl)-2,2-dimethylcyclopropane carboxylic acid (DBCA), 3-phenoxybenzoic acid (3-PBA) and 4-fluoro-3-phenoxybenzoic acid (FPBA). Two different sample collection methods were assessed: 1) passive flow and 2) the Salivette® collection device. Two solvents, toluene (MePh) and dichloromethane (DCM), and three time points, 20, 40 and 60 min, were evaluated for extraction efficiency. Liquid-liquid extraction (2 x 3 mL solvent) was performed to separate analytes from the aqueous layer. Metabolites were derivatized with hexa-2-fluoroisopropanol and N,N'-Diisopropylcarbodiimide, cleaned up with $NaHCO_3$ and isolated in hexane, then concentrated to 40 µL hexane prior to analysis by ion trap gas chromatography/mass spectrometry using the Agilent 7890B Gas Chromatograph and Agilent 240 Mass Spectrometer. MePh extracted all pyrethroid metabolites with 30% - 70% greater efficiency per one mL than DCM. The 40 min extraction period yielded greatest recoveries for both MePh and DCM: 109%, 88% and 104% recovery and 93%, 77% and 88% recovery for DBCA, FPBA and 3-PBA, respectively. Collection from the Salivette® cotton roll required three diH_2O washes to extract metabolites. Recoveries of over 150% for all analytes from the Salivette suggest interfering components from the collection device. A correlative study to quantify salivary, blood and urine concentrations of pyrethroid metabolites is underway.

Keywords: B-pesticides, A-biomonitoring, A-analytical methods, A - exposure measurement

TU-PO-393

A Pilot Study on Migrant Grape Workers Exposure to Pesticides in Sonora, Mexico

N. I. Lopez-Galvez; University of Arizona, Tucson, AZ

Abstract: Expanding agribusiness in Sonora, a state in Northern Mexico known for its cattle, tomato, and table grape production, has increased the demand for temporary migrant agricultural workers from the poorest regions of Southern Mexico. These migrant agricultural workers participate in strenuous tasks while exposed to a wide variety of occupational risks and hazards including pesticides. A cross-sectional observational study was conducted to assess the pesticide exposure of migrant grape workers. This is the first study to characterize exposure of migrant grape workers to pesticides in this region. A convenience sample of 20 participants were recruited from a large commercial grape farm employing approximately 2,000 workers during the harvest season near Hermosillo- Sonora, Mexico. A questionnaire was used to obtain information on working activities and demographics. Morning void urine samples were collected to assess pesticide exposure. Most participants were originally from the state of Chiapas-Mexico, none had completed high school, and half spoke an indigenous language. Pyrethroid and organophosphate urine metabolites were detected in the majority of workers. The creatinine-adjusted concentration for cyfluthrin, chlorpyrifos, and parathion metabolites in urine obtained in this study (geometric means: 0.942 µg/g, 3.559 µg/g and 1.630 µg/g, respectively) were higher than in the Mexican American population included in NHANES. Unfortunately, there is no study of the general population in Mexico for comparison. Our results suggest that grape workers in this region are exposed to high levels of pesticides, which may, eventually, affect their health. Additional research is needed to confirm these findings and to evaluate health outcomes associated to pesticide exposure in this region. Results from this pilot study can be used to conduct a larger pesticide study, create binational partnerships between researchers, and develop occupational health training resources in Mexico.

Keywords: A-biomonitoring, D-occupational, B-pesticides, A-environmental justice, A - population exposure

TU-PO-394

Construction of risk assessment system for chemicals in the agricultural environment

S. Choi; KnA consulting Co., Ltd, Yongin, Korea (the Republic of)

Abstract: Various standardized guidelines and systems have been developed both domestically and abroad to ensure the efficiency of risk assessment for the chemical residues in agricultural and food products. In this study, a risk assessment system, APRAS was developed to efficiently carry out risk assessment tasks for the RDA and support the storage and utilization of the previously accumulated monitoring data of hazardous chemical residues in the agricultural environment and products. APRAS consists of two database systems, the monitoring data in the NAAS and a food consumption database based on the NHNES data. This system is designed to extract data from major database systems relevant to a given exposure scenario so that proper exposure and risk assessment can be performed to achieve evaluation objectives. For the accumulation of the currently available data and efficient data connectivity, the system was developed to be web-based. The risk assessment system sets and associates parameters according to an exposure scenario constructed to evaluate the risk of agricultural chemical residues and predicts corresponding risk. In addition, APRAS was developed to incorporate monitoring databases of the residues and their quantities in the environmental media related to agricultural products, such as soil and water, and the risk assessment module can provide an estimated prediction through a probabilistic approach. Although the initial monitoring database of this system based on the hazardous chemical compounds database for agricultural products, it was upgraded to be a comprehensive agricultural environment monitoring database with the addition of hazardous chemical compounds databases for soil and ground water. Finally, the evaluation system was expanded to construct additional exposure routes of agricultural hazardous chemical residues in the agricultural environment to human body from soil and ground water and establish a necessary variable database.

Keywords: A-exposure models, A-risk assessment

TU-PO-395

Use of AHETF Data Outside North America

J. Holmsen; Agricultural Handler Exposure Task Force, Macon, MO

Abstract: The Agricultural Handler Exposure Task Force (AHETF) was established in 2001 to design and develop data on exposure of agricultural workers during the mixing, loading and/or application of pesticides in order to meet regulatory requirements imposed upon the member companies by EPA. The dermal and inhalation exposure data were intended to upgrade the data used in North American risk assessments based on PHED, and cover all important types of mixing/loading systems, differences in application equipment and diversity of formulations. The proprietary exposure data developed by the Task Force and submitted to North American regulators covers virtually all pesticide active ingredients and is being used by the U.S. EPA, California Department of Pesticide Regulation, and Health Canada's Pest Management Regulatory Agency to conduct agricultural handler risk assessments. Recently, the EPA's summary of AHETF data on EPA's web site has been used by CropLife International to develop a risk assessment tool for developing countries. As EPA has evaluated and in some respects proposed to modify aspects of the original AHETF data with different adjustment factors, it is important to understand the sources of these adjustment factors and recognize the impact of these modifications on the risk assessment outcomes. A comparison will be presented to highlight the differences in exposure endpoints that regulatory authorities have introduced following their own evaluations of the original AHETF data.

Keywords: B-pesticides, D-occupational, A - exposure measurement

TU-PO-396

The International Estimated Short-Term Intake: Conservatisms and Appropriateness for Estimating Acute Dietary Exposure to Pesticides

C. R. Fleming[1], C. Cleveland[2]; [1]Dow AgroSciences, Indianapolis, IN, [2]BASF Corporation, Research Triangle Park, NC

Abstract: The International Estimate of Short-Term Intake (IESTI) is used to estimate acute dietary exposures to pesticide residues as part of the process for establishing and reviewing maximum residue limits (MRLs) on foods in some regions (e.g., EU & Codex countries). The IESTI calculation is based on consumption of a large portion of a food item combined with a high-end estimate of the residues possible on that food item (the highest residue (or median for blended foods) from pesticide field trials). In late 2015, changes to the current IESTI calculations were proposed in a joint FAO/WHO workshop. Subsequent discussions at the Codex Committee on Pesticide Residues (CCPR) and an associated electronic working group focused on whether change is needed and potential impacts of proposed changes. No decision has been made on final proposed equations, but the appropriate level of conservatism has been a focus of discussion. The 2015 proposed equations included: use of the MRL in place of residue trial data, removal of consideration of unit weight, introduction of a new conversion factor accounting for metabolites not included in the MRL, and use of large portion consumption data on a body weight rather than per person basis. Preliminary assessments have projected an increase in calculated dietary intakes, despite statements ***that changes to the IESTI should not lead to substantial changes in the level of conservativeness*** (CX/PR 17/49/12; discussion paper from IESTI eWG at 49th CCPR meeting). The ultimate impact on MRLs critical for the free global trade of food remains unclear. Based on these events, it has become apparent that benchmarking the conservatisms in the IESTI approach will be necessary in order to better understand the need for any changes and the implications of those changes. In this presentation, we will provide a history of the proposed changes and provide case studies illustrating the conservatisms within the IESTI approach for estimating acute dietary exposures.

Keywords: A-risk assessment, A-exposure models, C-food

TU-PO-397

Characterization of Organophosphate Pesticides in Urine and Home Environment Dust in an Agricultural Community

C. Tamaro[1], M. Smith[1], T. Workman[1], W. Griffith[1], B. Thompson[2], E. Faustman[1]; [1]University of Washington, Seattle, WA, [2]Cancer Prevention Program, Fred Hutchinson Cancer Research Center, Seattle, WA

Abstract: Organophosphorus insecticides (OPs) are used on agricultural crops to control pests. Farmworkers (FW) have higher exposure to OPs than non-farmworkers (NFW), and FW children may have higher exposure than NFW children. A community-based participatory research strategy was used in the lower Yakima Valley of Washington state to identify 100 FW and 100 NFW adults, each with a referent child 2-6 years old. FW adults worked as thinners or harvesters in pome fruit orchards. Parents and children participated in three data collection periods over the course of one year. Urine samples were evaluated for the dialkylphosphate metabolites (DAP) characteristic of OP exposure, and dust from homes and vehicles was evaluated for intact OP residues. Geometric mean (GM) concentrations of OPs in house and vehicle dust were higher in FW households than NFW households across all three agricultural seasons. GM concentration of urinary DAPs was higher for children in FW households than NFW households. We found significant seasonal and occupational associations between urinary DAPs and intact OP residues in house dust among this cohort of FW and NFW adults and children. We found higher concentrations of intact OP residues in FW house dust and higher concentrations of OP metabolites in FW family urine. Results of our regression analysis showed a significant positive association between the concentrations of dimethyl OPs in household dust and dimethyl OP metabolites in urine, most notably during thinning season but also during harvest season when OPs are applied less frequently. These results provide support for the occupational take-home pathway of exposure by which children are exposed to agricultural pesticides used in the workplace. This project was made possible by The University of Washington Center for Child Environmental Health Risks Research supported by grant PO1 ES009601 from NIEHS, grants RD826886, RD83451401 and RD-83273301 from EPA, and NIH contract HHSN267200700023C.

Keywords: A - exposure measurement, B-pesticides, C-indoor, D-children, D-occupational

TU-PO-398

Left or Right: Where Should Personal Exposure Monitors Be Worn?

J. Thornburg[1], R. Chartier[1], S. N. Chillrud[2], D. W. Jack[2], S. Cho[1]; [1]RTI International, Research Triangle Park, NC, [2]Columbia University, Palisades, NY

Abstract: Widespread air pollution personal exposure measurements are becoming a reality as new sensor technologies become available or developed. The possibility of personal exposure measurements on larger cohorts of several hundreds or thousands of participants raises new questions about approaches for collecting the highest quality, most representative data to answer the research hypotheses. One such question is “where on the body should exposure monitors be worn?” The general consensus is that a person’s upper torso is the ideal location because of the proximity to the breathing zone. However, what side is best? Left? Right? Side of the participant’s dominant hand? Does exposure monitor placement defined in the data collection protocol depend on the study hypotheses? This work will consolidate air pollution personal exposure data from multiple studies to provide insight into these questions. The analysis will focus on particulate matter and gases. Multiple datasets that include duplicate, personal level integrated and real-time PM_{10}, $PM_{2.5}$, and $PM_{10\text{-}2.5}$ concentrations will be presented. Similarly, collocated exposure data for nitrogen oxides, ozone, volatile organic compounds, and aldehydes will be presented.

Keywords: A - exposure measurement, B-particulate matter, B-VOCs, C-air

TU-PO-399

Applications of Passive Silicone Wristband Samplers: Childhood Para-Occupational Exposures to Pesticide Mixtures

C. Poutasse[1], T. Arcury[2], S. Quandt[3], P. Laurienti[4], K. A. Anderson[1]; [1]Oregon State University, Corvallis, OR, [2]Wake Forest School of Medicine, Winston-Salem, NC, [3]Wake Forest School of Medicine, Winston-Salem, NC, [4]Wake Forest School of Medicine, Winston-Salem, NC

Abstract: Agricultural workers are exposed to a variety of pesticides at work and may put household members at risk through para-occupational exposures. However, self-reporting and biomonitoring samples may not fully capture total pesticide exposure. For a complementary assessment of external chemical exposure, silicone wristbands were used as passive sampling devices (PSDs) to sample complex environmental mixtures. In this pilot study, 20 children wore PSDs, which were subsequently analyzed for 72 pesticides by gas chromatography (GC) electron capture detection (ECD). It was hypothesized that farmworker household children (FHC) experience different pesticide exposures compared to non-farmworker household children (NFHC), which could lead to distinguishable neurological differences between study groups. FHC wristbands detected twice as many pesticides as the NFHC wristbands, and only FHC wristbands detected agricultural organophosphate pesticides. These preliminary data suggest an association between FHC group membership and an increased number of pesticide detections from para-occupational exposures. Future directions include investigating pesticide exposures as related to the agricultural season, incorporating data on participant neurological development, and expanding the study by increasing the sample size.

Keywords: B-pesticides, D-children, A - exposure measurement, A-epidemiology, A-sampling methods

TU-PO-400

Mobile and passive sampler measurements to identify spatial distributions of air pollution and its sources in Los Angeles

M. Tessum[1], T. Larson[2], S. Vedal[1]; [1]University of Washington, Seattle, WA, [2]University of Washington, Seattle, WA

Abstract: Mobile monitoring can measure small spatial scale variability in air pollutant concentrations. However, because spatial variability cannot be disentangled from temporal variability, it is not clear whether these can be used to estimate long-term concentrations. To assess the utility of short-term mobile monitoring measurements in estimating long-term exposure concentrations, 2-week mobile monitoring campaigns covering a large urban area combined with passive sampling campaigns in 43 intersections were conducted in Los Angeles in summer and winter. The mobile median concentration for each pollutant at each intersection across all days within a season was calculated to allow comparison with passive sampler detector (PSD) 2-week average concentrations for each pollutant at each intersection within a season. There were strong to moderate correlations between mobile and PSD NO_2 measurements in both seasons (Pearson's r = 0.62 and 0.79, respectively), and moderate correlations between mobile and PSD NO_x measurements in both seasons (r = 0.53 and 0.43). There were also correlations between mobile particulate matter, black carbon, CO, and CO_2 measurements and PSD NO_2 and/or NO_x measurements, indicating that mobile monitoring can be used to reflect long-term average concentrations of traffic-related and other air pollutants. A principal components analysis of combined mobile and PSD measurements identified emissions from light duty vehicles, diesel engines, crankcase venting, and aircraft as major pollution sources. Maps of absolute principal component scores identified intra-urban spatial distributions of those source emissions. Mobile monitoring combined with PSD sampling can provide good long-term estimates with high spatial resolution of traffic-related air pollution. These estimates can be used to identify source profiles of traffic-related and other emissions of multipollutant mixtures, as well as intra-urban spatial distributions of those source emissions.

Keywords: A - ambient monitoring, B-VOCs, C-air, B-particulate matter

TU-PO-401

A Novel Method for Characterizing Resident Behaviors and Housing Attributes using Photo Survey

P. Fabian, Z. Petropoulos, C. Scollaert, M. Scammell, J. I. Levy; Boston University School of Public Health, Boston, MA

Abstract: Exposure to air pollutants such as $PM_{2.5}$ is associated with a variety of adverse health outcomes including cardiovascular and respiratory diseases, preterm birth, and low birth weight. Accurately capturing an individual's exposure to ambient air pollutants requires focusing on the indoor environment, as this is where the average person spends the majority of their day. The impact of resident behaviors (i.e. window opening or air conditioning use) on infiltration of air pollutants into a home has been well-characterized. However, efforts to accurately predict these behaviors and their distribution (through field campaigns and mail-in surveys) have many limitations, including the ability to characterize these behaviors across entire communities. Using the ArcGIS Photo Survey tool and a GPS-enabled camera, we collected photos of homes in a community in Massachusetts, and created an online survey which was used to classify open windows and installation of window air conditioning units via crowdsourcing. Photos are also linked to parcel-level housing data and neighborhood-level sociodemographic data which will be used to develop predictive models of these behaviors. In our winter 2016-17 survey, we captured photos of 1,100 homes and classified the necessary parameters using crowdsourcing for image classification. These data will be used to identify the significant predictors—including housing characteristics, sociodemographic variables, and meteorological parameters—of window-opening and ownership of window air conditioning units. This approach to data collection can be used to concurrently capture information relevant to community stakeholders—which may include data on recycling and trash bins, trees, blighted or damaged homes, and graffiti—and can be implemented in any community to collect a wide variety of information. Photo survey techniques can complement parcel data and other geospatial information to enhance residential exposure assessment.

Keywords: A-behavior, A-activity patterns, A-built environment, A-indoor environment, C-air

TU-PO-402

Equines used as sentinels for human health - leveraging passive sampling and unique equine population exposures to assess negative post-natal health outcomes

L. G. Tidwell[1], C. E. Donald[1], K. R. Mullen[2], D. M. Ainsworth[2], K. A. Anderson[1]; [1]Oregon State University, Corvallis, OR, [2]Cornell University, Ithaca, NY

Abstract: In 2012 a high prevalence of neonatal foals born with neurological deficits (altered mentation and dysphagia) was observed at a farm located Pennsylvania (PA) near a number of unconventional natural gas drilling (UNGD) sites. The farm proprietor also owned another horse farm in New York (NY) that was not situated near UNGD sites. Newborn foals and mares on the NY farm did not exhibit similar health issues. From 2014 to 2016 passive sampling devices (PSDs) were deployed on brood mares at both farms continually, and PSDs were changed out every six weeks. PSDs were also deployed in the ambient air and in well water at each farm. Over the course of the study 19 foals that were at the PA farm during gestation/birth exhibited dysphagia while only 2 foals from the NY farm exhibited this deficit. Well water PSDs showed increased (2-20X greater concentration) levels of PAHs in PA. PSDs deployed in ambient air showed pyrogenic signatures of PAH ratios at both sites. However, the pyrogenic signature at the PA farm is associated with fossil fuel combustion, while the pyrogenic signature at the NY farm is more consistent with biomass combustion. Individual mare PSDs were recovered at rates of 95% and 92% from horses on the NY and PA farm respectively, showing that this adaptation of PSDs is robust enough to withstand the equine environment. Chemical analysis for PAHs on mare-deployed PSDs resulted in quantitative identification of 60 individual PAHs. Continued analysis of biological samples from each mare-foal pair and analysis of PSDs could reveal environmental stressors associated with this negative health outcome in a mammal model. These results may further inform human health concerns regarding the proximity of residences to UNGD facilities.

Keywords: A - ambient monitoring, A - population exposure, B-POPs,

TU-PO-403

Integrating Spatiotemporal Information System Approaches with Agent-Based Modeling for Studies of Human Exposures to Traffic Related Air Pollution (TRAP)

Z. Mi, R. Laumbach, H. Kipen, P. Georgopoulos; Environmental and Occupational Health Sciences Institute, Rutgers University, Piscataway, NJ

Abstract: Exposure to Traffic Related Air Pollution (TRAP) during commuting has been associated with various adverse cardiovascular and respiratory effects. A prototype Spatiotemporal Information System (STIS) was designed and implemented to support analysis of information collected through the Rutgers Commuter Community Cohort (RC3) study, providing data management, analytics, visualization, and screening exposure modeling capabilities. Using information collected via the RC3 study (i.e. spatiotemporal commuting trajectories and associated measured personal in-vehicle concentration profiles), the STIS was designed and implemented for direct integration with a comprehensive multiscale environmental and microenvironmental quality and human exposure modeling system, under continuing development at the Computational Chemodynamics Laboratory (CCL) of EOHSI. This integration can be used to inform and enhance the design of future exposure-relevant studies, while providing access to the field information collected via the RC3 study. Furthermore, an Agent-Based Model of Human Exposures has been under development as an option within the PRoTEGE (Prioritization and Ranking of Toxic Exposures with GIS Extension) modeling framework of CCL, to support further analyses of commuting behavior and of associated exposures, under both real and hypothetical scenarios that extend beyond the situations assessed and data collected via the RC3 study. The ABM presented here is being built on the REPAST (Recursive Porous Agent Simulation Toolkit) modeling platform (customizable via Python 2.7 scripts) and is used to simulate traveling patterns and behaviors of "virtual subjects" (represented as "agents" in the model) that allow the study of situations involving altered traffic patterns that reflect conditions corresponding to alternative planning options considered in urban and regional development plans.

Keywords: A-exposure models, A-geospatial analysis/GIS, A-risk assessment, C-air

TU-PO-404

Spatial variation of secondary inorganic $PM_{2.5}$ exposure: from exposure magnitude to exposure distance

K. Stylianou, O. Jolliet; University of Michigan School of Public Health, Ann Arbor, MI

Abstract: Secondary $PM_{2.5}$ exposure, particles formed from precursor emissions, is currently poorly characterized. We estimate secondary $PM_{2.5}$ exposure and exposure travel distances due to precursors such as NH_3, NOx, and SO_2 and evaluate the respective spatial variability within the US. We characterize $PM_{2.5}$ exposure due to NH_3, NOx, and SO_2 using intake fractions (iF), the fraction of $PM_{2.5}$ inhaled by population from a unitary precursor emission. We use a source-receptor (S-R) matrix approach based on the InMAP (Intervention Model for Air Pollution), a reduced-complexity air pollution model, to characterize exposure from ground level emissions in >40,000 source locations in the US. To investigate exposure travel distance we estimate *intake travel distance* of x (ITDx), distance from the source at which we reach x% of total cumulative iF, for each source. Results indicate substantial spatial variability for $PM_{2.5}$ exposure due to NH_3 emissions in the US. Estimates of $iF_{PM2.5,NH3}$ range between 0.01 and 41.5 ppm. Sources locations in close proximity to big cities (e.g. NY, LA) result in the highest $PM_{2.5}$ exposure from NH_3 emissions. Similar trends are observed for SO_2 and NOx emissions, with $iF_{PM2.5,SO2}$ and $iF_{PM2.5,NOx}$ estimates spanning between 0.009-3.0 ppm and 0.002-1.3 ppm, respectively. Our analysis produced higher secondary $iF_{PM2.5,NH3}$ estimates with higher variability compared to published estimates. In addition, our analysis suggest that exposure radius is precursor dependent with NH_3 having on average the shortest ($ITD_{50,NH3}$=70 km, 95% CI: 0-450) and SO_2 having on average the largest ($ITD_{50,SO2}$=200 km, 95% CI: 0-740) travel distance. Our results improve the exposure characterization of secondary $PM_{2.5}$, especially from NH_3 emissions. This S-R exposure assessment approach enables us to calculate sector specific (e.g. agriculture, electricity production) average iF at local, regional or nation level that can be used in alternative chemical assessment or life cycle assessment.

Keywords: A-exposure models, A-geospatial analysis/GIS, B-particulate matter, C-air

TU-PO-405

Applicability of mobile monitoring to high-resolution air quality mapping in New Delhi, India

A. Gurung, K. P. Messier, J. Apte; University of Texas at Austin, Austin, TX

Abstract: Aim:As Indian cities develop more robust air-quality monitoring infrastructure, a critical question is whether conventional ambient monitors can adequately characterize population exposure given the high density of primary combustion emissions. We analyze previously collected mobile monitoring data from New Delhi, India to assess use of routine mobile monitoring to reliably map spatial patterns of air pollution. **Methods**:Mobile monitoring was performed during morning and evening rush hours at 1s intervals for particulate matter with aerodynamic diameter less than 2.5μm ($PM_{2.5}$), black carbon (BC), and ultrafine particle number concentrations (PN) for 40 weekdays (Feb 22-May 26 2010) along a predetermined route. 1Hz measurements was snapped to the nearest 10-meter road segment and median pollutant concentrations estimated. Median precision for each road segment was estimated using bootstrap resampling. Intraclass correlation (ICC) for the 1Hz data grouped by road segment was estimated to investigate robustness of spatial patterns among repeated measures. **Results**:High concentrations of air pollutants with fine scale variability was observed along the route for ~3900 10m road segments. Mobile monitoring allowed repeated resampling to compute stable long-term median concentration. Road segment median concentrations were 195µg/m^3, 37µg/m^3, and 253×10^{3}cm^{-3} for $PM_{2.5}$, BC, and PN respectively. Low normalized standard error of individual 10m median concentrations of 4%, 11% and 9% for PM2.5, BC, and PN respectively suggested good overall precision in fine scale concentration estimates. Spatial concentration variability dominated the total variability and the observed long-term spatial pattern was robust to random variability among individual 1Hz samples, as noted by high ICC (0.90-095) for the pollutants. **Conclusion**:Findings demonstrate the applicability of mobile monitoring in Delhi as a cost-effective technique to characterize stable long-term spatial patterns of air pollution.

Keywords: A - exposure measurement, A - ambient monitoring, A - population exposure, A-global health

TU-PO-406

Meta-Analysis of Lead (Pb) in Multiple Environmental Media in the United States

J. J. Frank[1], A. Poulakos[2], J. Xue[3]; [1]Oak Ridge Institute for Science and Education, Oak Ridge, TN, [2]ASRC Federal Vistronix, Boston, MA, [3]U.S. Environmental Protection Agency, Durham, NC

Abstract: Introduction: The U.S. Environmental Protection Agency, Office of Research and Development, conducts probabilistic multimedia lead (Pb) exposure modeling to inform the development of health-based benchmarks for Pb in the environment. For this modeling, robust Pb concentration data are needed for developing distributional model inputs. **Objectives:** To create a database of Pb levels measured in environmental media within the United States using data from published literature; and to summarize Pb levels by media type to refine model inputs. **Methods:** A systematic literature search was conducted to identify qualifying articles meeting specified criteria published from 1996 to early 2016. Article screening and database creation were reviewed independently by two researchers. A random effects model was used to summarize data by media type. **Preliminary Results (reported as mean ± 95% CI for each single group summary):** Residential (RES) soil samples (50±24 ppm) from non-urbanized areas (rural) were ~8x lower than urbanized RES areas (383±74 ppm). RES sites on or nearby Pb Superfund locations were largely classified as rural (8 of 9 sites) with soil Pb levels of 267±56 ppm. Schoolyard and playground soil Pb levels were 54±22 ppm. Community and RES garden soil Pb levels were 160±37 ppm. Soil Pb from non-RES Pb Superfund sites (1316±402ppm) were ~2x lower than soil collected at outdoor shooting ranges (3137±136 ppm). Air Pb reported in the literature for sites classified as rural (0.0035±0.003 µg/m^3) were ~5x lower than urbanized areas (0.0169±0.014 µg/m^3). The results for Pb in drinking water for all areas reported in the literature was 2.25±1.79 ppb. **Conclusions:** The results from this analysis can inform future research and rule-making by providing insight into information gaps and key inputs in Pb multimedia modeling analyses, and by helping to identify potentially vulnerable groups. In addition, this information can be used to assess the effectiveness of remediation.

Keywords: B-metals, A-exposure factors, C-multimedia, D-community, A-environmental regulation

TU-PO-407

Characterization of total chromium and chromium-6 in UCMR3 drinking water monitoring data

C. L. Ring, F. Loko, M. A. Harris; ToxStrategies, Inc., Austin, TX

Abstract: We characterize the distributions of total chromium (Cr) and chromium-6 (Cr(VI)) in drinking water monitored by US EPA's Third Unregulated Contaminant Monitoring Rule (UCMR3) between January 2013 and December 2015 in all public water systems (PWSs) serving more than 10,000 people and 800 representative PWSs serving fewer than 10,000 people. Cr was below its Minimum Reporting Limit (MRL) (0.2 µg/L) in 49.5% of samples; Cr(VI) was below its MRL (0.03 µg/L) in 24.4%. With below-

MRL values substituted by the MRL, median concentrations nationwide were 0.2 µg/L Cr and 0.1 µg/L Cr(VI), and 95th percentile concentrations were 3.5 µg/L Cr and 3.4 µg/L Cr(VI). Levels of both Cr and Cr(VI) tended higher in PWSs with ground water sources (nationwide 95th percentile=5.6 µg/L for both Cr and Cr(VI)), compared to PWSs with surface water sources (nationwide 95th percentile=0.94 µg/L Cr and 0.69 µg/L Cr(VI)). To characterize geospatial distribution, samples were aggregated by ZIP code using UCMR3's list of ZIPs served by each PWS. Samples from each PWS were assigned to all ZIPs served by that PWS. Summary statistics for each ZIP are visualized on US and state maps. Cr and Cr(VI) levels exhibit some regional patterns; ZIPs with 95th percentile levels above 10 µg/L Cr or Cr(VI) tend to cluster in Arizona (104 ZIPs), California (54 ZIPs), and Oklahoma (16 ZIPs), with isolated occurrences in ten other states (1-2 ZIPs per state). This work provides a foundation for further geospatial analyses of Cr and Cr(VI) occurrence in drinking water, potentially including investigations into the sources of Cr and Cr(VI) in drinking water.

Keywords: C-water, B-metals

TU-PO-408

Microbiomes of the Built Environment: From Research to Application—a National Academies Study

E.B. Boyle[1], D. A. Butler[1], K. Bowman[2]; [1]National Academies of Sciences, Engineering, and Medicine, Washington, DC, [2]National Academies of Sciences, Engineering, and Medicine, Washington, DC

Abstract: Research has shown that built environments host a vast number and diversity of species of bacteria, viruses, fungi and protozoa in the air, water, and heating, ventilation, and air conditioning (HVAC) systems, and on surfaces. These constitute dynamic microbial communities or "microbiomes." The nature, composition, diversity, evolution, and growth of these microbiomes are influenced by interactions with humans, animals and plants; and occupant exposure is mediated by factors such as air flow, temperature, humidity, household chemicals, and building materials. These elements are, in turn, shaped by the design, construction, operation, occupation, and use of residences, commercial, office and other building types. The National Academies of Sciences, Engineering and Medicine convened a committee to examine the formation and function of microbiomes found in the interior of built environments. A report of its work will be released in summer 2017. It will explore the implications of this knowledge to positively impact sustainability and human health, and will • assess what is currently known about the interactions among microbial communities, humans, and built environments, and their relationship to indoor environmental quality; • articulate opportunities and challenges for the practical application of an improved understanding of indoor microbiomes; • dentify critical knowledge gaps and prioritized research goals to accelerate the application of knowledge about built environment microbiomes to improve built environment sustainability and human occupant health; and • discuss and recommend actions to advance understanding of microbiome-built environment interactions, including the potential impacts of building and health-related policies and practices, and social or public engagement dimensions. This presentation will summarize the report, focusing on how biological sciences, engineering, public health, and medical research are combining to provide insights on human exposures.

Keywords: A-built environment, B-microbial agents, C-indoor

TU-PO-409

Potential health risks of Polycyclic Aromatic Hydrocarbons (PAHs) associated with sediment and selected sea foods from a Ramsar site

D. S. Jyethi[1], P. Khillare[2]; [1]Indian Statistical Institute, Tezpur, India, [2]Jawaharlal Nehru University, New Delhi, India

Abstract: Persistent organic pollutants such as polycyclic aromatic hydrocarbons (PAHs) are ubiquitous in the environment. It is important to assess risk of PAHs exposure. Sixteen United States Environmental Protection Agency (USEPA) priority PAHs were determined in sediments and edible biota at a Ramsar site - Chilika lagoon, the largest brackish water lagoon in Asia. High Performance Liquid Chromatography was used to identify and quantify PAHs. Statistical tools were used for source apportionment and health risk assessment. Mean total PAH level in the sediments was 13674 ng/g dry weight, higher than other reported studies from the region. High molecular weight species dominated total PAH profile indicating pyrolytic origin. Diagnostic ratios and principal component analysis indicated diesel and coal/wood combustion as sources of PAHs in the sediments. Assessment of sediment associated individual PAHs effect on aquatic organisms of the lagoon revealed all the compounds, except Naphthalene and Anthracene, to be present above the lower range of concentrations related to toxicity. Risk quotient of PAHs revealed that Acenaphthene, Fluorene and Dibenz[a,h]anthracene require priority management concerns. PAH levels in crabs and prawns were 394.4 and 153.0 ng/g d.w. BaPeq concentrations were 42.9 and 15.2 ng/g d.w. in crabs and prawns respectively. Carcinogenic PAHs accounted for 33% of the PAHs concentrations in edible biota but consumption can be considered safe with respect to lifetime excess cancer risk guideline values. Low biota sediment accumulation factors in biota from the lagoon suggest that the studied PAHs have low bioavailability. The results of the study are useful for risk assessement due to dietary exposure of sea foods from the lagoon, which is consumed in domestic markets and also used for export to developed countries.

Keywords: B-POPs, C-food

TU-PO-410

Wi-Fi radiation exposure to children in kindergartens and schools in Australia

G. Benke[1], C. Bhatt[1], K. Karipidis[2], M. Abramson[1]; [1]Monash University, Melbourne, Australia, [2]Australian Radiation Protection and Nuclear Safety Agency, Yallambie, Australia

Abstract: Background: In 2010 the World Health Organization (WHO) published a research agenda for radiofrequency electromagnetic fields (RF-EMF). In the agenda was a call to focus future epidemiological research on the assessment of RF-EMF exposures to humans, especially children. Although there has been a significant amount of research regarding the cell phone frequencies (600 MHz to 2 GHz), little research on the exposures of children to Wi-Fi (2.4 GHz) has been published. Here we report on the results of two studies undertaken in Australia involving the measurement of Wi-Fi exposures in kindergartens and schools. **Methods:** In the first study, static and personal monitoring of Wi-Fi exposure was undertaken in 20 kindergartens. The ExpoM™ RF dosimeter was used to measure the personal exposure in 10 children over a four hour period. In the second study, undertaken in 23 schools, static measurements were undertaken in the class-rooms and school yards. **Results:** In the kindergartens, the maximum Wi-Fi levels were all below 0.05% of the ICNIRP limits. The contribution of Wi-Fi to total RF-EMF exposure was less than 10%. In the schools, the peak and median Wi-Fi levels in the class-rooms were in the order of 0.01 and 0.0001% of the ICNIRP limits, respectively. **Conclusions:** These findings should be reassuring to parents, teachers and students in schools and kindergartens, as the levels are very low compared to the ICNIRP limits and contribute only a small part of the overall RF-EMF exposure.

Keywords: A - exposure measurement, B-radiation, A-epidemiology

TU-PO-411

Addressing Toxicity of Water using the Challenge Process

E. Varughese, J. Garland; U.S. EPA, Cincinnati, OH

Abstract: Exposure concerns from an increasing number of pollutants in US waters requires the need for better monitoring. The pollutants are diverse and can include pesticides, heavy metals, personal care products, natural toxins such as cyanobacterial toxins, and a host of other organic and inorganic chemical pollutants and their transformation products. Current methods for detecting and identifying many of these contaminants are expensive, time-consuming, and require the use of specialized laboratories. Moreover, multiple methods are needed to detect the variety of contaminants of interest and, if the identity of the potential contaminants is unknown, water monitoring becomes even more complex. A chemically “agnostic” approach to water quality testing could allow for detection of multiple contaminants that are biologically active and trigger specific toxicity or adverse health outcome pathways. Biosensors, (i.e. those sensors which take advantage of biological phenomenon that are altered in the presence of contaminants), can potentially provide faster detection as well as portability, continuous monitoring, and/or detection in complex matrices using minimal sample preparation. The EPA has over the past years used a challenge approach that uses crowd-sourcing to find solutions to difficult problems. The Water Toxicity Biosensor Challenge is a recent task with the goal of producing design solutions for a biologically-based effects monitor/biosensor capable of responding to multiple environmental contaminant exposures that result in toxicity or adverse health effects when host organisms are exposed. This presentation will describe the process as well as provide information related to the challenge of designing a Water Biosensor to analyze the toxicity of water.

Keywords: A-sensor technology, A - exposure measurement, C-water, B-mixtures, A-analytical methods

TU-PO-412

A Double-Diffusion Model to Quantify the Sorption Effects of Indoor Surfaces on the Exposure to Chemicals Encapsulated in Products

L. Huang, O. Jolliet; University of Michigan, Ann Arbor, MI

Abstract: Chemicals encapsulated in products can constitute a major emission source in the indoor environment. Recently, increasing attention has been paid to the sorption effects of indoor surfaces as a sink for chemicals released to indoor air, especially for semi-volatile organic compounds (SVOCs). Previous studies have used simple assumptions to model the sorption effects which may not reflect reality. This study thus aims to use a more sophisticated model to better elucidate the role of sorption effects on the release of chemicals and near-field human exposures. A double-diffusion model was developed which describes the diffusive emissions of chemicals from materials, the subsequent loss by ventilation, and the loss by diffusive sorption by indoor surfaces. A typical North American house was modeled. Chemicals were assumed to originate from a 3mm-thick vinyl flooring, and the indoor surfaces considered were gypsum walls and ceilings. The model system was solved numerically using Method of Lines discretization. The model was tested for three chemicals, a typical VOC (ethylbenzene), a typical SVOC (DEHP) and one intermediate (tetradecane). Results show that sorption on walls and ceilings can significantly lower the peak air concentration of chemicals emitted from flooring, from a factor of 2 for ethylbenzene to a factor of 6 for DEHP. The predicted air concentration converges with that predicted by a model not accounting for sorption after several days for ethylbenzene and tetradecane, but the factor of 6 reduction for DEHP remains even after 15 years. Sorption also significantly increases the mass of DEHP released from flooring. Overall, this study shows that for VOCs the sorption effect is negligible, while for SVOCs the indoor surfaces can be viewed as infinite sinks in the time scale of consumer use. Simplified models will be presented on intermediate chemicals, for which the indoor surfaces are nonnegligible finite sinks and may serve as secondary sources.

Keywords: A-exposure models, A-indoor environment, B-SVOCs, B-VOCs, C-consumer products

TU-PO-413

Usage Patterns of Oral Care Products

M. Chiter[1], M. Gomez-Berrada[1], I. Boudieres-Laffont[1], A. Rielland[2], S. Guillou[2], D. De Javel[2], P. Ferret[1]; [1]Pierre-Fabre Dermo-Cosmétique, Toulouse, France, [2]Eurosafe, Saint Grégoire, France

Abstract: Purpose: An increase in the consumption of oral care products has been observed in the last decades. To conduct reliable safety assessment, accurate exposure information is needed. Very few studies focus on frequency data and usage patterns of oral care products. This information is essential for realistic and specific groups exposure assessment. The aim of this study was to obtain reliable data regarding usage patterns of oral care products by age groups and by sex. **Methods:** 1448 people (566 women, 386 men, 132 teenagers, 332 children, 32 babies) responded to a web based survey addressing the kind of oral care product they used, their frequency of use, the place of purchase etc. The questionnaire was made of 35 questions. This survey also permitted to select families for a real life exposure study. **Results:** **Kind of oral care product and frequency of use** Toothpaste is used by **100%** of adults and teenagers and **98.5**% of children. It's generally used twice or three times a day by adults (**86.6%**) and once or twice a day by teenagers (**91.3**%) and children (**89.0**%). Toothbrushes are used by **99.6** % of the respondents and **70.7**% of them use only a manual toothbrush. **88.1%** of children use an age-adapted toothpaste and **90%** of them an age-adapted toothbrush. **64.5%** of parents brush their baby's teeth with a toothbrush and half of them (**32.3%**) combine it with a toothpaste. Mouthwash and dental/gingival gel are mostly used occasionally. **Place of purchase** Most of the respondents buy their toothpaste and mouthwash at the supermarket (respectively **82.2%** and **66.4%**). Concerning dental/gingival gel, purchase at the pharmacy (**49.3%**) or para-pharmacy (**50.7%**) is more important than for the other products. **Conclusion:** This study provides frequency information and usage patterns for three kinds of oral care products. It also highlights the influence of age and sex on usage patterns. Data generated by this study provide an excellent basis for exposure assessment of oral care products.

Keywords: C-consumer/personal care products, A-behavior, D-children

TU-PO-414

An Overview of the NIEHS-EPA Pilot Study of Exposure to Chemicals in Consumer Products

D. Stout[1], L. Alston[1], F. Chen[1], P. P. Egeghy[1], S. Prince[1], R. A. Silva[1], K. Taylor[2], R. Walker[1], T. J. Buckley[1]; [1]USEPA, Durham, NC, [2]NIEHS, Durham, NC

Abstract: Consumer products provide a highly efficient means of exposure to thousands of chemicals for which risk is poorly understood. NIEHS conducted a large health study known as the Sister Study relying on questionnaires to assess exposure. The current pilot study is motivated by NIEHS's interest in evaluating their questionnaire-based exposure assessment against objective measures of exposure as well as EPA's interest in examining modeling approaches to exposure estimation. A small-scale, longitudinal study of women will investigate the performance of existing measurements and modeling methods to assess exposures to chemicals in personal care and household products. The overall goal of this study is to determine the effectiveness of those methods to assess exposures to chemicals in a number of consumer product categories. This will be achieved by collecting extensive and repeated exposure-related information and measurements from adult female participants (N=9) between the ages of 35 and 74 years over a consecutive 10- day period. Innovative tools for exposure assessment will be assessed for their utility through low burden, cost-efficient methods that simultaneously allow for the measurement of multiple chemical agents and the gathering of contextual information. Specifically, this study will obtain measurable concentrations of consumer product chemicals and metabolites from 24-h daily urine voids, blood serum, indoor surface sampling, airborne and particulate sampling, household vacuum sweepings, duplicate diet sampling, and silicon bracelets, to resolve potential chemical exposures. Individual behaviors and ethnographical data will be captured through food photo diaries, product use video journaling and daily web survey, GPS and accelerometers. These integrated tools and techniques will be used to refine measurement and predictive modeling methods that support high throughput chemical analysis and exposure assessment and inform exposure reduction.

Keywords: A-biomonitoring, C-consumer products, C-indoor, A-exposure factors, C-multimedia

TU-PO-415

Development of questionnaire and construct survey panels to obtain national exposure factors of children's goods in Korea

M. Lim[1], J. Park[2], S. Park[3], K. Lee[1]; [1]Graduate School of Public Health, Seoul National University, Seoul, Korea (the Republic of), [2]Seoul National University, Seoul, Korea (the Republic of), [3]World Research Inc., Seoul, Korea (the Republic of)

Abstract: As more consumer products for children have been used, we are more concerned about chemical burden in children's health. Children are more vulnerable to chemical exposure compared to adults because of their physiological characteristics. Children's behavioral characteristics and different usage patterns of children's goods make exposure and risk assessment more challenging. Accurately measuring the usage patterns of children's goods is important to conduct realistic exposure assessments to manage children's goods based on risk. The purpose of children risk and exposure assessment methodology (CREAM) project was to establish Korea national exposure factors of children's goods. In the first year, a total of 40,000 households as survey panels was set and recruited for surveys of children's goods from 17 metropolitan areas and provinces in Korea. The number of panels were determined by proportionate quota sampling based on the population composition ratio in each area and age distribution. In the second year, the systematic questionnaire was developed to investigate children's goods usage patterns and information regarding demographics and lifestyle. Surveyed products are classified into 4 categories as toys, household goods, stationery/books and playground equipment. Using the survey panel and questionnaire, the survey will be conducted face-to-face interviews as trained interviewers visit each household. We set 10,000 households as the target study population to ensure that we obtained a substantial number of children of rarely used products. The survey will be conducted two times for summer and winter to identify the seasonal variances of exposure factors. The study will provide national representative exposure factors of children's goods for accurate exposure and risk assessment. These exposure factor data will be useful as input data for exposure and risk assessments and setting safety guidelines to protect children's health.

Keywords: A-exposure factors, C-consumer products, D-children

TU-PO-416

Finding the Biologically Important Exposures in the Exposome Through Circulating Antigen-antibody Complexes

M. Chung, C. J. Patel; Harvard University, Boston, MA

Abstract: Background Finding the causes of chronic diseases is one of the key goals in public health. It is known that only a small fraction of chronic disease variability is ascribed to inherited factors. Consequently, many are turning their attentions to the exposome concept to enable discovery of exposures in the post genome-wide association study era. However, the exposure spectrum is so large and complex that many exposures remain unknown. Objective We proposed a novel, high risk yet high reward, receptor-based approach to this challenge. Specifically, we hypothesized that daily bioactive chemical exposures that may influence health and disease could be discovered by studying the circulating immune complexes (CICs). Methods In this proof-of-concept study, we developed a method that coupled an immunoprecipitation technique with a high-resolution metabolomics platform to query the CIC extracts for unknown exposures. Peak alignment, identification, and annotation were done with XCMS. Then, to demonstrate the real-world values, we tested the method with a blood sample from a healthy adult volunteer with only environmental level of exposures. Results For the top 20 features, half of them were assigned with a putative identity. These included 5 drugs (Remikiren, Loperamide, Quinidine Barbiturate, Fexofenadine, & Metergelone), a natural product from fruit (1,5-dibutylmethylhydroxycitrate), a tripeptide (Gln-Lys-His), and a likely microbial metabolite (all-trans-hexaprenyl diphosphate). Although individual identities were not experimentally confirmed, the results collectively pointed to an exogenous origin of the extracted molecules that is in line with our hypothesis. Conclusions: The methodology of our biology-driven approach is sound, and robust and sensitive enough to recover trace molecules from CICs. This revolutionary approach could open the possibility to investigate a series of hypotheses in exposomics and ultimately drive a new generation of exposome-wide association study.

Keywords: A-analytical methods, A-biomarkers, A-epidemiology, B-mixtures, A - exposure measurement

TU-PO-417

***In Silico* Prediction of Toxicokinetic Parameters for Environmentally Relevant Chemicals with Application to Risk-Based Prioritization**

B. Ingle[1], B. Veber[2], J. Nichols[3], J. Wambaugh[4], R. Tornero-Velez[1]; [1]Environmental Protection Agency, Research Triangle Park, NC, [2]Oak Ridge Institute for Science and Education, Oak Ridge, TN, [3]Environmental Protection Agency, Duluth, MN, [4]Environmental Protection Agency, Research Triangle Park, NC

Abstract: Toxicokinetic (TK) models can help bridge the gap between chemical exposure and measured toxicity endpoints, thereby addressing an important component of chemical risk assessments. The fraction of a chemical unbound by plasma proteins (F_{ub}) and metabolic clearance rate (CL_{int}) are critical TK parameters, accounting for aspects of the distribution, metabolism and excretion that determine *in vivo* tissue concentrations. Yet, limited TK data are available for environmentally relevant chemicals, including approximately 8000 chemicals with *in vitro* bioactivity data collected by Tox21. Quantitative structure-activity relationships (QSAR) for F_{ub} and CL_{int} were developed with *in vitro* assay data for both pharmaceuticals and chemicals in the ToxCast screening initiative using machine learning algorithms and open source descriptors. The models were shown to offer reliable *in silico* predictions of F_{ub} and CL_{int} for a diverse array of chemicals within the applicability domains. Incorporating the QSARs into TK models allowed a high throughput risk-based prioritization scheme informed by the margin between bioactive doses and human exposure. These QSAR models aid in the identification and prioritization of those chemicals with the highest probability of triggering adverse outcomes. *The presented work is that of the authors and does not necessarily represent U.S. EPA views or policies.*

Keywords: A-exposure models, A-risk assessment, A-chemical prioritization, A-statistical methods

TU-PO-418

Development of Pregnancy-PBPK models for bisphenol A and its metabolites.

V. Kumar, R. Prasad Sharma, M. Schuhmacher; Universitat Rovira i Virgili, Tarragona, Spain

Abstract: Bisphenol A (BPA) is produced at over 2 billion pounds/year and is found in wide variety of dietary and non-dietary products. Recent studies suggest that the universal fetal exposure to BPA imposes a high risk due to the underdeveloped physiology and metabolic capacity of the fetus. Prediction of the internal dosimetry profile of the fetal chemical exposure from transplacental transfer would be highly useful to predict the risk poses to a prenatal cohort from BPA exposure. This presentation will discuss the development a Pregnancy-Physiologically based pharmacokinetic (P-PBPK) model to predict toxicokinetic profile of BPA exposure during the gestational period in the growing fetus and to validate the model using BPA biomonitoring data in mothers and fetus from different cohorts' population. To achieve this objective, first, the adult PBPK model was developed and validated with human kinetic data. This human PBPK model was extended to pregnant women by including pregnancy physiology and fetus sub-model to develop mother-fetus or P-PBPK model that simulates both mother and fetal BPA pharmacokinetics (PKs). All physiological and enzymatic reactions that could impact BPA pharmacokinetics were taken into account. Metabolic kinetics from scaling of in-vitro experiments on metabolism for both mother and fetus were included and transplacental parameters were optimized for human from mice study. The P-PBPK model was validated with observed maternal and cord blood concentrations, and the impact of fetal conjugation and deconjugation of BPA and its metabolites on fetal PKs was investigated. The developed P-PBPK model predicts the BPA concentrations that matched observed cord blood concentrations, fetus liver concentration and amniotic fluid concentration using maternal blood concentration as an exposure source from different pregnancy cohort studies. Developed model helped in quantifying a priori the chemical exposure to the fetus for fast metabolizing chemical during pregnancy. Temporal internal dosimetry profile of BPA in fetus helped to identify second-trimester as a critical window of exposure. The model can be further extended to understand the toxicodynamic effect on the individual organ of the fetus by extending mechanism of toxicity.

Keywords: B-BPA, A-exposure models, D-prenatal

TH-PL-B2-642 ***(moved from Thursday oral presentation)***

Dietary and inhalation exposure to polycyclic aromatic hydrocarbons (PAHs) and monohydroxy metabolites in urine: A panel study for the elderly in Tianjin,

B. Han[1], P. Li[2], X. Qin[3], L. Zhang[3], T. Ni[4], J. Fan[3], N. Zhang[1], F. He[5], J. Xu[6], W. Yang[1], W. Zhang[1], X. Wang[1], Z. Bai[1]; [1]Chinese Research Academy of Environmental Sciences, Beijing, China, [2]Tianjin University of Technology, Tianjin, China, [3]Tianjin Medical University, Tianjin, China, [4]Tianjin University of Sport, Tianjin, China, [5]Hubei Meteorological Service Center, Wuhan, China, [6]University of Washington, Seattle, WA

Abstract: Objective: Our study aimed to analyze the levels of PAHs species from both respiration and diet pathway, as well as the levels of monohydroxy metabolites in the urine. Methods: GC-MS was used to analyze the concentrations of PAHs in personal exposed PM2.5 and food samples as well as the levels of OH-PAHs in urine collected from a elderly panel. We preliminarily analyzed the characterizations of PAHs in PM2.5 and food, and OH-PAHs in urine. Results 1. Season variation had significant influence on the concentrations of all PAHs species associated with PM2.5, and some PAHs species associated with food. Respiration pathway contributed a small fraction to the total exposure amount of PAHs. 2. 7 and 10 OH-PAH analytes were well detected respectively in summer and winter. Low molecular weight PAHs hydroxyl metabolites were the most abundant compounds. The concentrations of 1- and 2-OH-Naphthalene were much higher than other analytes, accounting for over 46% of the total. 3. Good correlations were found between 1-OHNaP, 2-OHNaP, ΣOHNaP and PM2.5-bound NaP, indicating the PM2.5 contribution to elderly exposure. However, no other significant correlations were observed between other urine metabolites and PM2.5-bound PAHs, indicating that ingestion could be the more exposure route for the general population. Conclusion: The contents of OH-PAHs in urine are easily to be affected by some factors, such as smoking, cooking, etc. Therefore, in this study, we found that simultaneous determination of multiple OH-PAHs species (1-OHPyr, 1-OHNap, 2-OHNap, 2-OHFlu and 2-OHBcPhe) should be implemented to accurately and comprehensively reflect the respiration and diet PAHs internal exposure level.

Keywords: B-particulate matter, A-biomarkers,

WE-PL-A4-510 ***(moved from Wednesday oral presentation)***

Exposure Assessment of $PM_{2.5}$ and Benzene in Bike Lanes in Taipei

C. Wu, T. Wu, S. Ho, C. Chan; National Taiwan University, Taipei, Taiwan

Abstract: Commuting by bicycle is encouraged in many urban areas. However, because of being close to roadways and traffic emissions, bike riders may be exposed to considerable amount of air pollutants. In this study, we aimed to characterize $PM_{2.5}$ and benzene exposures along the bike lanes in Taipei City. The monitoring period was from July to early August in 2016. $PM_{2.5}$ and benzene were monitored continuously at 12 different sites. Three sites were monitored simultaneously for one week. The results showed that the average concentrations ranged from 9.43 to 20.99 μg/m^3 and from 0.69 to 3.00 ppb for $PM_{2.5}$ and benzene, respectively. The differences of average concentrations across sites monitored in the same week were less than 7 μg/m^3 for $PM_{2.5}$ and about 1 ppb for benzene. Nevertheless, the correlations of the time series data have a wide range: R^2=0.02－0.71 and 0.07－0.55 for $PM_{2.5}$ and benzene, respectively, depending on the local characteristics and meteorological factors. In order to reduce the exposures for these commuters, the clean zone policy might be needed along the bike lanes.

Keywords: A - ambient monitoring, B-particulate matter, C-air

WEDNESDAY, OCTOBER 18, 2017

WE-PL-A1: Traffic Related Air Pollution - Part 1

WE-PL-A1-418

Measurement of Transportation Air Pollutant Exposure Concentrations

H. Frey[1], S. Singh[1], D. Gadre[2]; [1]North Carolina State University, Raleigh, NC, [2]Trinity Consultants, Phoenix, AZ

Abstract: Transportation-related air pollution (TRAP) is causally associated with asthma exacerbation and there is suggestive evidence that TRAP is causal for other adverse health endpoints. Given substantial spatial variability in the concentration gradient near roadways, the potential impact of signalized intersections, roundabouts, and bus stops on exposure concentrations, and variability in traffic flow with time of day, there is need for high resolution quantification of spatial and temporal variability in near-road TRAP exposure concentration. Furthermore, there is a need for data to relate in-vehicle exposures to ambient concentrations. The purpose of this study was to quantify the spatial and temporal variability in exposure concentrations to TRAP both on-road and near-road for four selected transportation modes: (1) pedestrian; (2) car; (3) transit bus; and (4) bicycle. Portable monitors for $PM_{2.5}$, CO, and O_3, and GPS receivers, deployed to measure exposure concentrations. Measurements were conducted during winter, spring period, and three summer periods in Raleigh, NC. Routes for each transportation mode were developed to allow side-by-side comparisons for several street segments. Measurements were made mid-day and during evening rush-hour. Spatial data were categorized as signalized intersection, roundabout, bus stop, and other. Statistical analysis was conducted using ANOVA with respect to intermodal variability, spatial variability, time of day variation, and seasonal (time period) variability. Relative variations in $PM_{2.5}$ concentration were statistically significant for season, location, time of day, and transportation mode. Results were also developed for CO and O_3. The results demonstrate that human exposure to TRAP in close proximity to roadways is highly variable over space and time.

Keywords: B-particulate matter, C-air, , ,

WE-PL-A1-419

Exposure to Air Pollutants in Selected High-rise Buildings in High Density Urban Areas in Hong Kong

T. Li[1], W. Che[1], A. K. Lau[1], H. Frey[2]; [1]Hong Kong University of Science and Techonology, Hong Kong, Hong Kong, [2]North Carolina State University, Raleigh, CA

Abstract: Air pollution is identified as the largest single global environmental health risk, contributing to 7 million annually. Half of the world's population lives in urban areas. Dense urban development characterized by high-rise buildings is becoming common, especially in Asian cities, to cope with population growth. Such urban morphology may trap air pollution in street canyons, resulting in elevated air pollutant concentrations. Urban citizens spend substantial time in high-rise buildings located in canyon-like areas; however, little is known about their exposure to air pollutants and how they are affected by outdoor air pollution. A field study was designed to quantify human exposure to fine particulate matter ($PM_{2.5}$) and carbon monoxide (CO) in three high-rise buildings in Hong Kong. Simultaneous indoor and outdoor concentrations were measured on multiple floors using portable air quality monitors. Variability in the $PM_{2.5}$ and CO concentrations was quantified. Factors affecting variability in exposure concentrations were identified. The associations between indoor and outdoor concentrations were examined. Preliminary results indicate that indoor $PM_{2.5}$ concentrations are sensitive to outdoor concentration, ventilation mode, height of air intake, and intensity of indoor sources. Based on linear regression, the infiltration of outdoor particles varies from 0.2 to 0.6 among selected locations, related to ventilation practice, choice of fresh air handling and maintenance of filtering system. Outdoor $PM_{2.5}$ concentrations are highly correlated (R^2 = 0.7~0.9) among multiple floors at the same building, but less correlated (R^2 =0.2~0.5) with $PM_{2.5}$ concentrations recorded at nearby fixed site monitors (FSMs), suggesting that FSMs may not be an appropriate surrogate for outdoor concentration in high-rise buildings. Such field studies may help better characterize exposure to air pollutants in high-rise buildings and parameterize exposure estimates in urban areas.

Keywords: A-exposure factors, A-built environment, A-sampling methods, C-air, B-particulate matter

WE-PL-A1-420

Characterizing Air Pollution Exposures in Complex Urban Environments: Experiences and Lessons Learned in Hong Kong

W. Che[1], A. K. Lau[1], H. Frey[2], Z. Li[1]; [1]The Hong Kong University of Science and Technology, Hong Kong, Hong Kong, [2]North Carolina State University, Raleigh, NC

Abstract: An estimated 54% of the world population resides in cities. In 2012, an estimated 6.5 million deaths (11.6% of all global deaths) were associated with air pollution, which is identified as the biggest environmental risk to health by the World Health Organization (WHO). The 194 WHO Member States developed a road map for "an enhanced global response to the adverse health effects of air pollution" in 2016. One of the main elements of the road map is the estimation of human exposure to pollutants. Exposure depends on the air quality in various microenvironments (e.g. school, offices) where people spend most of their daily time. However, such information is not well characterized, especially in urban areas. Over the past three years, we have carried out a range of exposure measurements in Hong Kong. Concentrations of fine particulate matter ($PM_{2.5}$) and carbon monoxide (CO) were measured using portable air quality monitors in several microenvironments including schools, offices, homes, shopping malls and popular transportation modes, such as the Mass Transit Railway (MTR) subway system, transit buses, trams, and minibuses. Field data collection was designed based on factors that may affect variability in pollutant concentrations in these microenvironments. A rigorous protocol was developed to enable accurate, reliable and valid measurement using portable monitors. The results demonstrate substantial variations in exposure concentration among microenvironments, affected by the differences in ventilation systems and ventilation practices, diversity in local emissions, and heterogeneity in outdoor concentrations caused by complex urban morphology, such as street canyons and high-rise buildings. Analysis results indicate that outdoor concentration can be a useful indicator for indoor exposure under certain conditions. The collected data will aid in developing estimates of human exposure in various microenvironments, and designing air pollution exposure management strategies.

Keywords: A-exposure factors, A-sampling methods, A-indoor environment, C-air, B-particulate matter

WE-PL-A1-421

Personal and occupational exposure to traffic-related fine particulate matter in Accra, Ghana

R. E. Arku[1], K. Dionosio[2], A. Hughes[3], J. Vallarino[4], J. Spengler[4], S. Agyei-Mensah[5], M. Ezzati[6]; [1]University of British Columbia, Vancouver, Canada, [2]United States Environmental Protection Agency, Washington, D.C., DC, [3]University of Ghana, Accra, Ghana, [4]Harvard T.H. Chan School of Public Health, Boston, MA, [5]University of Ghana, Accra, Ghana, [6]Imperial College London, London, United Kingdom

Abstract: Background/Aim Air pollution levels in fast-growing sub-Saharan African cities are among the highest in the world, but human exposure studies are limited, especially, traffic-related exposures. Our aim was to measure personal fine particulate matter ($PM_{2.5}$) exposure of commercial minibus and taxi drivers (the most popular means of public transportation in Ghana's capital), and street mobile vendors (hawkers) and street stationary vendors (vendors) in Accra, Ghana. Methods We measured 24-hour personal $PM_{2.5}$ exposure of 99 subjects, comprising 29 minibus drivers, 26 taxi drivers, 29 street vendors, and 15 street hawkers in the Accra metropolis. $PM_{2.5}$ was measured both gravimetrically and continuously, with time-matched global positioning system coordinates. The instruments were placed in backpacks, which were worn by the vendors and hawkers during the 24-hour measurement period. Taxi drivers had the backpacks placed beside them near the front passenger's seat. Field assistants carried the backpacks on their laps at the front passenger's seat and rode along the mini bus drivers. Results Across all four occupational groups, average (SD) personal $PM_{2.5}$ exposure was 56.4 (63.2) $\mu g/m^3$; group means ranged from 26.0 μg/m^3 for hawkers to 83.4 μg/m^3 for taxi drivers. Individual exposure was > 100 μg/m^3 for some drivers. Exposure was significantly higher among drivers than vendors (78 vs 29 μg/m3; 95% CI: 28, 70; p-value < 0.001). The highest exposure to the drivers occurred on major and secondary roads, although < 20% of their commute time was spent there compared to minor roads and alleys. Conclusions Our results support the need for urban air quality management plans that will address the role traffic to reduce exposure to the thousands who commute daily by minibuses and taxis or work near roadways in Accra, and when combined with policies to reduce urban biomass use, will greatly reduce urban population exposure in Accra.

Keywords: A-exposure factors, A-geospatial analysis/GIS, B-particulate matter, A-global health

WE-PL-A1-422

Exposure to Short Lived Air Pollutants on Public Transit in Brazil

E. Patrick[1], J. F. Segura[2], A. C. Targino[2], M. Gibson[1]; [1]Dalhousie University, Halifax, Canada, [2]Universidade Tecnológica Federal do Paraná, Londrina, Brazil

Abstract: Air quality is of particular concern in urban areas where high traffic and city design lead to accumulation of toxic air pollutants within street corridors. Urban aerosols are a heterogeneous mixture of solid particles and liquid droplets of varying physiochemical characteristics. Their concentration and composition are influenced by local and long range anthropogenic, geogenic, and biogenic emissions. Brazil in particular has grown rapidly in population, car ownership, and industry; it is currently the fifth most populous country in the world. Public transit is often chosen as the 'green' alternative to driving and helps alleviate congestion in cities. However, there is a gap in knowledge with regard to air pollution exposure on public transport in Brazil. The majority of buses in Brazil run on high sulfur diesel, do not have particle traps, and many cities contain old vehicle fleets. This contributes to higher emissions than buses that use cleaner fuels and have emission control. This study aims to quantify the bus passengers' exposure to air pollution in three varying sized cities in southern Brazil. Two different bus routes from each city were surveyed over a period of one week. Black carbon and particulate matter concentrations were measured using hand-held devices (an aethalometer model AE51, Aethlabs and a photometer model DustTrak 8520, TSI). Data between the cities will be compared to determine if city size and traffic variation affect the amount of air pollution aboard public transit. Highly temporally-resolved air pollution data will be linked with GPS data to create maps of air pollution observed during the bus trips. Preliminary results show that the average mass concentrations of fine particulate matter ≤ 2.5 μm ($PM_{2.5}$) and black carbon in a megacity, 32 μg/m^3 and 10 μg/m^3 respectively, are double the concentrations in a small city, 15 μg/m^3 and 4.5 μg/m^3 respectively.

Keywords: A - exposure measurement, A-geospatial analysis/GIS, B-particulate matter, C-air, D-environmental justice

WE-SY-B1: 2,4-D - A Case Study Of Decades Of Exposure Science; A Discussion of Quality, Quantity, and Harmonization

WE-SY-B1-423

Epidemiology and public health protection: the 2,4-D story

C. J. Burns; Burns Epidemiology Consulting, Sanford, MI

Abstract: Pesticide review and regulation in the US and globally is weighted heavily by prescribed guideline studies that adhere to Good Laboratory Practices. This is not to say that non-guideline human epidemiology studies have not been part of the discussion. As early as 1986, the US EPA initiated a Special Review of the herbicide, 2,4-D, due to the results of a published epidemiology study. Since then, findings of such human health studies have been considered as part of 2,4-D re-registration in the US,

Canada and Europe. In general, evaluations of epidemiology studies focus on the consistency of positive (or negative) associations with a specific outcome (i.e. 2,4-D and non-Hodgkin lymphoma). Checklists and meta-analyses are common tools to aid interpretation of a collection of studies. However, heterogeneity of study quality is an awkward reality. Assessing the quality of the underlying data and harmonizing the exposure and outcome measurements contribute to weight of evidence interpretation of epidemiology data. As regulatory decision makers move away from controlled animal data to more emphasis on human studies we will need to use transparent and well-accepted approaches to examine quality, harmonization, and bias, and their impact on the estimation of risk.

Keywords: A-biomonitoring, A-epidemiology, B-pesticides

WE-SY-B1-424

Critical and systematic evaluation of 2,4-dichlorophenoxyacetic acid (2,4-D) exposure data: quality and generalizability for human assessments

J. LaKind[1], C. J. Burns[3], D. Q. Naiman[2], C. O' Mahony[5], G. Vilone[5], A. J. Burns[4], J. S. Naiman[6]; [1]LaKind Associates, LLC, Catonsville, MD, [2]Johns Hopkins University, Baltimore, MD, [3]Burns Epidemiology Consulting, LLC, Sanford, MI, [4]Alma College, Alma, MI, [5]Creme Global, Dublin, Ireland, [6]University of Pennsylvania, Philadelphia, PA

Abstract: The herbicide 2,4-dichlorophenoxyacetic acid (2,4-D) has been commercially available since the 1940's and there are several decades' worth of data on levels of 2,4-D in food, air, soil, and water as well as in humans. Yet the robustness of these data for use in human exposure assessments has not been evaluated. Using three elements of the Biomonitoring, Environmental Epidemiology, and Short-lived Chemicals (BEES-C) instrument (avoidance of sample contamination, analyte stability, and urinary methods of matrix adjustment), we evaluate the quality of environmental- and biomonitoring-based 2,4-D data in the peer-reviewed published literature. We further explore issues of temporal variability in 2,4-D concentrations in environmental media and humans and the impact on our understanding of human exposures to 2,4-D. We present the results of our review of 156 publications using the BEES-C instrument and describe the need for greater transparency in quality control measures. We also discuss the impact of temporal variability and the sources of 2,4-D vary within the 2,4-D literature and the importance of distinguishing peak vs. background levels of exposure to 2,4-D. We finally describe the challenges for the exposure community associated with reaching consensus on how to address problems specific to short-lived chemical exposures in observational studies and the need for deeper conversations to advance our understanding of human exposures, and to allow interpretation of these data to catch up to analytical capabilities.

Keywords: A-biomonitoring, A-epidemiology, B-pesticides, 2,4-D

WE-SY-B1-425

2,4-D Human Exposure Data: Presentation of the results of a study on harmonisation of published Data

G. Vilone[1], C. O' Mahony[1], J. LaKind[3], C. J. Burns[2]; [1]Creme Global Ltd., Dublin, Ireland, [2]Burns Epidemiology Consulting, LLC, Sanford, MI, [3]LaKind Associates, LLC, Baltimore, MD

Abstract: The herbicide 2,4-dichlorophenoxyacetic acid (2,4-D) has been used for many decades around the world. Many scientific studies on its concentration in food, air, soil and water have been published, but concentration data in these media alone are not informative with regards to human health and risk assessment. Ideally, human exposure estimates in a common metric are required, permitting a comparison of doses across exposure datasets. In addition, numerous biomonitoring studies of 2,4-D exist, primarily in the form of measured levels in urine, but are not directly comparable to measurements in environmental media or calculated exposures. Building off the examination of quality and representativeness of environmental- and biomonitoring-based 2,4-D data, the next step is to develop a means to compare the data. The environmental data were input into several existing exposure models in order to convert them to a common exposure metric of mg/kg-day. The converted data were subsequently inputted into two Biomonitoring Equivalents (BEs) methods in order to estimate the related urinary concentrations, enabling comparison with published biomonitoring data. The BE method was chosen as it provides a deterministic estimate of the urinary concentrations that are easy to interpret and to compare with other data. Using this approach, the relative importance and magnitude of different media can be estimated, as well as their significance when compared with biomonitoring data.

Keywords: A-exposure models, A-statistical methods, A-risk assessment

WE-SY-B1-426

Advances in Methodologies For Conducting Systemic Reviews Within The Federal Government

K. Taylor[1], A. Boyles[2], A. Rooney[1]; [1]OHAT/NTP/NIEHS, Morrisville, NC, [2]DERT/NIEHS, Morrisville, NC

Abstract: Systematic review methodologies were first developed to assess the efficacy of health care interventions. More recently these approaches have been adapted and extended to address environmental health questions. The structured approach of a systematic review increases the objectivity and transparency in the process of collecting, synthesizing, and reaching conclusions based on the available scientific evidence. While studies of interventions rely primarily on clinical trials, extension of these methods to environmental questions, such as potential human health effects of 2,4-dichlorophenoxyacetic acid (2,4-D), requires synthesis of the human evidence from a broad range of epidemiological study designs as well as animal toxicology and mechanistic studies (evidence integration) to reach hazard identification conclusions. The National Toxicology Program's Office of Health Assessment and Translation (OHAT) established an approach for systematic review and evidence integration for conducting health assessments that begins with problem formulation and development of a project-specific protocol to guide the process of identifying, evaluating, and integrating the evidence to answer a specific research question. The presentation will outline the OHAT approach, highlight the transparency and increased rigor in the systematic review process and discuss OHAT's continued involvement in advances of methodologies for assessing the risk of bias in non-randomized observational studies of environmental exposures. These methodological advances stem from OHAT's international collaboration with leading experts in environmental epidemiology, exposure assessment, toxicology, and systematic review to adapt a tool that will facilitate assessment of potential for bias in environmental and occupational health study design and conduct.

Keywords: A-biomonitoring, A-epidemiology, B-pesticides, 2,4-D

WE-SY-B1-427

Ensuring Harmonized And Comparable Laboratory Measurements To Improve Public Health Programs

H. W. Vesper; Centers for Disease Control and Prevention, Atlanta, GA

Abstract: Measurements of exposure biomarkers are important for the assessment of health risks, the development of clinical and public health decision points, and for monitoring exposures over time. The non-comparability of biomarker measurements performed in different studies and at different times prevents the effective assessment, implementation and monitoring of public health activities. Procedures to determine, improve and track the analytical accuracy and variability of biomarker measurements are successfully used for certain clinical biomarkers. The same procedures and approaches can be applied to exposure biomarkers. Harmonization programs which create measurement results that are linked to one accuracy basis and thus are comparable across methods, locations and over time were successfully implemented for clinical analytes such as cholesterol, vitamin D and testosterone. These programs focus on the measurement accuracy and do not require laboratories to use the same analytical method. This is accomplished by providing specimens with reference values to routine laboratories. These specimens are used for assessment of accuracy and for identifying potential sources that cause inaccurate and unreliable results. For clinical analytes, these programs identified method calibration and specificity as two major sources of inaccuracy and between-laboratory variability. By continuously providing specimens with reference values and evaluating analytical performance of routine methods using consistent protocols and criteria, harmonized and comparable laboratory measurements are achieved.

Keywords: A-biomarkers, A-analytical methods, Harmonization

WE-SY-C1: Synthetic Biology and Exposure Science: Integrating Chemistry, Genetic Engineering, and Risk Management

WE-SY-C1-428

Synthetic biology: where we are, where we're going, and its impact on and need for exposure science

T. Kuiken; North Carolina State University, Raleigh, NC

Abstract: Innovations in gene editing technologies, like CRISPR-Cas9, are enabling advances in gene drive technologies and the overarching field of synthetic biology. These discoveries could have significant implications for exposure science, both in terms of how synthetic biology may advance exposure science (e.g. through development of advanced biological sensor/measurement systems) as well as how exposure science may improve our ability to understand risks from future synthetic biology applications. Introductions of synthetic biology related applications, or organisms containing gene drives, into the environment might require new measurement techniques in order to conduct ecological risk assessments and monitoring. Multiple reports that have examined the potential release of synthetic biology and gene drive applications have asked: how can existing systems of monitoring and surveillance be used in this effort. What new systems of monitoring and surveillance are needed; or whether it is even feasible to monitor these organisms and their ecological/evolutionary effects with our current methods. To help address these needs, competitions like iGEM, have created specialized tracks for measurement and the U.S. National Institute of Standards and Technology has collaborated with Stanford University to form the Joint Initiative for Metrology in Biology, which unites people, platforms, and projects to underpin standards-based research and innovation in biometrology. This talk will present an overview of the evolving state of synthetic biology science and discuss the needs and potential linkages to exposure science to benefit both disciplines.

Keywords: synthetic biology, gene drives, A-environmental policy, A-environmental regulation, A-ecological exposure

WE-SY-C1-429

Genetically engineered microbial sensors for remote mapping of environmental hazards

S. Belkin; Hebrew University of Jerusalem, Jerusalem, Israel

Abstract: The small size requirements, rapid responses and sensing versatility of bacterial bioreporters allow their integration into diverse types of devices, for laboratory as well as field applications. The relative ease by which molecular sensing and reporting elements can be fused together to generate dose-dependent quantifiable physical (luminescent, fluorescent, colorimetric, electrochemical) responses to pre-determined conditions allows the construction of diverse classes of sensors. Over the last two decades we and others have employed this principle to design and genetically engineer microbial bioreporter strains for the sensitive detection of (a) specific chemicals of environmental concern (heavy metals, halogenated organics etc.) or (b) their deleterious biological effects on living systems (such as toxicity or genotoxicity). In many of these cases, additional molecular manipulations beyond the initial sensor-reporter fusion may be highly beneficial for enhancing the performance of the engineered sensor systems. Recently, we have adapted this approach towards standoff detection of the sensors' responses, thus allowing remote "mapping" of environmental hazards. Recent results concerning the application of live *E. coli* bioreporters encapsulated in alginate microbeads for the remote detection of buried landmines and other explosive devices are presented.

Keywords: A-biomonitoring, A-sensor technology, C-soil, Explosives; Biosensors, Explosives; Biosensors

WE-SY-C1-430

Biosensors for water monitoring

D. Lloyd; FREDsense Technologies, Calgary, Canada

Abstract: Water quality analysis is a critical component of safeguarding our water supplies and maintaining a sustainable ecosystem; however, most of the highly analytical techniques to achieve this analysis require off-site labs and use expensive and highly technical equipment reserved for trained personnel. Biosensors using live or *in vitro* based detection methods represent an opportunity for comprehensive and highly modular field testing systems of target chemicals. FREDsense Technologies is designing live cell biosensor solutions to monitor various water chemistry parameters in the field. FREDsense is commercializing one of the first synthetic biology based electrochemical detection platforms for comprehensive water monitoring. Our goals is to produce sustainable, accurate and low cost solutions to move water monitoring out of the analytical laboratory environment into the field. This presentation will explore the challenges and opportunities biosensor solutions offer to the changing regulatory landscape within a North American context.

Keywords: A-analytical methods, C-water

WE-SY-C1-431

Synthetic Algae and Cyanobacteria: Great Potential but What Is the Exposure Risk?

M. Bagley[1], C. Penalva-Arana[2], A. Gillespie[3]; [1]US EPA, Cincinnati, OH, [2]US EPA, Washington, DC, [3]US EPA, Research Triangle Park, NC

Abstract: Green algae and cyanobacteria (hereafter, algae) have the advantageous properties of relatively simple genomes, rapid growth rates, and an ability to synthesize useful compounds using solar energy and carbon dioxide. They are attractive targets for applications of synthetic biology, as entrepreneurs seek to transform them into unicellular autotrophic "factories" for industrial chemical production, which could augment or potentially disrupt traditional production pipelines. Although initial genetic modifications were focused on optimizing the harvest of ethanol and other biofuels, commercial ventures are diversifying and climbing the value chain into industrial chemicals, specialty chemicals, and pharmaceuticals. Fueled by breakthroughs in molecular engineering, genomics, informatics and computer aided design, synthetic biology ventures have advanced rapidly and introductions of new products have the potential to present new challenges to regulators as they evaluate the associated environmental exposure risks. Ecologically, these exposure risks could be most profound for open water algal culture ventures that use new and untested containment controls that differ from those of closed systems. Ecological risk assessment for synthetic algae has some parallels with that of previous genetically modified organisms, but the open water nature of culturing algae presents new uncertainties that could greatly alter the risk calculus. In this talk, we will compare and contrast eco-evolutionary drivers of synthetic algae risk compared to previous genetically modified organisms and highlight key ecological exposure questions that must be evaluated for effective regulation.

Keywords: A-ecological exposure, B-microbial agents, C-water

WE-SY-C1-432

Facilitated Discussion on the Opportunities for Exposure Science to Inform Synthetic Biology Applications, and Vice Versa

A. Gillespie; US EPA ORD NERL, Research Triangle Park, NC

Abstract: The purpose of this Session is to increase awareness and call attention to the need and opportunities for integrating synthetic biology and exposure science. Synthetic biology combines new developments in genetic engineering, bioinformatics and DNA synthesis and is causing an exponential growth in new applications with potentially significant and novel risks and benefits for human health and the environment. Synthetic biology also presents new opportunities to develop tools and techniques which can advance the practice of exposure science. Mostly these applications are composed of genetically engineered organisms and ask for new tools and methods to evaluate their risks. The speakers in this session have discussed the state of synthetic biology and implications for exposure science; several current applications of synthetic biology for developing biosensors to measure environmental exposures; and a discussion of potential exposure considerations for synthetic biology applications in the context of chemical engineering. The final presentation will be a facilitated discussion amongst the presenters and the audience regarding potential avenues for synthetic biology to advance exposure science, and for exposure science to inform development and assessment of synthetic biology applications.

Keywords: A-biomonitoring, A-ecological exposure, B-microbial agents, A-sensor technology

WE-SY-D1: Social Determinants of Health, Environmental Exposures, and Disproportionately Impacted Communities: What We Know and How We Tell Others- Part 1

WE-SY-D1-433

Review of Non-Chemical Stressors from the Social Environment

K. Hibbert, N. S. Tulve; U.S. Environmental Protection Agency, Research Triangle Park, NC

Abstract: Recent studies have reported associations between exposure to chemical or non-chemical stressors in the social environment (e.g., SES, exposure to violence, acculturation) and adverse health outcomes in children. However, limited data are available on the interrelationships between chemical and non-chemical stressors and children's health outcomes. Children may be more vulnerable to exposures from stressors due to their developmental stage and lifestage-specific activities. The objectives of this review were to 1) examine the state-of-the-science of non-chemical stressors found in a child's social environment and 2) statistically rank and prioritize those stressors. Using the published literature, we performed a systematic review of non-chemical stressors found in a child's social environment. Combinations of search strings (i.e., overcrowding + health + child, or exposure to violence + health + child), were entered into PubMed, Google Scholar, and PsychInfo. Inclusion criteria resulted in 372 articles and 678 non-chemical stressors. From these articles, the data were extracted and stored in a searchable database for statistical analysis. The available non-chemical stressor data were classified into 11 different topic categories of non-chemical stressors: acculturation, adverse childhood experiences, economic, education, food, greenspace, overcrowding, social support, stress, urbanization, and exposure to violence. The most frequently analyzed non-chemical stressors in the literature were food (19%), economic (14%), acculturation (11%), and exposure to violence (10%). Analysis of the 11 topic areas suggested significant positive and negative impacts with children's health. In general, non-chemical stressors from the social environment are found in combination. Our analysis suggests that non-chemical stressors are associated with children's health. Additionally, chemical and non-chemical stressors should be considered together when evaluating children's health.

Keywords: non-chemical stressors, D-children

WE-SY-D1-434

The Role of Psychosocial Stressors in Mediating Socioeconomic Susceptibility to Air Pollution

J. E. Clougherty; Drexel University Dornsife School of Public Health, Philadelphia, PA

Abstract: Stronger health effects of air pollution have been observed in communities of lower socioeconomic position (SEP). The specific factors underlying this SEP-related susceptibility, however, remain unknown. Separate evidence suggests that chronic stress - shown to influence immune, endocrine, and metabolic function - may exacerbate the health effects of some pollutant exposures. As some chronic stressors (e.g., exposure to violence, noise, overcrowding) may be elevated in lower-SEP communities, it stands to reason that a substantial portion of SEP-related susceptibility may be mediated through chronic stress. More clearly elucidating the specific factors underlying susceptibility to pollution will improve our ability to identify and characterize at-risk populations, and to deliver targeted, cost-effective interventions to reduce health disparities. In this presentation, I will discuss the current state of the knowledge on combined effects of psychosocial factors in mediating SEP-related susceptibility. In addition, I will describe new methods for quantifying chronic stressor exposures across large urban cohorts, and for capturing both potential confounding and effect modification of air pollution effects by non-chemical stressors in environmental epidemiology.

Keywords: D-susceptible/vulnerable, C-air, D-community, A-epidemiology, D-environmental justice

WE-SY-D1-435

Identifying Neighborhood, Social determinants and factors that impact cardiorespiratory diseases

D. Diaz-Sanchez[3], J. Mirowsky[1], R. Devlin[3], W. Cascio[3], A. Weaver[3], C. Haynes[2], C. Blach[2], E. Hauser[2], S. Shah[2], W. Kraus[2], K. Olden[3], L. Neas[3]; [1]SUNY College of Environmental Science and Forestry, Syracuse, NY, [2]Duke Molecular Physiology Institute, Durham, NC, [3]Environmental Protection Agency, Chapel Hill, NC

Abstract: Individual-level characteristics, including socioeconomic status, have been associated with poor cardiopulmonary health. However, residential area-level characteristics may also independently contribute to health status. We examined a cohort of subjects with cardiovascular-related disease residing in Orange, Wake, and Durham counties in North Carolina. Addresses were geocoded and subjects assigned a block group according to the 2000 US Census. We generated 6 clusters comprising 444 Census block groups in these 3 counties. The cluster input variables were predominantly measures of neighborhood-level socio-economic factors and race. After controlling for individual-level demographic factors, significant differences in disease status were found based on a resident's neighborhood characteristics. Individuals residing in Cluster 2, which described urban, impoverished, and unemployed areas, exhibited an increased risk of obesity, coronary heart failure, diabetes and hypertension. When we examined their response to exposure to pollution, we again saw differences based on neighborhood characteristics. Exposure to PM2.5 was determined at a 1x1 km resolution using satellite imaging. Although the levels of ambient PM2.5 exposures were similar throughout the area, individuals in Cluster 2 had a greater PM2.5-associated increase in hypertension. While 82% of residents in this cluster are African-American, this finding held regardless of race. Matching clinical studies suggest that a biological basis for these observations may be through cumulative exposure from physical (e.g. pollution) and non-physical factors (eg psychological stress). These factors may all converge to alter the human epigenome and thereby determine responsiveness and future risk to environmental agents. The study highlights the importance of neighborhood characteristics on health even when individual risk factors are low.

Keywords: A - exposure measurement, B-particulate matter

WE-SY-D1-436

Working To Understand The Role Of Social Determinants Of Health For Minority And Vulnerable Communities In The United States

J. Lewis, J. Hoover; University of New Mexico, Albuquerque, NM

Abstract: Assessing social determinates of health is critical for predicting how environmental contaminants influence health outcomes. To this end five NIH-EPA Centers of Excellence on Environmental Health Disparities Research are working to understand the role of social determinants of health in toxicologic responses to environmental exposures. The Centers are located in distinct areas of the United States and represent a diversity of races and ethnicities, urban and rural communities, environmental contaminants, and health outcomes of concern. This diversity enables a comparative assessment of key social determinants across multiple exposure-outcome relationships among "minority" populations. The Centers are working together and using available data to determine similarities and distinctions between these partnering communities and the US population. Comparative results of key socio-economic and demographic variables will be summarized cartographically and tabularly. Examining the representativeness of Environmental Health Disparity Center partner communities provides a basis for generalization of results, improves our understanding of the key factors affecting exposure-response assumptions, and broadens the reach of results in informing health policy.

Keywords: D-susceptible/vulnerable, D-community, A-environmental justice

WE-SY-D1-437

Overview of Tribal Research in the Environmental Health Disparities Centers

M. Gonzales[1], C. Lefthand-Begay[2], C. DeVore[3]; [1]University of New Mexico, Albuquerque, NM, [2]University of Washington, Seattle, WA, [3]University of New Mexico, Albuquerque, NM

Abstract: The land-based nature of Native communities brings community members into direct and frequent contact with their environment, in which natural resources are relied upon to maintain traditional diets, customs and languages. The EPA recognizes the "unique need for tribal-focused research to identify impacts of pollution, dietary exposure, cumulative risk and climate change as well as to inform decisions to reduce health risks…" (EPA 2014). While sovereign tribal governments acknowledge that environmental disparities exist, they indicate that a lack of primary research in Native populations inhibits their understanding of the effects of environmental contamination on public health. The Native communities that live with human-made and natural contaminant sources know that these exposures are occurring on top of the impacts of social determinants of health, such as few employment opportunities, low wages, and inadequate or antiquated infrastructure, historical trauma, and the disappearance of Native languages and cultural practices. Addressing the pervasive environmental and social determinants of health is the focus of two NIH-EPA Centers of Excellence on Environmental Health Disparities Research: the Center for Native American Environmental Health Equity Research, based at the Univerisity of New Mexico, and the Center for Indigenous Environmental Health Research based at the University of Arizona . Both Centers are engaging with diverse native communities using a range of research study designs. Through Native-focused engagement, the Centers are building the capacity for research and data interpretation to characterize exposure pathways and define health not only reflective of tribal perceptions, but ultimately useful in informing regulatory decision-making.

Keywords: D-susceptible/vulnerable, A-exposure factors, D-First Nation, D-community

WE-PL-E1: Climate Change

WE-PL-E1-438

Drought and the risk of hospital admissions and mortality in older adults in the western USA from 2000 to 2013: a retrospective study

J. D. Berman[1], K. Ebisu[5], R. D. Peng[4], F. Dominici[3], M. L. Bell[2]; [1]Johns Hopkins School of Public Health, Newington, CT, [2]Yale School of Forestry and Environmental Studies, New Haven, CT, [3]Harvard T.H. Chan School of Public Health, Boston, MA, [4]Johns Hopkins School of Public Health, Baltimore, MD, [5]California EPA, Oakland, CA

Abstract: Occurrence, severity and geographic extent of droughts are anticipated to increase under climate change, but the health consequences of drought are unknown. For counties in the western U.S. (N=618) and for the period 2000 to 2013, we use data from the U.S. Drought Monitor to identify: 1) full drought periods; 2) non-drought periods; and 3) worsening drought periods stratified by low- and high-severity. We use Medicare claims to calculate daily rates of cardiovascular-related hospital admissions, respiratory-related admissions, and deaths among adults 65 years or older. Using a two-stage hierarchical model, we estimated the percentage change in health risks when comparing drought to non-drought period days controlling for daily weather and seasonal trends. On average there were 2.1 million days and 0.6 million days classified as non-drought periods and drought periods, respectively. Compared to non-drought periods, respiratory admissions significantly decreased by -1.99% (95% posterior interval (PI): -3.56, -0.38) during the full drought period, but not during worsening drought conditions. Mortality risk significantly increased by 1.55% (95% PI: 0.17, 2.95) during the high-severity worsening drought period, but not the full drought period. Cardiovascular admissions did not differ significantly during either full drought or worsening drought periods. In counties that experienced fewer drought events, we found risks for cardiovascular disease and mortality to significantly increase during worsening drought conditions. This research describes an understudied environmental association with both regional and global health significance.

Keywords: A-drought, A-epidemiology

WE-PL-E1-439

Assessing the potential impact of global warming on infiltration of outdoor air pollutants into residential indoor environments

D. Liang[1], K. Ono[1], W. Lee[2], J. Liao[1], J. Lawrence[3], S. E. Sarnat[1], J. A. Sarnat[1], P. Koutrakis[3]; [1]Emory University, Decatur, GA, [2]National Taiwan University, Taipei, Taiwan, [3]Harvard University, Boston, MA

Abstract: Aim. Rising temperatures due to climate change are expected to impact human adaptive response, including changes to home cooling and ventilation patterns. These change may affect potential air pollution exposures via alteration in residential air exchange rates, which impact contributions from indoor and outdoor particle sources to indoor air. We conducted a field study examining associations between particle infiltration and temperature to inform future studies of air pollution health effects. **Methods.** We measured indoor and outdoor fine particulate matter ($PM_{2.5}$) in Atlanta in 60 homes totaling 840 sampling-days. Indoor-outdoor sulfur ratios were used to estimate particle infiltration. Linear mixed-effects models were used to examine sulfur ratio-temperature relationships. Projected meteorological values were incorporated in the models to predict sulfur ratios in a 20 year future scenario (2046-2065) and past scenario (1981-2000). **Results.** The average sulfur ratio in Atlanta was 0.70 ± 0.30, with a 0.21 lower sulfur ratio in summer compared to transition seasons. Sulfur ratios were also 0.19 lower in houses using air conditioning (AC) compared to those without AC usage. We observed a negative linear relationship between temperature and sulfur ratio; a 1 degree Celsius increase in temperature was associated with a decrease in the sulfur ratio of 0.008. Future temperature was projected to increase by 2.1 degrees Celsius compared to the past, which would result in a corresponding 0.023 increase in sulfur ratio during cooler months and a 0.037 decrease during warmer months. **Conclusion.** We observed substantial variability in sulfur ratios in Atlanta by temperature, season, and AC usage, and projected changes to sulfur ratios in the future under a warming climate. Ongoing analyses will compare these relationships to a similar study in Boston. These analyses can help to provide a better understanding of the potential influence of climate change on $PM_{2.5}$ health effects.

Keywords: A-climate change, A - exposure measurement, B-particulate matter, A-indoor environment, C-air

WE-PL-E1-440

Modeling relationship between climate indicators and exposure to air pollutants for urban health

I. U. Din, A. Rashid; PMAS Arid Agriculture University, Rawalpindi, Pakistan

Withdrawn

WE-PL-E1-441

Accessing heat effects among migrant and seasonal farmworkers

K. Zhang, R. F. Arauz, S. Cooper; The University of Texas Health Science Center at Houston, Houston, TX

Abstract: Background: Although migrant and seasonal farmworkers are highly vulnerable to ambient heat because of their working conditions, heat effects in this population have been rarely studied. Objectives: We estimated effects of heat on mean daily counts of clinic visits among migrant and seasonal farmworkers by taking advantage of a unique longitudinal medical records database in the USA. Methods: We compiled a daily weather and clinic visit data set based on data from a health centre in Colorado for the summer of 2013. A total of 14 481 patients were included in our analysis, including 150 migrant farmworkers and 231 seasonal farmworkers with an average of 3 and 4 visits per day. We used Poisson regression to estimate the associations between heat and daily all-cause or cardiovascular-specific clinic visits among migrant or seasonal farmworkers or other stratified patients. We defined heat effects as the percentage difference in average daily counts of clinic visits, comparing 90-50th centiles of daily mean apparent temperature, a composite index accounting for both temperature and humidity. We conducted a sensitivity analysis to evaluate the impact of adjustment for ozone levels and different heat definitions. Results: Estimates of heat effects on average daily clinic visits among migrant farmworkers were positive (88.0%, 95% CI: 26.2% to 180.0%). We did not observe statistically significant associations between heat and clinic visits among other stratified groups. Conclusions: Our study appears to be the first to link heat effects with clinic data among migrant and seasonal farmworkers. This research suggests possible significant impact of heat on migrant farmworkers and provides justifications for further studies.

Keywords: A-climate change, A-epidemiology, D-occupational

WE-PL-E1-442

Personal Monitoring of Individual Temperature Experience in Outdoor Workers Using Wearable Sensors

J. Runkle[3], M. Sugg[1], C. Fuhrmann[2]; [1]Appalachian State University, Boone, NC, [2]Mississippi State University, Starksville, MS, [3]North Carolina State University, Raleigh, NC

Abstract: Background: Climate-related increases in the intensity and frequency of higher temperatures has and will continue to impact worker health and safety, particularly among workers who spend the majority of their workday outdoors. Wearable sensors offer a feasible strategy to collect personal monitoring data to enhance understanding of the individual patterns of workplace exposure to ambient temperatures. **Objective**: Examine the personal, occupational, and behavioral factors that influence physiologic response to heat strain. **Methods**: Outdoor workers from three geographic regions in the Southeast characterized by- high heat, high heat + high humidity, and moderate heat environments were recruited to participate during a 5-day work period in the summer of 2016 (July-August). Logistic regression models were estimated using generalized estimating equations. **Results**: In addition to wearing a temperature sensor, a total of 36 out of 66 workers consented to wearing a gps-enabled smartwatch to track heart rate, activity patterns, and location-based temperature measurements. Heat strain was more prevalent among workers in the high heat+humidity environment (55%) compared to those in the high heat (22%) and moderate heat (23%) environments. The risk of heat strain increased 16% for every 5°F increase in temperature, as well as with increasing BMI, higher education, and among Hispanics. In this study, heat prevention training, regular breaks, and access to shade did not reduce the risk of heat strain. However, a significant reduction in heat strain risk was observed in workers who perceived that time outside in the heat affects work productivity (34%) and among those who were told they had a heat-related illness in the past year (31%). **Discussion**: As wearable technology progresses, real-time temperature health and environmental monitoring at the individual-level across multiple occupational settings will become more feasible and inform targeted heat prevention strategies.

Keywords: A - exposure measurement, A-climate change, A-workplace, A-sensor technology, D-occupational

WE-SY-F1: Merging Non-Targeted Analytical Methods, Informatics, and Predictive Modeling to Advance Next-Generation Exposure Science

WE-SY-F1-443

Non-targeted analysis of chemicals in environmental media: methods and challenges

L. Ferguson; Duke University, Durham, NC

Abstract: The assessment of human and ecological exposure to environmental toxicants depends critically on our ability to detect, identify, and ultimately quantify the myriad organic and inorganic pollutants that constitute the "exposome". While conventional approaches to environmental analytical chemistry have yielded significant advancements in our understanding of toxicant exposure and effects for so called "priority pollutants", we remain far from achieving the ultimate goal of holistic exposure assessment for chemical mixtures. Advancements in analytical instrumentation and cheminformatics over the past decade have resulted in the emergence of non-targeted chemical analysis as a primary tool for pursuit of this goal. In this presentation, I will highlight the most recent and cutting-edge methods for conducting non-targeted analysis of organic contaminants in environmental and biological media. Examples will be provided demonstrating application of these methods to sewage-derived contaminants in ambient waters, semivolatile pollutants in house dust, and proxies of human exposure in the indoor environment. The role of advanced, high-resolution/accurate mass mass spectrometry coupled with multidimensional chromatographic separation will be examined, as will the role of high-performance computing and data processing workflows for organic chemical identification. Many of the methods and challenges faced by non-targeted analysis practitioners are shared with "omics" techniques pioneered in the first decade of the 21st century; parallels will be highlighted between the development and application of these approaches. Finally, the broad outlook for applying non-targeted analysis methods in context of human and ecological exposure science will be discussed, with particular focus on the challenges and opportunities within the field.

Keywords: A-analytical methods

WE-SY-F1-444

New Chemical Use Databases to Support Interpretation and Dissemination of Non-Targeted Analysis Exposure Data

K. Dionisio[1], K. Phillips[1], A. Williams[2], A. McEachran[2], J. R. Sobus[1], P. Price[1], J. Wambaugh[2], K. Isaacs[1]; [1]Environmental Protection Agency, Research Triangle Park, NC, [2]Environmental Protection Agency, Raleigh, NC

Abstract: Exposure to chemicals in consumer products has been identified as a significant source of human exposure. To predict such exposures, information about the ingredients and their quantities in consumer products is required, but is often not available. The Chemicals and Products Database (CPDat) has been developed under the U.S. EPA's Chemical Safety and Sustainability research program to fill these data gaps by using information from publicly-available sources. CPDat currently includes qualitative and quantitative information on reported product ingredients (from Safety Data Sheets or ingredient lists) in addition to multiple use categorizations (general, chemical function, and consumer product uses). Further, the CPDat database has been integrated with the Computational Toxicology Chemistry Dashboard (https://comptox.epa.gov/dashboard), which allows for linking of the product-specific (e.g. ingredients, weight fraction, use category) information in CPDat with chemical-specific information (e.g. structure, properties) and toxicity data (e.g. high-throughput screening and animal studies) available through the Dashboard. These data can aid in the interpretation or classification of accurate masses or molecular formulae identified in non-targeted analysis (NTA) studies. The Dashboard contains tools for searching for candidate chemical structures by mass or formulae, allowing selection among possible structures via investigation of CPDat information and other dashboard data sources. CPDat and the Dashboard are being updated to contain information about chemicals tentatively identified or confirmed in products in targeted and non-targeted analytical studies by EPA or others. As the database grows, it will increase the information available for rapidly and defensibly 1) evaluating the results of NTA of environmental or biological media or consumer products/articles and 2) characterizing chemicals in consumer products and articles for use in exposure and risk evaluations.

Keywords: Human Exposure Model, A-chemical prioritization, A-risk assessment, C-consumer products, C-consumer/personal care products

WE-SY-F1-445

Assessing Plausibility of Tentative Chemical Identifications from Suspect Screening Analyses via Chemical Function

K. Phillips[1], A. McEachran[2], K. Isaacs[1], J. R. Sobus[1], A. Williams[3], J. Wambaugh[3]; [1]U.S. Environmental Protection Agency, Research Triangle Park, NC, [2]US EPA / ORISE, Research Triangle Park, NC, [3]U.S. Environmental Protection Agency, Research Triangle Park, NC

Abstract: Suspect screening (SSA) and non-targeted analysis (NTA) have become increasingly useful methods for identifying chemicals in indoor environments, which is where many chemical exposures occur. However, the tentative chemical identifications from these analyses must be confirmed. Traditionally, pure chemical standards might be used for confirmation, but this testing is infeasible for the hundreds to thousands of chemicals indicated by these techniques due to both time constraints and availability of standards. Instead, a sometimes laborious process of matching molecular features to a chemical structure may be employed. Here chemical functional use (i.e., the end role of a chemical in a product) is used to filter out unlikely structural matches as part of SSA. In some cases, manufacturers report the functional use of a chemical in its products, but more often than not, chemicals of toxicological interest have little-to-no use information. To fill in these data gaps, quantitative-structure use relationship (QSUR) models have been applied to chemicals with no known use. These models use the structure and physicochemical properties of a chemical to provide a probability for that chemical serving a functional role. As a case study, in a recent SSA of 100 consumer products with 1639 identified chemicals, 1451 of those were not included in data sources listing chemicals associated with consumer products. Such a low number of hits from these data sources is not surprising as many of these sources would not contain information on chemicals used as fragrances, nor of chemicals contained in household articles. However, by examining functional uses, it was found that 696 of the tentatively identified chemicals had at least one reported functional use or had a valid QSUR prediction indicating that these chemicals could serve a functional role in a product and, therefore, should be considered for future confirmation. ***The views expressed here are those of the authors and do not necessarily reflect the views or policies of the U. S. EPA.***

Keywords: A-analytical methods, C-consumer products

WE-SY-F1-446

Development of Multi-Media Chemical Monitoring Database

A. Williams[2], C. Bevington[1], K. Isaacs[1], A. Williams[1], A. Pfahles-Hutchens[1], H. Hubbard[2], A. Varghese[2]; [1]Environmental Protection Agency, Washington, DC, DC, [2]ICF, RTP, NC

Abstract: There are many sources of data available that provide environmental and human biomonitoring data; however, there is not a source where the information has been collected and reviewed for reliability. In order to provide a better source for data, the multi-media monitoring database was compiled using two types of data sources. First, information from existing reputable monitoring databases was compiled using a combination of automated and manual curation approaches. Second, targeted information from the open literature was manually curated. The database strives for broad coverage of chemical-media concentrations, but does not systematically incorporate all reported monitoring values in the open literature. Instead, emphasis was placed on quality review, where criteria were defined and implemented in data evaluation. Monitoring studies from the open literature and database sources were reviewed using the following criteria: (1) a*ccurate-reliable* - source is reputable, defined as government entity, or an entity with documented credentials to discuss a particular topic; (2) a*pplicable* - source contains quantitative monitoring data for environmental samples and not spiked samples for the purpose of method validation or development; (3) *representative* - sample size must be greater than five with a preference for data from large surveys or studies; and (4) *accessible*- data sets must be freely available, generally on publicly available websites. Over time, additional chemical-specific monitoring data that have met the quality criteria, will be loaded into EPA's chemistry dashboard. This presentation will provide an overview of the status of database integration and quality assurance by noting the number of database and journal article sources reviewed and incorporated, the number of chemicals with available data, and the process for migration into the chemistry dashboard. The views expressed in this abstract are those of the authors and do not represent Agency policy or endorsement.

Keywords: A-biomonitoring, environmental monitoring

WE-SY-F1-447

Media-Specific Classification Models for Ranking Candidate Unknown Chemicals in Non-targeted Analyses

K. Isaacs[1], K. Phillips[1], A. McEachran[2], A. Williams[3], J. Wambaugh[3], J. R. Sobus[1]; [1]US Environmental Protection Agency, Durham, NC, [2]US EPA, Research Triangle Park, NC, [3]US EPA, Research Triangle Park, NC

Abstract: Monitoring of chemicals in environmental and biological media can inform estimates of exposure for supporting possible regulatory decisions and mitigation strategies. New technologies are allowing for the increased use of high-throughput (HT) monitoring approaches based on high-resolution mass spectrometry. However, HT approaches such as non-targeted (NTA) and suspect-screening (SSA) analyses require additional methods for identifying unknown chemical analytes from observed masses or molecular formulae. Current approaches for identifying "known unknowns" in NTA have relied on ranking candidate structures by number of literature sources or chemical database references. We now augment these approaches with chemical-specific predictions of the likelihood that a chemical might be found in an environmental medium. We use supervised machine learning (ML) approaches to classify chemicals on the basis of their unique properties, structures, uses, or other characteristics (collectively called "descriptors"). By analyzing a training set of chemicals known to be present or absent from a specific medium, the ML methods generate an empirical model that can be used to classify other chemicals. Here we develop ML models for likely media occurrence by analyzing chemicals previously identified in various environmental and biological media. The monitoring data, which were collected from public sources, were used to successfully generate 13 cross-validated medium-specific classification models, which can be used to predict probabilities of media occurrence for any unknown chemical. The models were applied to a set of chemicals that are poorly ranked using data-source based methods alone, and the value and limitations of the models for the purpose of chemical ranking in NTA were evaluated. *The views expressed here are those of the authors and do not necessarily reflect the views or policies of the U. S. EPA.*

Keywords: A - exposure measurement, A-analytical methods, A-exposure models, C-consumer products, A-biomonitoring

WE-SY-G1: Exposure sciences from east Asian perspectives

WE-SY-G1-448

Study on Phthalates, BPA and Antibiotics Exposure Assessment of Chinese School Children Based on Biomarkers Monitoring

Y. ZHOU, Q. JIANG; Fudan University, Shanghai, China

Abstract: A series of biomonitoring studies on internal exposure of Chinese School children to phthalates, BPA and antibiotics were investigated. In the first study, we analyzed 13 metabolites of 9 phthalates in morning urine of 782 Chinese school children aged 8−11 years collected from our cohorts of school children in China in 2012. Overall, 11 metabolites were found in at least 85% of the urine samples with the highest median concentration of 93.4 μg/g creatinine for MnBP. MOP and MiNP were not detectable. The cumulative risk assessment covering DEHP, DnBP, DiBP, and BBzP demonstrated that 19.8% (volume model-based) and 40.3% (creatinine model-based) of the children exceeded 1 for the HI based on TDI values. Furthermore, at least 36% of the children from the manufacturing-intensive region had a HI higher than 1. In addtition, total BPA was detected in 98.9% of urine samples from 666 school children aged 9-12 years with the geometric mean and median of BPA was 2.32 mg/g creatinine and 2.22 mg/g creatinine, respectively. The urinary BPA level was found decreased with decreasing age. The EDI of BPA (0.023 mg/kg bw/day) was much lower than the TDI and RD of 50 mg/kg bw/day recommended by either the EFSA or the US EPA. In another study, total urinary concentrations (free and conjugated) of 18 representative antibiotics (5 macrolides, 2 β-lactams, 3 tetracyclines, 4 quinolones, and 4 sulfonamides) were detected in morning urinary samples from 1064 Chinese school students in 2013. The detection frequencies of all 18 antibiotics ranged from 0.4% to 19.6%, and 58.3% of the whole urine samples. Of them, 47.8% of the urine samples had a sum of mass concentration of all antibiotics between 0.1 (minimum) and 20.0 ng/mL, and 8 antibiotics had their concentrations of above 1000 ng/mL in some urine samples. Three veterinary antibiotics (Tylosin, Chlortetracycline, Enrofloxacin) and 11 human/veterinary antibiotics were detected overall in 6.3 and 49.4% of urine samples, respectively.

Keywords: A - exposure measurement, A-biomonitoring, D-children, B-BPA, B-phthalates

WE-SY-G1-449

DEVELOPMENT OF LUR MODEL FOR NO_x EXPOSURE IN YOKOHAMA, JAPAN - A CASE STUDY

S. Nakai, K. Hanaoka, A. Fukuo; Yokohama National University, Yokohama, Japan

Abstract: Land Use regression (LUR) model should be the standard method for exposure assessment in air-pollution epidemiologic studies. However, few LUR models have been developed in Japan. To examine whether LUR model can be used for the exposure assessment in Japan, we developed a LUR model for NO_2 and NO_x in Yokohama. Annual mean of NO_x and NO_2 measured at 29 monitoring stations in Yokohama city in 2005 was regressed on land use data such as residential/industrial area, traffic volume, etc. with a manual stepwise method. Adjusted R^2 for NO_x and NO_2 regressions were 0.88 and 0.81, respectively. Distribution of NO_x and NO_2 concentrations were mapped by using Arc-GIS software. The LUR models we developed has a similar spatial resolution as the LUR models developed in the Western countries. However, it should be necessary to improve the distribution resolution of each pollutant at several places such as near expressways and near Tokyo bay. It may be necessary to increase the number of measurement points and to select the appropriate points to apply the developed LUR models for epidemiologic studies in Japan.

Keywords: A-exposure models, C-air, A-geospatial analysis/GIS

WE-SY-G1-450

Overview of VOC Monitoring in Taiwan - New Technology and Field Application Studies

L. Wang[1], P. Chung[1], C. Hsiao[1], S. Tsai[1], S. Lee[2], T. A. Chou[1]; [1]TricornTech Corporation, Taipei, Taiwan, [2]Hong Kong Polytechnic University, Hung Hom, Hong Kong

Abstract: According to World Health Organization (WHO) study in 2016, more than 90% of the world's population lives in places where the ambient air pollutions exceed WHO limits. The air pollutions are from sources, such as fuel burning, transportation exhausts, and emissions from industrial manufacturing processes. Among the air pollutants, volatile organic compound (VOC) is one of the contributors to PM2.5 and ozone formation. Furthermore, some VOCs have high toxicity, which can lead to serious adverse effects even under trace level exposure. In Asia, exposure of toxic VOCs from local industries has raised increasing public health concern due to densely mixing between industrial plants and residential areas under fast-growing economic developments. However, most of on-site monitoring works are only focusing on the VOCs contributing to ozone formation (photochemical assessment monitoring station, PAMS). There is no viable method to continuously monitoring receptor's exposure to non-PAMS air toxics due to the diversified VOCs from local pollution sources. Therefore, a solution to on-site continuously monitoring specific toxic VOCs is needed to effectively study the short-term and long-term exposure risks and assess the reduction of the air pollution. This presentation will address the current status of VOC monitoring implemented in Taiwan and the issues of lacking online reliable automatic VOC monitoring solution. A new technology and solution by Tricorntech MiTAP (Miniaturized Total Analysis Platform), online VOC monitor, to the current unmet application needs will be presented. Various field implementation and case studies utilizing automatic MiTAP for the continuous online VOC monitoring will be discussed in the applications of semiconductor facility air quality, petrochemical ambient air toxic monitoring, industrial park fenceline monitoring, urban traffic VOC pollution, and so on. Finally, the future trend of online VOC monitoring will be concluded.

Keywords: B-VOCs, C-air, A-sensor technology, A-analytical methods

WE-SY-G1-451

Recommendation of suitable sample size for exposure factors of consumer products

K. Lee, H. Ban; Seoul National University, Seoul, Korea (the Republic of)

Abstract: Introduction: Deterministic approach of exposure assessment for consumer products is simple to apply. However, there is no guideline for selecting a representative exposure factor. The purposes of this study were to determine the effect of different sample sizes on estimating exposure factors using a large database and recommend suitable sample size for acceptable margin of errors. **Methods:** This study used an existing database of 3,333 for three exposure factors (the frequency of use, the amount of use and the duration of use) of thirteen CPs including cosmetics, cleaning products and disinfectant products. In order to analyze the variation of exposure factors by different sample sizes, the exposure factor was randomly sampled by ten different small samples. The three exposure factors of the parent population were compared with a corresponding value from ten different smaller sample sizes. **Results:** For certain CPs, exposure factors determined by sample size of smaller than 50 were less than 10% different from those of parent population. When sample sizes were increased, exposure factors were similar to those of parent population for the most CPs. When sample size was 100, the 75th percentiles of exposure factors were less than or equal to 20% different from those of the parent population except for one product. **Conclusion:** This study provided the potential error using smaller sample sizes for 13 CPs. The sample size of 100 could provide exposure factors with less than 20% error for the most CPs.

Keywords: A-exposure factors, A-risk assessment, C-consumer products

WE-SY-G1-452

PANEL DISCUSSION

S. Nakai; Yokohama National University, Yokohama, Japan

Abstract: East Asian region has different aspects of problems and addresses unique approach to generate future solutions to complex environmental health problems. This symposium is organized by east Asian chapter (EAC) of ISES and we will provide a forum for the presentation and discussion of key environmental health issues facing researchers in East Asian countries. Participants will know the situation of environmental health, exposure sciences of other countries, discuss the difference and/or similarity among east Asian countries and give their perspectives/idea each other towards solutions of environmental problems.

Keywords: A-sustainability

WE-PL-A2-453

ASSESSMENT OF THE ORAL BIOACCESSIBILITY OF SEMI-VOLATILE ORGANIC COMPOUNDS IN INDOOR SETTLED DUST

G. Raffy[1], F. Mercier[1], P. Glorennec[1], C. Mandin[2], B. Le Bot[1]; [1]EHESP/Irset Inserm UMR 1085, Rennes, France, [2]CSTB, Marne-la-Vallée, France

Withdrawn

TH-PL-A2-638 *(being presented in place of WE-PL-A2-453)*

Diesel exposure in urban environment: evidence based approach to understand health risks

A. Rashid[1], A. Naseem[2], A. Kamal[1]; [1]PMAS Arid Agriculture University, Rawalpindi, Pakistan, [2]National University of Sciences and Technology, Islamabad, Pakistan

Abstract: We evaluated urban population with low socioeconomic status for ambient air particulate matter (PM) and diesel exhaust exposure effects. In general there seems no urban health policy for reducing environmental exposure and improving environmental health in Rawalpindi city. Three spots i.e. Faizabad, Shamsabad, and Saddar in Rawalpindi city were studied. People living and working along roads with dense traffic were compared with sub-urban areas for health outcomes. Hospital admissions, asthma, respiratory infections and cardiovascular hospital admission rates were higher. Five times higher PM concentration around roads and highways was noted. Stops for buses and public transport where people standing in queues have 10 times higher PM_{10} than urban ambient air. PM_{10} concentrations average ~ 2 mcg/m3 but have been detected at 125 mcg/m3 above background in urban public transport stops Faizabad, Shamsabad and Saddar. Significant effect modification by age, smoking status and poverty level was evident among those with higher frequency of respiratory infections, asthma and previous admissions. Based on logistic regression, we found that people living in neighborhoods with dense traffic and higher diesel exposure had high prevalence of respirator infections (OR=3.15, 95% CI=1.18-7.66) whereas location of occupational groups along roadsides have attributed substantial development of asthma (OR=4.02, 95% CI=1.59-9.61). Response of people towards exposure prevention such as use of masks was substantially low whereas their expectation from local government was very high to control smoke emission from public transport. We propose two-pronged approach to minimize health risks in Rawalpindi. First, urban governance should deliver by initiating strict action against smoke emitting vehicles. Exposure prevention by urban population at individual levels is suggested as an alternate approach through adopting safety measures such as use of masks and gloves.

Keywords: A-exposure factors, A-risk assessment, B-particulate matter

WE-PL-A2-454

Estimating Particle/Air Partition Coefficients for SVOCs on Airborne Particles

Y. Wu, C. Eichler, J. Little, L. Marr; Virginia Tech, Blacksburg, VA

Abstract: Semivolatile organic compounds (SVOCs) are ubiquitous in the indoor environment and may cause adverse health effects. Because SVOCs partition between the gaseous phase and airborne particles, the total airborne SVOC concentrations and resulting human exposure can be higher than expected. However, when it comes to understanding interactions between airborne particles and SVOCs, there are numerous models carrying different theories but significant challenges of experimentally measuring SVOC partitioning between the gaseous phase and particulate phase. In this study, a specially-designed tube chamber operating in the laminar flow regime was developed to investigate the particle-mediated mass transfer of SVOCs; this approach is advantageous in terms of both predictability and simplicity. Di-2-ethylhexyl phthalate (DEHP) was selected as the target SVOC due to its widespread use in consumer products and building materials. The particle/air partition coefficient (Kp) of DEHP was determined as 0.011 ± 0.004 m3/µg. The simulation by computational fluid dynamics model shows that the presence of particles increases the radial gas-phase concentration gradient, resulting in enhanced mass transfer from emission surface to the air. This novel experimental approach offers a rapid method for measuring Kp and a sound basis for characterizing gas-particle interactions of SVOCs.

Keywords: A-exposure models, B-SVOCs, C-indoor

WE-PL-A2-455

Characterization of water-soluble organic gases in 13 real homes

S. Duncan[1], B. Turpin[2]; [1]Rutgers University, New Brunswick, NJ, [2]University of North Carolina, Chapel Hill, NC

Abstract: Non-polar VOCs are well characterized indoors due to established analytical methods. However, besides a targeted few, polar (oxidized) VOCs indoors are not well understood. These polar VOCs are often water-soluble (deemed water-soluble oxygenated gases, WSOGs); they can be directly emitted or formed through gas phase or surface reactions. Characterizing these compounds is important to exposure assessment because people spend the majority of their time indoors. Also note that 18 to 50% of homes in the U.S. are considered "damp" (Mendell et al. 2011). In damp homes, WSOGs can partition into liquid water on surfaces, in wet particles and on skin. Subsequent aqueous chemistry could potentially alter dermal and inhalation exposures indoors. In this work, we aim to collect and identify WSOGs in real homes. Inside and directly outside 13 homes in total in New Jersey and North Carolina, we used mist chambers to collect WSOGs. Two mist chambers sampled in parallel inside and two outside, for two hours twice consecutively. Particles were removed at the mist chamber inlets using quartz filters. The flow rate was 25 L min-1, and the water collection volume was 25 mL. Total WSOG concentrations (µM - carbon) were 6 to 27 times greater in indoor samples than outdoor samples; thus most WSOG measured indoors originates indoors. Samples were also analyzed by ion chromatography for organic acids, which revealed that 33 to 53% and 8 to 24% of the total organic carbon in each indoor sample was acetic acid and formic acid, respectively. Finally, samples were analyzed by positive mode QTOF-ESI-MS and MS-MS to identify particularly reactive compounds indoors. The average N:C ratio of all identified compounds was 0.42, while the average O:C ratio was 0.26. 72% of compounds contained nitrogen. Compounds were identified as plasticizers, solvents, preservatives, pesticides, amino acids, pharmaceuticals, and others.

Keywords: A-built environment, C-indoor, A-indoor environment, C-air, B-VOCs

WE-SY-B2: Total Exposure Health - Advances in Exposure Sciences for Diverse Communities

WE-SY-B2-456

One Common Goal - Integrating Exposure Sciences through Total Exposure Health

K. Phillips, R. T. Hartman; Department of Defense, Falls Church, VA

Abstract: Background: Total Exposure Health (TEH) is the Air Force Medical Service (AFMS) solution to capture workplace, environmental and lifestyle exposures to the individual using advances in science, technology and informatics to prevent disease, improve health and well-being, and develop precision health solutions through effective risk assessment, management and communication. **Objectives**: The Air Force Medical Service Strategic Vision seeks to ensure that its "supported population is the healthiest and highest performing segment of the U.S. by 2025." Achieving this vision will require a paradigm shift in military medicine from its usual primary focus on diagnosis of disease and injury and treatment. TEH is an approach to primary prevention and precision health that seeks to evaluate the totality of an individual's exposure history through risk assessment and integrate that with an individual's genetic makeup and clinical disposition, in order to understand the root causes of wellness as well as that of injury and disease through risk management but ultimately communicate that risk to the individual. **Methods**: We are at the intersection of healthcare and technology. Currently, we have the ability to collect more refined information on the individual based on their genetics, which allows us to focus on unique interventions that will lead to better well-being. We will discuss how TEH will partner across multiple program areas and integrate occupational, lifestyle, and environmental exposure data from traditional and new exposure assessment technologies (including sensors and "-omics" based molecular biology) with clinical and genomic data. As a developing healthcare infrastructure, we will also show how TEH will position "exposure scientists" to improve the patient/provider experience with a focus on individual exposures (unique and targeted) through proper risk communication, which is often forgotten. **Results**: These capabilities will inform strategies by reducing the health impacts of exposures, hence improving human performance, readiness, and the resiliency of all Airmen and their beneficiaries. Moreover, by educating individuals about their exposure profile, TEH will enable individuals and families to make healthy lifestyle choices and become proactive about their own health.

Keywords: A - population exposure, A-exposure factors, A-sensor technology, A - exposure measurement, A-aggregate exposure

WE-SY-B2-457

We Are All Unique - Rethinking Risk Management

S. Lacey[2], R. T. Hartman[1]; [1]Department of Defense, Falls Church, VA, [2]Indiana University, Indianapolis, IN

Abstract: Background: Recent advances in science, medicine, technology, and informatics create a unique opportunity to pivot from population-based compliance regulations/standards to a more individualized approach to risk assessment and management in the prevention of disease and illness through precision health. **Objectives:** To introduce a comprehensive framework that describes a risk assessment process focused on individual versus population health risks based on occupational, lifestyle, and environmental exposures, genetic factors, and medical disposition, known as Individual Exposure Health Risk Profiles (IEHRP). **Methods:** The IEHRP will integrate occupational, lifestyle, and environmental exposure data from traditional and new exposure assessment technologies (including sensors) with clinical and genomic data. Leveraging a Big Data infrastructure and advanced analytics, the IEHRP will provide a new capability that will create risk management strategies based on more complete individual information and facilitate effective risk communication to inform lifestyle choices, enhance resilience and human performance, and reduce illness and injury. **Results:** This approach is intended to result in individual-level estimates of disease risk and therefore promote health and well-being through effective early intervention and risk mitigation. Conceptually, the IEHRP will provide health protection professionals a more informative set of data and recommendations regarding exposures to create "individual permissible exposure limits." Future work will further validate and refine the IEHRP models and processes to develop statistically validated algorithms to determine health risk profiles and drive prevention strategies for individuals.

Keywords: A - exposure measurement, A-exposure factors, A-risk assessment, A-cumulative exposure, A-exposure models

WE-SY-B2-458

Bringing It All Together - Noise, a Common Exposure

K. Montgomery[2], R. T. Hartman[1]; [1]Department of Defense, Falls Church, VA, [2]Intelesense Technologies, Milpitas, CA

Abstract: Background: To demonstrate the risk assessment, management and communication process through Total Exposure Health, we choose a common exposure, noise, which adversely affects over 1 million veterans, with an annual cost through compensation and treatment over $1.2B. **Objectives:** This presentation will provide an overview of the TEH "noise" proof of concept with the goals to: 1) demonstrate the ability to collect, store, and utilize unstructured "noise" exposure data to identify the unique health risk associated noise and manage NIHL, 2) create intervention strategies to protect beneficiaries (training, medical interventions such as functional hearing tests and customized protection) and 3) identify future research questions and research needs. **Methods:** To test the TEH concept, we used advances in sensors to collect total noise exposure (24/7), to include sound from media devices, along with medical disposition data from the Department of Defense Occupational and Environmental Health Readiness System (DOEHRS) Hearing Conservation (HC) and Industrial Hygiene (IH) data, and specific genetic data, single nucleotide polymorphisms (SNP), associated with noise induced hearing loss (NIHL) to identify active duty military members at risk for NIHL. Synthesizing this data into the IEHRP we translated these findings into valuable, actionable recommendations for individuals and their health care providers for illness/disease prevention and improving human performance. **Results:** The ultimate goal was to demonstrate the TEH concept and the ability to collect, store, and utilize disparate unstructured "noise" exposure data including sensor, clinical, survey and genetics data to identify the unique health risk associated with noise. Through this study, we showed how we can ingest exposure data, genomic data, and current medical disposition at the individual level to calculate personal risk profiles and provide information unique to the individual called, Course of Action (COA). These COA's can be used to inform and guide the individual to understand their risk profile and make healthy choices to reduce disease and illness. The aggregate information can also be used for policy and resource decision-making based on common exposures, genetic susceptibility and medical disposition vs. workplace or exposure.

Keywords: A - exposure measurement, A - ambient monitoring, A-cumulative exposure, A-sensor technology, A-risk assessment

WE-SY-C2: Exposure science - from research to report back: Working with your communities to communicate results appropriately

WE-SY-C2-459

Know Your Audience: Academic-Community Partnerships to Develop Data Disclosure Strategies

E. Haynes; University of Cincinnati, Cincinnati, OH

Abstract: Research results should be communicated effectively to the target population, yet the science of the communication strategies is lacking. The University of Cincinnati NIEHS Center's Community Engagement Core has worked with several unique populations to develop meaningful data disclosure strategies for different exposures: children's blood metal concentrations, community-level ambient volatile organic compounds near unconventional natural gas exaction waste disposal sites, and firefighter exposure and cancer prevention strategies. For each target population, members of the target audience were highly influential in the development of the risk communication message and media dissemination strategy. These methods will be presented.

Keywords: D-community, data disclosure, communication

WE-SY-C2-460

Improving Environmental Health Literacy and Justice Through Data Report Back and Free-Choice Learning

M. Ramirez-Andreotta; University of Arizona, Tucson, AZ

Abstract: Reporting back exposure data is carving out a new informal science education setting, particularly for those living in underserved areas. When applying a contextual model of learning, it has been observed that these communities are knowledgeable and intrinsically motivated to partner and inform communication strategies. Public participation, peer education, and information design are critical steps in data sharing. In this session, novel data visualizations and participatory approaches will be presented alongside learning and action-based evaluation outcomes. Selected community science and data sharing case studies will be presented where a mixed method approach was implemented to evaluate motivation, knowledge, and environmental self-efficacy and action as they relate to hazardous site-related exposures, environmental health risks, and intervention and prevention practices. Special attention will be given to the: data visualizations, role of knowledge mediators, and delivery of results (e.g. one-on-one home-based, community setting, web-based) to differentiate learning outcomes and advance the field of science, environment health, and risk communication.

Keywords: A-environmental justice, D-community, C-multimedia, A - exposure measurement

WE-SY-C2-461

DERBI: A digital framework for personalizing exposure reports at the individual and community levels

J. G. Brody, H. Susmann, K. Boronow, J. Ohayon; Silent Spring Institute, Newton, MA

Abstract: Personalized biomonitoring reports prepared in the Digital Exposure Report-back Interface (DERBI) can create tailored experiences for different audiences, draw attention to notable individual and study-wide results, and communicate strategies to reduce exposures. DERBI makes report-back practical in studies of any size by using a library of curated scientific information, presented in plain language, and a set of automated decision rules that personalize reports to draw attention to results most relevant to each participant. Reports include text and graphs that show an individual's results in relation to others in the study and national benchmarks or health guidelines, when they exist. Chemical exposure measurements are linked to contextual information about the sources of exposure, relevance to health, and strategies to reduce exposure through individual and community action. Through active collaboration between researchers and community representatives, reports can be tailored to the cultural context of each study. Reports can be produced on paper or online, and we are currently developing new smartphone-based DERBI reports in order to increase access in low-income communities where phones are commonly used to access the internet. DERBI has been used in economically and racially diverse studies to report results for pesticides, flame retardants, highly fluorinated compounds, phthalates, PAHs, and other chemicals. Participants who received their results were interviewed about their experience, and analytic data were collected about how users accessed and navigated online reports. Interviews with participants in studies that have reported individual exposures show that most want their results, and receiving results promotes environmental health literacy, understanding of science, and feelings of trust in the research team. Several major ethics statements now consider reporting personal and community-level research results as the norm for ethical research.

Keywords: A-biomonitoring, D-community, A - exposure measurement, A-behavior, A-epidemiology

WE-SY-D2: Social Determinants of Health, Environmental Exposures, and Disproportionately Impacted Communities: What We Know and How We Tell Others- Part 2

WE-SY-D2-462

Strategies for Community Engagement Related to Environmental Health Disparities

N. Teufel-Shone[1], M. Gonzales[4], J. Johnston[2], C. Scollaert[3]; [1]University of Arizona, Tucson, AZ, [2]University of Southern California, Los Angeles, CA, [3]Boston University, Boston, MA, [4]University of New Mexico, Albuquerque, NM

Abstract: The five NIH-EPA Centers of Excellence on Environmental Health Disparities Research are engaging with diverse communities across the United States, using a range of research study designs and engagement strategies. The strategies are guided by investigators' previous experience in the communities that provide insight into local social networks and stakeholder goals and community partners' guidance identifying existing coalitions, local formal and informal leaders and facilitating communication. Community engagement is vital to assure local concerns are examined using locally appropriate and evidence based methods to document the pathways and outcomes of the complex relationships between the environment and human health. The Centers work with communities to build local capacity to understand and apply the outcome of environmental health research. Local leaders and institutions need data collected using accepted scientific methods to weigh risks and benefits when making decisions to change practice and policy to create healthier communities. Centers are exploring locally meaningful ways to translate research outcomes using images created by local artists, info-graphics, video and even social media to support communication and learning. The goal is to empower communities to advocate for themselves. Throughout the course of their partnerships, these NIH-EPA Centers will describe effective processes for developing placed-based and culturally-relevant collaborative scientific projects in urban, rural and tribal communities. In this presentation, we will share community engagement strategies across the Centers and the lessons learned from experiences in the field, unique considerations in engaging with diverse communities, and approaches for meaningful community input in the scientific research.

Keywords: D-community, D-environmental justice, D-First Nation

WE-SY-D2-463

Communicating environmental risks with African-American communities: A Cognitive Psychology perspective

L. Rivers, D. McGregor; North Carolina State University, Raleigh, NC

Abstract: Poor urban stormwater management has been a consistent issue for African-American and underserved communities across the United States, particularly in the southeastern United States. Acute vulnerability to stormwater is often an artifact of the "classic southern" pattern of residential settlement that was established post-Reconstruction. Over the past two decades there has been growing efforts to use green infrastructure to address stormwater problems. However, this investment in community infrastructure has had at times the perverse effect of intensifying urban gentrification. Focus groups and a subsequent neighborhood survey in a predominantly African-American community in a large southeastern city in the United States are used to examine this phenomenon. This work suggests the need to address lingering questions of environmental equity and racism. A central concept key to addressing these questions is the role of trust between citizens and local governments in addressing stormwater concerns and other management issues. Our presentation makes use of this case study to examine the communication of environmental risks with African-American community partners making use of insights from the cognitive psychology and risk perception literature.

Keywords: A-environmental justice, A-environmental policy, A-environmental regulation, social trust

WE-SY-D2-464

Translation of EPA Research: Data Interpretation and Communication Strategies

A. G. Rappold[1], K. Benedict[2], D. Lobdell[1]; [1]US EPA, Durham, NC, [2]US EPA, Durham, NC

Abstract: Translation and communication of scientific findings will enable communities and individuals to limit exposure and reduce risks of adverse health effects. This presentation will provide examples of risk communication tools and indices being developed at the EPA such as communication strategies for wildfires and air sensors and indices incorporating various environmental domains (chemical and non-chemical stressors). Low-cost, portable air quality sensors that report air pollutant measurements in increments as short as one minute are becoming readily available. As availability and use increases, so does the need to help people interpret and communicate the relationship between a short-term sensor reading and health effects that may be experienced after longer exposure time periods. This presentation will provide a brief overview of a "sensor scale" designed to make instantaneous outdoor air quality data useful for the public pilot project and a summarize feedback received to date. The impacts of wildfire smoke exposure on the cardiopulmonary hospitalizations and doctor visits are well documented; but less is known about the health burden due to the less severe outcomes, which affect a wider population. This presentation will introduce the Smoke Sense Study - a crowd sourced study of wildfire smoke impacts on health. This study improves understanding of the low severity health impacts in populations and determines health risk communication strategies that reduce the public health burden. The EPA has also developed an Environmental Quality Index that includes five domains: air, water, land, sociodemographic and built environments. This presentation will include information on accessing the publically available data and how US EPA has used this index in research efforts. Disclaimer: The views expressed in this abstract do not necessarily reflect the views or policies of the U.S. Environmental Protection Agency

Keywords: A-sensor technology, D-wildlife, Environmental Quality Index

WE-SY-E2: The NTA Fractal: Metabolome, Exposome, and Biome

WE-SY-E2-465

Connecting genes to molecules: Identifying novel small molecules in microbe-plant interactions

E. M. O'Neill; University of North Carolina at Chapel Hill, Carrboro, NC

Abstract: Plant pathogens cause billions of dollars of crop damage worldwide annually. These bacteria produce many phytotoxins and antimicrobial molecules, which allow them to survive and flourish in competitive environments. While genomics-guided efforts may aid in the identification of small molecule products made by assembly-line enzyme pathways, such as non-ribosomal peptides and polyketides, the products of short and non-canonical pathways cannot be predicted. Here, we developed a dual approach that combines differential metabolomics with *in vitro* enzyme activity screening, which enabled us to elucidate the chemical structure of a structurally and functionally distinct molecule made by a small gene cluster in *Pseudomonas syringae* pv. tomato DC3000. We observed wide distribution of this three-gene pathway in more than one hundred other phytopathogens. Furthermore, gene expression of this cluster in DC3000 increased when a virulence factor master regulator was overexpressed, indicating this cluster is most likely linked to plant infection. Our work uncovered the molecular structure of an important molecule, which is a key step towards understanding the collection of small-molecule arsenals that *P. syringae* produces, with agricultural and biochemical impacts. The methods developed here will accelerate future identification of other short pathways and help us understand their roles in microbe-host interactions. Once these interactions are understood, we can engineer ways to prevent plant infection and minimize crop damage.

Keywords: A-analytical methods, B-microbial agents, metabolomics

WE-SY-E2-466

Exploring host-associated microbiota as mediators of neurobehavioral toxicity in zebrafish larvae developmentally exposed to triclosan

T. Tal[1], D. Phelps[2], A. Swank[3], T. Catron[2], C. Wood[1], M. Stryar[4], N. Brinkman[5], S. Keely[5]; [1]U.S. EPA, RTP, NC, [2]ORISE/U.S. EPA, RTP, NC, [3]U.S. EPA, RTP, NC, [4]U.S. EPA, RTP, NC, [5]U.S. EPA, RTP, NC

Abstract: Growing evidence indicates that host-associated microbiota modify the toxicokinetics and/or toxicodynamics of environmental chemicals; however, current risk assessment methods do not consider interactions between microbiota and chemical toxicity. We previously reported that microbial colonization is required for normal neurobehavioral development in zebrafish. We therefore hypothesized that neurobehavioral toxicity may be mediated by altered microbial colonization during development. We explored differences in swimming behavior, microbial community structure, and chemical metabolism in axenic (microbe-free) and conventionally colonized zebrafish larvae that were exposed to the antimicrobial triclosan (0.1-0.3 µM) or vehicle (0.4% DMSO) on 1, 6, 7, 8, and 9 days post fertilization (dpf). At 10 dpf, neurobehavioral function was assessed. Triclosan exposure had no effect on locomotor activity in axenic larvae. In comparison, locomotor hypoactivity was observed in conventionally colonized larvae exposed to 0.3 µM, but not 0.1 µM triclosan. Also on 10 dpf, triclosan exposure triggered concentration-dependent shifts in microbial community structure. To understand the temporal dynamics of triclosan-induced hypoactivity, conventionally colonized larvae were exposed to 0.3 µM triclosan in four scenarios: 1 dpf; 1 and 6 dpf; 1 and 9 dpf; or 1, 6, 7, 8, and 9 dpf. Triclosan exposure only caused hypoactivity at 10 dpf in larvae exposed on 1 and 9 dpf or 1, 6, 7, 8, and 9 dpf. As expected, these two groups contained elevated concentrations of triclosan (ng/larva) at 10 dpf compared to larvae exposed to triclosan on 1 dpf as measured by high resolution mass spectrometry. Ultimately, this study will serve as a test case to apply non-targeted chemical analyses to reveal unique biotransformation products in axenic and conventionally colonized zebrafish exposed to triclosan during development. In summary, these data suggest that triclosan may exert behavioral effects via dysregulation of microbial colonization during development. *This abstract does not necessarily reflect EPA policy.*

Keywords: B-microbial agents, D-children, Developmental neurotoxicology

WE-SY-E2-467

Exploring human exposure to organic micropollutants in the indoor environment by non-targeted analysis

L. Ferguson[1], B. Vogler[2], J. Cooper[2], H. M. Stapleton[1]; [1]Duke University, Durham, NC, [2]Duke University, Durham, NC

Abstract: Exposure to organic contaminants present in the indoor environment may be linked to adverse health outcomes including neurodevelopmental effects and obesity. Human exposure can occur through contact with consumer products, indoor furnishings, inhalation, and contact with indoor dust. Many indoor contaminants are endocrine disruptors, such as phthalates, the flame retardants, organophosphate polymer additives, and pesticides. We have utilized a non-targeted, HPLC-HRMS/MS analytical method to explore the identities and mixture composition of organic residues in household dust as well as to track human exposure to these compounds by analysis of hand-wipes. Our analytical workflow was based on use of accurate molecular mass, high-fidelity isotope measurements, and data-directed HRMS/MS spectra for querying public molecular databases (PubChem, Chemspider, and CAS), spectral libraries (NIST, MassBank, and local/proprietary libraries), identification of organic compounds. For analyses, hand wipe and dust extracts was analyzed using an LTQ-Orbitrap Velos mass spectrometer operated in either positive or negative ESI mode was programmed to acquire continuous high-resolution (R>100,000) full-scan data as well as data-dependent MS/MS spectra concurrently. Tentative compound identification was made using a customized data analysis workflow incorporating computational mass spectrometry and cheminformatics tools. Results indicate that nonionic, polyethoxylated surfactants (including nonylphenol ethoxylates and alcohol ethoxylates) are ubiquitous contaminants in household dust and that these compounds are also present in associated human hand wipes. Other compounds identified included polymer additives (including caprolactam oligomers and organophosphate plasticizers), azo dyes, pesticides (including imazalil and imidacloprid), and triglyceride breakdown products (e.g. dilinolein). These results demonstrate that non-targeted analytical approaches can enable a much more comprehensive exposure data set for organic micropollutants than can targeted approaches alone and can provide unique "leads" for further investigation of health effects.

Keywords: A-analytical methods

WE-SY-E2-468

Non-Targeted Screening and Identification of Chlorination Products in Reclaimed Water

J. McCord, M. Stryar; Oak Ridge Institute for Science and Education, Research Triangle Park, NC

Abstract: In recent years, total water demand and threats to natural water resources have driven interest in sustainable water practices, such as the reclamation and reuse of wastewater. The reclamation process can produce novel transformation products with limited structural or toxicological data. The application of high-resolution mass spectrometry and non-targeted screening techniques can be used to elucidate these unknown products of water reclamation as a step towards understanding the environmental impacts of reclaimed water usage. In this work, water obtained from wastewater treatment effluent, pre- and post- chlorination at North Carolina wastewater facilities, were examined using non-targeted analysis (NTA) approaches to identify novel products formed during the reclamation process. Concentrated water extracts were analyzed by liquid chromatography (LC) coupled to high resolution mass spectrometry (MS). Molecular features, consisting of accurate mass and chromatographic peaks resolved using MS vendor processing software, in these samples were isolated, and then filtered using a statistical data processing workflow in R. The remaining features were assigned tentative molecular formulae and searched against known compounds in the EPA's CompTox Chemistry Dashboard (https://comptox.epa.gov/dashboard). Assignments were confirmed using a semi-untargeted data-dependent MS/MS and independent targeted MS/MS characterization was carried out for compounds with no positive matches in the database. The identification of these unknown products is an important first step for future explorations of potential human exposure effects.

Keywords: A-analytical methods, A-sampling methods

WE-SY-F2: Chemical Prioritization via Computational Exposure and Hazard Screening

WE-SY-F2-469

Predicting Exposure Pathways with Machine Learning

J. Wambaugh[1], C. L. Ring[1], K. Isaacs[2], K. Phillips[2], P. P. Egeghy[2], R. W. Setzer[1]; [1]U.S. Environmental Protection Agency, Research Triangle Park, NC, [2]U.S. Environmental Protection Agency, Research Triangle Park, NC

Abstract: Prioritizing the risk posed to human health from the thousands of chemicals in the environment requires tools that can estimate exposure rates from limited information. High throughput models exist to make predictions of exposure via specific, important pathways such as residential product use, diet, and environmental fate and transport. These models can be parameterized in terms of physico-chemical properties that can be predicted with reasonable accuracy from chemical structure. However, as identified by Shin et al. (2015), there are extremely limited data available for identifying the relevant pathways for chemicals in a high throughput manner. Both expert opinion or conservative assumptions (i.e., all chemicals exposed by all pathways) have been considered but have obvious drawbacks. Here we examine the use of machine learning techniques to use structural features and physico-chemical properties to assess the probability that a chemical might be associated with exposure via different pathways. Estimating the relevant pathways using these techniques allows information (including model predictions and other exposure estimates) to be synthesized on a per pathway basis. For each pathway we evaluated the predictive ability of various sources of exposure information using inferred population exposure rates for only those chemicals relevant to the pathway. We can now synthesize exposure models and other predictions commensurate with the ability of those predictions to explain biomonitoring data. In addition, we can attribute the presence of a chemical to specific exposure pathways, potentially allowing structure-based forensic investigation of chemical exposure, and subsequent remediation. *The views expressed here are those of the authors and do not necessarily reflect the views or policies of the U. S. EPA.* Shin, H.-M *et al.*, Risk-based high throughput chemical screening and prioritization using exposure models and in vitro bioactivity assays. ES&T 2015, 49, 6760-6771

Keywords: A-exposure models, A-chemical prioritization, A - population exposure, A-statistical methods, A-aggregate exposure

WE-SY-F2-470

Using Chemical Space Analysis to Identify the Chemical Domain of Applicability for In Vitro Assays

C. Nicolas[1], K. Mansouri[1], S. Haider[1], P. McMullen[1], M. Yoon[1], J. Wambaugh[3], A. Tropsha[2], H. Clewell[1]; [1]ScitoVation, Research Triangle Park, NC, [2]University of North Carolina at Chapel Hill, Chapel Hill, NC, [3]USEPA, Research Triangle Park, NC

Abstract: High throughput in vitro hazard assessment is an integral part of prioritization in the proposed 21st century risk assessment framework. Dosimetry in in vitro systems can be impacted substantially by physicochemical properties of the test agent, leading to systematic bias in assays. Increased confidence in in vitro assay interpretation requires that assays are suitable for a compound's properties. Chemical space analysis interrogates chemical properties across multiple determinants of behaviors and informs their compatibility with current models. Current in vitro methods and models that are useful for predicting human relevant doses were designed with pharmaceutical compounds in mind. We can apply these methods to a set of environmental compounds that share similar chemical space in order to fill data gaps. We have identified an ensemble of active pharmaceutical ingredients among ~45,000 curated chemical structures from the Collaborative Estrogen Receptor Predictor Project (CERAPP), which includes compounds from Tox21, ToxCast, and the Endocrine Disruptor Screening Project (EDSP). This set of active pharmaceutical ingredients were then compared with a subset of CERAPP compounds that were identified as being pharmaceuticals, near-field, or far-field environmental chemicals, as defined by the USEPA Aggregated Computational Toxicology Resources (ACToR) database. CERAPP compounds were ranked by their chemical similarity to the top pharmaceutical actives, using physicochemical properties and ToxPrint chemotypes as descriptors. The resulting Jaccard-Tanimoto indices show that environmental compounds that were classified as pharmaceutical ingredients per ACToR are similar to top pharmaceutical ingredients. High throughput risk prioritization, via the incorporation of available hazard and exposure predictions, is then employed as a proof of concept for identifying the next opportunities for in vitro methods development.

Keywords: A-chemical prioritization

WE-SY-F2-471

QSAR-based Methods and Tools to Support HTP Risk Assessment

A. Tropsha; UNC-Chapel Hill, Chapel Hill, NC

Abstract: QSAR modeling is the major cheminformatics approach to predicting chemical compounds' biological and toxicological activities or properties. In silico modeling could provide a highly desirable alternative to animal testing via cost-efficient methodologies to screen chemicals that are potentially unsafe for the next tier of chemical testing. We will present and illustrate our approaches to toxicity and property prediction contrasting and combining statistical QSAR modeling and chemical alert based read-across methods using our recent studies on skin sensitization and skin permeability. Indeed, structural alerts are widely accepted in chemical toxicology and regulatory decision support as a simple and transparent means to flag potential chemical hazards or group compounds into categories for read-across. However, there has been a growing concern that alerts disproportionally flag too many chemicals as toxic, which questions their reliability as toxicity (or activity) markers. Conversely, the rigorously developed and properly validated statistical QSAR models can accurately and reliably predict the chemical bioactivity; however, their use in regulatory decision support has been hampered by the lack of transparency and interpretability. We demonstrate that contrary to the common perception of QSAR models as "black boxes" they can be used to identify statistically significant chemical substructures (QSAR-based alerts) that influence toxicity. We show through several case studies, specially, skin sensitization, that the mere presence of structural alerts in a chemical, irrespective of the derivation method (expert-based or QSAR-based), should be perceived only as hypotheses of possible toxicological effect but cannot be blindly relied on for reaching conclusions about chemical safety or lack thereof. We will propose and discuss a new approach that synergistically integrates structural alerts and rigorously validated QSAR models for a more transparent and accurate safety assessment of new chemicals.

Keywords: A-chemical prioritization, A-risk assessment, A-statistical methods, cheminformatics

WE-SY-G2: Exposure to Environmental Contaminants in Diverse Communities

WE-SY-G2-472

Quantification of Diné activity patterns with the San Juan River in the wake of the Gold King Mine Spill

P. I. Beamer[1], Y. Ornelas Van Horne[1], K. Chief[2], M. Begay[3], N. Lothrop[1], N. Teufel-Shone[4], M. Begay[5]; [1]The University of Arizona, Tucson, AZ, [2]The University of Arizona, Tucson, AZ, [3]Navajo Nation Department of Health, Windowrock, AZ, [4]The University of Arizona, Tucson, AZ, [5]Northern Arizona University, Flagstaff, AZ

Abstract: In August 2015, 3 million gallons of acid mine drainage was released from the Gold King Mine, eventually reaching the San Juan River. The river was reopened following a recreational risk assessment. However, this does not reflect the reality of the Diné (Navajo), who have a deep spiritual connection to the natural environment and rely heavily on the San Juan River for agricultural, spiritual and cultural practices. Investigating the impact of the Spill on the Diné's interaction with the River is crucial to understating the potential long-term health impacts. No activity pattern data exist for Diné that could be used to conduct a comprehensive residential risk assessment. Focus groups were used to identify 42 unique activities between the Diné and the River. The activities were grouped into one of four distinct categories: livelihood, recreational, spiritual and ceremonial, and arts and crafts activities. The activities were included in a questionnaire to collect pre- and post- Spill frequency and duration activity data. Within one year of the Spill, Navajo Community Health Representatives administered the questionnaire to adults living on the Navajo Nation. The 59 adults reported engaging in a combined total of 409 activities, with each individual engaging in an average of 7 (Range: 0- 29) activities before the Spill. After the Spill the 59 individuals reported engaging in a combined total of 177 activities, with each individual engaging in an average of 3 (Range: 0-23) activities. There was a 48.5 % decrease in livelihood activities, a 64.7% decrease in recreational activities, a 64.7% decrease in spiritual and ceremonial activities, and a 13.4% decrease in arts and crafts activities. Whether or not environmental contamination from the Spill persist in the River, the Spill has clearly impacted the Diné interactions with the River. There is likely to be long-term social-mental impacts from the trauma of the Spill.

Keywords: A-exposure factors, A-environmental justice, B-metals, C-water

WE-SY-G2-473

Exposure from solid fuel use: Case study of women in rural households of North India

D. S. Jyethi; Indian Statistical Institute, Tezpur, India

Abstract: Polycyclic aromatic hydrocarbons (PAHs) are by-products of incomplete combustion. Monitoring PAHs exposure in the environment can be useful in assessing health risks as several PAHs are carcinogenic in human beings. PAHs can be used as markers of sources. This study reports the profiles and source analysis of 16 US EPA priority polycyclic aromatic hydrocarbons (PAHs) associated with particulate matter at a rural site in North India. Weekly sampling was carried out at the location for 1 year. The annual mean particulate matter level was found to be ~9 times the World Health Organization limit. Seasonal variation of PAHs (range 37.2-74.0 ng m^{-3}) was significant with winter values being 72% and 68% higher than summer and monsoon respectively. Principal component analysis coupled with multiple linear regression identified wood and biomass combustion to contribute 25.4% and coal combustion contributed 9.6% to the observed PAHs. Domestic emissions from rural households using wood, biomass and charcoal for cooking in the vicinity of the site appeared to have significantly affected its air quality. A substantial portion (~55%) of the particulate PAH load was comprised of carcinogenic species, which yielded a considerably high lifetime inhalation cancer risk estimate (8.7E−04) in women.

Keywords: A - ambient monitoring, A-risk assessment, B-POPs, B-particulate matter, D-community

WE-SY-G2-474

A Pilot Study on Migrant Grape Workers Exposure to Pesticides in Sonora, Mexico

N. I. Lopez-Galvez[1], R. Wagoner[1], P. I. Beamer[1], J. de Zapien[1], C. Rosales[1], I. Ortega[2], P. Aranda[3]; [1]University of Arizona, Tucson, AZ, [2]Centro de Investigacion en Alimentacion y Desarollo, Hermosillo, Mexico, [3]Collegio de Sonora, Hermosillo, Mexico

Abstract: Expanding agribusiness in Sonora, a state in Northern Mexico known for its cattle, tomato, and table grape production, has increased the demand for temporary migrant agricultural workers from the poorest regions of Southern Mexico. These migrant agricultural workers participate in strenuous tasks while exposed to a wide variety of occupational risks and hazards including pesticides. A cross-sectional observational study was conducted to assess the pesticide exposure of migrant grape workers. This is the first study to characterize exposure of migrant grape workers to pesticides in this region. A convenience sample of 20 participants were recruited from a large commercial grape farm employing approximately 2,000 workers during the harvest season near Hermosillo- Sonora, Mexico. A questionnaire was used to obtain information on working activities and demographics. Morning void urine samples were collected to assess pesticide exposure. Most participants were originally from the state of Chiapas-Mexico, none had completed high school, and half spoke an indigenous language. Pyrethroid and organophosphate urine metabolites were detected in the majority of workers. The creatinine-adjusted concentration for cyfluthrin, chlorpyrifos, and parathion metabolites in urine obtained in this study (geometric means: 0.942 µg/g, 3.559 µg/g and 1.630 µg/g, respectively) were higher than in the Mexican American population included in NHANES. Unfortunately, there is no study of the general population in Mexico for comparison. Our results suggest that grape workers in this region are exposed to high levels of pesticides, which may, eventually, affect their health. Additional research is needed to confirm these findings and to evaluate health outcomes associated to pesticide exposure in this region. Results from this pilot study can be used to conduct a larger pesticide study, create binational partnerships between researchers, and develop occupational health training resources in Mexico.

Keywords: A-industrial hygiene, B-pesticides, A - exposure measurement, B-pesticides, A-environmental justice

WE-PL-A3: Air Pollution Geospatial Models

WE-PL-A3-475

Evaluation Of The Transferability Of Resolved Vs Unresolved Land Use Regression Models For Traffic-Related Air Pollutants

K. K. Shairsingh, C. Jeong, G. Evans; University of Tornto, Toronto, Canada

Abstract: Land use regression (LUR) models are statistical models used to predict intra-urban variability of air pollution and are frequently employed in mapping human exposure to traffic-related air pollutants. While LUR models developed for a specific city should be able to predict concentrations in other cities with similar infrastructure, land use, topography and climate; this transferability of models generally has been poor. In this study, we resolved mobile monitored high-resolution concentrations of black carbon (BC), ultrafine particles (UFP), nitric oxide (NO) and nitrogen dioxide (NO_2) into local and background signals, using time averaged minimum values, to investigate whether separation of the ambient measurements will improve LUR transferability. LUR models based on the unresolved (total) and resolved (local and background) input data were then developed for Toronto, Canada. These unresolved and resolved models were transferred to cities outside of the model domain, and their predictive performances were evaluated. Our results showed that resolved models moderately improved the transferability of UFP and NO_2. The resolved and unresolved models displayed R^2 of 0.54 and 0.54 for UFP and 0.60 and 0.62 for NO_2, respectively. When transferred to cities outside of the model domain, the local models displayed the lowest reduction in R^2 for all pollutants, while background models transferred better than total models for UFP and NO_2. This resulted in better transferability of the resolved models for NO_2 (R^2=0.36) and UFP (R^2=0.3) when compared to their unresolved models (NO_2:R^2=0.26, UFP: R^2=0.22).

Keywords: A - ambient monitoring, A-exposure models, A-geospatial analysis/GIS

WE-PL-A3-476

Improving ambient ultrafine particle concentration predictions at residences by adding central-site monitoring to a mobile monitoring campaign

M. C. Simon[1], A. P. Patton[2], D. Brugge[3], J. L. Durant[1]; [1]Tufts University, Medford, MA, [2]Health Effects Institute, Boston, MA, [3]Tufts University, Boston, MA

Abstract: Aim: Accurate exposure estimates of traffic-related ultrafine particles (UFP; <100 nm diameter) are needed to reduce exposure misclassification in health studies; however, because UFP concentrations are highly variable in space and time, accurate estimation is a considerable challenge. Our aim was to determine if hourly particle number concentration (PNC; a proxy for UFP) models built from intermittent, continuous mobile-monitoring data could be improved with the addition of data from a central-site monitor. **Methods:** PNC was measured in Boston (Dec 2011-Nov 2013) and Chelsea (Dec 2013-May 2015), Massachusetts (USA) by (1) mobile monitoring on ~45 days (3-6 hours/day), (2) continuous monitoring at central sites, and (3) continuous monitoring outside homes (6 weeks at 9 homes) (per city). Spatial-temporal models were built from the mobile-monitoring-only data set in each city. To incorporate central-site data, we built a time-series model from that data set and a spatial model based on the mobile-monitoring data standardized to the central site. Traffic, meteorology, indicators of secondary particle formation, and land use (e.g., distance from roads) were used as predictors in regression models of log-transformed PNC [ln(PNC)]. Models were evaluated by root-mean-square-error (RMSE) and validated by 10-fold cross-validation, and assessed for their ability to predict ambient concentrations made outside of the 18 homes using Pearson correlations. **Results:** Models showed generally good agreement with measurements, but overestimated PNC at homes 52-83% of hours. The addition of a central monitor increased the mean Pearson correlations between modeled and measured ln(PNC) at individual homes from 0.33 to 0.54 and decreased the mean RMSE by 7%. **Conclusions:** PNC models based on both mobile-monitoring and central-site data improved model predictions and reduced error of estimated ambient PNC at residences in two urban areas compared to models based on mobile-monitoring data alone.

Keywords: A-exposure models, B-particulate matter, C-air, A - ambient monitoring

WE-PL-A3-477

Errors associated with the use of roadside monitoring to study traffic pollutant-related health effects

D. Liang[1], J. L. Moutinho[2], H. Chang[3], R. Golan[4], D. Gao[5], S. E. Sarnat[1], R. Greenwald[6], R. Weber[5], A. Russell[2], J. A. Sarnat[1]; [1]Emory University, Atlanta, GA, [2]Georgia Institute of Technology, Atlanta, GA, [3]Emory University, Atlanta, GA, [4]Ben Gurion University of the Negev, Beer Sheva, Israel, [5]Georgia Institute of Technology, Atlanta, GA, [6]Georgia State University, Atlanta, GA

Abstract: Aim. Near-road monitoring creates opportunities to provide direct measurement of primary traffic emission and their impact on adjacent areas. The question remains, however, whether near-road monitoring accurately represents traffic-related pollution exposures for use in epidemiologic studies. A better understanding of potential exposure measurement error when utilizing near-road measurement is needed for the design and interpretation of the many observational studies linking traffic pollution and adverse health. **Methods.** We conducted an extensive field study to quantify the impact of exposure measurement error in time-series and small panel epidemiologic study designs when using near-road measurement as a surrogate of exposure to primary traffic emission. We measured several single traffic indicators with high spatial and temporal resolution at six indoor and outdoor sites ranging from 0.01 to 2.3 km away from a major highway artery. We computed exposure errors among five exposure metrics and calculated pollutant-specific effect attenuation and several acute respiratory endpoints. **Results.** Many of the primary pollutant species, including BC, CO, and NO, declined to near background levels by 20 m from the highway source, with relatively weak spatial concentration gradients among the various monitoring sites. We observed a moderate to strong attenuation in estimated pollutant health associations (35%-75%) when using near-road measurement as the proxy for residential location far away from the traffic hotspot. We also found that the magnitude of effect attenuation varied substantially over the course of the day for both time-series and small panel study. **Conclusions.** Use of near-road monitoring to investigate health effects of traffic pollutants may induce substantial measurement error, which should be considered when designing epidemiologic studies for traffic pollution.

Keywords: A - exposure measurement, A-epidemiology, C-air, A - ambient monitoring, A-statistical methods

WE-PL-A3-478

Satellite Predictions of $PM_{10-2.5}$ in Chicago

M. Pedde[1], I. Kloog[3], T. V. Larson[2], S. Adar[1]; [1]University of Michigan, Ann Arbor, MI, [2]University of Washington, Seattle, WA, [3]Ben Gurion University, Be'er Sheva, Israel

Abstract: Introduction: Few epidemiology studies have examined the health impacts of $PM_{10-2.5}$, largely due to an insufficient availability of exposure data (<1% of US communities have monitors). The availability of "big data" from satellites offers new opportunities to assess PM_{10-25} levels across space and time. Our project leverages NASA satellite-derived Aerosol Optical Depth (AOD) measurements to estimate daily concentrations of PM_{10-25} in Chicago for 2000-2009. **Methods:** We used mixed-models with random slopes for day to calibrate daily AOD measures (1 km resolution) from the MODIS satellite with EPA monitored PM_{10} and $PM_{2.5}$ concentrations in Chicago. We then used generalized additive mixed-models with spatial smoothing to predict daily PM_{10} and $PM_{2.5}$ for cells in the 1 x 1 km grid with missing AOD values, using regional measured PM_{10} and $PM_{2.5}$ and AOD measurements in neighboring cells. $PM_{10-2.5}$ concentrations were estimated using the difference of predicted PM_{10} and $PM_{2.5}$ at the same location. Results were externally validated against two-week average PM_{10-25} measurements in the Multi-Ethnic Study of Atherosclerosis and Coarse Airborne Particulate Matter (MESA Coarse) during the winter (33 sites) and summer (31 sites) of 2009. **Results:** Cross-validated results of the $PM_{2.5}$ model calibration had overall R^2=0.82 (spatial R^2=0.87, temporal R^2=0.82) while the PM_{10} model calibration had overall R^2=0.47 (spatial R^2=0.80, temporal R^2=0.43). We observed R^2=0.21 between satellite estimated PM_{10-25} and monitored PM_{10-25} during the winter, and R^2=0.56 during the summer. **Conclusions:** We had strong overall model performance for the satellite-predicted estimates of $PM_{2.5}$ and PM_{10}. We observed good out-of-sample comparisons between the satellite-estimated PM_{10-25} during the summer and moderate comparisons during the winter period. These results suggest that satellite data may be useful to predict PM_{10-25} for epidemiologic studies, though the performance may vary by season.

Keywords: A-exposure models, A-geospatial analysis/GIS, A - ambient monitoring, B-particulate matter, A - exposure measurement

WE-PL-A3-479

Exploring the Spatial Representativeness of NAAQS and Near Roadway Sites Using High-Spatial Resolution Air Pollution Maps Produced by A Mobile Mapping Platform

P. A. Solomon[1], A. K. Singh[2], A. Whitehill[3], S. Kaushik[3], M. M. Lunden[4], B. LaFranchi[4], D. Herzl[4]; [1]US Environmental Protection Agency, Las Vegas, NV, [2]University of Nevada Las Vegas, Las Vegas, NV, [3]US Environmental Protection Agency, Research Triangle Park, NC, [4]Acima, Inc., San Francisco, CA

Abstract: National Ambient Air Quality Standards (NAAQS) sites are located in urban areas to provide an estimate of air pollution at the neighborhood or community scale (1-5 km). EPA's near-roadway sites are designed and located to provide an estimate of pollutant concentrations emitted by motor vehicles on the road. These conventional, fixed-site monitors cannot capture the variability of pollutant concentrations that exist over small spatial scales in urban environments. However, mobile platforms equipped to measure air quality enable pollutant mapping at high spatial resolution that can greatly help to fill these spatial gaps and complement the data provided by stationary air monitoring sites. Understanding variability of pollutants on small spatial scales is important for exposure assessment, air quality model evaluation, and empowering individuals to reduce health risks from air pollution. In the current study, three Google Street View cars were equipped with the Aclima Environmental Intelligence ™ measurement and data acquisition platforms. The air pollutants of interest, including O_3, NO, NO_2, CO_2, black carbon, and particle number in several size ranges, were measured using a suite of fast time-response reference-grade equipment and combined with a data integration system to enable extensive, routine operation. Supplemental measurements that included temperature (T), relative humidity (RH) were also included. These three instrumented automobiles were driven concurrently in a variety of spatiotemporal patterns in Denver, Colorado over 22 days from mid-July to mid-August 2014, including dense driving around the NAAQS and near-roadway sites. Routes were designed to included interconnected and overlapping driving patterns. This presentation will examine pollution concentrations as observed at the stationary sites compared to those measured by the vehicles within a range of up to 2km from the NAAQS site and within a range ~500m of the near-roadway site.

Keywords: A - ambient monitoring, A-geospatial analysis/GIS, A - exposure measurement, A-sampling methods, B-particulate matter

WE-SY-B3: The Pyrethroids: Triangulating Exposure, Toxicology, and Epidemiology Part I

WE-SY-B3-480

Toxicology of the Pyrethroid Insecticides

T. P. Pastoor; Pastoor Science Communications, LLC, Greensboro, NC

Abstract: The principal effect of pyrethroids in insects and mammals involves interfering with the kinetics of voltage-sensitive sodium channel function in neurons. The associated perturbation of neuronal transmission produces transient neurotoxic signs of intoxication. Sufficient concentrations of the parent pyrethroid must be present at the neurological site for effects to manifest. The neurological effects are considered the most sensitive endpoint for human health risk assessment. For example, the USEPA grouped pyrethroids into a common mechanism group for assessment of cumulative risk based on neurological effects. No other effects were considered to be more potent or relevant. Nevertheless, other toxicological effects have been noted at higher doses and in conjunction with neurological effects. This presentation will discuss what was and was not observed in the extensive testing database that supports the registration and re-registration of each pyrethroid. This presentation will set the stage for the session by introducing the audience to the subject of pyrethroid toxicology by describing plausible adverse effects in humans based on animal data.

Keywords: B-pesticides, toxicology, pyrethroids

WE-SY-B3-481

Species and Age-Dependent Differences

T. G. Osimitz[1], L. Sheets[2], M. Creek[3]; [1]Science Strategies, LLC, Charlottesville, VA, [2]Bayer CropScience, Research Triangle Park, NC, [3]Valent U.S.A. LLC, Walnut Creek, CA

Abstract: Young rats are known to be much more sensitive to the acute toxic effects of certain pyrethroids than the adult. The available evidence suggests this differential sensitivity is specific to certain pyrethroids and high dose levels. In keeping with the goals of NAS's Toxicity Testing in the 21st Century (Tox21), we developed strategies to identify the adverse outcome pathway for pyrethroids and to design a multi-year research program to investigate the importance of both pharmacokinetic and pharmacodynamic factors associated with that pathway and their influence on species and age-dependent toxicity. This includes extensive characterization of the metabolism and the age-related ontogenetic differences between humans and rats. Most important, additional data were developed to better understand the relevance of the observations in rats to humans, particularly at real world exposure levels. The results from this research were incorporated into PBPK models for developing and adult rats and humans and to determine whether a safety factor is warranted to protect infants and children.

Keywords: pyrethroids, age-dependent toxicity, metabolism, PBPK models, Tox21

WE-SY-B3-482

Predicting Pyrethroid Internal Exposure in Different Populations Using Physiologically Based Pharmacokinetic Modeling

M. Yoon, G. Song, M. Moreau, P. Mallick, A. Efremenko, H. Clewell; ScitoVation, Research Triangle Park, NC

Abstract: A critical area in understanding the hazard profile of the pyrethroids has been modeling the fate of the pyrethroids in humans and determining the differences in concentrations to different physiological compartments at different stages in development. This effort has been integral in understanding the effects observed in animal toxicology studies and their relevance to humans at various stages of development. Physiologically based pharmacokinetic (PBPK) modeling provides an effective way to integrate three major determinants of internal exposure in different life stages; 1) age-dependent changes in physiology, 2) biochemical maturation processes, particularly metabolism, and 3) age-specific exposure characteristics. Development of a life stage PBPK model for deltamethrin based on a modern parameterization method of in vitro to in vivo extrapolation will be presented. Application of the life stage PBPK modeling to evaluate early life sensitivity to pyrethroid exposure will be discussed using the deltamethrin model as a proof of concept case study. Establishment of an in vitro and in silico-based evaluation strategy to predict internal exposure in conjunction with relevant exposure information in the population of concern in humans is of great importance in pesticide risk assessment for potentially vulnerable populations such as infants and children where the necessary information for risk-based decision making is limited.

Keywords: B-pesticides, D-children, A - population exposure, A-risk assessment, D-susceptible/vulnerable

WE-SY-B3-483

Epidemiology and Pyrethroid Insecticides: We Can Do Better

C. J. Burns; Burns Epidemiology Consulting, Sanford, MI

Abstract: Epidemiology studies published since 2000 specific to pyrethroid insecticide exposure were reviewed. Quality elements of outcome, exposure metric, exposure level, and study design were considered. Most of the pyrethroid epidemiology studies used single samples of pyrethroid urinary metabolites, which can only inform about current exposure. Many studies also relied upon a descriptive cross-sectional design, in which both outcome and exposure data are collected at the same time, preventing the investigator from evaluating the temporality of the exposure-disease relationship. None of the publications were of a quality that would provide robust evidence for a causal association between pyrethroid exposure and an adverse effect. As government regulators and policy makers increasingly look to epidemiology studies on human health, the status quo of the quality of epidemiology studies must change in order to contribute to our understanding of pesticide exposure and health. Opportunities for improvement will be discussed.

Keywords: A-biomonitoring, A-epidemiology, B-pesticides

WE-SY-C3: Biomonitoring in Action - Identifying and Remedying Harmful Chemical Exposures with Innovative Laboratory Tests, Surveillance, and Effective Communication

WE-SY-C3-484

National Trends in Exposure to Environmental Chemicals - An Update

L. Romanoff, K. A. Dortch, M. E. Mortensen; Centers for Disease Control and Prevention, Atlanta, GA

Abstract: Biomonitoring provides important data that helps assess human exposures to environmental chemicals. CDC's National Biomonitoring Program provides nationally representative data on Americans' exposure to environmental chemicals and supports state public health laboratories within the United States to assess human exposure in their communities. The *National Report on Human Exposure to Environmental Chemicals and Updated Tables* provide the most comprehensive assessment of Americans' exposure to environmental chemicals from CDC's National Health and Nutrition Examination Survey (NHANES). The data in the *Report* can be used to track trends over time, provide reference ranges for scientists and public health officials to identify unusual exposures, and assess effectiveness of interventions to reduce exposures. The *Updated Tables, January 2017* provides recent data for more than 300 chemicals. Of these, 96 chemicals have new data since the previous update. The January 2017 release reports 20 new chemicals measured for the first time, including atrazine metabolites, DEET metabolites, triclocarban, PFAS isomers, and six volatile organic compounds. Chemicals measured over several NHANES periods are useful to examine exposure trends over time. Serum PFOS concentrations have decreased more than 85 percent from 1999-2014; for serum PFOA, the decrease was more than 60 percent. The data can also help assess effectiveness of public health intervention. US EPA restricted use of the insecticide chlorpyrifos in 2001. Urinary concentrations of its metabolite, TCPy, have decreased more than 56 percent from 1999-2010. Public concerns about phthalates appears to have led to changes in consumer products and exposures in Americans. Urinary metabolite concentrations of diisononyl phthalate (DiNP) have increased more than 2.5 times from 2005-2012, while oxidative metabolites of di (2-ethylhexyl) phthalate (DEHP) have decreased more than 60 percent. CDC's continued assessment of environmental chemicals through biomonitoring will help set priorities for reducing exposure and focus research on chemicals of concern.

Keywords: A-biomonitoring

WE-SY-C3-485

Developing a Critical Laboratory Method for Assessing Arsenic Exposure from Private Wells

J. Schneider, C. Bean; New Hampshire Public Health Laboratories, Concord, NH

Abstract: Biomonitoring New Hampshire developed a targeted investigational study to determine the prevalence of elevated urinary arsenic and uranium concentrations in residents living in selected, high-risk areas who rely on private wells for drinking water. Arsenic concentrations above the limit for safe drinking water of 10 micrograms per liter affect twenty to thirty percent of all private bedrock wells in New Hampshire. Biomonitoring New Hampshire validated an analytical method for measuring urinary arsenic species to determine the amount of exposure to inorganic arsenic. Study participants with elevated levels of total arsenic in urine were tested for a panel of six arsenic species that have varying degrees of toxicity. The method uses high performance liquid chromatography-inductively coupled plasma-mass spectrometry (HPLC-ICP-MS) to separate and quantitate four inorganic-related arsenic species and two organic arsenic species. The method validation procedure followed guidelines to demonstrate accuracy, precision, linearity, and selectivity over twenty independent analytical runs. Matrix-matched quality control materials were prepared in-house and characterized during the validation. The method was validated in the linear range of 2 - 150 micrograms per liter; accuracy was within 20 percent recovery and precision was less than 10 percent relative standard deviation for each species spiked at a low level and high level. The laboratory results from urinary arsenic speciation testing provide a more complete picture for this targeted study because they can capture important information about the source in addition to the level of exposure.

Keywords: A-biomonitoring, A-analytical methods

WE-SY-C3-486

Utilizing Existing Public Health Capacities to Target Biomonitoring Investigations

S. Chaudhuri; Utah Public Health Laboratory, Taylorsville, UT

Abstract: State public health laboratories and epidemiologists in Utah, New Mexico, Colorado and Arizona formed a consortium to apply existing resources to address public health problems common to the four states region. The goal of the consortium is to collaborate among four states to develop new and expand laboratory based biomonitoring programs in this region and to assess the extent and nature of human exposure to environmental toxicants. The key outcomes include reduction or exposure elimination to environmental toxicants among the consortium residents. By leveraging resources from Environmental Public Health Tracking Networks (EPHTN), Public Health Emergency Preparedness (PHEP) programs and other health partners, the consortium was able to conduct a needs assessment to identify environmental chemicals of concern in the states'. The consortium developed three studies to investigate exposure and among the state's shared regional geophysical, cultural, economic, industrial, agricultural, and political environment. The Private Well Drinking Water and Metal's contamination study assess exposure for heavy metals of concern in well water. The Exposure of Four Corners States Residents to Pesticides and Consumer Products study assess exposure of herbicides, pesticides, and phthalates common in households in the consortium. The San Luis Valley Children's Study assess exposure of hazardous chemicals in children ages 3 to 13 years old. Each state develops laboratory capacity for at least one of the studies chemicals, spreading workload among four states while building capacity for the region. Routine conference calls, face to face meetings, and document sharing systems allow states to share challenges, successes, evaluate activities, help each other and make plans for future activities to build capacity and capability for biomonitoring. The Four Corners States Consortium's other successes include increased collaboration among states and development of communication resources for data reporting to study participants.

Keywords: A-biomonitoring, A - population exposure

WE-SY-C3-487

Communicating Biomonitoring Results to Promote Environmental Health Interventions

M. Nascarella; Commonwealth of Massachusetts, Boston, MA

Abstract: Communicating with the public on exposure to environmental chemicals is an important, technical, and complicated task. Communicating the scientific uncertainty associated with individual level biomonitoring results can be especially challenging. The Massachusetts Department of Public Health has developed and implemented a state-wide effort known as the Biomonitoring Massachusetts Study, focused on the surveillance of environmental chemical exposure. This effort includes the reporting-back of results to all study participants. The communication material that was developed for this effort draws from previous experience in communicating episodic results, refined through consultation with technical experts and community feedback in multiple focus group evaluations.

Keywords: A-biomonitoring, B-metals, D-community, A-emergency response, D-environmental justice

WE-SY-C3-488

Implementing Biomonitoring Studies to Assess Chemical Exposures in New York Communities

K. M. Aldous; Wadsworth Center, Albany, NY

Abstract: Public drinking water supplies were recently identified as sources of exposure to emerging industrial chemicals not targeted during routine water quality testing. Perfluoroalkyl substances (PFAS) have been the focus of a national drinking water supply study by EPA's Unregulated Contaminant Monitoring Rule (UCMR3). Public awareness and concern over possible exposures has resulted in efforts to eliminate sources from everyday items and firefighting foams. On January 27, 2016, NY Governor Andrew M. Cuomo announced actions to address PFAS contamination found in the Village of Hoosick Falls' water supply that served 3,600 residents. NY State Health Department began the task of collecting blood samples as part of a voluntary perfluorooctanoic acid (PFOA) biomonitoring program for all who live in or near Hoosick Falls. By leveraging the resources from the Public Health Emergency Preparedness (PHEP) Lab Response Network (LRN) program, the Wadsworth Center was able to develop a high throughput testing procedure for this large scale biomonitoring study and assess this and other community exposure from contaminated drinking water. The method developed for determination of perfluoroalkyl substances (PFAS) from human serum utilized liquid chromatography/tandem mass spectrometry (LC/MS/MS) and 96 well plate high-throughput sample preparation. The standard operating procedure was written, validated and CLIA approved. Over 3,000 serum samples were analyzed and reported in less than 3 months. The PFOA testing method was expanded to report the same 11 perfluoroalkyl compounds listed in the National Health and Nutrition Examination Survey (NHANES) that provides comparison levels for the general population. Improvements in logistics for serum sample collection, data transmission, sample shipment and reporting needed for a larger community (28,000 residents) will be discussed. Serum PFAS level found in residents from Hoosick Falls and other communities in New York will be presented.

Keywords: A-biomonitoring, A-analytical methods, A - exposure measurement, C-water

WE-SY-D3: Social Determinants of Health, Environmental Exposures, and Disproportionately Impacted Communities: What We Know and How We Tell Others- Part 3

WE-SY-D3-489

Modeling disparities in ambient air pollution exposure and residential air exchange rates across Massachusetts using publicly-available data

A. Rosofsky[1], J. I. Levy[1], A. Zanobetti[2], P. Fabian[1]; [1]Boston University School of Public Health, Boston, MA, [2]Harvard T.H. Chan School of Public Health, Boston, MA

Abstract: Housing is an important social determinant of health. Individual housing characteristics can modify outdoor ambient air pollution infiltration into the home environment through air exchange rate (AER). Due to time and labor-intensive methods needed to measure AER, few studies have characterized AER distributions and the implications for exposure disparities across large geographic areas. Using publicly available housing data and regression models associating AER with housing characteristics, we estimated AER for all Massachusetts residential parcels. We then conducted a disparities analysis of temporally- and spatially-resolved ambient $PM_{2.5}$ exposures, and residential AERs at an address-level. Across the state, mean ambient $PM_{2.5}$ (μg/m^3) concentrations in 2010 were 6.7 (range: 0.01-85.1). Median AERs (h^{-1}) with closed windows for winter and summer were 0.44 (IQR: 0.27-.65) and 0.26 (IQR: 0.16-.38), respectively. Overall, duplex and triplex homes, small apartment buildings and large apartment buildings had 2.3, 2.8, and 2.0 times higher median AERs than single family homes, respectively. Housing parcels in the 90th percentile distribution of both AER and $PM_{2.5}$ (i.e. the leakiest homes in areas of highest ambient air pollution) - versus the 10th percentile - were located in neighborhoods with higher proportions of ethnic/racial minorities (Hispanic 19.2% vs 2.1%), households with an annual income of less than $20,000 (26.0% vs. 7.7%) and low educational attainment populations (23.4% vs. 5.9% with less than a high school degree). We demonstrated a novel application of empirical AER models with high-resolution data. This approach can be applied in epidemiological studies to develop potential exposure modifiers, or to characterize exposure inequalities that are not solely based on ambient concentrations. This work additionally highlights the importance of considering both neighborhood- and housing-level factors as drivers of inequitable ambient air pollutant exposure.

Keywords: A-environmental justice, A-exposure models, A-built environment, C-air, D-community

WE-SY-D3-490

Community-Level Characteristics and Environmental Factors of Child Respiratory Illnesses in Southern Arizona

N. Lothrop, K. Hussaini, D. Billheimer, P. I. Beamer; University of Arizona, Tucson, AZ

Abstract: Lower respiratory illnesses (LRIs) and asthma are common diseases in children under 5 years of age. They are particularly susceptible to factors associated with their home, where they spend most of their time. No studies have analyzed the associations between multiple social and environmental risk factors and these disease outcomes in this population at the neighborhood level. We modeled relationships between emergency department visits and hospitalization rates for asthma and LRIs for these children and geographic factors, including socio-economic and housing characteristics, ambient air pollution levels, and population density in Maricopa and Pima Counties, Arizona, from 2005-2009. We used a generalized linear model with a negative binomial observation distribution and an offset for the population of children under 5 years in each census tract. Predictors exhibiting moderate-to-high correlations were combined into unitless indices using principal components analysis. Spatial autocorrelation among regression model residuals was assessed with the Global Moran's I test. Lower socio-economic status (SES) and reduced population density were associated with asthma hospitalization rates and both LRI outcomes (p values <0.001). Pima County residence was associated with lower asthma and LRI hospitalization rates (p values <0.001). Our study revealed complex, multi-factorial associations between community-level factors prevalence of asthma and LRIs. Rural areas with lower SES in Maricopa County require further investigation. Differences between counties in factors not investigated here (e.g., maternal characteristics, agricultural land use) need more study. By better understanding these relationships, more focused community-level, public health interventions could be developed to reduce and better control these diseases in very young children, who are more physiologically vulnerable.

Keywords: asthma, lower respiratory illness, children, neighborhood, socio-economic status, housing, air pollution

WE-SY-D3-491

Environmental Toxic Metals Exposures Across Disproportionately Impacted Communities

S. F. Farzan; University of Southern California, Los Angeles, CA

Abstract: Exposures to toxic metals, such as lead, arsenic, manganese, and cadmium, continue to be of public health concern. Potential sources of multiple metal exposures including food, drinking water, and industrial contamination have been described among low-income communities of color, as well as Native American tribal communities, but to date, these exposures have not been well characterized in these populations. Our work has begun to characterize metals exposures among participants in a primarily low-income Latina pregnancy cohort, the Maternal and Developmental Risks from Environmental and Social stressors (MADRES) Study. We measured 33 elements in 1st trimester hair samples collected from a subset of 41 MADRES participants, and observed detectable levels of 30 of these elements in >50% of the participants. Arsenic, manganese, and lead are contaminants of particular concern, and these three elements were detectable in 100% of tested participants. The average (range) exposures for arsenic, manganese, and lead were: 0.072 (0.012-1.281), 0.231 (0.053-2.439), and 0.766 (0.116, 6.499) micrograms per gram of hair, respectively. Ongoing analyses of metals in urine samples will provide information about more recent exposure patterns. Additionally we will present cross-cutting analyses of metals exposures across disproportionately impacted communities represented in the Environmental Health Disparities Centers. This approach will help to improve our understanding of the contribution of these harmful exposures to environmental health inequities.

Keywords: A - population exposure, A-biomarkers, A-epidemiology, B-metals

WE-SY-D3-492

Health and Well-being Impact of Contamination on Traditional Food on Navajo

J. C. Ingram[1], T. C. Rock[1], A. C. Lister[1], M. C. Lerma[2]; [1]Northern Arizona University, Flagstaff, AZ, [2]Northern Arizona University, Flagstaff, AZ

Abstract: The project goal is to determine health risks and community impacts (e.g., consumption issues, threats to cultural values, spiritual concerns, and public health impacts) from exposure to environmental uranium and arsenic contamination of traditional food and water for Navajo communities. The approach for this project is to utilize a community-engaged research model to explore exposure pathways and to identify culturally applicable and community-constructed models for mitigation of the impact of the identified chemical contaminants on Navajo communities. The Specific Aims of the project are to: 1) Characterize the extent of metal contamination in culturally significant food and water sources for two Navajo communities; 2) Model dietary metal exposure and utilize the Indigenous Health Indicator to assess health impact; and 3) Develop a Community-Participatory Multi-Level Policy Intervention Model. The policy work is being done in consultation with the Navajo Traditional Knowledge Holders to best shape policies around traditional food consumption.

Keywords: B-metals, A-exposure factors, C-food, C-water, D-Southwest-specific

WE-SY-E3: Per- and polyfluorinated substances (PFAS) in drinking water: Where exposure assessment, epidemiology, and communities meet

WE-SY-E3-494

PFAS in serum: national trends and interpretation

M. Mortensen; NCEH, Atlanta, GA

Abstract: Serum PFAS concentrations for individuals 12 years of age and older can be compared to U.S. population results from CDC's National Health and Nutrition Examination Survey (NHANES). Serum PFAS have been measured in NHANES since 1999. As part of the ongoing NHANES, serum PFAS are measured in a one third sample of participants, ages 12 and older. Population-based reference values are available by age group (12-19 years, 20+ years), sex, and race/ethnicity (non-Hispanic black, non-Hispanic white, Mexican American). Beginning in 2011, the racial/ethnic groups of Asian (e.g., non-Hispanic Asian) and all Hispanics were added. Typically, the 95th percentile is used as the upper end of the reference range for the U.S. population. For children younger than 12 years, no national reference values exist. This presentation will highlight observations in national PFAS serum trends and provide examples of how community-specific data can be compared with national results. Limitations of both individual and group (community) serum PFAS comparisons with NHANES data will be discussed.

Keywords: A-biomonitoring, A-biomarkers, A - exposure measurement

WE-SY-E3-495

PER- AND POLYFLUOROALKYL SUBSTANCE (PFAS) SITE CATEGORIZATION AND RECOMMENDATIONS

S. Moore, R. R. Worley, B. A. Anderson; ATSDR, Atlanta, GA

Abstract: Exposure to per- and polyfluoroalkyl substances (PFAS) is widespread and many of our local, tribal state, and federal partners may need help in addressing concerns over PFAS-contaminated water supplies. This tool should be used to categorize PFAS sites in order to determine recommendations for public health action. The algorithm is designed to ensure that responses at PFAS sites are consistent and result in the necessary action to mitigate exposure and prevent potential adverse health outcomes.

Keywords: A-risk assessment

WE-SY-E3-496

Pharmacokinetic modeling--exploring PFOA drinking water data and serum PFOA levels

R. R. Worley[2], X. Yang[1], J. Fisher[1]; [1]FDA/NCTR, Jefferson, AR, [2]ATSDR/DCHI, Atlanta, GA

Abstract: An individual's serum PFOA concentration is a product of a complex interaction of individual pharmacokinetic parameters and historical exposures. While drinking water is the most well characterized PFOA exposure pathway, past non-drinking water exposures may be important determinants of current serum PFOA concentrations. A novel model incorporating in vitro and in vivo extrapolation-derived parameters into a PBPK model successfully describes human PFOA serum kinetics. This model provides new information to improve risk assessment of human exposure to PFOA and inform decisions about exposure mitigation. Additionally, the model shows that past exposures are important for predicting current PFOA serum concentrations and that measured drinking water concentrations may not be ideal predictors of serum PFOA concentration.

Keywords: A-exposure models, A - exposure measurement, PBPK modeling

WE-SY-E3-497

Working with a community to assess the feasibility of a PFAS health study-an approach for incorporating community input

F. Bove, J. Stephens; ATSDR, Atlanta, GA

Abstract: In February 2016, ATSDR began recruiting community volunteers to advise ATSDR and other agencies in addressing PFAS drinking water contamination, exposure, and possible adverse health outcomes. ATSDR asked community representatives to serve as members of a community assistance panel (CAP). Technical advisors who could assist the CAP members in reviewing the scientific information on PFAS and proposed health activities were also recruited. The purpose of the CAP was to provide a mechanism for the community to participate directly in ATSDR's health activities related to the exposures to the contaminated drinking water. The CAP would provide input at the ground-floor concerning possible health activities proposed by ATSDR. The CAP members would also work with ATSDR to gather and review community health concerns, provide information on how people might have been exposed to hazardous substances, and inform ATSDR how to involve the community. The first public meeting of the CAP occurred in May 2016. Since that time, additional community meetings have occurred; monthly conference calls have been ongoing; and community representatives and ATSDR scientists have worked to determine if a study is feasible that will address some of their concerns. This presentation will outline how the CAP was formed, how concerns were incorporated into the feasibility assessment, and the ups and downs along the way.

Keywords: A - population exposure, community involvement, A-epidemiology

WE-SY-E3-498

Tools for assessing community exposures to PFAS

J. Schier; NCEH, Atlanta, GA

Abstract: Demand for serum testing in communities exposed to PFAS-contaminated drinking water has increased. In response, NCEH/ATSDR developed a set of exposure assessment tools that can be used by health officials if biomonitoring and exposure characterization are needed. The tools available include: Biomonitoring sampling and analysis protocol Laboratory biomonitoring sample collection and analysis protocols Water sampling protocols Exposure and health effects question bank Biomonitoring letters of interpretation, consent, and assent Communication materials Depending on community needs, some or all of the tools can be used to better understand exposures in a community or population, compare trends over time, provide a benchmark for activities that further reduce exposure, and monitor public health interventions.

Keywords: A - population exposure, A-biomonitoring, A-epidemiology

WE-SY-F3: The Tox21 Triangle: Toxicity Testing, Translation, and the Environment

WE-SY-F3-499

Integrative bioinformatics approaches for Tox21 data

D. Reif, M. Kosnik, A. Planchart, C. Mattingly; North Carolina State University, Raleigh, NC

Abstract: Tox21 has capitalized on recent advancements in high-throughput screening (HTS) technologies to generate volumes of toxicity data on thousands of chemicals. While the Tox21 data have been generated primarily using in vitro assays that have been assigned as targeting specific genes, assessing their disease-relevance remains challenging. Application of new bioinformatics approaches can integrate these in vitro data with HTS in vivo data, historical testing information, and public databases to provide additional context to individual assay results. We will detail examples of how integration across data sources can translate between HTS testing data and biomedical knowledge. These examples leverage data from the Comparative Toxicogenomics Database (CTD) and assay information from PubChem's Bioassay Database to map across common assays, genes, and chemicals from Tox21 data. The integrated data can then be assembled into pathways that corroborate (or oppose) screening data results. Applying a network approach to assess connectivity of genes, we can evaluate the disease coverage of HTS assays and even suggest additional assays that would optimize coverage.

Keywords: A-risk assessment, A-statistical methods, A-chemical prioritization

WE-SY-F3-500

Using ToxCast™ data to prioritize chemicals that impact biological processes related to obesity

S. Auerbach; National Toxicology Program at NIEHS, Durham, NC

Abstract: Obesity is a major threat to public health in the United States and abroad. Understanding the role that chemicals in our environment play in the development of obesity is an emerging issue in environmental health, although identifying and prioritizing chemicals for testing beyond those already implicated in the literature is challenging. Our first aim was to develop an approach to perform a screening level hazard assessment related to obesity-related outcomes using the ToxCast™ high-throughput screening data. We identified ToxCast™ assay targets relevant to several biological processes related to obesity (e.g., adipocyte differentiation and feeding behavior) and presented chemical screening data against those assay targets to identify chemicals of potential interest. The results of this screening-level hazard assessment suggest that the spectrum of environmental chemicals to consider in research related to obesity is much broader than indicated by research papers and reviews published in the peer-reviewed literature. Our second aim in the prioritization was to integrate exposure estimates from ExpoCast and the use of in vitro to in vivo extrapolation to perform a screening level margin of exposure (MOE) assessment. The MOE assessment suggested that only on a very limited number of environmental chemicals have high enough exposure to potentially impact biological processes related to obesity in the human population. Overall, the prioritization effort identified a broad array of chemicals that possess properties suggesting they pose an obesity-related hazard. Integration of a MOE assessment further helped refine the prioritization by identifying a small subset of chemicals that may have high enough human exposure that additional targeted testing may be justified.

Keywords: high throughput screening, obesity, diabetes, margin of exposure, prioritization

WE-SY-F3-501

Opportunities and Challenges in Employing *In Vitro-In Vivo* Extrapolation (IVIVE) to the Tox21 Dataset

B. A. Wetmore; US EPA, Research Triangle Park, NC

Abstract: *In vitro-in vivo* extrapolation (IVIVE), or the process of using *in vitro* data to predict *in vivo* phenomena, provides key opportunities to bridge the disconnect between high-throughput screening data and real-world human exposures and potential health effects. Strategies utilizing a combination of experimental and computational tools to predict chemical and drug toxicokinetics (TK) have made significant progress, as have evaluations of the uncertainty and variability of these predictions. A limited understanding of the linkage between *in vitro* assay bioactivity potencies, target site concentrations and potential downstream apical outcomes contribute to the lack of certainty in the assessment of *in vitro* toxicodynamics (TD), whether for chemicals or drugs. Data from the Tox21 program, designed to interrogate a discrete set of stress responses pathways but across a large swath of chemical space, can be mined to explore the predictivity of IVIVE in assessing the TK and TD across a broad range of chemicals with varying degrees of data availability. This talk will describe recent IVIVE efforts on this dataset, outlining both the opportunities and challenges that have been identified.

Keywords: A-chemical prioritization, A-exposure models, toxicokinetics

WE-SY-F3-502

Using High-throughput Screening Data to Guide Exposome Research

J. R. Sobus[1], A. Williams[2], A. McEachran[3], J. Grossman[5], L. Sarah[3], A. Marcotte[3], M. Russell[4], J. McCord[3], D. Mills[5], S. Newton[1], E. M. Ulrich[1], M. Stryar[1], J. Wambaugh[2], K. Isaacs[1], K. Phillips[1], G. Patlewicz[2], K. Mansouri[3], A. Richard[2], C. Grulke[2]; [1]US EPA, Research Triangle Park, NC, [2]US EPA, Research Triangle Park, NC, [3]US EPA, Research Triangle Park, NC, [4]US EPA, Research Triangle Park, NC, [5]US EPA, Research Triangle Park, NC

Abstract: Non-targeted analysis (NTA) methods serve as primary data collection tools for human exposome studies. Many NTA workflows utilize high-resolution mass spectrometry (HRMS), and proceed by first identifying thousands of features (i.e., unknown compounds identified by accurate mass, retention time, and mass spectra) in a sample set, and then proposing formulae, and ultimately structures for high-interest features. The process of correctly assigning structures to unknown features is challenging and cumbersome. As such, efforts should be focused on features that are most relevant from a human health perspective. The US EPA CompTox Chemistry Dashboard provides a web interface and public access point for high-quality chemistry, exposure, and bioactivity data. It also serves as a critical component of NTA workflows, and supports investigations into chemicals within consumer products, human biological media, and a variety of environmental media. Our workflow: 1) retrieves structures associated with masses/formulae detected in samples (based on HRMS); 2) identifies "probable" structures using a variety of diagnostic criteria; and 3) prioritizes compounds that are of highest interest for follow-up targeted analysis. Prioritization of high-interest compounds relies heavily on the availability of bioactivity data from high-throughput screening assays, such as those data generated by the Federal Tox21 consortium. In instances where these data are not yet available, QSAR and chemical read-across approaches are considered for candidate prioritization. This presentation will describe methods and tools being used by EPA to rapidly identify, prioritize, and quantify novel compounds in high-interest environmental and biological samples. Data generated from these tools will be used to prioritize parent compounds, metabolites, and chemical mixtures for future bioactivity screening and risk evaluation. *This abstract does not represent EPA policy.*

Keywords: A - exposure measurement, A-analytical methods, A-chemical prioritization, suspect screening

WE-SY-F3-503

Translating Tox21 Data to Risk Prioritization and Risk Assessment

C. L. Ring[1], J. Rager[1], C. Thompson[1], M. A. Harris[1], R. Pearce[2], R. W. Setzer[2], B. A. Wetmore[3], J. Wambaugh[2]; [1]ToxStrategies, Inc., Austin, TX, [2]US Environmental Protection Agency, Research Triangle Park, NC, [3]US Environmental Protection Agency, Research Triangle Park, NC

Abstract: Approaches to incorporate Tox21 data in risk-based chemical prioritization and decision making are currently under development, with case studies needed to evaluate methods and overall ranges of applicability. We present an example of a chemical risk prioritization approach that incorporated Tox21 data to identify chemical concentrations associated with *in vitro* bioactivity, which were then extrapolated to doses that produced equivalent concentrations in the body using a reverse TK approach. Generic TK models were parameterized using *in vitro* TK data and incorporated inter-individual variability through physiological parameters simulated for populations based on CDC NHANES data. For risk prioritization, equivalent doses were compared to population estimates of exposure rates inferred from NHANES urinary analyte biomonitoring data for 50 parent compounds. This approach allowed targeted risk prioritization for potentially sensitive sub-populations, including children 6-11 and people over 65. In an example of translating Tox21 data for risk assessment, *in vitro* and *in vivo* data were compared with respect to evidence for genotoxic versus non-genotoxic modes of action (MOAs) for hexavalent chromium (Cr[VI]). Similarities and differences between Tox21 *in vitro* data and *in vivo* data from the mouse small intestine (a sensitive target tissue) were observed. Specifically, genotoxicity appeared to predominate *in vitro* but not in the intestine. These findings have implications for chemical safety criteria, which are derived using different low-dose extrapolation methods for chemicals with genotoxic versus non-genotoxic MOAs, highlighting the need to define potential ranges of applicability for the use of Tox21 data in regulatory decisions. Together, these case studies support the utility of Tox21 data in chemical risk prioritization and lay the groundwork for future studies to assess ranges of applicability for incorporation into overall chemical risk assessment.

Keywords: A-chemical prioritization, A-risk assessment, high-throughput screening, in vitro-in vivo extrapolation, high-throughput screening, in vitro-in vivo extrapolation

WE-SY-G3: ZIKA VIRUS SECONDARY EXPOSURES, IMPACTS, and GUIDANCE

WE-SY-G3-504

ZIKA VIRUS SECONDARY EXPOSURES, IMPACTS, and GUIDANCE

F. Gould; NC State University, Raleigh, NC

Abstract: Biologists are on a steep learning curve when it comes to Zika. A couple of years ago we had little knowledge of either its within host behavior or its modes of transmission. From an entomological perspective there are many similarities to Dengue virus which is vectored by the mosquito, *Aedes* *aegypti*, that appear to also be the major vector of Zika. I will present an overview of what we now know about this virus/mosquito pair and the challenges that face us in developing control strategies. I will also take a look forward at novel genetic strategies for either suppressing the populations of the mosquito vectors or decreasing the ability of female mosquitoes to transmit the disease.

Keywords: A-Infectious disease, zika virus; mosquito vectors

WE-SY-G3-505

Congenital Zika syndrome in humans and non-human primate models

S. Permar; Duke University School of Medicine, Durham, NC

Abstract: Maternal infection with Zika virus (ZIKV) has been established to lead to congenital fetal anomalies, including microcephaly, sensorineural defects, motor and cognitive delays, and epilepsy, termed congenital Zika syndrome. To model fetal outcomes in nonhuman primates for studies of pathogenesis and prevention of disease, Asian-lineage ZIKV has been subcutaneously administered to pregnant rhesus macaques. While non-pregnant animals routinely clear viremia within 10-12 days, interestingly, maternal viremia was prolonged in 3 of 4 pregnancies. Of note, fetal head growth velocity in the last month of gestation determined by ultrasound assessment of head circumference was decreased in comparison with biparietal diameter and femur length within each fetus, both within normal range. ZIKV RNA was detected in tissues from all four rhesus monkey fetuses at term cesarean section. In all pregnancies, neutrophilic infiltration was present at the maternal-fetal interface (decidua, placenta, fetal membranes), in various fetal tissues, including fetal retina, choroid, and optic nerve. Consistent vertical transmission in a primate model may provide a platform to assess risk factors and test prophylactic and therapeutic interventions aimed at reducing fetal disease from this fetal neuropathogen.

Keywords: D-prenatal, D-neonatal, A-Infectious disease

WE-SY-G3-506

Implications of Congenital Zika Syndrome for Families

D. Bailey; RTI International, Research Triangle Park, NC

Abstract: The term "congenital Zika syndrome" (CZS) reflects the spectrum of symptoms observed in infants exposed in utero to maternal Zika infection. Although most news reports have focused on microcephaly, we now know that CZS has a more complicated picture and more variable set of problems for children, some of which only become apparent later in the first year of life. Parents who have a child with CZS face an immediate set of demands for caregiving and an anticipated lifelong responsibility for a child with a significant disability. This presentation describes the evolving knowledge about CZS and discusses some of the major implications for families.

Keywords: A-Infectious disease

WE-SY-G3-507

Title: Non-vertical transmission of Zika virus: exposure pathways and outcomes

J. Lebov, P. MacDonald; RTI International, Research Triangle Park, NC

Abstract: Zika virus (ZIKV) is transmitted primarily through mosquito bites of the *Aedes* genus, but other transmission routes have been observed, including vertical transmission (mother to fetus), and transmission through sexual contact, blood transfusion, and saliva. Additionally, other exposure pathways have been hypothesized, such as transmission to infants through breastfeeding and airborne transmission, but have not yet been confirmed. Though studies are underway to assess the association between adverse neonatal and pregnancy outcomes and ZIKV infection during pregnancy, little is known about the health impacts resulting from other routes of exposure to ZIKV. This talk will focus on describing the non-vertical transmission of ZIKV and the challenges to distinguishing these exposure routes in the general population. I will also discuss what is known and unknown about the impact that each type of exposure may have on human health outcomes, from an epidemiological perspective.

Keywords: A-Infectious disease, A - population exposure, A-epidemiology, A-global health, Zika virus

WE-SY-G3-508

U.S. Government Roles in the Zika Response

L. A. Colf, L. Walsh; U.S. Department of Health and Human Services, Washington, DC

Abstract: The U.S. government plays an instrumental role in the response to Zika virus. The Office of the Assistant Secretary for Preparedness and Response (ASPR) in the U.S. Department of Health and Human Services (HHS) leads the nation in preventing, preparing for, and responding to the adverse health effects of public health emergencies and disasters and serves as principal advisor to the Secretary on all matters related to the Zika response. Through a coordinating body called the Disaster Leadership Group (DLG), ASPR and HHS support a wide range of activities at federal, state, and local levels, including: surveillance and risk communication; sustained vector control measures; increased access to contraceptives and other physical barriers to infection; improved screening, diagnostic, and treatment services; and highly coordinated care for children and families affected by Zika. The DLG also addresses policy issues, which for Zika include: international sample collection; implementation of widespread domestic blood screening; standardization of clinical screening and laboratory testing protocols; and alignment of case definitions for birth outcomes and suspected, probable, and confirmed Zika cases. HHS also performs a number of operational Zika activities, including biomedical advanced research and development, downstream licensing and clinical trials, and deployment of medical personnel. In this session, we will share a number of previous and ongoing efforts and issues related to mitigating exposure and transmission of Zika virus.

Keywords: A-global health, A-Infectious disease, B-microbial agents, B-pharmaceuticals, D-neonatal

WE-PL-A4: Traffic Related Air Pollution - Part 2

WE-PL-A4-509

Characterizing trends in particulate matter concentration and size fraction distribution across a moderately populated metropolitan area on the coast: A pilot study in Charleston, SC

J. L. Pearce[1], A. Commodore[1], B. Neelon[1], R. Boaz[1], M. Bozigar[1], S. Wilson[2], E. Svendsen[1]; [1]Medical University of South Carolina, Charleston, SC, [2]University of Maryland, Baltimore, MD

Abstract: In this study, we aimed to improve understanding of ambient particulate matter in Charleston, SC, USA before anticipated port expansion by assessing the temporal and spatial distribution of particulate matter levels across multiple size fraction ranges. We monitored ambient air quality at five monitoring sites during February-July, 2016 by collecting real-time data on particulate matter (PM) concentrations at multiple size ranges: PM ≤ 15 μm, PM ≤ 10 μm, PM ≤ 4 μm, PM ≤ 2.5 μm, and PM ≤ 1 μm using light scattering laser photometers. Generalized additive model results identified that hour of day and location were both significant predictors of variation in hourly Federal Equivalent Method-adjusted PM2.5 after controlling for meteorology, instrument and temporal bias. Temporal peaks for PM2.5 were found to occur during the early morning hours (6-8am) and the marginal means for four of our inland sites were significantly different (higher) than a waterfront site. Self-organizing maps results identified six hourly size fraction compositions, ranging from hours when all size fractions were relatively low, to hours dominated by single size fractions (e.g., PM1), to hours when multiple size fractions were relatively high (e.g., PM15-10 and PM10-PM2.5). Frequency and duration distributions show marked variability in the occurrence and persistence of each 'hourly' type. Our findings illustrate the differential behavior of particulate matter over time and space across a relatively small geographic region. In addition, differences in size fraction distributions further suggest that the nature of particulate matter varies at our monitoring sites.

Keywords: A - ambient monitoring, B-particulate matter, A - population exposure, A-environmental justice, B-mixtures

WE-PL-A4-510

Exposure Assessment of $PM_{2.5}$ and Benzene in Bike Lanes in Taipei

C. Wu, T. Wu, S. Ho, C. Chan; National Taiwan University, Taipei, Taiwan

Moved to Tuesday Poster Sessions

WE-PL-A4-511

Incorporating Time Spent at Schools into Estimates of Traffic-Related Air Pollution Exposure

C. Wolfe[1], C. Brokamp[1], J. Burkle[2], G. LeMasters[2], P. Ryan[1]; [1]Cincinnati Children's Hospital Medical Center, Cincinnati, OH, [2]University of Cincinnati, Cincinnati, OH

Abstract: Background: Children spend a substantial amount of time at school, yet most epidemiologic studies investigating the effects of traffic related air pollution (TRAP) characterize exposure based solely on the participants' primary home. The objective of this analysis was to examine whether estimated TRAP exposures at school and estimates that integrate home and school are significantly different than exposures based on the home address alone. **Methods:** Complete address histories, including school locations, were collected for participants in the Cincinnati Childhood Allergy and Air Pollution Study, a longitudinal birth cohort. TRAP exposure was estimated at ages 4 and 7 for each participant's home and school using a validated land use regression model. Time-weighted average TRAP exposure was derived based on the proportion of hours spent at each location. Estimates of TRAP were log-transformed, and paired t-tests were used to evaluate the differences between exposures at home and school, as well as estimates at the home and time-weighted averages. **Results**: A smaller proportion of children reported attending school at age 4 (152, 22%) compared to age 7 (556, 91%), and children spent a correspondingly lower percentage of time at home per week (median: 93% vs 76% at ages 4 and 7, respectively). Estimated TRAP was, on average 6% higher at schools compared to homes at age 4 ($p = 0.02$), but there was no significant difference at age 7. Time-weighted estimates were significantly different from estimates based solely on the home at ages 4 ($p = 0.03$) and 7 ($p < 0.01$). However, time-weighted estimates were only 1% greater on average than exposures derived at the home at both ages. **Conclusion:** Incorporating school locations into time-weighted estimates of TRAP exposure results in small increases in estimated exposure, likely due to the greater proportion of time spent at the home. Future studies will investigate how geographic or socioeconomic characteristics influence these results.

Keywords: A - exposure measurement, A-epidemiology, B-particulate matter, C-air

WE-PL-A4-512

How much does ambient air pollution concentration increase because of private vehicle drop-offs at schools

M. Adams[1], W. Requia[2]; [1]University of Toronto, Mississauga, Canada, [2]McMaster University, Hamilton, Canada

Abstract: A child's exposure to environmental pollutants can have life-long health effects. Thus it is critical to understand the potential exposure pathways. In this paper, we examine the increase in ambient $PM_{2.5}$ concentrations at schools from private vehicle use for dropping children off at school. In North America, students are commonly driven to school in a private vehicle. Additionally, students walk or cycle, or take a school bus. Our results indicated that personal vehicles may increase ambient air pollution concentrations up to 177 μg/m^3 above baseline concentrations at schools. At the drop-off location in front of the school, we found ambient concentration increases of 5 μg/m^3, 10 μg/m^3, 25 μg/m^3 and 50 μg/m^3 during 16.8%, 7.6%, 2.0% and 0.5% of the mornings, respectively. Our vehicle surveys recorded between 23 and 116 personal vehicles at 25 schools, where enrolment ranged from 160 to 765 students. We fit a linear regression model to predict the number of vehicles at schools we did not observe within our study area, which explained 57% of the variation in our surveys. A microsimulation traffic model was created for each of the 86 schools we studied. Outputs from the traffic model were used to determine the emissions generated at each school. $PM_{2.5}$ emissions varied from 0.14 to 6.38 g. Lastly, we dispersed the emissions produced by private vehicles dropping off students, which are emissions generated by unnecessary trips because students further than walking distance are provided transportation by the school board. This research was conducted in a medium-sized North American city and should allow transferability to similar cities. We conclude that the use of private vehicles can significantly increase local concentrations, regardless of background conditions.

Keywords: A - exposure measurement, A - ambient monitoring, B-particulate matter, A-built environment, A-activity patterns

WE-PL-A4-513

Assessment of health effects of particulate pollution among traffic police in Kathmandu, Nepal using biomarkers

K. M. Shakya[2], R. E. Peltier[1]; [1]University of Massachusetts, Amherst, Amherst, MA, [2]Villanova University, Villanova, PA

Abstract: Serum samples collected from traffic police working on roads from six locations in Kathmandu, Nepal were analyzed for biomarkers such as CRP, SAA, ICAM-1, VCAM-1, IFN-γ, IL-1β, IL-2, IL-4, IL-6, IL-8, IL-10, IL-12p70, IL-13, and TNF-α. 81 serum samples were collected from 33 traffic police during the more polluted spring season (February 17 to April 4, 2014) and 62 serum samples were collected from 32 traffic police during relatively cleaner summer-monsoon season (July 20 to August 22, 2014). Samples were analyzed using proinflammatory panel 1 (human) kits and vascular injury 2 panel 2 (human) kits (MSD Multi-Spot Assay System, Meso Scale Discovery, MD, USA). Personal exposure levels of $PM_{2.5}$, black carbon, and chemical composition of $PM_{2.5}$ were also analyzed. $PM_{2.5}$ exposure levels were higher in spring season than in monsoon season. The effects of exposure seem to be more pronounced among women because several biomarkers such as IFN-γ, IL-2, IL-6, IL-10, IL-12, and IL-13 were found to have the highest mean level at the site with all female traffic police though this site did not have the highest $PM_{2.5}$ exposure levels. Association of biomarker levels with particulate pollution and several social variables will be presented.

Keywords: A - exposure measurement, A-biomarkers, B-particulate matter, D-occupational

WE-SY-B4: The Pyrethroids: Triangulating Exposure, Toxicology, and Epidemiology Part II

WE-SY-B4-514

Residential Exposure to Cyfluthrin and Deltamethrin

J. Thomasen; Bayer, RTP, NC

Abstract: Cyfluthrin and Deltamethrin are pyrethroid insecticides widely used in crop protection and to control various household pests. Residential exposure is a key source of exposure in understanding the aggregate exposure of children to pyrethroid insecticide use in a residential setting. The 2012 EPA Residential Exposure Standard Operating Procedures were used as the basis of determining the potential dermal and incidental oral exposures of children aged 1 to 2 years. Exposures were based on passive dosimetry and residue monitoring studies. These assessments incorporated extensive residential use pattern data obtained from a 12 month pesticide use survey developed by the Residential Exposure Joint Venture (REJV). Model inputs and results will be presented.

Keywords: B-pesticides, A-activity patterns, A-exposure models, D-children

WE-SY-B4-515

Aggregate Dietary Exposure to Cyfluthrin and Deltamethrin Using CARES NG

B. M. Young; Bayer CropScience, RTP, NC

Abstract: Human exposure based models are an estimate of the applied dose, however; linking with a physiologically-based pharmacokinetic (PBPK) model allows validation with data from biomonitoring. The Cumulative Aggregate Risk Evaluation System - Next Generation (CARES NG) is a cloud-based probabilistic exposure model based on publicly available data (consumption and residue monitoring) used to estimate aggregate dietary (food and water) exposure to pesticides. The CARES NG model is routinely used to generate longitudinal dietary exposure for a specific population over a defined timeframe (day, year) that relates to the timeframe of hazard endpoint. Current dietary intake data from "What We Eat In America" (WWEIA) records the time of eating/drinking events allowing CARES NG to generate exposures at each consumption occasion. This study focused on cyfluthrin and deltamethrin as representative pyrethroids because of their widespread use and potential exposure. The dietary exposure profiles were estimated based on recent monitoring or field trial data, current percent crop treated, and empirical processing factors. This presentation explores the different longitudinal approaches used by CARES NG to create temporal consumption diary for each individual: repeating diet (2-day food consumption diaries) or matching diet (nearest neighbor). The dietary exposure profiles from CARES NG can be linked to physiologically-based pharmacokinetic (PBPK) models to estimate internal dose and refine risk quantification.

Keywords: A-aggregate exposure, A-exposure models, B-pesticides

WE-SY-B4-516

Cyfluthrin and Deltamethrin Dose Reconstruction Based on Biological Monitoring Studies

J. H. Driver[1], J. Ross[1], C. Lunchick[2]; [1]risksciences.net, LLC, Manassas, VA, [2]Bayer Cropscience, Research Triangle Park, NC

Abstract: Biomonitoring data represent the "gold standard" for exposure assessment. Biological monitoring data available for cyfluthrin and deltamethrin, as exemplary pyrethroids are reviewed, and important considerations are highlighted, e.g., biomarker specificity. Supported by human studies on the metabolism and excretion kinetics of cyfluthrin and deltamethrin biomarkers, product-use scenario-specific and population-based biological monitoring studies results are summarized. Dose reconstruction based on available studies is presented, and the utility for comparison to model-based estimates of dose is discussed. Both cyfluthrin and deltamethrin have unique biomarkers that distinguish them using biomonitoring from any other pyrethroid registered in the U.S. Despite a relatively low limit of quantitation, the extant data indicate that the average exposure to these two pyrethroids is not measureable. As a result it is possible to estimate with some certainty the upper bound of exposure in the population to these two pyrethroids, but not averages. Comparison of the upper bound exposures estimated with biomonitoring with current Standard Operating Procedures (SOPs) of U.S. EPA reveals the overestimation bias inherent in the SOPs.

Keywords: A-biomonitoring, A-aggregate exposure, A - population exposure

WE-SY-B4-517

Aggregate and Cumulative Exposure Source Attribution for Pyrethroids: Consideration of Modeling and Biological Monitoring

R. Reiss[2], J. H. Driver[1], J. Ross[1], B. M. Young[3]; [1]risksciences.net, LLC, Manassas, VA, [2]Exponent, Arlington, VA, [3]Bayer Cropscience, Research Triangle Park, NC

Abstract: Source attribution for potential aggregate and cumulative exposures is important to inform risk management decision making. Evaluation of model-based exposure and dose estimates and source (dietary and non-dietary) attribution hypotheses can be facilitated using available population-based biological monitoring data. Pyrethroid-population-based biological monitoring data and source attribution will be presented. Considerations regarding biomarker interpretation, e.g. environmental levels, will be discussed. The Centers for Disease Control (CDC) has not released representative urine monitoring data for any pyrethroid since 2007-2008 despite the increased use of pyrethroids in and around homes following the elimination of organophosphates in those venues. Concomitantly there has been increased use of pyrethroids to control vectors for Zika, West Nile and Dengue in the U.S. As a result there is interest in cumulative exposure estimates of pyrethroids. For this purpose it is desirable to utilize a non-specific biomarker that captures as many pyrethroids as possible such as 3-phenoxybenzoic acid (3-PBA) which is generated by approximately half of all registered pyrethroids in the U.S.. Since 3-PBA itself forms in the environment and is bioavailable, there is an inherent overestimation bias of using 3-PBA to estimate cumulative pyrethroid exposure. Thus, use of 3-PBA as a pyrethroid biomarker is health protective.

Keywords: A-biomonitoring, A-aggregate exposure, A-cumulative exposure, A-exposure models

WE-SY-B4-518

Importance of Considering Non-Chemical Stressors in Interpreting Pesticide Exposures in Children

N. S. Tulve[1], J. D. Ruiz[2], K. Hibbert[1]; [1]United States Environmental Protection Agency, Research Triangle Park, NC, [2]ORISE, Research Triangle Park, NC

Abstract: Children are exposed to chemical and non-chemical stressors from their total environment, which is comprised of the built, natural, and social environments from places where they spend their time, including home, school, and daycare. Evidence in the literature suggests that the interrelationships between chemical and non-chemical stressors impact their health and well-being in ways that are different from exposure to a single stressor. Our objective was to explore the pesticide literature and evaluate the importance of non-chemical stressors in interpreting young children's exposures to pesticides and neurocognitive health. We identified publications containing pesticide data, information on non-chemical stressors, and children's neurocognitive health. We mined the literature, extracted relevant information, created a database, and conducted statistical analyses. We organized the information into stressors from the built, natural, and social environments. Various pesticides (e.g., mirex, chlorpyrifos, DDT) or pesticide metabolites (e.g., dialkyl phosphates) have been studied to understand the relationship between exposure and children's neurocognitive health. Our preliminary results showed inconsistent associations between pesticide exposure and children's neurocognitive health, suggesting that chemical exposures alone may not always be able to explain observed health effects. The complexities of multiple stressors suggest that the interrelationships between chemical and non-chemical stressors should be incorporated into studying children's neurocognitive health.

Keywords: B-pesticides, D-children

WE-PL-C4: Environmental Epidemiology - Birth Cohorts

WE-PL-C4-519

Evidence of Urine Glyphosate Exposure in Pregnant Women: A Prospective Midwest Birth Cohort

S. Parvez[1], R. R. Gerona[3], C. Proctor[2], J. Reiter[4], P. Winchester[2]; [1]Indiana University Fairbanks School of Public Health, Indianapolis, IN, [2]St. Francis Hospital, Indianapolis, IN, [3]University of California San Francisco, San Francisco, CA, [4]Indiana University School of Medicine, Indianapolis, IN

Abstract: Glyphosate (GLY) is the most heavily used herbicide worldwide. GLY is a known endocrine-disruptors and suspected to affect birth outcomes. However, there are currently no published estimates of GLY exposure in US pregnant women to establish whether environmental levels of GLY are safe for fetuses. Therefore, we designed a prospective and cross-sectional birth cohort study to measure GLY exposure in pregnant women and evaluate its association with birth outcomes. Urine and drinking water samples were obtained from 71 pregnant women with singleton births in Midwestern US. The mean age of women was 29 (range 18-39) and 67 (94.4%) pregnancies were Caucasians. The pregnancy outcomes, questionnaires relating to food and water intake and other demographic, and maternal risk factors were also collected. Urine and water samples were measured for GLY using mass spectrometry (Agilent LC 1260 Triple Quadrupole MS). Linear models were used to assess relationships between GLY levels and determinants of fetal risk (Gestational length and birthweight). 93% pregnancies tested positive for GLY. Mean GLY was 3.39ng/mL (range 0.5-7.2ng/mL). The GLY levels were found negatively related to gestation length ($r= -0.28$, $p=0.0202$) and birth weight ($r= -0.20$, $p=0.090$) respectively. In addition, GLY levels were found higher in women who lived in rural areas (means of GLY 4.15 vs. 3.47, $p=0.03$). All drinking water samples tested negative for GLY. Our study provides first direct measure of GLY exposure in US pregnant women and its association with gestational length and birth weight. Since water samples tested negative, food ingestion is assumed to be the main route of exposure, however, the contribution of inhalation route needs to be evaluated. These preliminary evidences are concerning, and a detailed research is warranted to establish their association with adverse fetal growth and underlying toxicity mechanism for risk characterization.

Keywords: A - exposure measurement, B-pesticides, A-epidemiology, A-biomonitoring, D-prenatal

WE-PL-C4-520

Investigation of Association between Environmental and Socioeconomic Factors and Preterm Birth in California

H. Huang[1], D. Woodruff[1], R. Baer[2], K. Bangia[3], L. August[3], L. Jelliffe-Pawlowski[2], A. Padula[1], M. Sirota[4]; [1]UCSF, San Francisco, CA, [2]UCSF, San Francisco, CA, [3]California Environmental Protection Agency, Oakland, CA, [4]UCSF, San Francisco, CA

Abstract: Background/Aim: Preterm birth (PTB), defined as gestational age < 37 weeks, is a significant cause of newborn morbidity and mortality. Relationships have been identified between PTB and environmental/chemical exposures, and social factors. However, fewer studies evaluate the joint effects of environmental exposure and social factors upon PTB. This study aims to 1) investigate relationships between singleton preterm birth and environmental exposures and socioeconomic factors across California; and 2) examine potential combined effects of environmental justice indicators upon PTB. **Methods**: We matched data from CalEnviroScreen 3.0 with the California's Office of Statewide Health Planning and Development (OSHPD) birth cohort datasets from 2009-2012 based on census tract geographic identifiers. We then statistically analysed associations between PTB and different tract-level environmental or social indicators including pollution exposures, environmental conditions, sensitive populations and socioeconomic status (SES), adjusted by birth race/ethnicity and maternal information such as age, education and insurance coverage. **Results**: We restricted our analysis to 1,432,627 eligible births with PTB rate 6.88%. We observed that some drinking water contaminants were positively associated with PTB. For example, the mean average arsenic concentration in drinking water of maternal residence for PTB was significantly higher than that for term birth (0.035 ppb, 95% C.I. [0.023, 0.048]). Tract-level SES also had significant association with PTB. Particularly, mothers who delivered preterm tended to reside in less educated neighbourhoods that had higher population percentage with < 12 years education (1.276 percent increase, 95% C.I. [1.168, 1.384]). Results from logistic regression models showed that individual race/ethnicity and insurance coverage could be potential effect modifiers. **Conclusion**: PTB is associated with both environmental exposure and socioeconomic factors in California.

Keywords: A-environmental justice, D-prenatal, D-susceptible/vulnerable, A - population exposure, A-cumulative exposure

WE-PL-C4-521

Prenatal Air Pollution Exposure and the Development of Allergic Sensitization in Early Life: A Randomized Controlled Trial

R. W. Allen[1], C. L. Bartolomeu[1], E. Gombojav[3], B. Boldbaatar[3], N. Gerel[3], C. Ochir[3], T. Byambaa[2], P. K. Barn[1], B. Lanphear[1], T. Takaro[1], S. A. Venners[1], G. M. Webster[1], W. Yuchi[1]; [1]Simon Fraser University, Surrey, Canada, [2]National Institute of Public Health, Ulaanbaatar, Mongolia, [3]Mongolian National University of Medical Sciences, Ulaanbaatar, Mongolia

Abstract: Background: Prenatal exposure to $PM_{2.5}$ has been associated with the development of allergic sensitization in children. We conducted a randomized controlled trial in Ulaanbaatar, Mongolia to test whether reducing $PM_{2.5}$ with HEPA filter air cleaners during pregnancy reduced the risk of allergic symptoms in infancy. **Methods:** Women were enrolled at 10.3 weeks gestation, on average. Mothers who were randomly assigned to the intervention (N = 217) received one or two air cleaners (depending on home size) to use from enrollment until childbirth. Women assigned to the control group (N = 187) received no air cleaners. We measured indoor $PM_{2.5}$ concentrations over 7-days at ~10 weeks gestation and again at ~30 weeks gestation with Dylos particle counters. We used questionnaires to ascertain the presence of eczema, wheeze, respiratory infections and otitis media in infancy. The efficacy of the intervention was analyzed in intention-to-treat analyses. **Results:** Baseline characteristics were similar between the two groups. $PM_{2.5}$ concentrations were 28% lower, on average, in intervention homes than control homes (GM: 17.3 vs. 24.4 μg/m^3). The prevalence of outcomes ranged from 5% for wheeze to 54% for eczema. Evidence of a protective effect was seen only for wheeze (OR: 0.62, 95% CI: 0.28 - 1.38), while the risk of eczema was slightly higher in the intervention group (OR: 1.43, 95% CI: 0.87 - 2.39). Otitis media and respiratory infections were similar between groups. **Conclusion:** HEPA air filter air cleaners significantly reduced indoor $PM_{2.5}$ concentrations during pregnancy. Our results do not provide strong support for the hypothesis that air cleaner use during pregnancy reduces risks of allergic symptoms in infancy.

Keywords: B-particulate matter, D-neonatal, A-epidemiology, D-children, C-air

WE-PL-C4-522

Mixtures of Early Pregnancy Environmental Exposures and Birth Size

L. B. Rokoff[1], S. Rifas-Shiman[1], B. A. Coull[2], A. Cardenas[1], A. M. Calafat[3], X. Ye[3], S. Sagiv[4], D. R. Gold[5], E. Oken[1], A. F. Fleisch[6]; [1]Harvard Medical School and Harvard Pilgrim Health Care Institute, Boston, MA, [2]Harvard T.H. Chan School of Public Health, Boston, MA, [3]Centers for Disease Control and Prevention, Atlanta, GA, [4]University of California, Berkeley School of Public Health, Berkeley, CA, [5]Harvard T.H. Chan School of Public Health, Boston, MA, [6]Maine Medical Center, Portland, ME

Abstract: Background: Low birth size increases risk for perinatal and later morbidity. Prenatal exposures to maternal cigarette smoking, air pollution, and perfluoroalkyl substances (PFAS) have been associated with low birth size, but the impact of combined exposures is unknown. **Methods:** In 1577 Project Viva cohort mother-infant pairs, at enrollment in the first trimester we assessed maternal smoking by questionnaire, residential black carbon (BC) exposure by land use regression, and perfluorooctane sulfonate (PFOS) plasma concentration. We calculated sex-specific birth weight-for-gestational age (BW/GA) z-scores based on national reference data. We fit multi-pollutant linear regression models adjusted for maternal and birth characteristics, and examined interactions between exposures, using a likelihood ratio test to identify a best fit model. **Results:** 206 (13%) mothers smoked during early pregnancy. Median [interquartile range (IQR)] BC was 0.8 (0.4) μg/m^3, PFOS was 25.7 (16.1) ng/mL, and BW/GA z-score was 0.2 (1.3) units. In the best fit model, BW/GA z-score (95% CI) was lower in infants of mothers who smoked [-0.09 (-0.23, 0.06)] or were exposed to greater BC [-0.07 (-0.14, -0.001) per IQR] or PFOS [-0.03 (-0.07, 0.02) per IQR], without any interactions between exposures. In a secondary analysis, we examined perfluorononanoate (PFNA) [median (IQR): 0.7 (0.4) ng/mL], a PFAS more closely linked to low birth size in our cohort, instead of PFOS; and the best fit multi-pollutant model included positive two-way interactions between PFNA and both smoking and BC (p-interaction=0.03 for each term). **Conclusions:** We found that concurrent prenatal exposures to maternal smoking, BC, and PFOS act independently to lower fetal growth, whereas PFNA may counteract the impact of smoking and BC on lower birth size. It is important to examine interactions between multiple exposures in relation to health outcomes, as effects may not always be additive and may shed light on biological pathways.

Keywords: B-mixtures, D-prenatal, A-epidemiology

WE-PL-C4-523

Patterns and Predictors of Environmental Chemical Mixtures Among Pregnant Women: The HOME Study

G. Kalloo[1], A. M. Calafat[2], A. Sjodin[2], A. Chen[3], K. Yolton[4], B. Lanphear[5], J. M. Braun[1]; [1]Brown University School of Public Health, Providence, RI, [2]Centers for Disease Control and Prevention, Atlanta, GA, [3]University of Cincinnati, Cincinnati, OH, [4]Cincinnati Children's Hospital Medical Center, Cincinnati, OH, [5]Simon Fraser University, Burnaby, Canada

Abstract: Introduction: Pregnant women in the United States are exposed to a multitude of environmental chemicals, but there is a limited understanding of the patterns and predictors of mixture exposures. **Methods:** In a prospective cohort of 389 pregnant women from Cincinnati, OH (enrolled 2003-2006), we estimated exposures to 26 environmental chemicals including metals, phthalates, phenols, polybrominated diphenyl ethers, organophosphate (OP) pesticides, polychlorinated biphenyls (PCBs), and perfluoroalkyl substances (PFAS) by measuring chemical biomarker concentrations in urine and blood samples. We calculated pairwise correlations to examine bivariate relationships between biomarkers. Then, we used k-means clustering to classify women into clusters based on biomarker concentrations and used generalized logistic regression to identify predictors of cluster membership. **Results:** The median correlation between exposure biomarkers was 0.07 (Range: -0.44 to 0.89). Chemicals within structurally, commercially, or industrially related families (e.g., PFASs) were moderately to strongly correlated compared to chemicals unrelated in structure or use (e.g. pesticides and phthalates). We identified three clusters of women (n's=106, 158, and 125). Women in cluster 1 had on average higher concentrations of benzophenone-3, triclosan, Σ_4PCBs, OP pesticides, As, and Cd compared to women in clusters 2 and 3. For instance, women in cluster 1 had a higher average OP pesticide concentration (63 vs. 50, 39 ug/g creatinine) compared to women in clusters 2 and 3. Women in cluster 1 were more likely to be white and consume fruits and vegetables daily compared to women in cluster 3. **Discussion:** We identified three clusters of women in this cohort based on profiles of their chemical exposure biomarkers; cluster membership was associated with sociodemographic and lifestyle factors. Future studies may use these clusters to investigate the cumulative impact of chemical exposures on children's health.

Keywords: A-cumulative exposure, D-prenatal, A-statistical methods, A-epidemiology

WE-PL-D4: Indoor Cooking Exposure

WE-PL-D4-524

Exposure to cooking emitted PM- Reveiw of previous research and future needs

M. Amouei Torkmahalleh[1], S. Gorjinezhad[1], P. Hopke[2]; [1]Nazarbayev University, Astana, Kazakhstan, [2]University of Rochester, Rochester, NY

Abstract: Scientific evidences show that exposure to cooking particulate matter (PM) is correlated with adverse human health effects. The level of PM emission during cooking varies with several factors. This study reviewed cooking PM literature and found that cooking method, type and quality of the energy (heating) source, burner size, cooking oil, food, additives, source surface area, ventilation and position of the cooking pan on the stove are among influential factors affecting cooking PM emission rate and concentration. Conclusions to reduce indoor PM concentrations during cooking have been proposed in this study, and possible future studies have been discussed.

Keywords: A-indoor environment, A - exposure measurement

WE-PL-D4-525

Impact of wearing compliance on CO-PM2.5 relationships for personal exposure in cookstove studies

K. Burkart[2], S. N. Chillrud[1], E. Boamah-Kaali[3], Z. Zhou[2], K. Ayuurebobi Ae-Ngibis[3], M. Mujtaba[3], A. Quinn[2], P. Kinney[4], D. W. Jack[2], K. Asante[3]; [1]LDEO of Columbia University, Palisades, NY, [2]Mailman School of Public Health of Columbia University, New York, NY, [3]Kintampo Health Research Centre, Kintampo, Ghana, [4]Boston University School of Public Health, Boston, MA

Abstract: Cookstove exposure response studies have often used personal carbon monoxide (CO) measurements as proxies for predicting personal PM2.5 exposures, but mixed results have been obtained leading to controversy on when it is appropriate. Here we report modeling results from the subset of participants (N= 980) within the Ghana Randomized Air Pollution and Health Study (GRAPHS) that measured both CO and PM2.5 over 72 hrs with good data. Questionnaire data collected included both general description of home characteristics as well as a recall questionnaire detailing cooking and other sources of exposure at the end of each 24 hr period during the monitoring window. Lascar CO monitors and the RTI microPEMs were both worn on a locally made harness, and participants were instructed to not wear the devices during bathing or sleeping periods. In addition to collecting a filter for gravimetric analysis, the microPEM records real-time PM2.5 and 3-axis accelerometer data, which we have used to determine the # of hours the monitoring harness was worn out of each 24 hr period of the 3 day deployment. The median wearing compliance was 33% of the 24 hr period (or 8 hrs). Binning all the data into deciles of wearing compliance (< 10% to > 70% of the 24 hr period), resulted in the distribution of exposure levels for both CO and PM2.5 being positively associated with wearing compliance indicating potential for exposure misclassification. Furthermore, excluding poor wearing compliance, increases the power of of our predictive models. Regression models that include all observations (i.e., compliance blind) and household- and cooking session specific covariates explain around 30% of the variance; while accounting for compliance results in over 80% of the variance being explained. As far as we know, this is the first study to identify wearing compliance as an important issue to consider when considering using CO as a proxy for PM2.5.

Keywords: A - exposure measurement, B-particulate matter, A-statistical methods, A-exposure models, A-activity patterns

WE-PL-D4-526

Exploring emission and concentration linkages across multiple seasons during a cookstove intervention trial in rural India

T. W. Aung[3], A. Grieshop[1], M. Kelp[2], J. Marshall[2]; [1]North Carolina State University, Raleigh, NC, [2]University of Washington, Seattle, WA, [3]University of North Carolina at Chapel Hill, Chapel Hill, NC

Abstract: Burning solid fuel (i.e. wood) in traditional stoves for cooking and heating is a major public health concern. Air quality guidelines for household fuel combustion and emission standards for alternative cookstoves were recently developed but limited empirical-based assessment exists on how emissions relate to exposure, and factors that may modify this relationship. Emission factors (EF) and concentration of fine particles ($PM_{2.5}$) and black carbon (BC) were derived from a randomized controlled trial implemented in a rural Indian village from 2011-2012. A subset of households (n=31) had $PM_{2.5}$ and particle absorbance/BC measurements for: 1) integrated kitchen-area concentrations; 2) kitchen-area real-time concentrations; 3) plume measurements with complementary gases to yield fuel-based EF; along with household fuelwood usage in pre- and post-intervention seasons. Ambient temperature, and wind speed and direction were measured in the village center. Significant seasonality (baseline > follow-up) effect was observed; mean PM concentration/EF ratios were higher by a factor of 2-5; and BC concentration/elemental carbon EF rations were lower by half in the control households. Ongoing analysis will compare paired data from the same households (both daily average and meal-averaged) to confirm whether the observations hold across households and time of day. Analysis of HH air exchange rates will be explored to determine if concentration decay values reflect the changing linkage between emissions and concentrations. The observed seasonality suggests that seasonally varying conditions (e.g.

meteorology, HH ventilation/air exchange rates) may be important modifiers for the linkages between indoor emissions, concentrations and exposures. This highlights the importance of designing evaluations to separate the influence of time-varying processes and properties, and ensuring emission standards meet the air quality guidelines.

Keywords: A - exposure measurement, A-indoor environment, A-global health, B-particulate matter

WE-PL-D4-527

Household air pollution exposure and nasopharyngeal carriage of streptococcus pneumoniae in Malawian infants

R. Chartier[1], M. Dherani[2], T. Tafatatha[3], S. Cho[1]; [1]RTI International, Research Triangle Park, NC, [2]University of Liverpool, Liverpool, United Kingdom, [3]The Karonga Prevention Study, Chilumba, Malawi

Abstract: Approximately 3 billion people rely on the burning of solid fuel for everyday cooking. Children can be exposed to harmful household air pollution (HAP) generated during the combustion of these solid fuels. These exposures have been linked to many adverse health outcomes including pneumonia. The mechanism linking exposure to pneumonia has been poorly characterized. Here we aim to quantify the relationship between HAP exposure and prevalence of nasopharyngeal carriage (NPC) of streptococcus pneumoniae (SP), which is considered a necessary step to developing pneumonia. This study was conducted in the Karonga district in northern Malawi and focused on 6-week and 6-month old children. 48-hour $PM_{2.5}$ exposures were measured using the RTI MicroPEM and nasal swabs were collected at the end of exposure sampling. This study was embedded in the Cooking and Pneumonia Study (CAPS) to take advantage of the available intervention stoves (Philips HD4012LS). We collected 673 valid exposure measurements and found no significant difference in personal $PM_{2.5}$ exposures of children in homes using an open fire or the Philips stove (p=0.1525; 53.4 [95%CI 49.9, 57.1] µg/m^3; 48.0 [42.1, 54.6] µg/m^3; respectively). However, exposures did vary significantly by age ($p<0.0001$), presence of cigarette smoker (p=0.0181), and kitchen location (p=0.0080). We also found that personal $PM_{2.5}$ exposures in 6-month old children that tested positive for SP-NPC were significantly higher than children that tested negative (p=0.0020; 60.6 [56.1, 65.6] µg/m^3; 43.4 [35.3, 53.4] µg/m^3; respectively). We observed that 6-week old children living with an active cigarette smoker were 70% more likely to test positive for SP-NPC than their peers. This is the first study in which MicroPEMs were used to measure the personal exposures of infants on a large scale. To our knowledge this is also the first study to show a direct link between HAP exposure and prevalence of nasopharyngeal carriage of streptococcus pneumoniae.

Keywords: A - exposure measurement, D-children, B-particulate matter, A-sensor technology

WE-PL-D4-528

Wood-smoke Inhalation Exposures from Traditional Hunting Practices in a First Nations Community

R. E. Peltier[1], P. Pant[1], R. J. Moriarty[2], M. Wilton[2], E. Liberda[3], L. Tsuji[2]; [1]University of Massachusetts, Amherst, MA, [2]University of Toronto Scarborough, Scarborough, Canada, [3]Ryerson University, Toronto, Canada

Abstract: Exposure to wood smoke is a significant health risk for tribal populations who participate in seasonal subsistence hunting and wood burning is an integral part of this culturally-relevant activity. Entire families travel to remote hunting camps, and use wood as an energy source for cooking and heating. In addition, game meat is smoked in specially constructed teepees with limited ventilation. The objective of this study was to measure indoor air quality in tents used for subsistence hunting activities in a population of Native North American hunters by characterizing wood-smoke aerosol components through microenvironment sampling. Indoor air was monitored for black carbon (BC), carbon monoxide (CO) and $PM_{2.5}$ using a microAethalometer (AE51, AethLabs, USA), and a portable nephelometer (pDR 1500). Time-integrated samples were also collected using Teflon and quartz filters for chemical speciation (ions, metals and carbon). The study protocol was reviewed and approved by the IRB at the University of

Massachusetts, Amherst (USA), the University of Toronto and Ryerson University (Canada), and Tribal representatives. Average indoor $PM_{2.5}$ and BC concentrations were significantly higher during meat smoking compared to the in-cabin concentrations. A strong correlation was observed between OC and EC during both phases (hunting and meat smoking), but the concentrations were several fold higher for the meat smoking phase. For short periods of time, concentrations as high as 1000 $\mu g/m^3$ were observed for $PM_{2.5}$, and 300-500 $\mu g/m^3$ for BC.

Keywords: A-indoor environment, D-community, D-First Nation

WE-SY-E4: Improving Our Understanding of Exposures to Tire Crumb Rubber Used on Playing Fields and Playgrounds

WE-SY-E4-529

The Federal Research Action Plan on Recycled Tire Crumb Used on Playing Fields and Playgrounds - Background and Exposure Research Goals

A. Guiseppi-Elie[1], A. D. Ragin[2], E. Hooker[3], K. Thomas[1], E. A. Irvin-Barnwell[2], J. L. Zambrana[1]; [1]United States Environmental Protection Agency, Los Angeles, CA, [2]Centers for Disease Control and Prevention, Atlanta, GA, [3]United States Consumer Product Safety Commission, Rockville, MD

Abstract: Concerns have been raised by the public about the safety of recycled tire crumb rubber used in synthetic turf fields and playgrounds in the United States (U.S.). Several studies have been identified that examine potential exposure to tire crumb rubber infill in these settings. The existing studies do not comprehensively evaluate the potential exposures associated with these use scenarios. Additional research is needed to help fill important data gaps leading to improved exposure assessment for children and adults using synthetic turf fields and playgrounds with tire crumb rubber. In response, the U.S. Environmental Protection Agency (EPA), the Centers for Disease Control and Prevention/Agency for Toxic Substances and Disease Registry (CDC/ATSDR), and the U.S. Consumer Product Safety Commission (CPSC) launched a multi-agency federal research action plan to study key environmental human health questions associated with tire crumb rubber on synthetic turf fields and playgrounds. The "Federal Research Action Plan on Recycled Tire Crumb Used on Playing Fields and Playgrounds" (referred to as the FRAP) was finalized in February 2016. The U.S. EPA and CDC/ATSDR prepared a research protocol to implement portions of the research activities outlined under the FRAP, including 1) conduct a literature review and data gaps analysis; 2) perform tire crumb rubber characterization, and 3) perform human exposure characterization for synthetic turf field users. CPSC is conducting research for assessing the potential risks to children associated with the use of recycled tire crumb rubber in playground surfaces. This presentation will provide the context for this exposure issue, an overview of the federal research efforts, and results of the literature review and data gaps analysis. The overall aim of these efforts to be presented will be to improve our understanding of child and adult exposures at fields and playgrounds with recycled tire crumb rubber.

Keywords: A - exposure measurement, A-analytical methods, A-exposure factors, tire crumb rubber, D-community

WE-SY-E4-530

Characterizing Tire Crumb Rubber for Exposure Assessment

K. Thomas[1], E. A. Irvin-Barnwell[2], A. Guiseppi-Elie[1], A. Ragin-Wilson[2]; [1]U.S. EPA, Research Triangle Park, NC, [2]Centers for Disease Control and Prevention/Agency for Toxic Substances and Disease Registry, Atlanta, GA

Abstract: Tire crumb rubber derived from recycled tires is widely used as infill material in synthetic turf fields in the United States. An estimated 95% of the over 12,000 installed fields in the U.S. use tire crumb rubber infill alone or mixed with sand or alternative materials. Concerns have been raised about exposures of field users to the many potential tire chemical constituents. Most previous U.S. research studies examining tire crumb rubber at synthetic fields have been relatively small, restricted to a few fields or material sources, and limited chemical constituents measured. Characterizing chemical and physical constituents and properties for tire crumb rubber is needed to improve human exposure assessment. Working under the U.S. Federal Research Action Plan, researchers collected tire crumb samples from nine tire recycling plants and 25 outdoor and 15 indoor synthetic turf fields across the U.S. Field ages ranged from new installations to 12 years old. This presentation will focus on methods employed the characterize the tire crumb rubber material. Tire crumb samples were analyzed for metals using acid digestion and ICP/MS. SVOCs were extracted with 1:1 hexane/acetone followed by GC/MS and LC/MS analyses. Dynamic chamber tests measured VOC and SVOC emissions at 25° and 60°C. SVOC and VOC analyses included both targeted analyses for chemicals of interest (e.g. polyaromatic hydrocarbons) and non-targeted analyses to more fully characterize chemical constituents. Particle size, moisture content, and sand fraction were characterized. Finally, bioaccessibility tests were performed for metals and SVOCs using simulated saliva, sweat, and gastric fluids. It is important to recognize that chemical constituent information is necessary but not sufficient for assessing human exposures. Material variability, environmental conditions, bioaccessibility, and human activity factors are among the complex parameters needed to understand exposures at synthetic turf fields.

Keywords: tire crumb rubber, multi-residue analyses, tire crumb rubber, multi-residue analyses, A-analytical methods, A-exposure factors, D-community

WE-SY-E4-531

Characterization of Exposure Potential during Activities on Synthetic Turf Fields with Recycled Tire Crumb Rubber Infill

E. A. Irvin-Barnwell[1], K. Thomas[2], K. Benson[1], Z. Li[1], A. D. Ragin[1]; [1]Centers for Disease Control and Prevention, Atlanta, GA, [2]U.S. EPA, Research Triangle Park, NC

Abstract: The Federal Research Action Plan on Recycled Tire Crumb Used on Playing Fields and Playgrounds (FRAP), released in February 2016, is a multi-agency research plan in response to concerns over the use of tire crumb rubber as infill on synthetic turf fields. The FRAP outlines specific research objectives, including characterizing tire crumb rubber and implementing a pilot-scale observational exposure characterization study. In December 2016, ATSDR and the USEPA released a FRAP Status Report. The report included an in-depth literature review and data gaps analysis. Key data gaps included limited biomonitoring data and exposure information. Specifically, there was limited data on exposure factors, ingestion and dermal routes of exposure, and exposures to tire crumb particles. Additionally, no epidemiological studies were identified. The Status Report also described tire crumb characterization activities. Data from the tire crumb characterization study is being used to identify key chemical and physical property information and to inform implementation of the exposure characterization study. Data from the pilot-scale exposure characterization study, while limited, will allow for further exploration of activities and use patterns that could result in exposure to chemicals from the tire crumb rubber infill by children and athletes who have the potential for high-end exposures. Additionally, the data gained from the study will inform biomonitoring approaches for exposure assessment and elucidate the key information needs which would be required for the development of an epidemiological study. The current presentation will focus on the exposure characterization component of the FRAP and how the ongoing activities, along with exposure information from previous studies, will be used to better characterize synthetic turf field user exposures and inform future research activities. An update on the exposure characterization study will be presented.

Keywords: A-exposure factors

WE-SY-E4-532

Characterizing Children's Exposure to Recycled Tire Rubber Used in Playground Surfacing

E. Hooker, K. Philips, K. Talcott, H. Nesteruk, R. Balci-Sinha; United States Consumer Product Safety Commission, Rockville, MD

Abstract: Recycled tire rubber is used in unitary and loose-fill surfaces to reduce impact-related injuries on playgrounds. Parents and playground owners have expressed concerns regarding the possible chemical hazards of recycled tire rubber in the playground environment. Children can be exposed to playground surfacing materials by oral, dermal, and inhalation routes. As a partner in the Federal Research Action Plan, CPSC is assessing the potential risks to children associated with the use of recycled tire crumb rubber in playground surfaces. The scientific literature lacks information to estimate children's exposure to chemicals from playing on recycled tire rubber playground surfaces. CPSC is using a combination of field observations, focus groups, and a national survey of parents and child care providers to collect information on children's behaviors on playgrounds and to identify exposure factors that may be useful in an exposure assessment. Examples of information sought include frequency and duration of visits to playgrounds, identification and frequency of behaviors and situations that can promote exposure, and supervising adults' awareness of potential hazards and their strategies to reduce exposure. CPSC will use exposure modeling to identify bioavailable substances in recycled tire rubber that may pose a health hazard to children using playgrounds. Oral, dermal, and inhalation routes of exposure will be considered individually and in combination. Subsequent risk assessment strategies will be determined based on review of new data, including the tire crumb rubber characterization and bioavailability studies currently being performed by EPA and CDC/ATSDR, and likely will focus on the substances of highest concern. CPSC staff will present the methods used and any preliminary results. *This work is that of CPSC staff and has not been reviewed or approved by, and may not necessarily represent the views of, the Commission.*

Keywords: A-activity patterns, playground, A-exposure factors, D-children, C-consumer products

WE-SY-E4-533

Improving our understanding of exposures at synthetic turf fields and playgrounds using recycled tire materials

K. Thomas[1], A. D. Ragin[2]; [1]United States Environmental Protection Agency, Research Triangle Park, GA, [2]Agency for Toxic Substances and Disease Registry, Atlanta, GA

Abstract: In recent years, there has been growing concern from the public over the use of recycled tire crumb rubber in synthetic turf fields. In response to these concerns, ATSDR, CPSC, and USEPA developed and implemented a Federal Research Action Plan on Recycled Tire Crumb Used on Playing Fields and Playgrounds (FRAP). The research conducted under the FRAP is intended to provide information about the composition of tire crumb rubber, human exposure activities for field and playground users, and exposure characterization and personal exposure measurements, including personal air monitoring and dermal wipe sampling, among high-end field users. Moreover, previous and ongoing research in the U.S. and Europe provide additional insights for human exposure assessment. The panel will bring together researchers from the US and Europe, and the discussion will provide an opportunity to explore strengths and weaknesses of existing information, future research directions, and the adequacy of exposure data for risk evaluations. Insights on challenges associated with conducting research on the topic area will be discussed. Panel members will also briefly discuss ongoing research activities, including community outreach and engagement efforts and areas for potential collaboration across agencies.

Keywords: A-exposure factors

WE-SY-F4: Integrating exposure and hazard in risk assessment using computational tools

WE-SY-F4-534

Frameworks for organizing exposure and toxicity data - the Aggregate Exposure Pathway (AEP) and the Adverse Outcome Pathway (AOP)

C. Tan; US EPA, Durham, NC

Abstract: Recent risk assessment approaches involve applying *in vitro* dose-response data to estimate potential health risks that arise from exposure to exogenous chemicals. One main challenge involved with these approaches is relating bioactivities observed in an *in vitro* system to the perturbations of biological mechanisms that lead to adverse outcomes (AOs) in living organisms. The Adverse Outcome Pathway (AOP) framework addresses this challenge by linking the binding of a stressor with the technical target of an *in vitro* assay to a molecular initiating event (MIE), which connects to an AO via measurable key events occurring *in vivo.* Another challenge involves converting concentrations that are sufficient to produce bioactivity *in vitro* with *in vivo* concentrations that can trigger an MIE at the relevant biological target. These internal concentrations also should be compared to realistic exposure concentrations and pathways for risk assessment and risk mitigation purposes. The Aggregate Exposure Pathway (AEP) framework addresses this challenge by organizing diverse exposure data and predictions contained within multiple repositories, starting with a source and ending at a target site of exposure (TSE). The TSE serves as the bridge between an AEP and the AOP, as it describes the state of a stressor at a target site that corresponds to an MIE. Thus, integration of these two frameworks provides a natural linkage to processes that move a stressor from its source to its TSE, and subsequently, to the processes that link exposure to AOs. Disclaimer: This abstract has been cleared by the EPA but solely expresses the view of the author.

Keywords: A-aggregate exposure, A-risk assessment

WE-SY-F4-535

Quantitative *in vitro* to *in vivo* extrapolation (QIVIVE) and physiologically based modelling of human toxicokinetics for AOP-based risk assessment

M. Yoon[1], P. McMullen[2], H. Clewell[1]; [1]ScitoVation, Research Triangle Park, NC, [2]ScitoVation, Research Triangle Park, NC

Abstract: For future risk assessments based on in vitro data it will be necessary to perform quantitative in vitro to in vivo extrapolation (QIVIVE) - translation of the in vitro data on bioactive concentrations into estimates of safe in vivo exposures, or extrapolation of the in vitro concentration-effect relationship to an in vivo dose-response, for regulatory use in safety assessments. QIVIVE approaches with differing degrees of complexity can support different aspects of safety assessment, from prioritization to safe exposure estimation. In particular, physiologically based pharmacokinetic (PBPK) modeling can be used to integrate data collected based on aggregate exposure pathways (AEPs) and adverse outcome pathways (AOPs) to support risk-based decisions. Their physiological structure facilitates the incorporation of in silico- and in vitro-derived chemical-specific parameters in order to predict in vivo kinetic behaviors of the test chemicals. With the use of modern parameterization methods based on in silico and in vitro data, PBPK modeling will provide the quantitative bridge from bioactive concentrations in vitro to equivalent in vivo exposure conditions, using the internal exposure at the target tissue as the linking point. Together with the USEPA, we are currently developing an open source PBPK modelling platform, the Population Lifecourse Exposure-To-Health-Effects Model (PLETHEM) suite, to efficiently describe the exposure to health effect continuum using a suite of PBPK models as the integration point.

To support high-throughput PBPK modeling, the platform includes capabilities for rapid parameterization of PBPK models, including a database of physiological parameters, automated IVIVE for in vitro metabolism data, and in silico prediction models for tissue partitioning. This freely available PBPK platform will increase the applicability of PBPK models to support the new safety assessment paradigm.

Keywords: A-risk assessment, A-chemical prioritization, A - population exposure, Toxicokinetics

WE-SY-F4-536

The influence of a surfactant's polar head group on its in vitro distribution and cytotoxic potency

N. Kramer[1], S. T. Droge[2]; [1]Utrecht University, Utrecht, Netherlands, [2]University of Amsterdam, Amsterdam, Netherlands

Abstract: Despite similar nominal effect concentrations, the biologically effective concentration in cells may vary significantly between chemicals and between *in vitro* and *in vivo* assay setups. Not considering differences in biologically effective concentrations between assays hampers quantitative *in vitro-in vivo* extrapolations (QIVIVE). The *in vitro* distribution and the effect of this distribution on *in vitro* bioassay readout have been studied for neutral organic chemicals. The octanol water partition coefficient (K_{OW}) and Henry's law constant correlate with a chemical's bioavailable fraction *in vitro*. The partitioning behaviour of charged, complex chemicals like surfactants in *in vitro* assays has not been well studied. The aim of this study was to measure the fraction of cationic and anionic surfactants with alkyl chains of eight and twelve carbons in exposure medium and well plate plastic over time in an RTgill-W1 basal cytotoxicity assay. The extent to which a surfactant's head group, charge, charge shielding, pKa and membrane-water partition coefficient (K_{MW}), as well the extent to which pH and serum levels of the exposure medium influence the partitioning behaviour and cytotoxicity readout was assessed. Results indicate that there was a large variation in cytotoxic potency between surfactants with varying head groups and alkyl chain length, where cationic surfactants, especially with a permanently charged head group, were found to be most cytotoxic, which was likely due, in part, to the mechanism of membrane insertion. Cytotoxic effect concentrations were dependent on the head group, chosen dose metric (measured medium concentrations versus nominal) and exposure time. Given the variation in partitioning behaviour, the nominal effect concentrations used to rank surfactants according to their cytotoxic potency across assays will not reflect the ranking of toxic potencies based on the freely available or cell-associated effect concentrations.

Keywords: A-exposure models, A-risk assessment, in vitro

WE-SY-F4-537

Biologically Based Dose Response modeling and its application in risk assessment: A case study of thyroid-active compounds

A. Lumen; National Center for Toxicological Research, Jefferson, AR

Abstract: Computational approaches are increasingly being considered in risk assessment to guide regulatory decision-making. Biologically Based Dose Response (BBDR) model comprises of both physiologically based pharmacokinetic model (to link external exposure to internal dose) and mode-of-action based model (to link internal dose to corresponding response). Such integrated models allow for the use of available kinetic and dose-response data in animal models and non-pregnant adults and extrapolate to data-scarce sensitive sub-populations. A case study on the development and application of a BBDR pregnancy model for thyroid-active compounds, such as perchlorate and thiocyanate, will be discussed. A deterministic model for perchlorate was first developed in late-gestation pregnant women and extended to a population-based model. Reverse dosimetry approaches were implemented to utilize urinary biomarker data to reconstruct the daily iodine intake levels for late-gestation pregnant women in the U.S., which subsequently allowed for the identification of the prevalence of iodine inadequacy. Furthermore, probabilistic exposure-based risk assessment for perchlorate was also conducted. Our current efforts aim to continue the expansion of this computational pregnancy modeling framework to address the issue of co-exposure to mixtures of thyroid-active chemicals that reflects better real world exposure scenarios. Specifically, thiocyanate, a thyroid-disruptor with multiple modes of action, predominantly found in food is being evaluated in our ongoing work as a constituent of a ternary mixture. The recent findings in the modeling of thyroid-chemical mixtures and its implications in risk assessment will be summarized. The overall population-based risk assessment framework will provide regulatory agencies with a valuable and robust tool for the integrative assessment of the risks of adverse effects due to exposure to thyroid-active chemicals by a population of pregnant women.

Keywords: B-perchlorate, B-mixtures, A-biomonitoring, A-cumulative exposure, A-risk assessment

WE-SY-G4: Global harmonization of exposure science data

WE-SY-G4-538

Harmonization of Exposure Science through Reference Materials and Data

C. A. Gonzalez; National Institute of Standards and Technology, Gaithersburg, MD

Abstract: For over three decades, the Chemical Sciences Division (CSD) at the National Institute of Standards and Technology (NIST) has been providing metrological tools for human and animal health-based measurements in the form of higher-order analytical methods, reference materials and data, and interlaboratory comparisons programs. The CSD also maintains the Marine Environmental Specimen Bank (ESB), an important national resource of research materials that are leveraged to advance measurement science for emerging contaminants, and reveal spatial and temporal changes in environmental exposure and health outcomes. Computational tools for enhanced data analysis are also being developed within CSD to improve the identification of unknown compounds, further streamlining non-targeted screening processes of biological fluid reference materials. These measurement services have directly supported the environmental exposure assessment of organic pollutants and toxic metals, as well as promoted and ensured quality assurance for metabolomics and lipidomics, nutritional assessment and clinical marker measurements. Additionally, multimodal, nonlinear models are being developed to couple these environmental, chemical and biological data, and determine the minimum set of parameters necessary to describe the behavior of the whole exposome. This metrological infrastructure of chemical science, data and informatics, and accuracy-based standards is critical to understand, harmonize and validate the links between environmental conditions and human health and disease status.

Keywords: A - exposure measurement, A-exposure models

WE-SY-G4-539

Integrative Mass Spectrometry to Sequence the Human Exposome

D. I. Walker[1], Y. Liang[1], J. Fernandes[1], A. Esper[1], G. Martin[1], G. Miller[2], D. P. Jones[1]; [1]Emory University, Atlanta, GA, [2]Emory University, Atlanta, GA

Abstract: Over a lifetime, humans are exposed to a wide range of chemicals from multiple sources. A more complete understanding of how environment contributes to disease is required for mitigating risk and developing treatment strategies; however, no unified method exists to characterize the sum involvement of lifestyle and environment in disease. As the high-resolution metabolomics (HRM) core for the HERCULES Exposome Research Center at Emory University, we have developed a multi-platform, mass spectrometry-based framework providing in-depth coverage of the exposome, metabolome and metallome that uses less than 500 µL of biological material. Results show gas-chromatography (GC) with ultra-high resolution mass spectrometry (UHRMS) makes possible routine detection of 75,000 molecular features in plasma, which includes over 250 confirmed environmental pollutants. HRM by dual column liquid chromatography (LC) with UHRMS detects over 25,000 chemical signals that arise from core nutrient metabolism, lipids, the microbiome, diet-derived chemicals, pharmaceuticals and environmental contaminants, providing measures for evaluating biological response, metabolic health and toxicodynamics. The addition of inductively coupled plasma MS (ICP-MS) now provides capabilities that include measure of toxic metals and micronutrients, improving characterization of nutritional status and metal exposure. Using the combined analytical capabilities of GC-UHRMS, LC-UHRMS and ICP-MS, it is now possible to provide extensive measures of environmental chemicals, nutrients and endogenous metabolites in a cost-effective manner. The resulting data can be used to estimate exposure in human populations, evaluate disease-exposure associations and provides systematic measures of environment and biological response for exposome research. Demonstration in a critically ill population shows this approach makes possible a systems-biology understanding of chemical toxicology and environmental health in humans.

Keywords: A-biomonitoring, A-biomarkers, Metabolomics, Exposome, High-resolution mass spectrometry

WE-SY-G4-540

The European Human Biomonitoring Initiative - HBM4EU

M. Kolossa-Gehring, U. Doyle, U. Fiddicke, A. Conrad, K. Pack, L. Reiber; German Environment Agency (UBA), Berlin, Germany

Abstract: The overarching goal of HBM4EU is to improve the protection of human health in Europe by generation of knowledge and informing chemicals policy. Human Biomonitoring (HBM) will be the basis to assess the risks from human exposure to chemicals and its health impacts. Intensive communication with policy makers from the very start of the programme and onwards will ensure that HBM4EU results are used in the further development of chemicals policies and the evaluation of existing measures. HBM4EU is organised around 16 work packages led by key players of national HBM studies and research programmes. They represent science policy transfer, HBM studies, and exposure and health. One work package is specifically devoted to the investigation of mixtures because HBM reveals the multiple exposure and the need for health risk assessments concepts beyond traditional single substance evaluation approaches. Key objectives of HBM4EU are: 1. Harmonizing procedures for human biomonitoring across 26 countries, to provide policy with comparable data on internal exposure to chemicals. 2. Linking data on internal exposure to chemicals to aggregate external exposure and identifying exposure pathways and upstream sources. Information on exposure pathways is vital for targeted policy measures aiming for exposure reduction. 3. Generating scientific evidence on the causal links between human exposure to chemicals and health impacts. 4. Adapting chemical risk assessment methodologies to use HBM data. We will achieve these objectives by building upon knowledge from national HBM initiatives, and building new capacities. By establishing National Hubs in each country, we will create a robust HBM network at the European level. This initiative contributes directly to the improvement of health and well-being. We will also investigate how factors like behavior, lifestyle and socio-economic status influence exposure to chemicals across Europe on different population sub-groups.

Keywords: A-biomonitoring, A-biomarkers, A-chemical prioritization, B-mixtures, A-risk assessment

WE-SY-G4-541

Quality control, SOPs and data harmonization in metabolomics: the good, the bad, and the ludicrous

C. E. Wheelock[1], I. Meister[2], R. Chaleckis[2]; [1]Karolinska Institute, Stockholm, Sweden, [2]Gunma University, Maebashi, Japan

Abstract: Environmental exposures are a leading cause of morbidity and mortality worldwide. There are >80,000 chemicals on the EPA's inventory of compounds used in commercial and consumer products. There is an unmet need to quantify environmental exposures and elucidate their collective effect upon incidence of disease. The sheer numbers and chemical complexity call for an omics-based approach to data acquisition. The sum of one's environmental exposure can be described as the exposome, while the metabolic profile is the metabolome. Implementing the exposome concept poses a number of challenges. Full measurement of the exposome is not yet possible analytically. While measurement of exposome signals using omics technologies addresses this issue to some extent, there is a need to significantly improve the methodologies used to perform exposome science. Progress is being made with the use of high-resolution mass spectrometry, but additional effort is required to acquire a full exposome. This talk will present an overview of metabolomics followed by in-depth discussion of its application to molecular phenotyping and the exposome. Emphasis will be placed on the design of experiments, choice of clinical material, confounding experimental factors as well as metabolite annotation and data analysis. Particular attention will be placed on the vital role of quality control samples, standard operating procedures and data harmonization to ensure overall data quality. The issue of speed *vs.* specificity in large-scale data acquisition will also be addressed. Examples will be given using urine, which is a common biofluid for assessing environmental exposure. While readily available and non-invasive, the heterogeneity in collected urine can have a large impact on the resulting analyses and must be accounted for in the experimental design. Particular challenges in the field will be presented and placed within the context of the future of applying omics approaches to the study of the exposome.

Keywords: A-analytical methods, A-biomarkers, metabolomics, exposome

WE-SY-G4-542

Data users view: What data quality is required?

R. C. Vermeulen; Utrecht University, Utrecht, Netherlands

Abstract: Major scientific advancements in the last decade have resulted in extensive new insights in the role of genomics in health. This has been achieved in part by standardization of techniques, and data quality and reporting guidelines allowing the pooling of large datasets with genomics data. While the genomic advances are impressive, they have not necessarily produced insights that contribute to major public health improvements. At the same time, there has been increasing recognition of the crucial and significant role that the environment, alone and in interaction with genomics, plays in health. As a result of this renewed appreciation, multiple countries over the past decades have launched large scale biomonitoring and cohort studies of their populations to monitor their exposure status and to find new exposure - disease associations. However, procedures on standardization, quality control, and reporting guidelines are much less developed as in genomics resulting in problems when comparing or pooling different datasets between laboratories and over time within laboratories. From a measurement error perspective random laboratory variance results in nuisance variance that leads to a decrease in power to find differences in exposure experiences between individuals or populations. If the measurement error is however differential or if the measurements are biased than such comparisons are not only comprised in the sense of statistical power but can also lead to spurious results. For example, in a cross-sectional study among 602 men from Greenland, Poland, and Ukraine who had recently fathered a pregnancy, associations between four chemical classes and markers of male reproductive function were evaluated. As outcome parameters by country could differ (either biologically or due to the use of different labs) there was a high potential for spurious results due to so called ecological fallacies. In this presentation, we will explore what statistical options there are to control and correct for laboratory measurement error in the comparison of individuals and populations, and what the impact of measurement error can be on exposure-disease associations.

Keywords: A - exposure measurement, A-epidemiology, measurement error

WE-PO-543

Linking aviation emissions to residential and personal ultrafine particle exposures

N. Hudda, M. C. Simon, W. Zamore, D. Brugge, J. L. Durant; Tufts University, Medford, MA

Abstract: Aviation emissions adversely impact air quality as far as 7.3 kilometers downwind of Logan International Airport, Boston, MA (about 1000 flights/day). During 1.67 and 3.67 years of monitoring at central-sites in Chelsea and Boston, respectively, concentrations of ultrafine particles (UFP) were 100% and 33% higher when winds were from the direction of the airport. Our new analysis investigates exposures in greater detail at individual residences. Specific objectives were as follows: 1. Extend previous central site UFP investigations to individual residential locations within each community. 2. Determine if airport related elevated UFP infiltrate homes and likely impact personal/residential exposures. Outdoor and indoor UFP monitoring was successfully completed at 17 of the 24 homes and at two central sites. Chelsea homes were 2-6 km downwind of the airport during southeast to south winds. Boston homes were 4-12 km downwind during northeast to east winds. We analyzed UFP concentrations, taking flight activity at Logan airport, meteorology and local traffic volumes into account. Results confirm our findings at the central-sites and underscore the pervasive footprint of the impact of Logan-related emissions, particularly in Chelsea. Highest concentrations at 7 of 8 homes in Chelsea (population: 39,000) occurred during winds from the direction of the airport. Higher temporal correlations were observed between (hourly average) concentrations at residences and the central-site during such winds; Pearson correlation coefficients ranged from 0.65-0.97 versus 0.44-0.83 for the entire dataset. Indoor infiltration was evident in the data, with a concurrent increase in both outdoor and indoor concentrations during winds from the direction of the airport indicating that airport-related emissions are likely to have a significant impact on personal exposures. Our results provide compelling evidence for the impacts of aviation emissions on personal ultrafine particle exposures.

Keywords: A - ambient monitoring, A - exposure measurement, A-built environment, A-environmental justice, A-indoor environment

WE-PO-544

A Field Study to Validate Inhalation Exposure Factors Used to Create the Particle Inhalation Rate Metric

L. Corlin[1], H. Amaravadi[2], N. Henderson[1], J. Cordova[2], S. Ball[2], M. Woodin[1], J. L. Durant[1], D. Gute[1], D. Brugge[2]; [1]Tufts University, Boston, MA, [2]Tufts University, Boston, MA

Abstract: Exposure assessment in air pollution epidemiology usually focuses on ambient concentrations, ignoring inhalation as part of the exposure pathway. To account for inhalation patterns, we developed a new exposure metric that multiples ambient particle concentrations by minute respiratory volume (MRV). The MRV estimates were derived from fairly homogenous, healthy populations. We sought to test whether the published MRV estimates were valid among a vulnerable population representative of participants in the Boston Puerto Rican Health Study (BPRHS). We recruited 38 adults (79% Puerto Rican, 68% female, mean age=60.9 years) to participate in a Spanish-English bilingual health fair/research day. Participants reported health and demographic characteristics that affect inhalation patterns. We measured participants' height, weight, and peak expiratory volume. We also measured participants' pulse, oxygen saturation, and inhalation patterns while they walked, stood, lied down, and sat. From participants' respiration rate (RR) and inspiratory volume (IV), we calculated the average age, sex, and weight-adjusted MRV for each activity level. The effect of using these calculated values instead of published values on the PIR was assessed. Demographic and health characteristics were similar between our sample and the larger BPRHS cohort (34% of our participants had also participated in the BPRHS). About one-third of our participants were smokers, had asthma, and had low peak expiratory volume. Pulse, RR, IV, and MRV were significantly greater when participants walked or stood compared to sat or lied down. The MRVs were 1.9-4.9 times higher than the published estimates. Applying these coefficients within the BPRHS cohort increased the variability of the PIR distribution and doubled the mean and median PIR. Using the new values strengthened the effect estimates for PIR on blood pressure. Population-specific estimates of inhalation metrics are important for accurate exposure assessment.

Keywords: A-exposure factors, D-susceptible/vulnerable, C-air, B-particulate matter

WE-PO-545

Comparison of Bioaerosol Samplers and Media for the Collection of Aerosolized Norovirus

C. L. Boles, M. W. Nonnenmann; University of Iowa, Iowa City, IA

Abstract: The United States (US) has an estimated 179 million cases of acute gastroenteritis (AGE) each year. Of the total cases of AGE more than 39 million are caused be an infectious agent, such as bacteria and virus. One of the most common infectious agents that result in AGE (iAGE) is norovirus. Norovirus causes greater than 90% of iAGE outbreaks, and results in 56,000-71,000 hospitalizations in the US each year. Route of transmission is known to occur by direct and indirect contact with contaminated surfaces; however, evidence has suggested that the virus can be transmitted via aerosol. Determining routes of exposure is important as the infectious dose for norovirus has been reported as low as 10 virions. Little research has been performed measuring aerosolized norovirus in the field and laboratory, which may be a significant route of transmission. Therefore, a developed methodology is needed for norovirus aerosol sampling so that inhalation exposure to norovirus can be determined. This study compared two samplers and two liquid sampling medias in the collection of airborne norovirus. The SKC BioSampler and NIOSH Bioaerosol sampler containing either Phosphate-Buffered Saline or Hanks Balanced Salt Solution were compared for collection of murine norovirus (MNV:CW3). Ten 30-minute trials were completed by aerosolizing MNV:CW3 in a laboratory bioaerosol chamber to compare concentrations of norovirus collected across two types of bioaerosol samplers using two liquid sampling medias. Samples were analyzed using RT-qPCR, as well as PMA:RT-qPCR to evaluate virus capsid integrity. The mean total virus particles per cubic meter of sampled air were compared between samplers and liquid sampling medias. Preliminary results (n=40) indicate no difference between media type, however there was a difference between the percentage of virus with intact capsids before and after sampling. Future investigations targeting aerosolized norovirus can incorporate the findings of this study.

Keywords: A-industrial hygiene, A-Infectious disease, D-occupational, B-microbial agents, A-workplace

WE-PO-546

Investigation and modelling the residential infiltration of fine particulate matter in Beijing

C. Xu, N. Li, Q. Wang, D. Xu; National Institute of Environmental Health, Chinese Center for Disease Control and Prevention, Beijing, China

Abstract: Aim The objective of this study was to estimate the residential infiltration factor (Finf) of fine particulate matter ($PM_{2.5}$) and to develop models to predict $PM_{2.5}$ Finf in Beijing. **Methods** 88 paired indoor-outdoor $PM_{2.5}$ samples were collected by Teflon filters for seven consecutive days during both non-heating and heating seasons (from a total of 55 families between August, 2013 and February, 2014). The mass concentrations of $PM_{2.5}$ were measured by gravimetric method, and elemental concentrations of sulfur in filter deposits were determined by energy-dispersive x-ray fluorescence (ED-XRF) spectrometry. $PM_{2.5}$ Finf was estimated as the indoor/outdoor sulfur ratio. Multiple linear regression was used to construct Finf predicting models. **Results** The residential $PM_{2.5}$ Finf in non-heating season (0.70±0.21, median=0.78, n=43) was significantly greater than in heating season (0.54±0.18, median=0.52, n=45, $P<0.001$). Outdoor temperature, window width, frequency of window opening, and air conditioner use were the most important predictors during non-heating season, which could explain 57% variations across residences, while the outdoor temperature was the only predictor identified in heating season, which could explain 18% variations across residences. **Conclusions** The substantial variations of $PM_{2.5}$ Finf between seasons and among residences found in this study highlight the importance of incorporating Finf into exposure assessment in epidemiological studies of air pollution and human health in Beijing. The Finf predicting models developed in this study hold promise for incorporating $PM_{2.5}$ Finf into large epidemiology studies, thereby reducing exposure misclassification.

Keywords: A-indoor environment, A - exposure measurement, B-particulate matter, C-air, A - population exposure

WE-PO-547

Controlled Dermal Exposure to Thirdhand Cigarette Smoke

S. F. Schick; University of California, San Francisco, San Francisco, CA

Abstract: Thirdhand cigarette smoke is a term for the chemicals released during smoking that linger in the indoor environment. Nicotine and many of the other semivolatile organic compounds in sidestream cigarette smoke sorb to surfaces before they can be removed by ventilation. Under normal ventilation conditions, approximately 70% of the nicotine released during smoking sorbs to indoor surfaces. Human beings can be exposed to thirdhand cigarette smoke by the respiratory, oral and dermal routes. Transdermal nicotine patches demonstrate that nicotine can be absorbed through the skin. Research suggests that nicotine can also be absorbed from clothing. To test whether other cigarette smoke chemicals can be absorbed from smoke-contaminated clothing and to estimate the dosage, we are performing controlled human exposure studies. Using a crossover study design, healthy nonsmokers wear cotton clothing for three hours. The clothing is either clean or exposed to a controlled amount of cigarette smoke. The cigarette smoke exposure of the clothing is designed to mimic the exposure in the home of a pack-a-day smoker. We collect blood and urine samples from the study participants before and after wearing the clothing and test for metabolites of nicotine and the tobacco-specific nitrosamine, NNK. Exposures take place in a chamber with 0.85 air changes per minute, to minimize respiratory exposure. The order of exposures is randomized and the study visits are separated by three weeks to minimize crossover effects. Our data show that both nicotine and NNK are absorbed dermally from smoke-contaminated clothing. These findings, combined with previous research showing that NNK forms from nicotine on indoor surface materials, suggests that dermal uptake of NNK from thirdhand cigarette smoke constitutes a significant and hitherto unappreciated exposure to the carcinogens created by smoking cigarettes.

Keywords: A - exposure measurement, A-indoor environment, B-particulate matter, B-SVOCs, C-consumer products

WE-PO-548

Potential Inhaled dose and cardiovascular indicators - initial pilot results

D. W. Jack[1], P. Kinney[2], X. Lui[1], D. Shimbo[1], R. Sloan[1], C. M. Smith[1], J. Thornburg[3], Q. Yang[1], S. N. Chillrud[1]; [1]Columbia University, New York, NY, [2]Boston University, Boston, MA, [3]RTI International, Triangle Research Park, NC

Abstract: Exposure to air pollutants while commuting to work on a bicycle in New York City may affect blood pressure (BP) and heart rate variability (HRV) in the hours following exposure. We carried out a pilot study in NYC with 43 healthy commuting bikers for testing feasibility of different aspects of the study protocol. The aim was to have each subject self deploy exposure, GPS, and physiological monitors for five 24 hr periods over two weeks. Black carbon and $PM_{2.5}$ monitors were used to assess exposure. Minute ventilation and heart rate variability were assessed using a biometric shirt. An ambulatory blood pressure monitor logged BP automatically. BP and HRV in the hours post exposure were compared to pre-exposure baseline readings. Utilizing potential inhaled dose (product of minute ventilation and concentration in the breathing zone) of $PM_{2.5}$ and black carbon may enable more accurate analysis of the association between exposure and health outcome than air concentrations. Pilot results showed a positive correlation between increased inhaled dose of black carbon and systolic blood pressure following exposure (R = 0.12 to 0.41). Correlations were strongest on a 2 hour lag (R = 0.38 to 0.41). Results also showed a negative correlation between inhaled dose of black carbon and HRV following exposure (R= -0.13 to -0.48). Correlations were strongest on a 4 hour lag (R = 0.42 to 0.48). Based on these pilot results, we hypothesize that increased BC and $PM_{2.5}$ potential inhaled dose incurred during morning bicycle commutes will predict increased blood pressure 2 hours after exercise ends. Additionally, we predict an increase in BC and $PM_{2.5}$ potential inhaled dose will decrease HRV after exercise ends. This hypothesis will be tested in a cohort of approximately 150 cyclists over the next 3 years.

Keywords: A - exposure measurement, A-biomonitoring, A-epidemiology, A-sensor technology, C-air

WE-PO-549

Impact of High-Efficiency Air Filtration on Indoor Particle Levels: The AIRE Study

D. Bennett[1], P. Krakowiak[2], D. Tancredi[1], N. Kenyon[1], R. Moran[1], K. Roudneva[1]; [1]University of California, Davis, Davis, CA, [2]University of California, Davis, Davis, CA

Abstract: The AIRE study enrolled 172 non-smoking homes in the Fresno and Riverside regions of California. High-efficiency filtration was installed, either through stand-alone air cleaners (N=140) or through high-efficiency filters placed in the air intake of the HVAC system (N=32). 136 homes completed the study, which included one year of true filtration and one year of sham filtration. Concentrations of PM were monitored for one week every six months, for a total of five weeks. Generalized linear mixed-effects regression models were used to determine the differences between true and sham filtration and the impacts of potential modifiers. All measured indoor PM concentrations (i.e., $PM_{0.2}$, $PM_{2.5}$, and PM_{10}) were significantly lower during true compared with sham filtration periods (all $p<0.0001$). The geometric mean $PM_{0.2}$ concentrations during sham and true filtration periods were 2.33 mg/m^3 [SE= 1.04] and 1.21 mg/m^3 [1.05], respectively; mean $PM_{2.5}$ concentrations were 6.76 mg/m^3 [1.05] and 3.49 mg/m^3 [1.05]; and mean PM_{10} concentrations were 12.79 [1.04] and 8.79 [1.04]. The mean sham vs. true differences in PM concentrations were significantly greater in homes with air cleaners versus those with HVAC filtration ($p<0.05$), and levels with true filtration were lower using stand-alone air cleaners. When compared with pre-intervention conditions, particulate levels were lower in both the true and sham post-intervention periods. Heterogeneity of treatment effects associated with candidate effect modifiers was assessed using interaction terms. We found that the effectiveness of filtration was modified by the age of the home and how many days during the sampling week windows were open, both thought to affect air exchange rates. The impacts of various particulate sources (cooking, smoking, burning of candles or fires) were also considered. The statements and conclusions in this abstract are those of the contractor and not necessarily those of the California Air Resources Board.

Keywords: A-indoor environment, B-particulate matter, A - exposure measurement, A-built environment,

WE-PO-550

Fluorescent Biological Aerosol Particle Exposure in a Northern California Residence

Y. Tian[1], Y. Liu[2], P. K. Misztal[2], J. Xiong[2], C. M. Arata[3], A. Goldstein[2], N. W. William[1]; [1]UC Berkeley, Berkeley, CA, [2]UC Berkeley, Berkeley, CA, [3]University of California, Berkeley, CA

Abstract: Residences, where Americans spend about 70 percent of their time, represent an important site of bioaerosol exposure. We studied bioaerosol exposure in a single-family residence in northern California with two occupants using real-time instruments in two multiweek monitoring campaigns (summer 2016 and winter 2017). Information regarding occupancy and activities was obtained from occupant-maintained logs. An ultraviolet aerodynamic particle sizer was used to measure size-resolved fluorescent biological aerosol particles (FBAP) and total particles in real-time in the kitchen. Wireless environmental monitors and co-pollutant sensors were distributed throughout the house. During the summer and winter monitoring periods, occupants spent an average of 10 and 8.5 hours per day, respectively, awake and at home. The geometric mean time-averaged FBAP exposure concentration (1-10 micrometer diameter) was 42 particles/L for summer (GSD=1.7) and 29 particles/L (GSD=1.3) for winter. During awake hours, human occupancy strongly influenced indoor FBAP levels. Indoor FBAP concentrations were an order of magnitude higher when the house was occupied than when the house was vacant. FBAP exposure concentration increased with occupancy level. FBAP size distribution peaked in 2-4 micrometer diameter range for occupied condition, while a dominant mode at 1-2 micrometer was observed when the house was vacant.

Keywords: A-indoor environment, A-built environment, A - exposure measurement, B-microbial agents, B-particulate matter

WE-PO-551

High-Throughput, Simultaneous Quantitation of Hemoglobin Adducts of Acrylamide, Glycidamide, and Ethylene Oxide Using UPLC-MS/MS

C. Tse[1], M. Yang[2], T. Frame[1], H. Vesper[1]; [1]Centers for Disease Control and Prevention, Atlanta, GA, [2]Battelle Memorial Institute, Atlanta, GA

Abstract: Exposure to ethylene oxide (EO), acrylamide (AA) and its primary metabolite glycidamide (GA) is associated with mammary tumors in animals. People are exposed to these chemicals through certain occupational activities, tobacco smoke and food. These chemicals are highly reactive and form covalent reaction products (adducts) with DNA and proteins. There is a need to reliably assess long-term human exposure to these chemicals and potential health effects associated with the exposure. We developed a sensitive and high-throughput ultra-high performance liquid chromatography-tandem mass spectrometry (UPLC-MS/MS) method to simultaneously quantitate hemoglobin (Hb) adducts of AA (HbAA), GA (HbGA), and EO (HbEO) from the N-terminal valine of the Hb beta-chain in blood samples. Based on a modified Edman reaction, these Hb adducts were cleaved from the Hb chain, extracted and analyzed as pentafluorophenyl thiohydantoin derivatives. The limits of detection for HbAA, HbGA, and HbEO were 3.9, 4.9 and 30.1 in pmol/g (Hb) respectively, which were sufficient to determine the background levels of these adducts in the general population. The intra-day and inter-day precision, expressed as percent coefficient of variation, for all analytes using quality control materials ranged from 3.8 to 14.0%. The mean accuracy (SD) for all analytes was 100.8% (0.52). Blood bank samples from 105 individuals with self-reported smoking status were analyzed. The median adduct concentrations for HbAA, HbGA, and HbEO were 64.9, 45.3 and 113.6 pmol/g in self-reported non-smokers and 126.3, 68.2, and 241.1 pmol/g in self-reported smokers, respectively. These findings indicated that higher concentration of HbAA and HbEO levels were found in smokers and suggested that elevated exposure to AA and EO are associated with smoking status. The described method is suitable for measuring HbAA, HbGA, and HbEO in the general population and can process 296 samples per day.

Keywords: A-biomonitoring, A-analytical methods, A-biomarkers

WE-PO-552

New, Highly Specific UPLC-MS/MS Method for Analyzing Steroid Hormones and their Conjugates in Human Serum

L. F. Duke[1], P. Kim[2], J. Botelho[1], H. Vesper[1]; [1]CDC Atlanta, Atlanta, GA, [2]Battelle Memorial Institute, Atlanta, GA

Abstract: Steroid hormones are critical for human development and health. Endocrine disruptor chemicals (EDCs) are substances of public health concern known to interfere with biosynthesis, pathways, and action of steroid hormones and are suspected to affect human reproduction, development, and metabolism. To support the need for investigating the association between EDCs and multiple steroids, CDC developed an isotope dilution liquid chromatography tandem mass spectrometry (ID LC-MS/MS) method capable of measuring 8 hormones, including some of their conjugates in human serum. The method will help elucidate endocrine diseases and may provide information concerning the impact of EDCs on the metabolism of these hormones. This method uses sequential liquid-liquid extractions for isolation of progesterone (P4), estradiol (E2), estrone (E1), estrone sulfate (ES), total testosterone (TT), androstenedione (AD), dehydroepiandrosterone sulfate (DHEAS), and 17-hydroxyprogesterone (17-OHP) in 200 mL of human serum without derivatization or hydrolysis. Chromatographic separation of the steroids is carried out using a C18 guard column and a phenyl-hexyl HPLC column with a methanol gradient. Scheduled selective reaction monitoring (SRM) by electrospray ionization (ESI) in the positive and negative ion modes is used to quantify the analytes. The specificity is enhanced compared to methods using derivatization, because the selected mass transitions are specific to the fragmentation of the analytes and not to that of the derivative. The method was developed to quantitate concentrations over four orders of magnitude, with limits of detection (LOD) in matrix-based materials reported as: P4 0.01 ng/mL, E2 1.6 pg/mL, E1 1.5 ng/dL, ES 90 pg/mL, TT 1.0 ng/dL, AD 6.0 ng/dL, DHEAS 2000 pg/mL, and 17-OHP 10 ng/dL. This new, highly specific method has improved LODs compared to methods described in the literature and is suitable for detecting these steroid hormones in men and pre- and postmenopausal women.

Keywords: A-biomonitoring, A-exposure factors, A-epidemiology, B-phthalates

WE-PO-553

Electrochemical Sensing of Nitrite, a Biomarker of Oxidative Stress, in Exhaled Breath Condensate

C. P. Weisel[1], A. Gholizadeh[2], M. Javanmard[2], D. Voiry[3], A. Gow[1], M. Chhowalla[3], R. Laumbach[1]; [1]Rutgers University, Piscataway, NJ, [2]Rutgers University, Piscataway, NJ, [3]Rutgers University, Piscataway, NJ

Abstract: We have developed a portable sensor for measuring nitrite levels in exhaled breath condensate (EBC), a biomarker of inflammation and oxidative stress in the respiratory tract. EBC is collected in a non-invasively by breathing into cold tube. The miniaturized electrochemical sensor for detecting nitrite was fabricated using reduced graphene oxide. The unique properties of reduced Graphene Oxide (rGO) is its being resilient to corrosion while exhibiting rapid electron transfer with electrolytes allowing for highly sensitive electrochemical detection and minimal fouling. The rGO sensor was housed in an electrochemical cell fabricated from Polydimethyl Siloxane (PDMS), to facilitate its ability to analyze small EBC sample volumes (<0.1 mL). The sensor is capable of detecting nitrite at a low over-potential of 0.7 V with respect to Ag/AgCl reference electrode. The sensor demonstrated a sensitivity of 0.21 $\mu A\ \mu M^{-1}\ cm^{-2}$ in the range of 20-100 μM and 0.1 $\mu A\ \mu M^{-1}\ cm^{-2}$ in the range of 100-1000 μM nitrite concentration with a low detection limit of 830 nM in the EBC matrix. The performance of the sensors was characterized using standard nitrite solutions in buffer, nitrite spiked into EBC, and also clinical EBC samples as well as compared to nitrite measured using a colometric method. The two colormetric and sensor method characterized the same trend in nitrite concentration in both fresh and frozen EBC samples, with the results being within a factor of two across nitrite concentrations ranging between 0.14-6.5 μM. This enzyme-free and label-free method of detecting biomarkers in EBC can pave the way for development of portable breath analyzers for use in studies of the effect of air pollution on oxidative stress and respiratory inflammation.

Keywords: A-biomarkers, A-sensor technology, B - ozone

WE-PO-554

Does subcooled solubility play any role in dermal absorption?

R. Tibaldi[1], W. ten Berge[3], D. Drolet[2], A. Redman[1]; [1]ExxonMobil Biomedical Sciences, Annandale, NJ, [2]volunteer EASC, Longueuil, Canada, [3]Santoxar, Westervoort, Netherlands

Abstract: Emergency incidents can present a risk for skin absorption from accidental contact to a neat chemical substance or mixture of contaminants in water. Workers normally take steps to prevent exposure, however unplanned contact may occur during response or cleanup operations. Dermal exposure measurement following such events is often not possible and standard methods not available. Skin permeation models can be a practical solution in these circumstances to evaluate risk for systemic effects. The assessment of skin absorption to components in a mixture remains challenging. Many chemical specific factors must be known and considered to arrive at reasonable estimates. It is especially difficult to estimate the physiochemical properties and behavior of every constituent in case of substances with unknown or highly variable composition. The hydrocarbon block approach is often used to reduce the complexity of UVCBs which models partitioning as a function of abundance in the oil phase (mole fraction) and the subcooled solublity of the block. The aim of this presentation is to illustrate the combination of these concepts into a screening level approach using the IH SkinPerm tool to estimate dermal absorption of specific contaminants from an aqueous solution containing fuel or heavy oil.

Keywords: A-exposure models, A-risk assessment, A-industrial hygiene

WE-PO-555

Bioavailability of POPs: method development for physiologically based in vitro extractions

X. Cui; Nanjing University, Nanjing, China

Abstract: Risk assessment for persistent organic pollutants (POPs) are often based on total contaminant concentrations. However, mounting evidence suggests that understanding contaminant bioavailability in soil, food, or dust is necessary for accurate assessment of exposure to humans via the ingestion pathway. Animal-based in vivo tests have been used to assess contaminant bioavailability. However, due to ethical issues and high cost, it is desirable to use physiologically based in vitro methods as alternatives. Although in vitro assays offer an attractive alternative to predict bioavailability, their application for refining POP exposure is still in the developmental phase. Here we first present a work about bioavailability of one emerging contaminant, i.e., perfluorooctanoic acid (PFOA), in food. An in vivo mouse model and three in vitro methods (unified BARGE method-UBM; physiologically based extraction test-PBET; and in vitro digestion method-IVD) were used to determine bioavailability and bioaccessibility of PFOA in 17 types of food. The feasibility of in vitro method was validated by correlation with in vivo results. In the second part, we will focus on legacy POPs in soil, such as PAHs and DDTs. Most in vitro methods can not provide sufficient sorption capacity for POPs with low solubility. Tenax with infinite affinity for POPs was included as sorption sink to optimize the in vitro methods for hydrophobic POPs.

Keywords: A-risk assessment

WE-PO-556

Quantification of formaldehyde hemoglobin adducts in blood with protein calibrators

M. Yang[1], C. Tse[2], K. Omri[2], D. Wirtz[2], T. Nabors[2], A. Gostilean[2], H. W. Vesper[2]; [1]Battelle, Atlanta, GA, [2]Centers for Disease Control and Prevention, Atlanta, GA

Abstract: Formaldehyde (FA) is classified as carcinogenic to humans by WHO/IARC thus creating the need to determine human exposure to FA. Reaction products of FA with hemoglobin (FA-Hb adducts) are promising biomarkers for assessing the internal exposure to this chemical. Analytical methods have been described for measuring FA-Hb adducts using enzymatic digestion and quantitation with stable isotope labeled peptide standards and liquid chromatography-tandem mass spectrometry (LC-MS/MS). However, peptide standards do not compensate for variation in enzymatic digestion and subsequent sample preparation caused by enzymes with different digestion activities. To address this challenge, we investigated different strategies using in-house prepared FA-Hb adducts, commercial available ^{15}N labeled Hb, chemically synthesized peptides with multiple cut-points for the enzyme, and native FA-Hb adducts as calibrators. Quality control samples and different calibrators were digested for 1 to 48 hours using trypsin at 48C. We stopped the digestion reaction with formic acid and analyzed the resulting N-terminal peptide containing the adduct (FA-VHLTPEEK) using LC-MS/MS. Results suggest that the efficiency of enzymatic digestion with trypsin depends on both protein sequences and folding structures. In-house prepared FA-Hb adducts, commercial available ^{15}N labeled Hb, and the chemically synthesized peptide showed different digestion efficiency than the native FA-Hb adducts. We prepared calibration curves by diluting native FA-Hb adducts. When using these calibrators, the variability of measurement results obtained with different enzyme lots was reduced to within 0.1% to 7%, compared to 3.5% to 17%

as observed with peptide calibrators. Using native FA-Hb adducts as calibrators provided consistent measurement results that are independent with the enzyme digestion efficiency. These calibrators will be used in our analytical method for measuring human exposure to FA.

Keywords: A-biomonitoring, A - population exposure, A - exposure measurement, A-analytical methods, A-biomarkers

WE-PO-557

Susceptibility to xenobiotics toxication is dictated by the time of the day that exposures occur

P. Charisiadis[1], S. Gaengler[1], R. Seth[2], S. Chatterjee[2], K. C. Makris[1]; [1]Cyprus University of Technology, LImassol, Cyprus, [2]University of South Carolina, Columbia, SC

Abstract: Aim: Non-persistent environmental chemicals (NOPEC) are xenobiotics with short half-lives of elimination (<7h). NOPEC metabolism may follow diurnal hepatic patterns of cytochrome P450 activity, which may be under the liver clock influence. Metabolic activation (toxication) of trihalomethanes (THM) by CYP2E1 leads to harmful metabolites, enhancing susceptibility to xenobiotic toxication. We investigated the diurnal patterns of exposures to THM (NOPEC class of disinfection byproducts) and their metabolism catalyzed by circadian patterns of hepatic CYP2E1 enzyme activity (4-hydroxynonenal (4HNE). Methods: We implemented three time-series studies with adult volunteers conducting specific cleaning activities at predefined times of the day. An activity day was designed where each participant conducted four well-controlled activities that could generate THM in the surrounding indoor environment (showering, hand dishwashing, mopping, and bathroom cleaning); another sampling day was included with none of the aforementioned activities (control). Each participant collected spot urine samples at predetermined intervals on both days. Urinary THM and HNE-His levels were measured using gas chromatography tandem mass spectrometry (GC-MS/MS) and immunoassay test, respectively. Nychthemeral variation in lipoperoxidation rhythm of 4HNE patterns was assessed with a cosinor model. Results: The time of exposure in the day dictated the magnitude of urinary THM levels and their toxication effect; in all three studies and relative to urinary THM before the activity, lower and higher median urinary THM were measured after the activity in morning and afternoon/night, respectively, consistent with higher diurnal CYP2E1 activity in light/active phase. Conclusions: The inclusion of diurnal measurements in exposome studies is warranted to better describe the dynamic changes in exposure patterns between and within subjects that are prospectively followed during critical life stages.

Keywords: A-biomarkers, A-biomonitoring, D-susceptible/vulnerable

WE-PO-558

Crowdsourced and Crowdfunded Biomonitoring

R. Dodson, H. Susmann, K. M. Rodgers, J. G. Brody, R. A. Rudel; Silent Spring Institute, Newton, MA

Abstract: Communities and individuals may be interested in knowing about their exposures, but biomonitoring for most chemicals requires costly and specialized expertise, and so is not routinely available. Instead, biomonitoring is typically conducted as part of research studies or population surveillance, and in many cases participants do not receive their personal results. In addition, research funding for exposure studies, particularly community-initiated studies, is limited. Crowdsourcing and crowdfunding provide a novel mechanism to recruit people who are interested in their exposures into research, to ask them to support the project financially, and to develop a large research cohort of biomonitored individuals. We have launched a first-ever project to build a cohort of several hundred biomonitored individuals using this approach. After providing informed consent, participants are sent a kit for urine sample collection, called the Detox Me Action Kit, and asked to provide information about exposure-related behaviors. A main goal of our project is to identify influential sources of exposure. In this phase of the study, analytes include consumer product chemicals such as disinfectants, parabens, a UV-filter, flame retardants, and BPA, many of which are endocrine active. Participants receive personalized results using an NIH-funded Digital Exposure Report-Back Interface (DERBI). Participants also may use a smartphone app—Detox Me—that we developed to provide information on how to reduce exposure to common environmental pollutants and consumer product chemicals. To date, we have enrolled >250 participants. This project also provides a platform for community-initiated biomonitoring studies by removing barriers to access, taking advantage of efficiency of scale. This program offers a flexible and scalable model for participatory community biomonitoring for environmental chemicals. All study protocols have been reviewed by an independent IRB to ensure human subject protection.

Keywords: A-biomonitoring, C-consumer/personal care products,

WE-PO-559

Exposure data for sunscreen products

M. Gomez-Berrada[1], S. Rakotomalala[1], M. Bellec[2], S. Guillou[2], D. De Javel[2], P. Ferret[1]; [1]Safety Assessment Department, Pierre Fabre Dermo-cosmétique, Toulouse, France, [2]Eurosafe, Saint Grégoire, France

Abstract: Purpose: In Europe, sunscreen products are cosmetics used during a limited period by wide range of population to protect skin from radiations. However, to conduct reliable safety assessment, accurate exposure information is required. Most of the available data are resulting from SCCS guidelines (18g/day) or clinical studies in which subjects are submitted to a specific protocol of use. The aim of this study was to determine the daily quantity used of a SPF50 face and body sunscreen product according to subject's daily habits and to compare the quantity used depending on the age and the gender. **Methods**: This study was carried out on 109 families who are usual consumers of sunscreen product: 212 adults (from 23 to 79 years old) and 179 children (from 3 to 12 years old). Subjects were provided with a SPF50 face and body sunscreen spray product and were asked to use it according to their habits during their summer holidays. Products were weighed at the start and completion of each study in order to determine the total amount of product used and detailed daily usage information was recorded. **Results**: Mean, standard deviation and 90th percentile were as follows: - Daily quantities used: Adults=3.07±2.33g(5.96g);men=3.06±2.05g(6.06g);women=3.07±2.59g(5.98g) Children=3.24±2.18g(6.29g);boys=3.18±2.38g(6.25g);girls=3.29±1.96g(6.29g) - Daily frequency of use: Adults=1.69±0.97(2.91);men=1.58±0.93(2.48);women=1.79±1.00(3.43) Children=1.93±1.08(3.28);males=1.86±1.06(3.33);females=2.00±1.10(2.95) **Conclusion**: This study provides relevant information on exposure for face and body sunscreen SPF50 which will be useful for risk assessment purposes. We also highlighted a difference in product uptake between adults and children and a similarity between women and children.

Keywords: A-risk assessment,

WE-PO-560

Usage patterns of sunscreen products: a key point for safety assessment

M. Gomez-Berrada[1], S. Rakotomalala[1], M. Bellec[2], S. Guillou[2], D. De Javel[2], P. Ferret[1]; [1]Pierre Fabre Dermo-cosmétique, Toulouse, France, [2]Eurosafe, Saint Grégoire, France

Abstract: Purpose: In Europe, sunscreen products are cosmetics used during a limited period by wide range of population to protect skin from UV radiations. Thus, due to their extensive skin contact combined with direct exposure to UV radiation, specific safety assessment is required. The aim of this study was to obtain reliable information regarding patterns of sunscreen products usage that will be useful for safety evaluation. **Methods**: A web-based questionnaire was developed in March 2014 in order to select families for a real-life exposure study of a SPF50 spray. It contained questions related to frequency of use, sun exposure conditions, type of product used (SPF, galenic form) and body surface exposed (face;body). 1064 participants completed the questionnaire. 1054 respondents are used to applying sunscreen products during their holidays. Finally, 109 families were chosen to participate to the real-life conditions study at the end of which they were questioned about the use of the product. **Results**: The 1054 respondents mainly indicated applying sunscreen product daily and several times per day during their holidays. Most women answered using SPF50 products on face (37%) and SPF30-40 on body (39%); men, SPF30-40 on face and body (37%); children, SPF50 on face (67%) and body (64%). Respondents indicated favouring spray (>46%) and cream (>24%) forms. At the end of the real-life conditions study, families reported the influence of tan and sunlight intensity on product use. **Conclusion**: This study provides current sunscreen exposure information for most commonly used products which will be useful for risk assessment purposes.

Keywords: A-behavior, A-activity patterns

WE-PO-561

Trace chemicals in consumer goods: differentiating analytical presence from human health risk

D. Schowanek[1], K. D. Acuff[3], S. A. Tozer[4], K. L. Krivos[2], S. P. Felter[3]; [1]Procter & Gamble, Brussels, Belgium, [2]Procter & Gamble, Cincinnati, OH, [3]Procter & Gamble, Cincinnati, OH, [4]Procter & Gamble, Egham, United Kingdom

Abstract: Increasingly, both industry and regulatory agencies are challenged to address low-level human exposures to chemicals in consumer products. The chemicals of concern may be impurities, or traces of omnipresent environmental contaminants. Current analytical methods allow detection and quantification of chemicals at extremely low part per trillion levels, the so-called 'race towards zero'. Even where risk assessment science would indicate no objective safety concerns -since actual exposure under use conditions is negligible- examples are plenty where reporting the mere presence of certain compounds resulted in a spread of unfounded chemophobia with consumers and the general public. A 'flat world' with real time social media usually plays an important role in this process. In the end, such cases often lead to economic damage, effects on company reputation and brand equity, and dedication of resources to address issues that do not translate to tangible gains in public health. This poster focuses on approaches and suitable methodologies to be able to decide whether detection of trace compounds requires further risk management. A very important element to consider is the extraction step prior to the analytical measurements. While the presence of a dissolved chemical in a liquid product allows a relatively straightforward human exposure assessment, it is more challenging to quantify exposure from solid matrices such as articles and assembled consumer products. To be useful in a safety assessment, realistic bioavailability and physiology considerations should guide the extraction approaches used and the final data interpretation. This poster reviews some case studies in this regard, and explains the principles followed by P&G for consumer products.

Keywords: A-analytical methods, C-consumer products, A-exposure models

WE-PO-562

Modeling use phase chemical releases, fate, and disposal for modeling longitudinal human exposures to consumer products

G. Glen[1], R. Avanasi[1], J. Levasseur[1], H. Hubbard[1], P. Price[2], K. Isaacs[3]; [1]ICF, Durham, NC, [2]US EPA, Durham, NC, [3]US EPA, Durham, NC

Abstract: The US EPA's Human Exposure Model (HEM) is an integrated modeling system to estimate human exposure to chemicals in household consumer products. HEM consists of multiple modules, which may be run either together, or independently. The Source-to-Dose (S2D) module in HEM uses a compartmental mass partitioning approach that models the release and transport of chemicals during product use as a series of stages including the initial use phase release, transfers during the use phase, end of the direct use phase, and direct exposure to residual chemicals. In the first stage, the fractional mass releases of chemicals that occur during the use of a product are determined for 11 possible compartments (indoor/outdoor air, indoor/outdoor surface, skin, gut, other body parts, solid waste, drain, appliance and retained solid product). The mass releases to each compartment vary by the type of product. Products are categorized into 17 different product categories based on the physical processes involved in their use (sprays, liquids, dusts etc.) and on their application location (body, indoor, outdoor surfaces, among others). The second stage determines the transfers between compartments during the time the product is used by processes such as, but not limited to, volatilization and settling. The third stage determines the movement of chemicals to household trash and residential waste water from wipe-off and rinse-off events. Some products have a fourth stage for modeling direct exposure to any residual chemical left on the skin or in the nearby area. The mass of chemicals in the indoor air and surface compartments at the end of the fourth stage are used as inputs to a fugacity module to determine the time series of post-use chemical concentrations in the home's indoor air and surfaces. This approach allows for the efficient calculation of population-based residential human exposures and releases in support of mid-tier assessments and life-cycle analyses.

Keywords: A-indoor environment, A-exposure models, C-consumer/personal care products, A-exposure factors, A-life cycle analysis

WE-PO-563

Understanding the endocrine disruption potential of hazardous chemicals in children's products and their proposed alternatives using in vitro data

M. N. Smith[1], R. M. Shaffer[1], E. Cohen Hubal[2], E. Faustman[1]; [1]University of Washington, Seattle, WA, [2]Environmental Protection Agency, Research Triangle Park, NC

Abstract: Children's consumer products represent an important exposure source for many toxicants due to their intended uses, which lead to direct contact with children. Examples of chemicals found in children's products include phthalates, bisphenols and parabens. As regulation and reporting requirements increase, demand for safer alternatives to these chemicals has also increased. Alternatives assessments are used to identify safer chemical alternatives, however, many times these assessments are limited by lack of toxicity data. This biases the alternatives assessment to conclude that the new chemical is safer, when in fact it may be just less well studied. This project examines how in vitro data can fill gaps in alternatives assessments for hazardous chemicals found in children's consumer products. Formal national and international lists, such as the European Chemical Agency's (ECHA) Endocrine Disruptor Substances of Concern classification are compared with the toxicological prioritization index (ToxPi) score calculated based on the United Stated Environmental Protection Agency's ToxCast Database. Formal lists are primarily derived using data from human and animal studies, and the ToxPi score is based on in vitro assay results. Three chemical groups were considered: phthalates, parabens and bisphenols. Though alternative chemicals were rarely classified as endocrine disruptors by the ECHA, the in vitro ToxPi scores for alterative chemicals were slightly higher for bisphenols and phthalates. The results from this case study suggest that in vitro data can help fill gaps when existing classifications are incomplete. This project is supported by the Environmental Protection Agency (FP-91779601-0, RD 83573801, RD 83451401) and the National Institute of Environmental Health Sciences (5P01ES009601).

Keywords: A-chemical alternatives, A-life cycle analysis, C-consumer products, A-green chemistry, D-children

WE-PO-564

Systematic Review and Meta-Analysis of Occupational Styrene-Induced Dyschromatopsia

A. R. Choi[1], J. M. Braun[2], G. D. Papandonatos[3], P. B. Greenberg[1]; [1]The Warren Alpert Medical School of Brown University, Providence, RI, [2]Brown University School of Public Health, Providence, RI, [3]Brown University School of Public Health, Providence, RI

Abstract: AIM: Styrene is an industrial solvent commonly utilized in the manufacture of plastics-based products. Observational human studies have indicated that prolonged occupational exposure to this chemical may result in acquired dyschromatopsia, or defective color discrimination. However, reviews and meta-analyses on this topic are conflicting and exclude studies from the past decade. To address this information gap, we systematically reviewed all studies of occupational styrene-induced dyschromatopsia.

Using a subset of identified studies, we estimated the styrene-induced effect through meta-analysis. METHODS: PubMed, EMBASE, Web of Science electronic databases were systematically queried for eligible studies. We employed a random effects model to compare measures of dyschromatopsia between exposed and non-exposed workers, from which we calculated the standardized mean difference (Hedges' *g*). Assessments of between-study heterogeneity and publication bias were also conducted. RESULTS: All but one study presented findings in support of styrene-induced dyschromatopsia. Meta-analysis demonstrated significantly greater dyschromatopsia among exposed workers relative to their non-exposed counterparts (standardized mean = 0.56; 95% CI: 0.37, 0.76; p < 0.0001). Furthermore, a non-significant Cochran's Q test result (Q = 23.2; df = 10; $p = 0.171$) and an I^2 of 32.2% (95% CI: 0.0%, 69.9%) indicated low-to-moderate heterogeneity between studies. Funnel plot and trim-and-fill analyses showed evidence of publication bias. CONCLUSIONS: Qualitative synthesis of the evidence supports the putative exposure-outcome relationship, at levels well below regulatory agency-prescribed exposure limits. Additionally, meta-analysis indicates a moderate effect size with low-to-moderate heterogeneity. Our review suggests a need to re-assess current styrene exposure limits through further investigation. To this end, current work conditions and practices at high-exposure sites should be better elucidated.

Keywords: D-occupational, A-workplace

WE-PO-565

Exposure to Organophosphate Flame Retardants and Thyroid Cancer Risk in Women

N. Deziel[1], H. Yi[1], H. M. Stapleton[2], H. Huang[1], N. Zhao[1], Y. Zhang[1]; [1]Yale School of Public Health, New Haven, CT, [2]Duke University, Durham, NC

Abstract: Background Exposure to organophosphate flame retardants (PFRs) occurs worldwide; growing evidence demonstrates that these chemicals can alter thyroid hormone regulation and function. We investigated the relationship between PFR exposure and thyroid cancer and examined whether individual characteristics or temporal factors predicted PFR exposures. **Methods** Our analysis included 100 incident female, papillary thyroid cancer cases and 100 female controls of a Connecticut-based thyroid cancer case-control study. Interviews and spot urine samples were collected from 2010-2013. We used mass spectrometry to measure concentrations of six PFR metabolites. We used unconditional logistic regression to estimate odds ratios (OR) and 95% confidence intervals (95% CI) for thyroid cancer risk for categories (low, medium, high) of concentrations of individual and summed metabolites, adjusting for potential confounders. We used multiple linear regression models to examine relationships between concentrations of PFR metabolites and individual characteristics (age, smoking, alcohol consumption, body mass index [BMI], income, education) and temporal factors (season, year). **Results** None of the individual or summed PFRs were significantly positively associated with risk of papillary thyroid cancer. However, the odds ratios of the PRF metabolite bis(1,3-dichloro-2-propyl) phosphate was elevated; the OR (95%CI) for the high versus low category was 1.78 (95%CI: 0.83-3.8). The metabolite bis(1-chloro-2-propyl) phosphate was inversely related to thyroid cancer risk; the OR (95%CI) for the high versus low category was 0.32 (0.20-1.00). Our exposure determinants analysis observed higher urinary PFR concentrations with increasing BMI and in the summer season. **Conclusions** Our results do not support an increased risk of thyroid cancer with exposure to PFRs. Given the modest sample size and use of spot urine samples to represent long-term exposure, the research question warrants additional study.

Keywords: B-flame retardants, A-epidemiology

WE-PO-566

Mixtures of PBDEs and thyroid hormones: Critical importance of exposure correlations, novel methods and causal assumptions

T. F. Webster; Boston University School of Public Health, Boston, MA

Abstract: Previous studies of the associations between PBDEs and thyroid hormones have typically analyzed one PBDE congener at a time due to concerns about colinearity. This ignores at least two issues: potential confounding by co-exposures and cumulative effects. We examined a small longitudinal cohort of 52 office workers, sampled 3 times at 6 month intervals; serum was analyzed for various thyroid hormones, PBDEs, PCBs and organochlorine pesticides. The correlation matrix for the exposure biomarkers showed a roughly block diagonal structure between the PBDEs and other persistent organic compounds, implying little likelihood of confounding between the two groups. Correlations between several PBDEs were high. In single exposure analyses, a number of congeners were negatively associated with total T4 (TT4), indicating the possibility of confounding by co-exposure. However, including all of these congeners in the regression model produced high variance inflation factors. Weighted quantile sum regression (WQS) assumes the outcome is linearly related to a weighted sum of quantiles of the exposures; it can be used to analyze small data sets with correlated exposures. As WQS is not currently set up for longitudinal data, we used the estimated weights to construct the weighted sum which was then used as the exposure in a mixed effects model. The WQS index was negatively associated with TT4, similar to single congener analysis. BDE-99 had the largest weight, with less weight estimated for BDEs 47, 153, 28, 100. Co-exposure amplification bias has recently been shown to be a potential problem for highly correlated exposures when an uncontrolled factor is related to the outcome and one but not all exposures. Confounding by an enzyme related to a PBDE congener and TT4, both measured via biomarkers, is at least possible in this situation. Careful consideration of the causal assumptions are therefore needed for interpreting the results.

Keywords: A-statistical methods, B-mixtures, A-epidemiology, B-flame retardants, A-cumulative exposure

WE-PO-567

Association between Pollen Risk Indexes, Air Pollutants, and Allergic Diseases in Korea

H. KIM[1], Y. Park[1], K. Park[2], B. Yoo[1]; [1]Soonchunhyang University, Cheonan, Korea (the Democratic People's Republic of), [2]Ewha Womans University Mokdong Hospital, Seoul, Korea (the Democratic People's Republic of)

Abstract: Objectives: This study, different from the past researches, has been conducted in all age groups to understand the association between air pollutants, pollen risk indexes, and outpatients with allergic rhinitis (AR), asthma, and atopic dermatitis (AD). Methods: Data on air pollutants, pollen risk indexes, and outpatients with each disease were collected from 2003 through 2011 to verify the association between them. All data are time-series materials that have been observed by time (day) and region, and are in a nonlinear shape. In particular, the total number of outpatients per day is a count data that had a Poisson distribution as the response variable. SAS 9.3 was used to make a statistical model, generalized additive model, with lag effects for the analysis. Results: For allergic diseases during spring (April-May) and fall (September-October), a significant association was shown between the variables of air pollutants, pollens, and the number of outpatients. Especially, the estimates of NO2 [AR (43.00967_ 0.11284), asthma (52.01837_ 0.06452), AD (52.01837_ 0.06452), $p < 0.001$] in spring and SO2 [AR (43.00967_ 0.11284), asthma (52.01837_ 0.06452), AD (52.01837_ 0.06452), $p < 0.001$] in fall were highly significant and showed a positive association with all diseases. Conclusion: Domestically and even internationally, various studies on the allergic diseases are being conducted. However, not many studies related to similar studies. In the need of creating grounds to back up these efforts, additional studies on allergic diseases, as well as researches utilizing pollen data, air pollution data, and claims data provided by the Health Insurance Corporation that has no problem in the representativeness of the data that have close relationships to the allergic disease will be needed.

Keywords: A-climate change, A-exposure factors, A-exposure models

WE-PO-568

Assessing the Role of Caregiver Stress and Cockroach Allergen in Asthma Related Healthcare Utilization

D. Werthmann, F. Rabito; Tulane University SPHTM, New Orleans, LA

Withdrawn

WE-PO-569

Early-life exposure to household chemical products with relation to patterns of whistling and wheezing in children up to 5 years old and their association to asthma in school age

O. Mikes[1], M. Klugova[1], J. Klanova[1], P. Cupr[1], J. Svancara[1], H. Pikhart[2]; [1]Masaryk university, Brno, Czech Republic, [2]University College London, London, United Kingdom

Abstract: The prevalence of the asthmatic symptoms and asthma among children increases globally over the time. Few hypotheses about the causes of such increase have been proposed including the hygiene hypothesis saying that reduced exposure to pathogens in early childhood and increased exposure to anthropogenic irritants results in increased risk of whistling and wheezing symptoms in children. Evaluation of aspects related to asthmatic symptoms is of high importance in order to possibly reduce these inconvenient life obstacles. Of importance is to study the effects of such exposures longitudinally including many potential confounders. The Czech part of the European Longitudinal Study of Pregnancy and Childhood (ELSPAC) which was established in the early nineties in the Czech Republic with recruitment of thousands of families and follow up until the 19 years of children, has been used to evaluate the hypothesis. The Central European region is currently understudied and its results may support the results from other similar studies. Composite household chemical exposure scores have been used in line with sister study ALSPAC and they were standardized after and expressed as a z-score. Whistling and wheezing patterns were based on the reported onsets and their persistency or transiency from birth until five years of age. Social, medical and environmental factors were taken into account for analysis. Multivariable multinomial logistic regression analysis was performed using patterns of wheezing/whistling as a study outcome. We were able to determine the association between several whistling and wheezing childhood patterns and the frequent usage of household chemicals in the fully adjusted model, which supports the hygiene hypothesis. Moreover, the persistent and late onset patterns were highly significantly associated with the school age asthma. Acknowledgements: Supported by the Czech Ministry of Education, Youth, and Sports (LM2015051 and CZ.02.1.01/0.0/0.0/15_003/0000469).

Keywords: A - population exposure, A-epidemiology, C-consumer products, D-children

WE-PO-570

Evaluation of the impact of exposure to air pollution change on variation of lung function among refractory asthmatics in urban areas of South Korea

S. Kim[1], S. Yu[1], C. Park[2]; [1]Soonchunhyang University, Asan, Korea (the Republic of), [2]Soonchunhyang Bocheon Hospital, Bucheon, Korea (the Republic of)

Abstract: In this study, we examined the association of variation of lung function (FEV1/FVC) with the change of air pollution level among refractory asthmatics living in metropolitan cities of South Korea, Seoul, Incheon or Bucheon. We retrospectively studied the spirometry data of adult patients, aged 19-87 years, who had been registered since 2005. All subjects were followed for a minimum of 2 years in an asthma cohort from the Genome Research Center for Allergy and Respiratory Diseases, Soonchunhyang University Bucheon Hospital, South Korea. The diagnosis of asthma was based on the Global Initiative for Asthma (GINA) guidelines. In our study, we included only the 82 patients living in the metropolitan city of Seoul, Incheon or Bucheon, South Korea, who presented at the Soonchunhyang Hospital between 2005 and 2015 with multiple episodes of exacerbations. Our final data comprised 2310 episodes from 214 RA nonsmoking patients whose ratio of FEV1 divided by FVC was < 80%. We constrained an "episode" to have a frequency no greater than one visit per month to avoid attributing follow-up visits to new independent episodes. Air pollution measurements were obtained from the national ambient monitoring station located closest to each patient's home address. Multilevel model analyses were conducted using the SAS 9.1 package. The data stratified by season and smoking status showed that in multi-pollutant models with smoking RA patients, in winter, a 1^{O}C decrease in ambient temperature and a 1 ppb increase in SO2 concentration significantly decreased lung function (FEV1/FVC) on Lag3 and Lag4 after controlling for other explanatory variables: 0.3 % for temperature and 2.0% for SO2. Exposure to temperature drops and increased sulfur dioxide concentrations are positively associated with decrease of lung function during winter with 3 or 4 day lags.

Keywords: A - population exposure, A-epidemiology, C-air

WE-PO-571

Improving the Accuracy and Usability of Residential Histories for Exposure Assessment

H. Minor, F. W. Lurmann, B. Penfold; Sonoma Technology, Inc., Petaluma, CA

Abstract: Residential histories are essential for retrospective environmental exposure assignments and are usually collected by questionnaire; however, they often contain incomplete or inconsistent addresses and dates. Addresses may be missing, misspelled, or have erroneous street numbers and names, city names, states, and postal codes. Start and end dates may be incomplete or overlapping. Although such problems are understandable, given the difficulty of recalling precisely where and when a person lived over long periods, the errors and omissions in addresses lead to poor geocoding results. Likewise, gaps and overlaps of dates limit the usability of residential histories. For environmental health studies with high geocoding accuracy requirements, such as those examining near-road environmental exposure, residential history problems can cause significant reduction in the number of subjects and reduced statistical power. Interviewers often have only one opportunity to collect residential histories; follow-up with subjects regarding problematic addresses and dates may not be feasible when address verification and geocoding occur months after the data collection. Real-time interactive systems can improve the accuracy and usability of residential histories. We describe the elements of a computerized survey system that incorporates real-time address verification, geocoding, mapping, and date checking to provide immediate feedback on each address collected. Maps afford the opportunity to correct address errors, and address verification can improve address geocodability. The system checks the consistency of start and end dates for each location and across locations, and prompts the user to verify or correct gaps and resolve overlapping dates. The ability to correct residential history information during data collection will improve overall quality and reduce the number of subjects excluded from health analyses for missing or inaccurate locations.

Keywords: A-epidemiology, C-multimedia,

WE-PO-572

Health Risk Ranking Assessment of Human Exposure to Multiple Air Pollutants Emitted from Municipal Solid Waste Incineration in China

Q. Zhou, M. Liu, J. Bi; Nanjing University, Nanjing, China

Abstract: Municipal Solid Waste Incineration (MSWI) has developed rapidly with doubled increase in China during 2010-2015. However, risk level of MSWI as well as its spatial distribution were still unknown. A systematic risk identification and ranking assessment is essential to improve effectiveness and pertinence of risk management. In this study, we developed a health risk ranking framework and applied it to assess the MSWI related risk across China. Firstly, a bottom-up multiple-pollutant emission inventory was calculated based on detailed information of 222 MSWIs in mainland China and localized emission factors from literature investigation. Then a Gaussian Plume Model was used to simulate the dispersion of pollutants emitted by each MSWI, considering the perennial dominant wind direction and mean wind speed. Finally, spatialized and population-weighted Hazard Index (HI) and Relative Risk (RR) were calculated to evaluate both non-carcinogenic and carcinogenic health risks caused by individual respiratory exposure from heavy metals and PCDD/Fs. Results show that, in 2015, 59651.6 t of NOx, 20072.2 t of SO_2, 13215.3 t of CO, 7240.0 t of HCl, 4392.5 t of particulate matter (PM), 26.0 t of Pb, 25.8 t of Cr, 15.5 g of Hg, 5.0 t of Cd, 3.4 t of Ni, 273.2 kg of As, and 73.2 g-TEQ of polychlorinated dibenzo-p-dioxins and polychlorinated dibenzofurans (PCDD/Fs) were emitted by MSWI in mainland China. The biggest HI and RR at county-level were 2.75E-3 (<1) and 9.11E-7 (<1E-6), and the relatively high-risk regions mainly concentrated in the eastern coastal areas (Shanghai, Jiangsu), and southwestern inland (Sichuan, Chongqing). Besides, the greatest contributor to HI and RR were Pb and PCDD/Fs separately. The identified relatively high-risk areas and pollutants above need to be set as the priority control areas or pollutants in the risk management. Certainly, more field tests are suggested to be conducted in these places to provide more exact information.

Keywords: A-risk assessment, A-geospatial analysis/GIS, C-air, A-environmental policy

WE-PO-573

Applying the Source to Outcome Approach for Refining Inhalation Risk Assessment

S. Flack[1], P. Hinderliter[1], M. T. Ledson[1], A. Szarka[1], A. Charlton[2], B. Parr-Dobrzanski[2], T. Ramanarayanan[1], D. Wolf[1]; [1]Syngenta, Greensboro, NC, [2]Syngenta, Bracknell, United Kingdom

Abstract: The Source to Outcome Approach provides a framework for improving and modernizing human health risk assessments where uncertainty is reduced through integrating hazard and exposure characterization. For non-volatile pesticides, using human-relevant particle size distributions from application scenarios with refined inhalation dosimetry models more accurately estimates a Human Equivalent Concentration (HEC) for human health risk assessments. Inhalation exposure data used by EPA in human health risk assessments is typically measured with personal air monitors and environmental sampling without consideration of particle size distributions. Side-by-side air sampling with a Respicon TM particle sampler characterized the size distribution of aerosols captured on personal air monitors (i.e., OVS tubes) during spraying to derive a particle size distribution for respiratory dosimetry modelling. Computational Fluid Dynamics (CFD) modelling further link this exposure information with target site-specific dosimetry for the human to calculate surface concentrations of deposited aerosol formulations in discrete regions of the respiratory tract. CFD models are especially useful to provide a more accurate reflection of the deposition necessary to initiate the cascade of events resulting from irritant mediated response in the upper respiratory tract. Coupling the human dosimetry model with a 3D *in vitro* model of the human respiratory epithelium was used to calculate an HEC for risk assessments.

The Source to Outcome Approach was applied, using chlorothalonil, by integrating exposure characterization and respiratory dosimetry modelling to refine the human health risk assessment of an irritant aerosol.

Keywords: A-risk assessment, B-pesticides

WE-PO-574

Introduction of the Environmental Exposure Related Activity Patterns Research for the Chinese population (Children)

X. Duan; University of Science and Technology Beijing, Beijing, China

Abstract: With regard to the 12th five-year plan for the environmental health work of national environmental protection, the Chinese Research Academy of Environmental Sciences has conducted the environmental exposure related activity patterns research for the Chinese population (CEERAPR) in two steps (Adult and children, respectively). The research on the adults (>18 yrs) was conducted from 2011-2012, and the Exposure Factors Handbook for the Chinese population (Adults) was published based on the first step study. The CEERAPR for the children (CEERAPR-C) was the second step of the research program, which targeted the population in 0~5 and 6~17 yrs, respectively, and was carried out from 2013-2014. In the CEERAPR-C research, a total of 75519 subjects from 316 schools in 165 villages, 55 counties, 30 provinces in China were covered. This following research was conducted on the basis of face-to- face questionnaire and filed measurements, and focused on the exposure factors including the inhalation factors, ingestion factors, time-activity factors, and other exposure factors such as body weight and surface area. Based on the survey and other data from scientific literatures, the Exposure Factors Handbook for the Chinese population (Children: 0-5 years and 6-17 years, respectively) were published, considering the variance on the activity patterns for the children in different ages. The above-mentioned two handbooks both contain 12 chapters, which introduced the research background and applicable scope of the handbook, the method and procedure of survey, the samples and collection principle of the subjects, and showed the above-mentioned exposure factors in the ways of regionalizing, age-specific, sex-specific, province-specific, seasons, etc. These handbooks are the first exposure factors handbooks for the Chinese children, which are now largely applied in the risk assessment in lots of scenarios, and the risk management and the criterions and benchmarks revision, etc.

Keywords: A-activity patterns, A-exposure factors, A - population exposure, D-children

WE-PO-575

Canada's Exposure Toolbox: Resources and Tools Used to Characterize Consumer Exposure in Human Health Risk Assessments Conducted under Canada's Chemicals Management Program

Y. Zhang, A. Zidek, L. MacKinnon, V. Bergeron; Existing Substances Risk Assessment Bureau, Ottawa, Canada

Abstract: The general population can be exposed to chemical substances present or released from consumer products every day. Since the launch of the Canada's Chemicals Management Plan (CMP) in 2006, over 3000 substances have been assessed in approximately 260 published assessment reports, more than 50% of which include a quantitative assessment on exposure of the general population from the use of consumer products. This presentation will describe Canada's exposure toolbox, with a focus on exposures from consumer products, covering resources and tools used for problem formulation, exposure source identification, and exposure estimation using modelling, measurement and/or biomonitoring data. Web-based automated tools for exposure data gathering will be highlighted. An exposure-oriented problem formulation database, used primarily for assessment prioritization, grouping and initial information gathering on exposure potential, will be introduced. Various in-house exposure databases and guidance documents will also be presented, including personal care products and cosmetics, dermal absorption, product migration, cleaning products, mouthing exposure, as well as models typically used in deriving consumer product exposure estimates. Case studies will be presented on human exposure to chemicals via various product uses, including children's exposure through post-application and/or mouthing behaviour. The case studies will highlight the use of the toolbox in regulatory decision making and priority setting for chemical substances manufactured, used or imported into Canada. Through the illustrative case studies, we will also discuss uncertainties and challenges encountered within the context of regulatory risk assessment, such as data availability, combined exposure to multiple chemicals, and model validation.

Keywords: A-risk assessment, C-consumer products, A-exposure models, A-exposure factors, D-children

WE-PO-576

A Systems Framework for Exposure Science to Inform Environmental and Public Health Decision-Making

R. Araujo[2], M. Bagley[1], J. Garland[1], L. Stanek[2], P. Whung[2], V. G. Zartarian[3]; [1]US EPA, Cincinnati, OH, [2]US EPA, Research Triangle Park, NC, [3]US EPA, Boston, MA

Withdrawn

WE-PO-577

Health Benefits of GHG policies in Suzhou, China, and Kuopio, Finland

M. J. Jantunen[1], M. Liu[2], A. Asikainen[1], Y. Huang[2], E. Pärjälä[3], Z. Jin[2], J. Tuomisto[1], J. Bi[2], C. Sabel[4]; [1]National Institute for Health and Welfare, Kuopio, Finland, [2]Nanjing University, Nanjing, China, [3]City of Kuopio, Kuopio, Finland, [4]Aarhus University, Aarhus, Denmark

Abstract: Background: GHG abatement and health benefits of policies adopted by two municipalities in China and Finland were assessed by the GHG-PAM model. **Material & Results:** In Suzhou (70 km West of Shanghai, China, pop. 10.6 M) the 2020-ICR scenario vs. the 2020-BAU increases the tertiary industry from 37 to 60%, reduces the automobile fleet from 240 to 180/1000, and increases the industrial and motor transport efficiencies to state of the art. 2020-ICR reduces the CO_2 emission from 297 to 173 Mt/a. Respectively the PM2.5 emission is reduced from 2326 to 1279 kt/a and the burden of disease (BoD) from 117 to 65 kDALY. In Suzhou emissions of the rapidly expanding industries dominate and all 2020-ICR impacts are still marginally higher than in 2010. In Suzhou, and also in China more broadly, the role of industries is critical for both GHG abatement and health cobenefits. In Kuopio (East Finland, pop. 112 000) the 2020-Policy scenario, vs. the 2010-Baseline replaces 50% of peat in combined heat and power generation and all heating oil with wood, renovates 3% of residences per year targeting at 25 $kWh/m^2/a$, and increases liquid biofuels in traffic from 5 to 30%. The GHG-PAM assessment of Kuopio (a) is compared to another one (b) based on same data and scenario, using dispersion and exposure rather than intake fraction modelling, and limited to the policy affected sources, fossil CO_2 and local health outcomes, summed as burden of disease (BoD) in DALYs. Replacing fossil fuels by biofuels in Kuopio reduces fossil but not the total CO2 emission, it does, however reduce secondary PM and thereby the BoD impact. In Kuopio, as one European city, the roles of industries, built urban environments and transportation are all of importance for both GHG abatement and health cobenefits. **Acknowledgement:** URGENCHE funded by EU FP 7 RTD Grant 265114.

Keywords: A-environmental policy, A-built environment, A-risk assessment, B-particulate matter

WE-PO-578

The Role of Refined Exposure Modeling in Risk Communication: Balancing Efficiency with Accuracy

C. R. Fleming; Dow AgroSciences, Indianapolis, IN

Abstract: Often in risk assessment, exposure assessments are performed in a tiered manner, starting with modeling using a very general set of assumptions, incorporating more refined exposure information only as it is needed to reach an acceptable risk conclusion. This approach maximizes the efficiency of the risk assessment process, incorporating only the amount of effort required to ensure that exposure levels are below established safety thresholds. However, the exposure estimates in these assessments are sometimes orders of magnitude above the true exposure level. As such, tiered exposure assessments present a trade-off between efficiency and accuracy. The tiered approach is quite valuable in a regulatory setting where often the number of chemicals in need of assessment far outweighs the capacity to perform those assessments. Nevertheless, this approach can present unique challenges for risk communication where quantitative estimates of risk (e.g., margin of exposure, or % of a reference dose) are generated, but may not have any direct quantitative meaning regarding the actual risk of an effect in the population being assessed. In this way, it may be valuable to invest resources in more refined exposure assessments, consequently providing more realistic representations of actual risk. This presentation will provide case studies of depiction of pesticide risk to the general public (e.g. through the media) discussing the role of refined exposure evaluation in aiding public understanding of regulatory decisions on chemical risk. Based on these case studies, generation of refined and accurate estimates of exposure for chemicals of high public interest would be valuable to improve risk communication even if not needed to reach a reasonable certainty of no harm conclusion in a risk assessment.

Keywords: A-risk assessment, A-exposure models

WE-PO-579

Harmonizing Life Cycle Human Exposure and Toxicity Assessment: A Roadmap

P. Fantke[1], O. Jolliet[2], T. McKone[3]; [1]Technical University of Denmark, Kgs. Lyngby, Denmark, [2]University of Michigan, Ann Arbor, MI, [3]University of California, Berkeley, Berkeley, CA

Withdrawn

WE-PO-580

Assessing human exposure to flame retardants in the indoor environment

C. F. Chaisson[2], J. Patterson[3], K. Diskin[2], A. Parker[3], M. Babich[1], M. Biggs[1]; [1]Consumer Product Safety Commission, Rockville, MD, [2]The LifeLine Group, Annandale, VA, [3]University of Cincinnati, formally Toxicology Excellence for Risk Assessment, Cincinnati, OH

Abstract: Flame retardants (FRs) are chemicals that are added to natural and synthetic materials to improve their resistance to ignition or reduce flame spread after ignition occurs and may result in the reduction of injuries or deaths due to fire. FRs are used in a variety of consumer products and have been detected in many media, such as indoor air, house dust, and human body fluids. Incidental ingestion of household dust is believed to be a major source of human exposure to FRs. FRs have been under scrutiny due to their health effects in animal studies, which include reproductive and developmental toxicity, chronic organ toxicity, and cancer. An exposure assessment was performed to estimate human exposure to select FR chemicals in the indoor environment using available indoor air and household dust data from the home, office, child care, and car environments. These chemicals include (tris(1,3-dichloro-2-propyl) phosphate (TDCPP), tris(chloropropyl) phosphate (TCPP), tris(2-chloroethyl) phosphate (TCEP), triethyl phosphate (TEP), triphenyl phosphate (TPP), 2-ethylhexyl 2,3,4,5-tetrabromobenzoate (TBB), di(2-ethylhexyl) tetrabromophthalate (TBPH), tetrabromobisphenol A (TBBPA), and antimony trioxide (ATO). Using a probabilistic exposure assessment methodology, an aggregate exposure assessment was performed resulting in an exposure assessment for each chemical, considering multiple routes of exposure (i.e., oral, dermal, inhalation). Infants and children 1 to <3 years experienced their highest exposures from TCPP levels in the home environment (indoor air and dust). The highest contributor for adult combined exposure was air and dust TCEP concentrations in the office environment. The FR studies and publications used in this assessment presented substantial information on exposure concentrations but often lacked information on experimental design, collection methodologies, and description of environments and FR sources, which limited the reliability of any conclusions.

Keywords: B-flame retardants, A-aggregate exposure, A-indoor environment, C-consumer products

WE-PO-581

Quantitative Property-Property Relationships for Estimating Diffusion Coefficients in Solid Materials and Solid Material-Air Partition Coefficients of Organic Compounds

L. Huang, O. Jolliet; University of Michigan, Ann Arbor, MI

Abstract: Indoor releases of organic chemicals encapsulated in solid materials are major contributors to human exposures and are directly related to the internal diffusion coefficient in solid materials (D) and material-air partition coefficient (K_{ma}). Existing correlations to estimate D and K_{ma} are only valid for a limited number of chemical-material combinations. The aim of the present study is to develop quantitative property-property relationships (QPPRs) valid for a large number of chemical-material combinations. We compiled datasets of measured D (K_{ma}) from 68 (36) literature references, which contained 1103 (911) data points for 158 (164) chemicals in 32 (21) consolidated material types. Multiple linear regression (MLR) models were developed to predict D (K_{ma}) as a function of the chemical's MW (K_{oa}), temperature and the material type. Both the D and K_{ma} models show good fitting performance of the experimental datasets with adjusted R-squares of 0.93 and 0.92, respectively. The models have been verified by internal validation (cross-validation and Y-scrambling) and external validation to be robust, internally stable and have good predicting ability ($R^2_{ext} > 0.8$ for D and $R^2_{ext} >0.9$ for K_{ma}). Since chemical properties are more influential than material attributes for K_{ma}, a generic QPPR is also developed to predict K_{ma} from K_{oa} only, with adjusted R^2 of 0.81, for use with material types without experimental data. These QPPRs provide a comprehensive correlation method to estimate the D and K_{ma} as they cover a much wider range of organic chemicals and solid materials compared to previous studies, and provide a relatively accurate generic correlation of K_{ma} for materials without experimental data. The QPPRs will thus facilitate high-throughput estimates of human exposures for chemicals encapsulated in solid materials, particularly in building materials and food contact materials.

Keywords: A-indoor environment, A-statistical methods, C-consumer products, C-indoor

WE-PO-582

Indoor dust ingestion rate for Japanese children

M. TAKAGI, T. Isobe, M. IWAI-SHIMADA, S. F. Nakayama; National Institute for Environmental Studies, Japan, Tsukuba, Japan

Abstract: Indoor dust is one of the potential route of exposure to chemical substances such as lead, frame retardants and phthalates. As shown in the previous studies, children are more likely to have contact with dust and thus ingest larger amount. Dust ingestion rates are regularly used for exposure assessment. However, indoor dust ingestion data for Japanese children is scares. In this study, we estimated the daily indoor dust ingestion for Japanese children using phthalates and their metabolites as tracers. Preschool children at the age of 3-6 (n = 104) were recruited. Indoor dust and indoor air (24 hours) were collected from their house. Duplicate diet (one day) and 24-hour pooled urine were sampled. Urine pooling cycle were from the 2nd urine on the day which duplicate diet was collected to the first morning void on the next day. Phthalates (di(2-ethylhexyl) phthalate (DEHP), dibutyl phthalate (DBP) and butyl benzyl phthalate (BBzP)) in indoor dust, indoor air and diet samples were measured by gas chromatography tandem mass spectrometry (GC-MS/MS). Urine samples were analysed for phthalate metabolites using liquid chromatography tandem mass spectrometry (LC-MS/MS). As a feasibility study, the first five sample data were analysed. DEHP daily intake was estimated to be 37-122 μg day^{-1} from urine metabolites. DEHP intake via indoor dust was calculated by subtracting the intake via indoor air and diet from the total daily intake. It was 16-58 μg day^{-1}. DEHP concentrations in indoor dust were 640-2800 μg g^{-1}. Daily indoor dust ingestion was thus (DEHP Intake via indoor dust) / (DEHP concentrations in indoor dust), ranging 9-250 mg day^{-1}. Those are comparable to previous reports from other countries. Further details of the method and its full result will be shown in the presentation.

Keywords: A-exposure factors, C-indoor, D-children, B-phthalates

WE-PO-583

Indicators of human indoor exposure: thyroid hormone disruptors identified in cat blood and house dust by Effect-Directed Analysis

M. Lamoree[1], L. Lucattini[1], N. Zwart[1], P. Cenijn[1], J. Weiss[2], T. Hamers[1]; [1]Vrije Universiteit Amsterdam, Amsterdam, Netherlands, [2]Swedish University of Agricultural Sciences, Uppsala, Sweden

Abstract: Indoor pet cats are exposed to thyroid hormone disrupting compounds (THDCs) through dust. We have investigated cat blood plasma as model for indoor human exposure to THDCs. Effect-Directed Analysis (EDA) combines biotesting, fractionation and chemical analysis to identify toxicants in complex matrices. EDA has already demonstrated to be a valid tool for identifying THDCs in blood serum [Simon et al., 2013] and environmental matrices [Ouyang et al., 2017]. For cat blood and house dust, extracts were prepared for EDA. The samples causing a response in the fluorescence T4-TTR binding assay were further fractionated into 96 well plates and tested in the bioassay again as well as screened for the presence of THDCs using liquid chromatography coupled to QuadrupoleTime-of-Flight Mass Spectrometry (LC-QToF-MS). The microfractionation was done using Ultra performance liquid chromatography (UPLC) into 96-well plates, thus avoiding time-consuming evaporation steps of the excess solvent and enabling high throughput. Prior to the fractionation of the dust and cat blood extracts, the procedure was optimized by the evaluation of the fractionation pattern of a test mixture of known thyroid hormone disrupting compounds, such as the OH-PCBs and several flameretardants. LC-QToF-MS using electrospray ionization (ESI) was used for suspect screening and nontarget analysis. The identification of suspects was done by sequential use of several instrument (Bruker Daltonics) software tools. For suspect screening, an 'in house' database was used containing around 11000 chemicals including chemicals reported in dust and consumer products (e.g. plasticizers and flame retardants), potential persistent and bioaccumulative (P&B) compounds, P&B transformation and by-products, impurities and pharmaceuticals. In addition, non-target screening was performed for the identification of the compounds responsible for the observed thyroid hormone disruptive effects.

Keywords: A-analytical methods, A-indoor environment, C-indoor, D-children, C-consumer products

WE-PO-584

Persistent Organic Pollutants in Dust Collected from Elderly Care Facilities in Portugal and the United States

A. Salamova[1], K. Arnold[1], J. Teixeira[2]; [1]Indiana University, Bloomington, IN, [2]National Institute of Health, Porto, Portugal

Abstract: In this study, forty-two indoor dust samples were collected from nursing homes in Portugal and in the United States, and the concentrations of organophosphate esters (OPEs), polybrominated diphenyl ethers (PBDEs) and other brominated flame retardants (BFRs), polycyclic aromatic hydrocarbons (PAHs), organochlorine pesticides (OCPs), and polychlorinated biphenyls (PCBs) were measured in these dust samples. Measured concentrations were compared based on the sampling location (Portugal vs. United States) and the microenvironment type in which the samples were collected (bedrooms vs. drawing rooms vs. corridors). Total daily intake rates of selected target chemicals were estimated based on dust ingestion and compared with established reference doses. Overall, OPEs and PAHs were the most abundant and OCPs and PCBs were the least abundant chemical groups in both Portuguese and United States' dust. ∑OPE, ∑PAH, ∑BFR, and ∑PBDE concentrations were significantly higher in United States' dust than those in Portuguese dust ($P < 0.001$), while ∑OCP and ∑PCB concentrations were not different between the two countries (at $P < 0.05$). ∑OPE, ∑PAH, ∑BFR, and ∑PBDE concentrations were up to 26 times higher in corridors compared to bedrooms and drawing rooms. ∑OCP and ∑PCB concentrations were overall higher in bedrooms and in drawing rooms and lower in corridors. Total daily intake estimates were based on the measured dust concentrations in bedrooms, drawing rooms, and corridors, and on the amount of time spent in each of these microenvironments. All total daily intake rates were lower than corresponding reference doses.

Keywords: A-indoor environment, B-flame retardants, B-POPs, A - ambient monitoring, A - population exposure

WE-PO-585

Using SHEDS-S/D to Estimate Soil and Dust Ingestion Rates for Children

H. Hubbard[1], G. Glen[1], H. Ozkaynak[1], N. S. Tulve[2], L. Phillips[2], J. Cohen[1], K. Thomas[2], J. Moya[2]; [1]ICF, Durham, NC, [2]US EPA, Durham, NC

Abstract: Soil and dust ingestion can be the primary pathway for environmental exposure to some pollutants. Studies have shown that young children, due to greater mouthing behavior than older children or adults, are more vulnerable to incidental ingestion of soil and dust. However, available data to support the development of age-specific soil and dust ingestion rates are either limited or uncertain for most age groups. Our objective was to use the Stochastic Human Exposure and Dose Simulation Soil and Dust (SHEDS-S/D) model to estimate distributions of soil and dust ingestion rates for infants and children to determine if data gaps could be filled using this modeling approach. We developed a new exposure scenario in SHEDS-S/D to capture exposures to indoor dust via pacifier use. This exposure scenario accounts for the use of blankets or similar surfaces that prevent direct contact of the pacifier with the floor. Although the inputs for this scenario were uncertain, the scenario was estimated to contribute about 20 mg/day to the median dust ingestion rate, whereas the overall mean soil and dust ingestion rate ranged from 30 mg/day (infants) to 60 mg/day (toddlers). We also conducted a sensitivity analysis to determine key drivers of exposure. For infants and younger children, pacifier use drove exposure with the most sensitive variables being carpet dust loading, pacifier drop frequency, and floor-to-pacifier transfer fraction. For older children, key variables were carpet dust loading, soil adherence, and four hand properties (frequency of hand-to-mouth contact, area of the hand mouthed, amount of floor contacted, and fraction of dust removed by each hand-to-mouth event). These results can be used to identify data limitations in developing robust soil and dust ingestion rates and to focus resources on variables that will provide the most insight into developing age-specific rates, as well as describing variation in ingestion rates.

Keywords: A-exposure models, A - population exposure, D-children, C-indoor, C-soil

WE-PO-586

Estimating Surface/Air Partition Coefficients for SVOCs on Interior Surfaces

C. Eichler, Y. Wu, L. Marr, J. Little; Virginia Tech, Blacksburg, VA

Abstract: Sorption of semi-volatile organic compounds (SVOCs) onto interior surfaces, often referred to as the “sink effect”, and their subsequent re-emission significantly affect the fate and transport of indoor SVOCs and the resulting human exposure. Unfortunately, experimental challenges and the large number of SVOC/surface combinations have impeded progress in understanding sorption of SVOCs on indoor surfaces. An experimental approach based on a diffusion model was thus developed to determine the surface/air partition coefficient K of di-2-ethylhexyl phthalate (DEHP) on typical impervious surfaces including aluminum, steel, glass, and acrylic. The results indicate that surface roughness plays an important role in the adsorption process. Although larger data sets are needed, the ability to predict K could be greatly improved by establishing the nature of the relationship between surface roughness and K for clean indoor surfaces. Furthermore, different surfaces exhibit nearly identical K values after being exposed to kitchen grime with values that are close to those reported for the octanol/air partition coefficient. This strongly supports the idea that interactions between gas-phase DEHP and soiled surfaces have been reduced to interactions with an organic film. Collectively, the results provide an improved understanding of equilibrium partitioning of SVOCs on impervious surfaces.

Keywords: A-exposure models, B-SVOCs, A-indoor environment

WE-PO-587

Sex Differences in the Impact of Sire Exposure to BPA on Weight Gain, Glucose Tolerance, and Metabolic Endpoints in Offspring

S. J. Sumner[1], T. Fennell[2], S. McRitchie[1], H. Allardice[3], A. Mason[3], E. Rissman[3]; [1]University of North Carolina at Chapel Hill, Chapel Hill, NC, [2]RTI International, RTP, NC, [3]North Carolina State University, Raleigh, NC

Abstract: Studies using model systems, and epidemiology investigations, have raised concerns regarding the potential for bisphenol A (BPA) to disrupt the endocrine system and cause weight gain, diabetes, behavioral disorders, and cardiovascular disease. Our study investigated the impact of sire exposure to BPA on weight gain, glucose tolerance, and metabolic endpoints in male and female offspring. Adult male CD1 mice were fed diets containing 0 mg/kg BPA (control sires) or 5 mg/kg BPA (BPA exposed sires) for 50 days before mating. The F1 offspring were tested for glucose tolerance. Male mice born to dams who mated with BPA exposed sires had an increase in body weight starting at 49 days and continuing to 160 days of age at which time they were less glucose tolerant than their controls. Females sired by BPA-exposed males were heavier than their controls starting at 35 days of age, but were not different at 160 days old. Serum, liver, and brain from 60 day old mice were analyzed via a targeted LC-MS/MS method for 188 low molecular weight metabolites. In general, female mice born to dams mated with BPA exposed sires (compared with controls) showed an increase in serum acylcarnitines, amino acids and biogenic amines, and a decrease in sphingolipids and glycerophospholipids, while metabolites from each of these classifications increased in the serum of male mouse offspring (compared with controls). Perturbations in metabolites of these classifications were also observed in brain and liver of BPA exposed groups compared with control. This study demonstrates sex differences in the impact of sire exposure to BPA on metabolic endpoints in male and female offspring.

Keywords: B-BPA, A-biomarkers

WE-PO-588

Health effect of different gaseous formaldehyde fluctuation forms on mice: a preliminary study

X. Zhang[1], Y. Zhao[2], X. Yang[2], Y. Zhang[1], R. Li[2]; [1]Tsinghua University, Beijing, China, [2]Central China Normal University, Wuhan, China

Abstract: Formaldehyde (FA), among the most hazardous air pollutants in indoor environment, will result in diseases such as asthma, lung cancer and leukemia at certain dose. However, so far very few experiments covering its health effects of different fluctuation forms have been reported, especially those existing in the actual residence environment. The aim of this study is to compare the health damage difference between FA at fluctuant exposure (based on the real monitoring result) and at constant exposure.Mice were divided into 3 groups for fluctuant exposure (12 hours 1 mg/m^3 and 12 hours 0 mg/m^3), constant exposure (24 hours 0.5 mg/m^3) and zero exposure (no formaldehyde) per day for 7, 14 and 28 days, respectively. After exposure, the bronchoalveolar lavage fluid (BALF), lung tissue and its homogenate as well as marrow samples were prepared to examine the relevant biomarkers. It was shown that exposure to FA could induce elevated level of both oxidative stress and inflammation in the lung tissue of mice. Additionally, the inflammatory and pro-inflammatory cytokines like IL-4, IL-5, IL-13, IL-6, IL-17 and IL-1β were found to be up-regulated by FA exposure. What's more, total number of inflammatory cells and the number of some specific inflammatory cells as well both exhibited an increase in the BALF when exposed to fluctuating FA. Interestingly, our study has demonstrated that the fluctuating FA exposure rather than the constant FA exposureplayed a more intensified role in promoting the above biological alterations and even the apoptosis. This study has confirmed from several aspects that fluctuating FA exposure could be more likely to exert an adverse influence on the health of mice, leading to some guidance information for the construction of healthy indoor environment.

Keywords: C-air, A-epidemiology, B-VOCs, C-indoor

WE-PO-589

Expression of genes involved in stress, toxicity, inflammation, and autoimmunity, and levels of heavy metals in human blood

R. Monastero[1], C. Vacchi-Suzzi[1], C. Marsit[2], B. Demple[1], J. Meliker[1]; [1]Stony Brook University, Stony Brook, NY, [2]Emory University, Atlanta, GA

Abstract: Objectives: Exposure to heavy metals is associated with a range of diseases, but the pathways involved are not certain. Abundance of gene (mRNA) transcripts can be measured in participants' blood and contribute to our understanding of how the body reacts to metals. This study aimed to correlate blood levels of three metals (mercury (Hg), cadmium (Cd), and lead (Pb)) with gene expression in 24 participants from the Long Island Study of Seafood Consumption. Methods: We measured Hg, Cd, and Pb in blood, and peripheral blood mRNA expression of 98 genes implicated in stress, toxicity, inflammation, and autoimmunity. We fit multiple linear regression models with multiple testing correction to correlate metals with mRNA abundance. Mean blood Hg was 16.1 μg/L, nearly three times the EPA reference dose (5.8 μg/L); levels of other metals were consistent with background levels (Mean Pb: 2.7 μg/dL; Mean Cd: 0.4 μg/L). Results: Three genes were significantly associated with Hg, four with Cd, and five with Pb, though none were significant after multiple testing correction. We examined whether genes involved in similar/interacting functional pathways, what we refer to as "related genes", showed similar directions of association. IL1RAP, CXCR1, and ITGB2 and their related genes showed negative directions of association with Hg. MMP9 and VEGFA, along with their related genes were positively correlated with Pb. No other related genes showed consistent directions of association. Conclusions: Overall, in this cross-sectional study with one-time measures of metal levels and gene expression, little evidence was found to associate metal exposure with mRNA abundance involved in stress, toxicity, inflammation, and autoimmunity. Potential genes and related genes were suggested for association with specific metals; future work should include larger sample sizes and a wider array of analyzed genes for a more complete picture of physiological reactions to metals exposure.

Keywords: B-metals

WE-PO-590

Comparison of Genetic and Environmental Mechanisms of Motor Neuron Death by Contemporary Proteomics

J. Beri[1], T. Nash[2], M. Bereman[3]; [1]NC State University, Raleigh, NC, [2]NC State University, Raleigh, NC, [3]NC State University, Raleigh, NC

Abstract: Amyotrophic lateral sclerosis (ALS) is a fatal disease characterized by motor neuron death with a median survival time from symptom onset of 3 years. Evidence suggests that environmental exposures play a significant role in the development of ALS. Beta-methylamino-L-alanine (BMAA) has been investigated as a potential environmental link to ALS. Though the toxic mechanisms of BMAA exposure remain poorly understood, BMAA is thought to cause cellular stress via misincorporation into cellular protein. In this study, we perform shotgun proteomics to investigate the cellular pathways perturbed by in-vitro BMAA exposure using the NSC-34 cell line. In addition, we explore genetic mechanisms by proteomic comparison of NSC-34 cells expressing wild type and mutant (G85R and G93A) human SOD1. We also investigate the possibility of gene-environment interaction by exposure of NSC-34 cells expressing mutant human SOD1 to BMAA. FInally, we investigate the hypothesis that BMAA becomes misincorporated into cellular protein via amino acid analysis and through use of an in-vitro cell-free protein synthesis kit.

Keywords: A-analytical methods, A-biomarkers

WE-PO-591

Differences in biological activity among household dust samples detected by changes in expression of biomarker genes in zebrafish *Danio rerio*

T. Henry[2], C. Papageorgakopoulou[2], M. Loh[1]; [1]Institute of Occupational Medicine, Edinburgh, United Kingdom, [2]Heriot-Watt University, Edinburgh, United Kingdom

Abstract: Household dust is a complex mixture of particulates with sorbed chemicals. Since assessment of all complex mixture substances by analytical chemistry is not practical, this study investigated relative biological activity among dust samples by *in vivo* assessment of biomarkers of toxicant exposure with the zebrafish *Danio rerio* model. Samples were collected from vacuum cleaners of 6 households in Edinburgh UK, sieved to < 63 µm and extracted using *n*-hexane:acetone (3:1). Three independent replicates of zebrafish larvae were exposed to each dust extract, spiked positive controls, or vehicle controls. Total RNA extracted from the larvae was reverse transcribed to generate cDNA for quantitative real-time PCR. Relative changes in gene expression of *cyp1a* (a biomarker of oxidative stress and metabolism of toxicants such as polycyclic aromatic hydrocarbons) and *vtg* (a biomarker of exposure to estrogenic substances) were compared among dust samples and positive controls. Significant induction of *cyp1a* ($p<0.05$) was observed in two household dust extracts compared to the other extracts and the vehicle control. Although the spiked positive control induced *vtg* expression, there was no induction of *vtg* transcripts with exposure to un-spiked samples. Results indicate the presence of organic toxicants in some households at sufficient quantities to induce *cyp1a* activity, and that households differ in the presence of bioactive substances associated with dust. While numerous substances present within households are suspected of having estrogenic activity, none of the dust samples had sufficient activity to induce *vtg* expression in larval zebrafish. This study demonstrates that the zebrafish model is an effective screening tool to identify biological activity in household dust samples. Dust samples with biological activity can be prioritized for further examination of compounds present, or for examination of factors that contribute to biological activity within specific households.

Keywords: A-biomarkers, B-SVOCs, C-indoor, B-mixtures

WE-PO-592

Evaluating the synergistic effects of cyanotoxic mixtures on the ALS pathway using targeted proteomics and statistical design of experiments

R. Martin, M. Bereman; NCSU, Raleigh, NC

Abstract: Amyotrophic lateral sclerosis (ALS) is a devastating fatal disease characterized by the death of motor neurons which leads to the loss of voluntary muscle control - ultimately robbing the individual of the capability to walk, eat and eventually breathe. ALS is classified in to two categories: idiopathic (sporadic ALS) and heritable (familial ALS). Familial forms of ALS have autosomal inheritance patterns and account for about 10% of all ALS cases. The cause for the other 90% of sporadic ALS cases remains poorly understood. Current research suggests that environmental stressors may contribute to the development of sporadic ALS. β-methylamino-L-alanine (BMAA) is a neurotoxin produced ubiquitously by cyanobacteria and has been linked to various neurodegenerative disorders, including sporadic ALS. In addition to BMAA, cyanobacteria produce other known neurotoxins such as anatoxin, mycrocystins, and 2,4 diaminobutyric acid (2,4 DAB). To further investigate if cyanotoxins mixtures could potentially act in a synergistic manner to enhance neurodegeneration and promote the onset of sporadic ALS, we performed targeted mass spectrometry (MS) based proteomics of a preselected group of proteins that are associated with motor neuron degeneration in the ALS pathway. We used the NSC-34 cell line, which is an increasingly used in vitro model for neurodegenerative diseases, as it shares many properties of motor neurons cells. We employ statistical design of experiments, a powerful approach in process optimization, to screen for synergistic effects of mixtures of cyanotoxins. Taken together, we intend to develop new methodologies to screen for interactions amongst chemicals and apply these methods to elucidate the mechanisms at which the environment initiates motor neuron degeneration.

Keywords: A - exposure measurement, A-analytical methods, A-statistical methods, B-mixtures, B-microbial agents

WE-PO-593

Aldehyde dehydrogenase 2 (ALDH2) polymorphism and metabolism of aldehydes

T. Kawamoto, M. Tsuji, R. Tanaka; University of Occupational and Environmental Health, Kitakyushu, Japan

Abstract: It is well known that many Asian people, especially half of Japanese people, genetically lack Aldehyde dehydrogenase (ALDH) 2 which is an isozyme of ALDH family. However, the influence of the ALDH2 polymorphism on the metabolite production of the chemicals which undergo ALDH metabolism is not clear, even though the metabolites are used as exposure markers. We investigated the metabolism of propionaldehyde and dodecyl aldehyde in Aldh2 knockout mice, which have been developed in our laboratory. The ALDH activity toward propionaldehyde was measured by production of NADH. At low concentration (50mM), ALDH activity of mitochondrial fraction was more than 10 times lower in the Aldh2 knockout mouse than in the control mouse. Enzymatic activity was also studied by PAGE-isoelectric focusing (IEF) and zymogram. When propionaldehyde is used as substrate, an activity band of dehydrogenation was found at pI 6.0 only in the control mouse liver fraction, but not in the Aldh2 knockout mouse liver fraction. This band was reacted with anti-ALDH2 antibody by the Western blot analysis. When dodecylaldehyde was used as substrate, the difference between the knockout mouse and control mouse was not significant in NADH production or zymogram. This result suggests the ALDH2 polymorphism affects metabolism of some aldehydes.

Keywords: A-biomarkers, A-activity patterns, D-susceptible/vulnerable

WE-PO-594

Proteomic analysis of cerebrospinal fluid and plasma in patients with Amyotrophic Lateral Sclerosis

T. Nash[1], J. Beri[3], M. Bereman[2]; [1]North Carolina State University, Raleigh, NC, [2]North Carolina State University, Raleigh, NC, [3]North Carolina State University, Raleigh, NC

Abstract: Amytrophic Lateral Sclerosis (ALS) is a fatal illness that is characterized by the progressive loss of upper and lower motor neurons. Greater than 90% of ALS incidence is not attributable to known genetic causes and recent studies suggest that environmental factors may play a significant role in disease pathogenesis. Currently, clinical evaluation is the main form of diagnosis leaving a delay between symptom onset and diagnosis, effectively reducing the therapeutic window. Identification of biomarkers for the disease will aid in early diagnosis, assessment of disease progression and may facilitate the advancement of drug therapeutics. In this study, we use LC-MS/MS to evaluate changes in protein expression in cerebrospinal fluid (CSF) and plasma from 20 ALS and 20 healthy patients. Due to highly abundant proteins in CSF and plasma, we increased proteome coverage through optimization of protein depletion, allowing for a more comprehensive evaluation of differential protein expression profiles. The work detailed herein serves as an initial step toward identification of clinical biomarkers for use in ALS diagnosis.

Keywords: A-biomarkers

WE-PO-595

Optimizing a Sensor Network with Data from Hazard Mapping Demonstrated in a Heavy-Vehicle Manufacturing Facility

J. D. Berman[1], T. M. Peters[2], K. Koehler[3]; [1]Johns Hopkins School of Public Health, Newington, CT, [2]University of Iowa School of Public Health, Iowa City, IA, [3]Johns Hopkins School of Public Health, Baltimore, MD

Abstract: Hazard mapping is an increasingly popular technique to enhance occupational exposure assessments and risk communication. In an alternate application, we use preliminary hazard mapping data to optimize the number and location of sensors within a network for a long-term assessment of occupational concentrations. Our goal is to reduce sensor locations, while preserving temporal variability, accuracy, and precision of predicted hazards. Particle number concentrations (PNCs) and respirable mass concentrations (RMCs) were measured with direct-reading instruments in a large heavy-vehicle manufacturing facility at 82 locations during 7 mapping events, stratified by day and season. Using kriged hazard mapping, a statistical Monte Carlo approach identified optimal orders for removing locations to capture temporal variability and high prediction precision of PNC and RMC concentrations. We compared optimal-removal, random-removal, and least-optimal-removal orders to bound prediction performance. Temporal variability of PNC was found to be higher than RMC with low correlation between the hazards (ρ=0.30). Optimal-removal orders resulted in more accurate PNC kriged estimates (Root Mean Square Error, RMSE=49.2) compared to random-removal order (RMSE=55.7). For estimates at locations having concentrations in the upper 10th percentile, the optimal-removal order preserved estimated concentrations better than random- or least-optimal-removal orders ($p<0.01$). However, the optimal-removal was not statistically different than random-removal when averaged over the entire facility. No statistical difference was observed for optimal- and random-removal methods for RMCs that were more spatially and temporally homogeneous than PNCs. These results provide evidence that the number of locations used in a network of static sensors for long-term monitoring of hazards in the workplace can be reduced, without sacrificing prediction performance.

Keywords: A-industrial hygiene, A-geospatial analysis/GIS, A-sampling methods, D-occupational, C-air

WE-PO-596

Historical Population Exposure to Fine Particulate Matter Extracted by Spatiotemporal Interpolation from 2005 to 2015 across the Contiguous U.S.

X. Zhou[1], L. Li[2], W. Tong[2], G. Besenyi[3]; [1]Georgia Southern University, Statesboro, GA, [2]Georgia Southern University, Statesboro, GA, [3]Augusta University, Augusta, GA

Abstract: Fine particulate matter ($PM_{2.5}$) isassociated with increased risk of mortality and respiratory diseases. Epidemiological studies consistently show an association between atmospheric particle pollution and the number of deaths from cancer, and cardiovascular and respiratory diseases. To further investigate the links between $PM_{2.5}$ and adverse health effects, it is imperative to estimate $PM_{2.5}$ exposure in a continuous space-time domain. To estimate $PM_{2.5}$ density and distribution, spatial interpolation methods have been well developed to estimate values at unknown locations, but few approaches have concurrently considered the contribution of data in the time dimension. Integrating space and time simultaneously is anticipated to yield better interpolation results than treating them separately for typical GIS applications (Li *et al.*, 2012). Unfortunately, there are far fewer models for spatiotemporal interpolation compared with spatial interpolation (Li *et al.*, 2004, Li *et al.*, 2016), especially in the application of air pollution where data varies within a large time domain. In addition, continuous exposure to a higher level of $PM_{2.5}$ may have a much severer impact on public health than intermittent exposure. Hence, it is important to consider long-term, temporal variation of air pollution exposure across multiple years. In this study, we applied innovative spatiotemporal interpolation techniques based on cloud computing to process large amount of historical $PM_{2.5}$ data. Results include monthly $PM_{2.5}$ values at census block group level in the past 10 years from 2005 to 2015. We linked the pollution levels with the demographics across the U.S. The spatiotemporal exposure levels to population across the contiguous U.S. were quantified and visualized in our results. In summary, this study explored population exposure to

$PM_{2.5}$ in the contiguous U.S. from 2005 to 2015 and provided public health implications based on $PM_{2.5}$ exposure at both geography and time dimensions.

Keywords: A-geospatial analysis/GIS, A-environmental justice, A-exposure models

WE-PO-597

Spatial point pattern analysis of congenital heart defects in Lanzhou, China

L. Jin[1], J. Reuning-Scherer[1], J. Qiu[2], Q. Liu[2], Y. Zhang[1], M. L. Bell[1]; [1]Yale University, New Haven, CT, [2]Gansu Provincial Maternity and Child Care Hospital, Lanzhou, China

Abstract: Analysis of spatial point processes has been used to investigate spatial variation of disease risks, and their relationships with environmental factors; however, few studies have been conducted in China, where emissions from industries and heavy traffic pose serious public health concerns. The objective of this work is to investigate the impacts of major point sources and road networks on congenital heart defects (CHD) risks in Lanzhou, China, where maternal exposure to ambient air pollution was linked to increased risks of CHD. From a Lanzhou birth cohort during 2010-2012, 8,227 singleton live births with home addresses in the city urban area were included in this study. K, L, and Kcross functions were used to detect clustering tendency of the CHD (n=65) and healthy infants (n=8162). A Kernel density ratio was used to identify potential clusters of CHD cases adjusting for the distribution of healthy infants. A Poisson point process model was used to model the intensity of CHD cases as a function of emission-weighted distance to major point sources (power plants and cement factories), road length density within 100m buffers, the estimated intensity of healthy infants, maternal income, and education. Similar distributional patterns were identified for CHD cases and healthy infants. Adjusting for the distribution of healthy infants, maternal income and education, CHD risks were significantly associated with increased road length density within 100 m buffer (RR: 1.07) and decreased emission-weighted distance from major point sources (RR: 0.07). Chi-square tests of quadrat counts and Kolmogorov-Smirnov tests validated the model and showed no spatial clustering of residuals. Results indicate proximity to major point sources and road networks have adverse impacts on newborn's health in Lanzhou, China. The identified clusters of CHD cases may allow policy makers or future mothers to make informed decisions regarding exposure control and risk management.

Keywords: A-geospatial analysis/GIS, A-epidemiology, D-prenatal

WE-PO-598

3-Dimensional Paper Microfluidic Devices for the Detection of Toxins in Environmental Matrices by Paper Spray MS

R. Carter[1], M. Bereman[2]; [1]NC State University, Raleigh, NC, [2]NC State University, Raleigh, NC

Abstract: Gas or liquid chromatography coupled to mass spectrometry is the gold standard analytical tool for traditional toxicity testing in a variety of environmental matrices. These instruments are both sensitive and specific and thus are powerful tools for detection of environmental chemicals. However, they are expensive, require a high level of expertise/training, and are not easily field deployable. Paper-based tests are a promising avenue for environmental monitoring owing to the substrate's inexpensive nature and various physical properties (light weight, porous), and have been used for centuries in analytical testing. The Bereman laboratory at NC State University was the first to combine paper spray mass spectrometry (PS-MS), an ambient ionization technique for mass spectrometry, to paper microfluidics for improved versatility and analytical performance. In constructing a 3-dimensional paper based microfluidic device, wax-designed paper with cationic exchange and physical filters were layered to extract and enrich for both validation and quantitation of known environmental toxins: beta-methylamino-L-Alanine (BMAA), atrazine, and propazine. In this study, sample loss per paper layer, the efficacy of cationic exchange, and utilizing the constructed the device for internal-standard quantitation has been evaluated for BMAA, atrazine, and propazine. Preliminary results show an average sample loss of 53 ± 10% between each layer of the device. Results also show the average peak area for each of the toxins was significantly larger while utilizing cationic exchange when compared to only using physical filter layers as a means of extraction.

Keywords: A-analytical methods, A - ambient monitoring, A-sampling methods, C-soil, C-water

THURSDAY, OCTOBER 19, 2017

TH-PL-A1: Behavioral & Policy Interventions for Traffic Related Air Pollution

TH-PL-A1-599

Study on Environmental Health Effects of Public Bicycles

M. Liu, Z. Jin, X. Liu, Y. Huang, J. Bi; Nanjing University, Nanjing, China

Abstract: Many studies have assessed environmental health effects of shifting from other transport modes to Public Bicycles and showed that the benefits are far greater than the risks. However, existing studies have two common limitations: (1) The pattern of transport mode substitution is mostly determined based on assumption; (2) Few of localized studies have explored health risk of Public Bicycles in China. Therefore, we assessed environmental health effects of Public Bicycles Program in Nanjing, China based on actual usage data. The data of actual usage and transport mode substitution are obtained from Nanjing Public Bicycle Co., Ltd. and local questionnaire survey was conducted from September to November 2016, respectively. 461 valid questionnaires were collected. COPERT-4 model is applied to calculate the emissions reductions and the linear dose-response functions are used to calculate health effects. The results show that Public Bicycles program in Nanjing brings 13.65 tons of CO, 1.10 tons of NOx, 1.29 tons of VOC, 6.84 kilograms of particulate matter and 626 tons of CO_2 emissions reductions in 2015. According to our survey, if Public Bicycles is not available, 29.93%, 24.51%, 3.90%, 2.39%, 35.14%, and 4.12% of all respondents would use bus, metro, taxi, car, walking, and private bike as a substitution for particular trip made by Public Bicycles. If the users shift to use bikes 13 minutes per day from cars, buses, subways, the average $PM_{2.5}$ exposure concentration would increase about 4 μg/m^3 in Nanjing. And if shifting from walking, it will reduce 1 μg/m^3. 17.48 deaths were avoided due to increased physical activity, while 3.21 deaths were caused by changes of air pollution exposure in Nanjing in 2015. Overall, the benefits of Public Bicycles program in China are much smaller than what they are supposed to be.

Keywords: A-activity patterns, A-risk assessment, C-air

TH-PL-A1-600

The effectiveness of facemasks used to protect Beijing residents against particulate air pollution

M. Loh[1], J. W. Cherrie[2], A. Apsley[1], H. Cowie[3], W. Mueller[3], A. Sleeuwenhoek[1], S. Steinle[1]; [1]Institute of Occupational Medicine, Edinburgh, United Kingdom, [2]Heriot-Watt University, Edinburgh, United Kingdom, [3]Institute of Occupational Medicine, Edinburgh, United Kingdom

Abstract: Due to high particulate air pollution concentrations in Beijing, many residents use disposable facemasks in an attempt to protect their health. The aim of this study was to evaluate the effectiveness of a range of facemasks commercially available in China. Nine masks were purchased from consumer outlets in Beijing. The filtration efficiency of each mask was tested by drawing an aerosol of diesel exhaust through the material and measuring the black carbon aerosol concentration upstream and downstream of the material. The mean filter penetration for each mask material ranged from 0.26% to 21% under a flow rate of 40 l/min and from 0.71% to 29% under 80 l/min. Based on these test results four masks were selected for further evaluation. Volunteers who were generally not experienced at wearing respirators were exposed to diesel exhaust inside an experimental chamber for approximately 20 minutes, on two occasions (one with sedentary tasks and one with active tasks). Black carbon air concentrations were continuously monitored inside and outside the mask throughout the test. The average total inward leakage (TIL) of black carbon aerosol in the volunteer tests ranged from 3% to 68% in the sedentary tests and from 7% to 66% in the active tests. Only one of the four masks tested showed an average TIL of less than 10% under both test conditions. Our results demonstrate that TIL should be measured in health effects studies of facemasks, as exposure may vary greatly between individuals. Furthermore, our results imply that there should be greater governmental oversight on mask regulation and communication of mask effectiveness and proper fit for public consumption.

Keywords: B-particulate matter, D-community, C-air

TH-PL-A1-601

Public's Risk Perception and Willingness-to-Pay for Air Pollution

D. Bi[2], S. Pu[1], J. Liu[2], S. Wu[2], J. Ding[2], M. Fang[1], X. Lu[2], M. Liu[1]; [1]Nanjing University, Nanjing, China, [2]Nanjing foreign language school, Nanjing, China

Abstract: Frequent haze event in China raised great public concerns on air pollution issue. This work analyzed temporal and spatial difference of four risk perception factors (risk cognition, worry, perceived benefit and trust), risk acceptance and willingness to pay (WTP) (shorted as "five variables and WTP"), so as to provide decision-makers science-based evidence. Specifically, Likert scale and contingent value method (CVM) were applied to quantitatively measure five variables and WTP. We carried out 3-years questionnaire survey from November to December in Nanjing from 2014 to 2016, and collected 570, 598 and 595 valid questionnaires respectively. To better capture spatial dynamics, we collected another 10,646 valid questionnaires in 2016 from 31 provinces in China through Internet platform. T-test and one-way ANOVA were used to examine significant difference of five variables among different demographical groups, years and provinces to explore temporal and spatial variations. In addition, average value of WTP in Nanjing and China were calculated. Result show that females , environmental professionals, highly educated people, and medium to high income groups are sensitive population. Risk cognition in 2015 was significantly higher than that in 2014($p<.000$) and 2016($p<.000$), degree of worry in 2016 was significantly higher than that in 2014($p<.000$) and 2015($p<.000$), perceived benefit was increasing year by year (4.14<4.18<4.45), while trust was declining year by year (3.11>3.00>2.84). Moreover, we didn't observe any significant heterogeneity among regions in terms of risk perception. This result indicate that in a society with increasing air pollution risk, government should pay attention to guide public' risk perception, especially for sensitive population. Additionally, public's WTP for 50% reduction of heavy air pollution days within two years in Nanjing and China are 364 and 283 RMB per year, respectively. When making policy, government should also take it into account.

Keywords: A-behavior, C-air, A-statistical methods, ,

TH-PL-A1-602

A randomized, blinded, crossover intervention study of traffic-related air pollution (TRAP) and heart rate variability in healthy adults

B. Han[1], R. Zhao[1], N. Zhang[1], L. Zhang[2], J. Xu[3], Z. Bai[1], S. Vedal[3]; [1]Chinese Research Academy of Environmental Sciences, Beijing, China, [2]Tianjin Medical University, Tianjin, China, [3]University of Washington, Seattle, WA

Abstract: Background/Aim: Traffic related air pollution (TRAP) may increase the risk of cardiovascular disease. This study aimed to assess short-term cardiovascular effects of TRAP, using heart rate variability (HRV) and randomized respirator interventions. **Methods:**A randomized, blinded, crossover intervention study was conducted in which 39 healthy adults spent 2 hrs next to a busy road. Participants wore either a powered air-purifying respirator or an N95 (mask mode). PAPRs were fitted with a PM2.5 filter (PM mode), a PM2.5 and VOCs filter (PM/G mode) or a sham filter (Sham mode); participants were blinded to filter type. The four interventions were determined randomly for each participant and separated by at least one week. HRV (SDNN and rMSSD) was measured by Holter monitor every 15 min from before the roadside exposure to 6 hrs after. Linear mixed-effect models were used to evaluate the effects of the interventions. **Results:**There was no significant change in either SDNN or rMSSD during the 2-hr roadside exposure, and for the 6 hrs following. Both SDNN and rMSSD generally increased in all three intervention modes compared to the sham mode, both during and following exposure. For SDNN, there were no substantial differences across the three intervention modes during exposure, but with the mask mode, statistically significant increases of SDNN were observed, with increases of 5.97ms (95%CI:0.34~11.60) in the 2 hr time period after exposure, and 6.51ms (95%CI:1.09~11.93) during the 2-6 hr time period after exposure. In the PM mode, there was a significant increase in rMSSD relative to the sham mode of 8.65ms (95%CI:2.70-14.58) during exposure, and 8.36ms (95%CI:2.03~14.66) in the 2 hrs time period after exposure. No significant differences in rMSSD increases were observed across three modes during the 2-6 hr time period after exposure. **Conclusions:**SDNN and rMSSD were not significantly changed with roadside exposure, but increased by preventing inhalation of TRAP-related Particles

Keywords: C-air, A-biomarkers,

TH-PL-A1-603

Modeling Air Pollution Exposure Metrics for the Coronary Artery Disease and Environmental Exposure (CADEE) Health Study

M. Breen[1], S. Chang[2], S. Arunachalam[2], V. Isakov[1], R. Devlin[1]; [1]US Environmental Protection Agency, Durham, NC, [2]University of North Carolina, Chapel Hill, NC

Abstract: Air pollution health studies often use outdoor concentrations from a central-site monitor as exposure surrogates. To improve exposure assessments, we previously developed and evaluated an exposure model for individuals (EMI), which predicts five tiers of individual-level exposure metrics for ambient air pollutants using outdoor concentrations, questionnaires, weather, and time-location information. We linked a mechanistic air exchange rate (AER) model to a mass-balance air pollutant infiltration model to predict residential AER (Tier 1), infiltration factors (Finf, Tier 2), indoor concentrations (Cin, Tier 3), personal exposure factors (Fpex, Tier 4), and personal exposures (E, Tier 5) for ambient air pollutants. In this study, we developed a novel model, called Exposure Tracker (ETrac), which extends EMI by including: (1) an air quality model to predict hourly census-block outdoor concentrations for four pollutants (PM, NOx, CO, EC), (2) a GPS-based microenvironment tracker (MicroTrac) model to predict time spent by individuals in various microenvironments. Using ETrac, we predicted daily exposure metrics (Tiers 1-5) for the 15 participants across 10 consecutive weeks in a cohort health study in central North Carolina called Coronary Artery Disease and Environmental Exposure (CADEE). Our modeled predictions for a total of 708 participant-days showed substantial house-to-house and temporal variability of AER, Finf, and Cin (Tiers 1-3); and subject-to-subject variability of Fpex and E (Tiers 4-5) for the four pollutants. The capability of ETrac could help reduce uncertainty of ambient pollutant exposure metrics used in health studies, such as CADEE, in support of improving health risk estimates.

Keywords: A-exposure models, A-built environment, B-particulate matter, A-activity patterns, C-air

TH-SY-B1: ENVIRONMENTAL, OCCUPATIONAL, AND COMMUNITY EXPOSURES TO LIVESTOCK AGRICULTURE

TH-SY-B1-604

Use of source-tracking methods to understand antimicrobial-resistant *S. aureus* exposure pathways among industrial hog operation workers and community residents

C. Heaney; Johns Hopkins University, Baltimore, MD

Abstract: Background: Administration of antimicrobials to food animals, including non-therapeutically, can select for antimicrobial-resistant (AMR) bacterial pathogens, such as methicillin-resistant *Staphylococcus aureus* (MRSA), and influence microbial communities due to selective pressure. In the United States, understanding of occupational and community AMR *S. aureus* exposure pathways related to industrial hog operations (IHOs) is improving, but significant knowledge gaps remain due to challenges of gaining access to IHOs. Few studies have explored how microbial source-tracking methods (swine-specific and/or microbiome-based) might advance understanding of AMR *S. aureus* exposure sources and help identify signatures of microbial contamination, colonization, and infection status. Aims: To summarize results of recent studies that integrated AMR *S. aureus*, microbial source-tracking, and microbiome measurements among IHO workers and community residents. Methods & Results: Presence (culture-based) and abundance (qPCR-based) of AMR *S. aureus* was assessed in IHO worker and community resident nasal swabs. Abundance of a swine-specific fecal microbial source-tracking marker (qPCR assay for Pig-2-bac) and the nasal microbiome will be examined in relation to measures of AMR *S. aureus*. A positive relation of Pig-2-Bac DNA copy number was observed with *fem*A DNA copy number (beta per $\log_{10}$ unit=0.21, 95% confidence interval [CI]=0.10, 0.32), *S. aureus* CC398 DNA copy number (beta per $\log_{10}$ unit=0.21, 95% CI=0.11, 0.30), *mec*A DNA copy number (beta per $\log_{10}$ unit=0.33, 95% CI=0.24, 0.42), *scn* negative *S. aureus* nasal carriage (odds ratio [OR]=1.52, 95% CI=1.17, 1.97) and multidrug-resistant *S. aureus* nasal carriage (OR=1.84, 95% CI=1.32, 2.56) in IHO worker and community residents' nasal swabs. IHO workers demonstrated higher alpha diversity than community residents (rank sum statistic of Simpson diversity index=3.9, p<0.0476). Conclusions: Source-tracking methods may improve understanding of sources of AMR *S. aureus* exposure among IHO workers and community residents.

Keywords: A - exposure measurement, A - population exposure, A-biomonitoring, B-microbial agents, Agricultural Exposures

TH-SY-B1-605

Bioaerosol emissions, transport, deposition and risk downwind of CAFO (dairy) manure application areas.

S. Rogers; Clarkson University, Potsdam, NY

Abstract: Manure application to land yields benefits of fertilization and waste disposal. However, aerosolization and subsequent deposition of manure pathogens poses infection risks to downwind receptors and food safety concerns for crops grown for human consumption. A field sampling campaign at dairy manure application sites was used to support emission, transport, and deposition modeling of bioaerosols following manure application to land using the US EPA's AERMOD dispersion model. Results were coupled with a quantitative microbial risk assessment model to estimate infection risks due to inhalation of bioaerosols, and to consumption of leafy green vegetable crops grown downwind from manure application areas upon which bioaerosols have deposited. Median 8-h (post-application) infection risks due to inhalation decreased exponentially with distance from a median of 1to2700 at the downwind edge-of-field to 1to13,000 at 100 m and 1to200,000 at 1000 m; peak risks (95th percentiles) were considerably greater (1to33, 1to170, and 1to2500, respectively). Overall median one-time infection risks due to consumption of leafy green vegetables at the time of maximum loading (0.7 d - 2.6 d) decreased from 1to1300 at 0 m downwind the field to 1to6700 at 100 m and 1to92,000 at 1000 m; peak risks were 1to18, 1to89, and 1to1200, respectively. Considering, bioaerosols emitted following manure application present significant public health risks to downwind receptors and to consumers of fresh leafy green vegetables grown in proximity to manure application areas. Improved practices for pathogen reduction prior to application or to reduce resuspension from land-applied manures will reduce risks of disease transmission. In the case of vegetable crops, a minimum 160-m setback distance between manure application areas and crop production areas based on median infection risk of 1to10,000 is recommended. Additional distance or delay before harvest will provide further protection of public health.

Keywords: A - ambient monitoring, A - exposure measurement, B-microbial agents, Agricultural Aerosols

TH-SY-B1-606

Inhalable bioaerosols, bacterial microbiome, and inflammation: A day in the life of a dairy worker

J. Schaeffer, S. Reynolds; Colorado State University, Fort Collins, CO

Abstract: Dairy workers are at the nexus of bioaerosol exposures, which contain microbial communities and associated pro-inflammatory components. These workers experience adverse respiratory health outcomes. We evaluated the full size distribution (from 0 to 100 µm in aerodynamic diameter) and biological composition of dairy milking parlor bioaerosols across four aerodynamic size fractions (<3, 3-

10, 10-30, and >30µm). Our focus was on modeling microbial communities using high-throughput DNA sequencing. In a related study, we conducted a panel study measuring cytokines in sequential nasal lavages from dairy workers over repeated workdays to model the inflammatory response from bioaerosol exposure. Cytokines were measured using Luminex, a multiplexing assay. We used linear mixed models to understand the association between inhalable dust and select constituents (endotoxin and β-glucans) and transcripts coding for pro-inflammatory (TNF-α) and anti-inflammatory (IL-10) cytokines. Airborne particulate mass followed a bimodal size distribution (one mode at 3 µm and a second above 30 µm), which also correlated with the relative concentrations of the following microbiological markers: bacterial endotoxin, 3-hydroxy fatty acids, and muramic acid. DNA sequencing of these aerosols revealed a diverse microbiome containing potentially hazardous constituents and opportunistic pathogens. Findings from our repeated measures affirm that workplace endotoxin exposures play a meaningful role in subclinical inflammation. A total of 36 workers (7 female, 29 male) participated for a median of 3 days (range: 1 - 3). Controlling for gender, smoking status, and length of employment, we found that a 10 EU increase in endotoxin was associated with a 0.12 pg/mL increase in TNF-α (95% CI: 0.0007, 0.244) and a 0.073 pg/mL decrease in IL-10 (95% CI: -0.142, -0.004). These studies are the first of their kind and are helping to inform a current intervention study to develop effective control strategies.

Keywords: A - exposure measurement, A-industrial hygiene, C-air, B-particulate matter, B-microbial agents

TH-SY-B1-607

Characterization of Personal Inhalation Exposure to Bioaerosols using Metagenomic DNA Sequencing Tools in Broiler Chicken Production

M. W. Nonnenmann[1], K. O'Brien[1], M. Farnell[2], T. Tabler[3], J. Bray[4]; [1]University of Iowa, Iowa City, IA, [2]Texas A&M University, College Station, TX, [3]Mississippi State University, Mississippi State, MS, [4]Stephen F. Austin, Nacogdoches, TX

Abstract: Poultry production is the largest meat producing industry in the United States and employs about 200,000 workers. Workers are exposed to dust containing organic components, ammonia and bioaerosols while completing tasks in production houses. Inhalation exposure to dust has been associated with respiratory disease in agricultural workers and animals. However, personal exposure to bioaerosols during work in broiler chicken production is relatively unknown. This study characterized the microbial composition using next generation sequencing technology among personal exposure samples of inhalable aerosols during work in commercial broiler chicken production. Personal inhalable dust samples were collected during the work tasks of littler sampling and mortality collection at broiler chicken farms. Genomic DNA was extracted from settled dust (n=3) and personal inhalable dust (n=18). DNA libraries were sequenced using a paired-end sequencing-by-synthesis approach on an Illumina HiSeq 2500. Geometric mean inhalable dust concentrations were 3.9 mg/m^3 (SD=2.8). Poultry dust was predominantly composed of microorganisms (98%) with a small quantity of avian, human and feed DNA (<2% of total reads). *Staphylococcus sp AL1*, *Salinicoccus carnicancri*, and *Lactobacillus crispatus* were the most abundant bacterial species and genera of *Penicillium* and *Aspergillus* are highly abundant among personal exposure samples of inhalable dust. However, the diversity of microbial species were reduced significantly in settled dust. Unlike settled dust composition, aerosolized dust composition had little variance between samples. Although inhalable dust exposure concentrations exceeded recommended exposure limits, future work is needed to understand the impact of exposure to multiple inhalation hazards. The use of cost-effective engineering, administrative and personal exposure controls is needed to effectively reduce inhalation exposures during work in broiler chicken production.

Keywords: A - exposure measurement, A-indoor environment, B-microbial agents, C-air, metagenomics

TH-SY-B1-608

A One Health Approach to Agricultural Exposures

M. Davis[1], S. C. Rankin[2], N. Pisanic[1], A. Brown[1], S. Cole[3], S. Rhodes[3], J. Schurer[4], A. M. Rule[1], L. Conti[5], C. Heaney[1], P. Rabinowitz[4]; [1]Johns Hopkins University, Baltimore, MD, [2]University of Pennsylvania, Philadelphia, PA, [3]University of North Carolina at Chapel Hill, Chapel Hill, NC, [4]University of Washington, Seattle, WA, [5]Florida Department of Agriculture and Consumer Services, Tallahassee, FL

Abstract: The Checklist for One Health Epidemiological Reporting of Evidence (COHERE) was created to improve the quality of epidemiological studies that integrate data from human, animal and environmental domains. One Health emphasizes multidisciplinary research teams with expertise from all three disciplines related to human, animal and environmental health. Assessment of the three domains for agricultural exposures and the integration of data collection, analysis and interpretation provides scientific, logistic and economic benefits over approaches that do not use a harmonized One Health approach. COHERE guidelines were applied to the conduct of a pilot study of worker exposure to antibiotic-resistant *Staphylococcus aureus* in hog production environments. Four hog operations were evaluated: 1 that raised swine in confinement with antibiotics (industrial hog operation: IHO) and 3 that raised swine on pasture without antibiotics (antibiotic-free hog operation: AFHO). Hog, ambient air and worker surrogate personal air samples were analyzed for *S. aureus*. Confirmed isolates were characterized by antimicrobial susceptibility phenotype, *spa* type and absence of the *scn* gene. At the IHO, *S. aureus* was recovered from 17/20 (85%) hogs, 8/14 (67%) ambient air and both (100%) personal air samples from worker surrogates. *S. aureus* was not recovered from 30 swine, 19 ambient and 6 personal air samples at all 3 AFHOs. All *S. aureus* recovered from the IHO operation were *spa* type t337, *scn* negative, and 62/63 (98%) were multidrug resistant. Concurrent sampling of direct (hog) and indirect (environmental) pathways of potential worker exposure had lower travel costs and improved time efficiencies for researchers and farmers compared to separate sampling. This work informs future studies to relate operation practices and environmental exposures to animal and worker health outcomes and shows how COHERE can be applied to improve reporting of epidemiological studies.

Keywords: A - exposure measurement, B-microbial agents, A - ambient monitoring, Agricultural aerosols

TH-PL-C1: Temporal Trends from Biomonitoring

TH-PL-C1-609

Trends in Human Exposure to Bisphenol A and other Bisphenols

A. M. Calafat, L. Wong, P. Dwivedi, X. Zhou, T. Powell, T. Jia, X. Ye; Centers for Disease Control and Prevention, Atlanta, GA

Abstract: Background: Because of bisphenol A (BPA) potential adverse effects on health, regulatory actions and public concerns, use of BPA may be decreasing while use of other bisphenols as substitutes may be increasing; little is known about consequent trends in population-level exposures. We investigated Americans' exposure to BPA, bisphenol S (BPS) and bisphenol F (BPF) using data from the National Health Nutrition Examination Survey (NHANES). Methods: We combined data on BPA for 15,593 NHANES 2003-2014 participants and calculated least square geometric means (GMs) from multivariate regression models. For NHANES 2013-2014, we calculated GM and distribution percentiles of urinary concentrations for all bisphenols, and conducted weighted multivariate ANCOVA to examine associations by sociodemographic variables (e.g., sex, age, race/ethnicity). Results: BPA adjusted GMs from 2003 to 2014 showed a significant downward trend (beta=-0.05, $p< 0.001$). GMs (95% confidence interval [CI], in µg/g creatinine) of BPA in 2013-2014 (1.26 (1.17-1.35) were significantly higher than those of BPS (0.428 (0.471-0.606)) and BPF (0.534 (0.387-0.474)). By contrast, 95th concentrations (95% CI, in µg/g creatinine) were highest for BPF (8.39 (5.90-12.0)) followed by BPA (5.09 (4.65-5.96)), and lowest for BPS (3.33 (2.60-4.68)). Race/ethnicity was significantly associated with adjusted GMs of BPF ($p=0.03$) and BPS ($p=0.002$), age with GMs of BPS ($p=0.0001$) and BPA ($p<0.0001$), sex was only associated with

BPA GM (p=0.048). Conclusions: Differences in current exposure determinants suggest that chemical-dependent sources or lifestyle-related activities impact exposures, but exposures to all three bisphenols remain widespread. Nonetheless, legislative activity and consumer advocacy campaigns may have contributed to a significant downward exposure trend to BPA since 2003. Additional NHANES will be useful to better understand the general public's exposure to these chemicals and monitor future exposure trends.

Keywords: A-biomonitoring, A-exposure factors, B-BPA, C-consumer/personal care products, A-chemical alternatives

TH-PL-C1-610

Trends in Cumulative Exposures of Six Phthalates in the United States from 2005 to 2014

J. Reyes[1], P. Price[2]; [1]ORISE at US EPA, Research Triangle Park, NC, [2]US EPA, Research Triangle Park, NC

Abstract: Phthalates are utilized in a wide range of consumer goods and are common contaminants in food. Exposures and resulting dosages of individual phthalates vary over time as a result of changes in their use in consumer products and food contact materials. We calculated the trends in screening estimates of daily dose, Hazard Quotient (HQ), Hazard Index (HI), and Maximum Cumulative Ratio (MCR) for a group of six phthalates using the tolerable daily intakes for the compounds and biomonitoring data collected from 2005 to 2014 under the National Health and Nutrition Examination Survey. HQ is the ratio of an individual's dose of a chemical and the chemical's permitted dose. HI is the summation of an individual's HQs. The MCR is the ratio of the HI to the largest HQ for each individual. There was a 2.2-fold decrease in the mean HI over this period (0.34 to 0.15) and a 7.2-fold decrease in the percentage of participants with an HI greater than 1 (5.7% to 0.8%). Decreases in HI were due to the decreases of diethylhexyl phthalate (DEHP) and dibutyl phthalates (DBP). Diisononyl phthalate (DINP) exposure increased between 2005 and 2014 and the remaining three phthalates remained approximately constant during this period. The phthalate with the greatest frequency of maximum HQ among individuals with HI values greater than 1 in 2005 was DEHP and in 2014 was DINP. While the individual HI values decreased over time, the values of MCRs increased, indicating that the need to consider cumulative exposures has become more important. These findings suggest that there has been a shift in phthalate exposures in the US population, leading to a decrease in HI values. These secular decreases were most pronounced in the groups experiencing the highest phthalates exposures and occurred due to the declines in HQs for DEHP and DBP out weighing the increases in DINP HQs. The views expressed in this abstract are those of the authors and do not necessarily reflect the views or policies of the U.S. EPA.

Keywords: B-mixtures, B-phthalates, A-cumulative exposure, A-biomonitoring

TH-PL-C1-611

Time Trend of the Internal Exposure to Lead in Young Adults: 35 Years of Human Biomonitoring with Data from the German Environmental Specimen Bank (ESB)

D. Lermen[2], T. Göen[3], M. Bartel-Steinbach[2], A. Conrad[1], T. Weber[1], M. Rüther[1], H. von Briesen[2], M. Kolossa-Gehring[1]; [1]German Environment Agency (UBA), Berlin, Germany, [2]Fraunhofer Institute for Biomedical Engineering (IBMT), Sulzbach, Germany, [3]Institute and Outpatient Clinic of Occupational, Social and Environmental Medicine Friedrich-Alexander-Universität, Erlangen, Germany

Abstract: Background Lead is used since ancient times by mankind in various ways. Its extensive use and emission also of its compounds resulted in considerable exposure of the environment and the human population. Due to its known toxic potential lead exposure has been routinely monitored by the German ESB since 1981. **Methods** Blood samples from 10,257 participants of the German ESB from 1981 to 2015 were analyzed for this study. From a subgroup of participants covering the years 2010 to 2015 data on sex, smoking, alcohol consumption, and housing situation were investigated on their impact on lead exposure in young adults. **Results** Blood lead levels (BLL) of young adults decreased from 1981 to 2015 about 85.5 % as demonstrated by geometric mean values from the sampling site Muenster. A similar time trend is found for all four sampling sites from 1997 to 2015. Since 2008 a decreasing trend is not observed anymore. In general male participants have higher BLLs than females. Both, male and female smoker have significantly higher BLLs than non-smoker. Alcohol consumption significantly increased BLLs. An association with housing situation on BLLs is only revealed for male participants. **Discussion** The results reflect regulation measures undertaken to decrease human exposure to lead which started in Germany in 1971 with the regulation of leaded gasoline. The observed associations with sex, alcohol and tobacco consumption are in line with data from comparable studies. A decrease of 85.5% from 1981 to 2015 is remarkable but still it is not known if the currently observed lower BLLs can be considered as harmless. Further monitoring activities also for the most vulnerable groups - new-borns and children - are warranted in order to evaluate currents risks of lead exposure in Germany. **Acknowledgement** The ESB is funded by the German Ministry for the Environment, Nature Conservation, Building and Nuclear Safety (BMUB).

Keywords: A-biomonitoring, A-biomarkers, B-metals

TH-PL-C1-612

Time and Age-Based Trends of Chemical Exposure Biomarkers in the United States from 1999-2014

V. K. Nguyen[1], J. Colacino[2], J. Arnot[3], O. Jolliet[2]; [1]University of Michigan, Ann Arbor, MI, [2]University of Michigan, Ann Arbor, MI, [3]Arnot Research & Consulting Inc., Toronto, Canada

Abstract: Individuals are exposed to complex chemical mixtures, driven by behavioral, socioeconomic, and demographic factors. To systematically address the interactions between chemical exposure and demographic and lifestyle factors, we have developed an untargeted approach to study the 1999-2014 National Health and Nutrition Examination Survey (NHANES) data of over 300 chemical biomarkers in 82,000 study participants. The present study aims 1) to determine chemical exposure differences between young and older individuals; 2) to identify chemicals with increasing exposure in the US population by determining biomarker time trends; and 3) to evaluate key determinants that explain these patterns, particularly considering half-lives in human. We use a series of multivariate linear regression models that include age, sex, study years, ethnicity, smoking behaviors, poverty income ratio, and creatinine levels as covariates and chemical biomarker concentration as the outcome. Metabolites from chemicals in personal care products such as benzophenone-3, parabens, and triclosan along with arsenobetaine and enterodiol are higher in young individuals, while 2,5-dichlorophenol, mono-benzyl phthalate, and NNAL are elevated in the older population. Concentrations of 1-pyrene, mono-isobutyl phthalate, 2-phenanthrene, 2-napthol, and perfluorononanoic acid are significantly increasing over time. Correlation analysis between the half-lives in human and age coefficient (yearly increase in the log-transformed biomarker concentration) demonstrates a positive correlation with $R^2 = 0.5$ when considering over 50 chemicals with half-lives of more than 1000 hours. Systematically studying the NHANES biomarker dataset by applying an untargeted approach allows for the identification of expected and unexpected exposure trends by age and over time. These findings can be utilized to prioritize chemicals for toxicological evaluation or targeted environmental health interventions.

Keywords: A-biomarkers, A-exposure factors, A-chemical prioritization

TH-PL-C1-613

German Environmental Specimen Bank (ESB): Time Trend of the Internal Exposure to Glyphosate

A. Conrad[1], M. Rüther[1], H. Hoppe[2], S. Pieper[3], T. Weber[1], D. Lermen[4], M. Kolossa-Gehring[1]; [1]German Environment Agency (UBA), Berlin, Germany, [2]Medical Laboratory Bremen, Bremen, Germany, [3]German Environment Agency (UBA), Dessau-Roßlau, Germany, [4]Fraunhofer Institute for Biomedical Engineering (IBMT), Sulzbach, Germany

Abstract: Background Being increasingly applied, glyphosate became one of the most important professionally and non-professionally used pesticides in Germany. The health-relevance of this broadband herbicide is subject to broad debate. Against this background, the German ESB initiated a retrospective study on changes in the population's internal exposure to glyphosate over time. **Methods** Human ESB samples are collected annually in four German cities, most of them being cryo-archived for later analysis. For this study, 24 h-urine samples from 399 non-vegetarian/-vegan males and females collected from 2001-2015 at the sampling location Greifswald were analyzed for glyphosate and its main metabolite aminomethylphosphonic acid (AMPA) by GC-MS-MS (LOQ=0.1 µg/L). Results were related to physiological and anthropometrical parameters recorded for all participants. **Results** 127 samples (32%) exceeded the LOQ for glyphosate (AMPA: 160 samples, 40%). The fraction of glyphosate levels at or above LOQ increased from 10% in 2001 to nearly 60% in 2012 and 2013 before decreasing to 33% and 40% in 2014 and 2015, respectively. AMPA levels generally followed the same trend. Glyphosate and AMPA levels tend to be higher in males. Concentrations of both compounds are significantly rank-correlated (r_{SP}=0.51). A first comparison with a smaller data set from 20 self-reported vegetarians/vegans does not hint to differences in glyphosate and AMPA levels due to this diet. **Discussion** Results are in agreement with German glyphosate sales data. Additional measurements are warranted for confirming the observed reduction in glyphosate exposure since 2012 and contributing to the further discussion on risks and benefits of this pesticide. Data from the cross-sectional German Environmental Survey (GerES 2014-2017) will provide further insight into inter-individual variation in exposure. **Acknowledgement** The ESB is funded by the German Ministry for the Environment, Nature Conservation, Building and Nuclear Safety (BMUB).

Keywords: A-biomarkers, A-biomonitoring, B-pesticides, C-food, A-risk assessment

TH-PL-D1: Consumer Product Exposure

TH-PL-D1-614

Health Risks from Chemicals during the Use Phase of Consumer Products - Current Knowledge and Information Gaps

D. Li, S. Suh; University of California Santa Barbara, Santa Barbara, CA

Abstract: The number of chemicals in consumer products is increasing rapidly, while our understanding on their impact on humans is limited. Recent cases show that chemicals in consumer products may pose health risks to consumers. This study aims at a systematic review of the literature on human health risks of chemicals in consumer products. We analyze the functional uses, product applications, exposure routes, exposure mediums, toxicity endpoints and their combinations that are frequently reported in the literature. In addition, research needs and knowledge gaps are discussed. A total of 207 published peer-reviewed journal articles published between 2006 and 2016 are reviewed. The body of literature covers 116 chemicals, which are analyzed in terms of their characteristics including their functional uses, product applications, exposure routes, exposure mediums, and toxicity endpoints as they appear in the literature. Plasticizers, polymers, and flame retardants used in food contact products, personal care products, cosmetics, furniture, flooring, and electronics are the most frequent functional use-product application combinations in the literature with over 20 reports in the literature. Phthalates, bisphenol-A, and polybrominated diphenyl ethers are the most discussed chemical groups with 53, 28, and 26 papers, respectively, out of the 207 papers reviewed. Chemicals used in food contact products and those producing house dust are most frequently reported in the literature. Few papers focus on predicting the health risks of chemicals, the risk of novel chemicals, or the risk of alternative chemicals that displaces existing chemicals. We highlight the need to develop more comprehensive risk assessment methods with capability to predict the risks of chemicals used in consumer products.

Keywords: A-indoor environment, C-consumer products

TH-PL-D1-615

Exposure to Preservatives in Personal Care Products: Case Study Comparing Biomonitoring Data to Model Predictions

L. Aylward[1], G. Vilone[2], C. Cowan-Ellsberry[3], S. Hays[4], C. O' Mahony[2]; [1]Summit Toxicology, LLP, Falls Church, VA, [2]Creme Global, Dublin, Ireland, [3]CE2 Consulting, LLC, Cincinnati, OH, [4]Summit Toxicology, LLP, Bozeman, MT

Abstract: Exposure models are critical to the evaluation of the safety of personal care product ingredients. However, very few studies are available comparing model predictions to exposure data. Data from a recent intensive observational biomonitoring study of eight volunteers using personal care products were used to estimate minimum and maximum absorbed doses of three parabens (methyl-, ethyl-, and n-propyl [MP, EP, and PP]) and triclosan over four days of recorded use of toothpaste, shampoo, shower gel, shaving cream, face cream, and sunscreen with the target preservative ingredients. Four available models were run in default mode assuming 0.1% ingredient concentration to estimate average daily absorbed doses for each ingredient and participant based on the participant-specific combination of product uses: ConsExpo, run in deterministic mode; the European Commission Scientific Committee on Consumer Safety (SCCS) risk assessment screening model algorithms; SHEDS-HT; and Creme Care & Cosmetics (Creme CC). Minimum estimated absorbed doses across volunteers based on biomonitoring data ranged from 0.2 to 12.1 ug/kg-d for MP, 0.05 to 2.4 ug/kg-d for EP, 0.02 to 5.0 ug/kg-d for PP, and 0.02 to 26.9 ug/kg-d for triclosan. With a few exceptions, the results of the exposure models fell within a factor of 10 of the biomonitoring data depending on the chemical modeled. ConsExpo and SCCS point estimates were often the most conservative, but often fell within the range of minimum to maximum estimated exposure levels. The Creme CC and SHEDS-HT distributions of exposure estimates often encompassed the estimated exposure levels based on biomonitoring data. Specific patterns of under- and over- estimation of actual exposures will be presented, including comparison of estimated and measured product use amounts and frequencies. A sensitivity analysis on the assumed concentration of preservatives in the products and other factors will also be presented.

Keywords: A-aggregate exposure, C-consumer/personal care products, A-biomonitoring, A-exposure models

TH-PL-D1-616

Consumer Product Use and Suspect Screening of Phthalate/Phthalate Metabolites and Environmental Phenols in Maternal Serum

A. Wang[1], J. M. Schwartz[2], R. R. Gerona[3], T. Lin[3], R. Morello-Frosch[4], M. Sirota[5], D. Woodruff[2]; [1]University of California, San Francisco, San Francisco, CA, [2]University of California, San Francisco, San Francisco, CA, [3]University of California, San Francisco, San Francisco, CA, [4]University of California, Berkeley, Berkeley, CA, [5]University of California, San Francisco, San Francisco, CA

Abstract: Consumer products are important sources of exposure to endocrine disrupting chemicals, such as phthalates and environmental phenols (including parabens, triclosan, and bisphenol-A). Our study sought to identify the relationship between the presence of phthalates/phthalate metabolites and environmental phenols in serum, and demographic characteristics and consumer product use (including use of personal care products (PCPs) or household cleaning products (HCPs)). We used liquid chromatography-quadrupole time-of-flight mass spectrometry to perform a suspect screen for 68 phthalates/phthalate metabolites and 168 environmental phenols in serum collected at delivery from 83 pregnant women in Northern California and obtain both detection frequency and relative concentration (peak areas) for each suspect (matched) compound. Statistical analysis included Poisson regression, multivariable linear regression, and Chi-squared test. Asians and Whites (compared to Latinas) and women with at least some college education or household income ≥ $80,000 used 1-2 fewer HCP daily on average. We detected, on average, three fewer phthalate/phthalate metabolite suspects in women with some college education and six more phenol suspects in African Americans (all $P<0.05$). Daily use of lotion was associated with the number of phenol suspects detected after race/ethnicity adjustment ($b=3.3$, adj$P=0.10$). Use of lotion, deodorant or scented candles was positively associated with the detection and the relative concentrations of several phthalate and phenol suspects, although these associations were not significant after multiple comparison adjustment. Our results suggest potential demographic differences in consumer product use and the presence of phthalate and phenol suspects in maternal serum. Daily consumer product use may be associated with the presence of phthalates and phenols in maternal serum, indicating higher exposures to these compounds among certain race/ethnicity and income groups.

Keywords: A-biomonitoring, C-consumer/personal care products, A-exposure factors, D-prenatal, A-analytical methods

TH-PL-D1-617

Assessment of Aggregate Consumer Exposure to Isothiazolinones via Cosmetics and Detergents Using the Newly Developed model PACEM-KD

E. Garcia Hidalgo, N. von Goetz, K. Hungerbuehler; ETH Zurich, Zurich, Switzerland

Abstract: Isothiazoliones can cause allergic contact dermatitis and are present in a variety of consumer products, such as cosmetics, detergents, and do-it-yourself products. Allergic contact dermatitis is induced following dermal exposure to a sensitizer in an amount exceeding the sensitization threshold. The critical determinant of exposure for evaluating skin sensitization risks is dose per unit area of exposed skin. Thus, aggregate exposure per body site should be assessed as the basis for a quantitative risk assessment (QRA). Currently, aggregate (non-food) consumer exposure is not routinely evaluated and most available models are not suited for estimating aggregate exposure. Recently, a few models were developed, which are able to perform probabilistic aggregate consumer exposure, but the risk assessment of sensitizing substances (including isothiazolinones) is hampered by the lack of relevant exposure data like use frequency, duration, dilution and on chemical concentrations in products. We assessed aggregate exposures to four isothiazolinones (BIT, OIT, MI, CMI) by using a newly proposed individual-based Probabilistic Aggregated Consumer Exposure Model - Kinetic, Dermal (PACEM-KD) by combining the reported individual use-patterns in Switzerland (N=669, ages 0-91) with the isothiazolinone concentrations in the products used by the individual person. These were measured in a parallel analytical study. Thus, for each substance we provide realistic distributions of exposure for both genders and across all age groups. PACEM-KD-based higher tier predicted maximum exposure was equal to 0.04 $\mu g/cm^2$, 0.3 $\mu g/cm^2$, 0.003 $\mu g/cm^2$, and 0.003 $\mu g/cm^2$ for BIT, OIT, MI, and CMI, respectively. We presented a detailed population level cosmetic and detergent exposure model, which is able to predict the aggregated internal dermal load of sensitizing agents on specific body sites. This study can serve as a basis for exposure assessments for other sensitizing substances.

Keywords: A-exposure models, A-aggregate exposure, C-consumer/personal care products, A-exposure factors

TH-PL-D1-618

Study Design using Video Ethnography to Assess Role of Risk Perception in Use of Consumer Products

R. A. Silva[1], S. Prince[3], K. Taylor[2], D. Stout[3], L. Alston[3], P. P. Egeghy[3], T. J. Buckley[3]; [1]U.S. Environmental Protection Agency, Research Triangle Park, NC, [2]NIEHS, Research Triangle Park, NC, [3]U.S. EPA, Research Triangle Park, NC

Abstract: The nature and extent of human exposure to chemicals found in consumer products is strongly influenced by personal choices and product use. However, little is known about the extent to which risk perception factors into product selection and usage. As part of a feasibility study, we will evaluate video ethnography as a way to improve our understanding of exposures to chemicals from household consumer products and the modifying effect of risk perception. By recording individuals in their home environment and capturing their behaviors, attitudes, and opinions through a series of "in-the-moment" questions and prompts, we hope to better understand both the motivations behind specific behaviors and differences between reported and actual behaviors. Nine women between the ages of 35 and 74 years will be recruited for a 10-day monitoring period. Prior to the monitoring period, the field team will train the participants to facilitate familiarity with a video diary guide and to decrease any potential unease with video recording. Using an iPad and the video diary guide, participants will record a set of videos focused on one regularly used consumer product from each of three general product categories: personal care products, home cleaning products, and consumer-applied pesticides. They will demonstrate technique of use and quantity applied and answer questions designed to probe their knowledge and perceptions of exposure and hazard related to the chemical ingredients of those products. Video recordings will be analyzed to gain insight into motivations behind specific behaviors, with particular attention to risk perception. Self-administered video recording may reduce field deployment costs while increasing a sense of privacy and a more accurate disclosure of behaviors than questionnaires alone. Field deployment will occur during the Spring and Summer of 2017.

Keywords: C-consumer/personal care products, C-consumer products, A-behavior, A-indoor environment

TH-PL-E1: Metals - Human Health

TH-PL-E1-619

Hopi Environmental Health Project: Initial development and implementation.

G. S. Honanie[2], L. Joshweseoma[2], M. K. O'Rourke[1], R. Harris[1], M. Brian[1], J. L. Burgess[1]; [1]University of Arizona, Tucson, AZ, [2]Hopi Tribe, Kykotsmovi, AZ

Abstract: As part of the "Center for Indigenous Environmental Health Research (CIEHR)," the Hopi Environmental Health Project employs CBPR strategies in its design and implementation. We seek to (1) characterize the magnitude of environmental exposures to particulate matter (PM) and selected metals from multimedia; (2) evaluate social determinants of health and community resilience; and (3) expand tribal capacity to address areas of environmental concern that may inform programs and policy. As an initial step, we identified members of a Community Advisory Board (CAB); these members exhibit expertise related to community values, housing needs, and environmental health disparities. This active CAB critiqued the study design, suggested additional analytes, and altered questionnaires to meet project needs and reflect appropriate community conditions/interests. The project required extensive approval from numerous Hopi agencies, Hopi Tribal Council, and university IRB. CAB and tribal researchers recommended that Community Health Representatives (CHRs) play an active role in the project. By visiting homes with the field teams, the CHRs expand their reach in the community while providing spirometry, FeNO, body measures & food/beverage intake assessments. During the fall of 2016, CAB members volunteered their homes for training visits. Concurrent with questionnaire development & training, the research team identified a recruitment strategy that met concerns of Tribal Council for representation of all villages. A regional air monitoring program was implemented in January 2017 to collect $PM_{10 \& 2.5}$. Focus groups conducted in March of 2017 sought input to explore community resilience when facing environmental issues. A questionnaire module derived from these focus groups will be employed during the second phase of household visits. We are now collecting data from homes. CBPR projects are complex; other presentations at this meeting show preliminary results from this study.

Keywords: A - population exposure, D-First Nation, D-Southwest-specific, C-multimedia, B-particulate matter

TH-PL-E1-620

The Pregnancy Exposome: Multiple Environmental Exposures in the Iranian Environment and Neurodevelopmental Disorders (TEND) Birth Cohort

M. Zare Jeddi[1], M. Shamsipour[1], M. Yunesian[1], Y. Bruinen de Bruin[2], S. Hassanvand[1]; [1]Institute for Environmental Research (IER), Tehran University of Medical Sciences, Tehran, Iran (the Islamic Republic of), [2]Joint Research Centre of the European Commission, Ispra, Italy

Withdrawn

TH-PL-E1-621

Early-life Manganese Exposure and Intrinsic Functional Connectivity of the Developing Brain

E. de Water[1], E. Proal[3], V. Wang[2], S. Martínez Medina[3], L. Schnaas[3], M. M. Téllez-Rojo[4], R. O. Wright[1], C. Y. Tang[2], M. K. Horton[1]; [1]Icahn School of Medicine at Mount Sinai, New York, NY, [2]Icahn School of Medicine at Mount Sinai, New York, NY, [3]National Institute of Perinatology (INPer), Mexico City, Mexico, [4]National Institute of Public Health (INSP), Mexico City, Mexico

Abstract: Manganese (Mn) is an essential trace metal that is neurotoxic at high levels of exposure. Disruption of brain maturation processes during the prenatal period may have lasting consequences. During this critical period, the developing human brain is uniquely vulnerable to exposure to environmental toxicants such as Mn, and prenatal Mn exposure has been associated with changes in brain areas involved in emotion processing and regulation. The goal of the present study was to examine whether prenatal Mn exposure is associated with changes in the intrinsic functional connectivity (iFC) of the brain in childhood, focusing on changes in emotional brain areas. In this pilot study, 15 subjects (aged 6-7 years) were selected from an ongoing longitudinal birth cohort study to participate in a resting state functional magnetic resonance imaging (fMRI) study. Prenatal Mn exposure was determined from maternal blood collected during the 2nd and 3rd trimesters of pregnancy. We used seed-based correlation analyses and independent component analyses to examine whether prenatal Mn exposure was associated with the iFC of the brain in children. We found that the bilateral anterior cingulate cortex and right globus pallidus showed reduced iFC with medial and lateral prefrontal areas in children who were exposed to higher prenatal Mn levels, and these children further showed reduced iFC between the bilateral insula and occipito-temporal areas. These findings indicate that prenatal Mn exposure is associated with reduced iFC of brain areas involved in emotion processing and regulation in children.

Future studies should investigate whether this reduced iFC mediates the association between prenatal Mn exposure and emotional dysfunction in childhood.

Keywords: B-metals, D-children

TH-PL-E1-622

The joint effects of metals on general cognitive ability in adolescents living near ferromanganese industry

J. A. Bauer[1], B. Coull[2], J. Bobb[3], S. Guazzetti[4], S. Zoni[5], C. Fedrighi[5], D. Placidi[5], G. Cagna[5], R. White[1], M. Arora[6], R. O. Wright[6], D. R. Smith[7], R. G. Lucchini[6], B. Claus Henn[1]; [1]Boston University School of Public Health, Boston, MA, [2]Harvard T.H. Chan School of Public Health, Boston, MA, [3]University of Washington, Seattle, WA, [4]Azienda Unita Sanitaria Locale, Reggio Emilia, Italy, [5]University of Brescia, Brescia, Italy, [6]Icahn School of Medicine at Mount Sinai, New York, NY, [7]University of California, Santa Cruz, CA

Abstract: Introduction: Although mixtures research is increasingly popular, there are virtually no data on chemical mixture effects in adolescents. We explored associations of exposure to a metal mixture with general cognitive ability in adolescents living proximal to ferromanganese industry, a source of multiple airborne metals emissions. **Methods**: We measured manganese (Mn), lead (Pb), copper (Cu), and chromium (Cr) in hair and blood samples collected from 608 Italian adolescents ages 10-14 years. Concurrent with biosampling, trained neuropsychologists administered the Wechsler Intelligence Scale for Children, 3rd edition (WISC-III) to measure full-scale, verbal and performance IQ (FSIQ, VIQ, PIQ) scores. Bayesian kernel machine regression, which allows for higher order interactions and nonlinear effects, was used to estimate associations of FSIQ, VIQ, or PIQ scores with the metal mixture and its individual components. **Results:** Median metal concentrations were: hair Mn, 0.08 ug/g; hair Cu, 9.6 ug/g; hair Cr, 0.05 ug/g; blood Pb, 1.3 ug/dL. Adjusting for sex, age and socioeconomic status, we observed an inverse association between the metal mixture and VIQ scores (e.g. when all metals concentrations were at their 55th percentile compared to 50th percentile: -0.3 [95% credible interval: -0.5, -0.1]). Within the mixture, when other metals were fixed at their 25th percentile, VIQ scores were inversely associated with an IQR increase in hair Mn (-1.9 [-3.6, -0.2]) and with an IQR increase in blood Pb (-1.5 [-3.0, -0.1]). Inverted u-shaped associations between hair Cu and FSIQ and VIQ were found, consistent with Cu as an essential nutrient. There was suggestive evidence of interactions between Mn and Cu. **Conclusion**: Adolescent exposure to a mixture of metals was inversely associated with verbal cognitive ability, suggesting that metals from ferromanganese emissions jointly affect neurobehavior.

Keywords: A-epidemiology, B-metals, D-children

TH-PL-E1-623

Working Memory and the Delayed Spatial Alternation Test in a Population of Children from Mexico City

L. Zheng[1], A. Williams[2], J. Doucette[1], K. Svensson[1], D. Cory-Slechta[3], C. Gennings[1], M. Tomayo Y Ortiz[5], M. Torres Calapiz[4], M. M. Téllez-Rojo[5], L. Schnaas[4], R. O. Wright[1], H. Megan[1]; [1]Icahn School of Medicine at Mount Sinai, New York, NY, [2]Augusta University, Augusta, GA, [3]University of Rochester, Rochester, NY, [4]National Institute of Perinatology, Mexico City, Mexico, [5]National Institute of Public Health, Cuernavaca, Mexico

Abstract: Background: Working memory (WM), a key component of executive function (EF) is critical for problem solving and reasoning. The Delayed Spatial Alternation Task (DSAT) is one of few standardized assessments to evaluate WM/EF among preliterate children but little normative data on this task exists. We evaluated DSAT performance among 18 and 24 month old children and tested whether prenatal lead exposure was associated with DSAT performance. **Methods:** The DSAT is a two-choice task in which the subject must alternately retrieve a reward hidden on the right or left side of a tray. As the subject progresses through the test, the time delay between hiding and retrieval increases. We administered the

DSAT to 286 children enrolled in a longitudinal birth cohort study, PROGRESS. We evaluated the highest level achieved (max = 5), the # correct retrievals ("correct") and the # successful retrievals before the first error ("error"). We examined the impact of 3rd trimester maternal blood lead (MBPb) on DSAT performance controlling for sex, maternal education, maternal SES, birth weight percentile, household smoking, and depression. **Results:** 24m children perform better on the DSAT than 18m children; 24m subjects reached a higher level (3.3 vs 2.4, $p<0.01$), had higher "correct" (20.3 vs 17.2, $p<0.01$) and fewer "errors" (2.5 vs 3.6, $p<0.01$). Females performed better than males. Maternal education predicted better DSAT performance; household smoking predicted worse DSAT performance. At 24 months of age, In MBPb exposure was associated with fewer correct retrievals (β, (95%CI) = 0.16 (-0.35, 0.68), more errors ((β, (95%CI) = -2.44 (-4.67, -0.21)) and lower odds of achieving higher levels (OR (95%CI) = 0.38 (0.02, 0.73). **Conclusion:** In our study, 24m children perform better on the DSAT than 18m. Maternal Pb was associated with poorer performance on the DSAT parameters we explored. Further study will investigate other parameters of the DSAT as well as other potentially toxic exposures.

Keywords: A-biomarkers, B-metals, D-children, A-epidemiology

TH-SY-F1: Modeling Longitudinal Patterns of Exposure Using Agent Based Modeling and Related Techniques

TH-SY-F1-624

Characterizing Exposure-Related Behaviors Using Agent-Based Models Embedded with Needs-Based Artificial Intelligence

N. Brandon, P. Price, K. Dionisio, K. Isaacs, D. Kapraun, R. W. Setzer, R. Tornero-Velez; US Environmental Protection Agency, Durham, NC

Abstract: Information on where and how individuals spend their time is important for characterizing exposures to chemicals in consumer products and in indoor environments. Traditionally, exposure assessors have relied on time-use surveys in order to obtain information on exposure-related behavior. In lieu of using surveys, we create an agent-based model (ABM) that is able to simulate longitudinal patterns in human behavior. By basing our ABM upon a needs-based artificial intelligence (AI) system, we create autonomous agents that mimic human decisions on residential exposure-relevant behaviors. The model predicts the behavior patterns for the following actions: sleeping, eating, commuting, and working/schooling. The model uses four different types of agents parameterized to represent the following U.S. demographic groups: working adults, non-working adults, school-aged children, and pre-school children. The parameters for the model are calibrated using survey data from the US Environmental Protection Agency's Consolidated Human Activity Database (CHAD). The results demonstrate that the ABM can capture both inter-individual and intra-individual variation in the aforementioned behaviors as well as providing a needs-based rational as to how decisions on one's behavior can affect subsequent behaviors. A key advantage of the needs-based AI is the possibility to synthesize plausible time-activity diaries *de novo* where this information is absent. We propose that by simulating human behavior, this ABM may allow exposure-assessors and other scientists to characterize exposure-related behavior quicker and in ways not possible with traditional survey methods.

Keywords: A-exposure models, A-risk assessment, A-activity patterns

TH-SY-F1-625

High-resolution scheduling model of consumer product use by individuals and households

T. Hong[1], J. Levasseur[1], A. Varghese[1], H. Hubbard[1], K. Dionosio[2], N. Brandon[2], P. Price[2]; [1]ICF, Durham, NC, [2]EPA, Research Triangle Park, NC

Abstract: The scheduling model (SM) was developed for arranging the use of consumer products in the U.S. EPA's Human Exposure Model (HEM), an integrated modeling system to estimate human exposure to chemicals in household consumer products. The SM input is a set of year-long daily activity patterns which are produced by a separate module in HEM. These patterns specify the start times and durations for five macro behaviors: sleep, eat, work/school, commute, and idle time (time not spent in the other activities). The SM schedules the use of more than 300 types of products during idle times. The SM determines which of the product types are used in a household based on the characteristics of the household (e.g., house size and type), the characteristics of individuals living in the house (e.g., gender, age, and work status), and the prevalence of product use. The predictions of product use on a given day are then determined for each product type based on seasonality of product use and estimates of frequency and duration taken from the SHEDS-HT model. The SM considers whether products are used to satisfy personal or communal needs (frequency is based on the person or household, respectively) and whether a period of idle time is sufficiently long to allow the use of product, and also clusters products typically used in tandem (e.g., toothpaste and mouthwash, paint primer and paint, car wash and car wax). The SM output is a set of activity patterns that specifies the products used on each day of a year by each individual in a household, the durations of product use, and the mass of product used. These estimates are temporally consistent (individuals do not do two things at once), are consistent with longitudinal predictions of macro behaviors, reflect demographic information, and are consistent with the day of the week and season of the year. The results are used to model human exposure to chemicals in products and the releases of the chemicals to the environment.

Keywords: A-activity patterns, A-exposure models, C-consumer/personal care products, A-exposure factors, Human Exposure Model

TH-SY-F1-626

The CARES NG Temporal Residential Model

G. Vilone; Creme Global Ltd., Dublin, Ireland

Abstract: CARES NG is a cloud-based probabilistic model that provides an intuitive and efficient tool to estimate aggregate and cumulative exposure to pesticides from foods, drinking water and consumer products. The CARES NG residential module estimates exposure to pesticides from products used in and around houses, such as insect repellents and indoor/outdoor foggers. The model follows the algorithms described in the EPA's standard operating procedures for residential pesticide exposure assessment published in 2012. The CARES NG model generates a calendar-based series of residential events for each person in a given population. The longitudinal pattern (1 to 365 days) of pesticide product use is based on product application information from product label and conditional probabilities from consumer product use surveys. Residential pesticide use surveys (like the Residential Exposure Joint Venture-REJV) provides data on the likelihood of someone making an application in a given scenario on a specific day, as well as, co-occurrence of pesticide product use across residential scenarios on the same day. Using the calendar-based approach allows applying a probabilistic model to the EPA algorithms to estimate a temporal exposure pattern. This talk will focus on the methodologies and assumptions used in the CARES NG residential model to create a longitudinal pattern of residential exposures based on residential pesticide use survey data.

Keywords: A-exposure models, A-statistical methods, A-longitudinal metrics, B-pesticides

TH-SY-F1-627

Integrated use of Agent Based Modelling with sensor webs for personal exposure assessment

D. Chapizanis, S. Karakitsios, D. Sarigiannis; Aristotle University of Thessaloniki, Thessaloniki, Greece

Abstract: Innovations in sensor technology create possibilities to collect environmental data at unprecedented depth and breadth. In this study, human movement and interaction behaviour were simulated using Agent Based Modelling (ABM). A model was developed for cities that can feed into population-based exposure assessment without imposing prior bias, but rather basing its estimations onto emerging properties of the behaviour of the computerised autonomous decision makers (agents) that compose the modelled system. Survey outputs with data on lifestyle/behavioural patterns were associated with human agent behavioural rules to model realistic conditions. Observations on spatiotemporal behaviours, derived from a portable sensor campaign on 150 households were extrapolated to the larger population of the city and translated into human agent rules, considering sociodemographic variations. Human agents differentiated for age, gender or income follow different rules, express different behaviours leading to an agent-specific exposure profile. At the end of a model run, activity patterns are determined for every individual based on the prevalence of specific preferences and decision-making. Coupling position information with spatially resolved pollution levels allows us to assign pollutant concentrations to an individual. PM levels and size distribution varied among different parts of the urban agglomeration and hours of the day. Personal exposure results were between up to 20% more accurate than the equivalent estimate using ambient air concentration of PM as exposure proxy. This approach permits the cost-effective construction of refined time-activity diaries and diurnal exposure profiles. Our method leads to a refined exposure assessment model addressing effectively vulnerable population sub-groups integrating socio-economic status indicators. The computational platform developed can be used to study the effect of different measures to reduce pollution levels in cities, as well as the effects of different public health strategies prior to implementation. Thus, they reduce the time and cost required to identify effective public health protection interventions.

Keywords: A-exposure models, A-sensor technology, C-air

TH-SY-F1-628

APEX5: The Latest Refinements and Added Features of U.S. EPA's Air Pollutants Exposure Model

C. Holder[1], G. Glen[1], J. Levasseur[1], J. Cohen[4], J. Hader[1], H. Hubbard[1], J. Langstaff[2], S. Graham[2], K. Isaacs[3]; [1]ICF, Durham, NC, [2]U.S. Environmental Protection Agency, Research Triangle Park, NC, [3]U.S. Environmental Protection Agency, Research Triangle Park, NC, [4]ICF, Rohnert Park, CA

Abstract: U.S. EPA's Air Pollution Exposure (APEX) model is a flexible and powerful inhalation exposure model. Over decades, APEX has been adapted to advances in exposure science and computational power. APEX5, released in 2017 and the first update since 2012, employs 2010 Census data, expanded activity and energy-expenditure data, and several new algorithm refinements. APEX5 is being used in the SO_2 National Ambient Air Quality Standards review and a high-profile evaluation of potential inhalation exposure to VOCs near oil and gas wells. APEX5 includes new multiple linear regression models for resting metabolic rate (RMR) and ventilation rate (VE). New RMR calculations are customized by age and sex, resulting in overall $r^2>0.8$ and a 10-30 kcal/day smaller bias than previous versions. New VE calculations introduce a relationship between an individual's maximum and activity-specific oxygen consumption, improving r^2 to 0.94 and improving trends with age. APEX5 includes updated body-weight distributions, and simulated heights have better correlations with body weight. Census tract-level data on age, sex, employment, and commuting now reflect the 2010 U.S. Census, and activities are sampled from an expanded Consolidated Human Activity Database and linked to more specific activity codes with updated distributions of relative energy expenditure. APEX5 also includes a Sobol analysis—a sensitivity analysis partitioning the variance in exposure among groups of stochastic input variables. APEX5 is easy to use and computationally efficient for local and regional inhalation exposure assessments. It allows users to control random seeding, customize the contents of output tables (including summaries for sensitive individuals), define microenvironments, and evaluate multiple pollutants, using a choice of time steps. A new APEX Introduction Document provides an overview of the model and contains detailed instructions on conducting an example APEX run and common simulation variations.

Keywords: A-exposure models, C-air, A-activity patterns, A-statistical methods

TH-PL-G1: Sensor Validation

TH-PL-G1-629

Quality Assurance and Quality Control of Portable Devices using a Standard Reference Material

J. L. Reiner; NIST, Charleston, SC

Abstract: Standard Reference Materials (SRMs) are homogeneous, well-characterized materials that are used to validate measurements and improve the quality of analytical data (www.nist.gov/srm). The National Institute of Standards and Technology (NIST) has a wide range of SRMs that have values assigned for clinically important analytes, legacy organic pollutants, and toxic metals in human matrices. Examples of some SRMs include organic contaminants in human serum, human milk, and human urine, lead in caprine blood, arsenic species in human urine, and toxic elements in human urine. These SRMs can serve as materials for quality control when developing methods on a portable devise. Currently NIST has looked at different sampling techniques, including dried blood spot cards (DBS), and different portable devises, including CardioChek and VetScan. SRM 1958 Organic Contaminants in Fortified Human Serum (Freeze-Dried), SRM 955c Toxic Metals in Caprine Blood, and a solution of 25-hydroxyvitamin D_3 were spotted onto DBS cards to test the feasibility of using DBS to screen for analytes of interest. SRM 1950 Metabolites in Frozen Human Plasma was used to assess the precision and accuracy of the CardioChek and VetScan devises for field research. Results indicate that precise and accurate measurements for many compounds can be made using both devises.

Keywords: A-biomonitoring, A-analytical methods

TH-PL-G1-630

Performance of Consumer-Grade Air Pollution Measurement Devices in Residential Environments

G. Mainelis, S. N. Manibusan; Rutgers University, Edison, NJ

Abstract: A developing trend in consumer electronics is production and use of inexpensive (<$300) devices to monitor airborne contaminants of health concern. Most of these devices monitor the mass concentration of $PM_{2.5}$ and PM_{10} particles, temperature, relative humidity, and some include volatile organic compounds or other pollutants. This study focused on four consumer devices: the Air Quality Egg 2, Blueair Aware, Foobot, and Speck that use optical sensors to measure airborne particulate matter concentrations. The devices were collocated in three residences over 7 consecutive days in each residence. Household measurements were compared against established optical sensing devices including two Personal Data RAMs (passive pDR-1000, Thermo Fisher), a DustTrak DRX (TSI Inc.), and gravimetric mass measurements using two Personal Modular Impactors (PMI, SKC, Inc.). A follow-up study in one residence included a collocation of Foobot, Speck, and the pDR-1000 in the kitchen (near source cooking exposure) and bedroom (residential exposure away from a source) combined with time-activity behaviors in the residence over 14-days. Overall, the results show inconsistent performance of consumer grade devices relative to reference methods. For example, in one household, the Foobot registered a Pearson's R^2 of 0.565 when compared to the DustTrak DRX for 1-hour averages, while the Speck showed poor associations (R^2=0.206). Similar results were shown in the cooking study with poor associations from the Speck compared with the pDR-1000 (R^2=0.003), while the Foobot had a strong association with the pDR (R^2=0.830). Among the investigated sensors, the Foobot was most correlated with the reference devices, but the cooking study indicated its limitations regarding maximum loading capacity. Overall, the data show the utility of the consumer-grade air quality measurement devices, allowing for wide-scale deployment; however, further testing is needed to determine their applicability in field studies.

Keywords: A-sensor technology, B-particulate matter, A-indoor environment, A-sampling methods

TH-PL-G1-631

Using low-cost temperature sensors to monitor the use of domestic woodstoves and their potential for indoor smoke emission

I. Longley, A. Kachhara, S. Edwards, E. Somervell; NIWA Ltd, Auckland, New Zealand

Abstract: Domestic woodstoves can be a major source of local $PM_{2.5}$ emissions. However, managing these emissions is inhibited by large uncertainties in spatio-temporal variability in emission activity. Home-heating behaviour is a complex response to meteorology, climate, thermal performance of the appliance and the home, as well as habit, lifestyle, fuel affordability and accessibility, and subjective thermal tolerance/comfort. Some studies have suggested woodstoves can also be intense sources of indoor $PM_{2.5}$ through direct leakage. Trials were conducted to investigate the feasibility of using low-cost temperature sensors to continuously monitor woodstove use in a manner that could easily be scaled-up. In a controlled trial 14 "iButton" devices were distributed at various locations around a single woodstove. This test showed a strong relationship between device placement and the interpretability of data collected. This showed that the criteria for device placement need to be carefully defined in response to study objectives. In a cohort of 10 homes iButtons were deployed in pairs (one on stove, one in main living area) alongside a reference thermocouple during the winter and spring of 2015. The iButtons performed as well as the thermocouples for the purposes of monitoring stove us. In a follow-up study during the following winter iButtons were deployed in triplicate (stove, living area and kitchen). This allowed the main features of heating activity to be observed. The data were compared with indoor measurements of particulate matter and time-activity diaries maintained by householders. In this small dataset we observed a possible reduction in heating activity related to changes in meteorology. We detected no significant correlation between events related to woodstove use (lighting, stoking, re-fuelling) and observable increases in indoor particulate matter, although smoke leakage cannot be ruled out. A larger study aimed at approximately 50 homes is planned for winter 2017.

Keywords: A-activity patterns, B-particulate matter, C-air, C-indoor, A-sensor technology

TH-PL-G1-632

Evaluating Real-time Air Quality Monitors Prior to Deployment as Community-level Monitors

A. Commodore, R. Boaz, J. L. Pearce; Medical University of South Carolina, Charleston, SC

Abstract: The increasing availability of low cost and easy-to-use air pollution monitors has created new opportunities for local community groups to collect air pollution using Citizen Science. An important step for this area of research is to determine the reliability and comparability of the data. Therefore, we aimed to evaluate the DustTrak DRX (hereafter DRX) as a non-federal reference monitor in order to determine which Citizen Science monitoring tier our instruments are suitable for. In addition, we aim to develop a 'calibration model' that removes observed biases and improves comparability of our data with federal equivalent methods (FEM). To achieve our aims, we first co-located five DRX units with an FEM $PM_{2.5}$ monitor from April 3rd to 20th 2017. We will develop a calibration model to adjust raw $PM_{2.5}$ measurements using the framework of a generalized additive mixed model with the FEM and DRX measurements as dependent and predictor variables respectively. All measurements were collected at one minute intervals. Data were trimmed in order to retain the central 99% (0.5%-99.5%) for each DRX to avoid potential skewing by outliers. Once trimmed, hourly data summaries were calculated for each DRX if measurements greater than or equal to 80% of the minute data for a particular hour block were available, per Environmental Protection Agency's citizen science recommendations. Preliminary results indicate that the mean $PM_{2.5}$ concentrations during the evaluation period (n=390 hours) ranged between 4 and 8 $\mu g/m^3$. The correlation coefficients (r) between raw DRX measurements were between 0.97 and 0.99

(p<0.0001), among all DRX units, demonstrating strong inter-instrument agreement. Comparisons to FEM measurements will be presented. Our methods highlight the importance of adopting quality control and quality assurance methods capable of removing observed instrument biases and improving the comparability and reliability of data collected for community-level air monitoring projects.

Keywords: A - ambient monitoring, A-exposure models, D-community, A - exposure measurement, A-sensor technology

TH-PL-G1-633

Validation of Hexoskin Biometric Shirt at Rest, Submaximal Exercise, and Maximal Exercise

C. M. Smith[1], S. N. Chillrud[1], D. W. Jack[1], P. Kinney[2], Q. Yang[1], A. M. Layton[1]; [1]Columbia University, New York, NY, [2]Boston University, Boston, MA

Abstract: Minute ventilation (V_E) is difficult to measure in the field, but is of considerable interest as it allows researchers to estimate potential inhaled dose of air pollutants. Advances in commercially available wearable technology make it easier to measure V_E in mobile field studies, but the reliability of these systems is unknown. The Hexoskin Biometric Shirt continuously records heart rate through an ECG. Respiratory rate (RR), tidal volume (V_T), and V_E are recorded through a dual band respiratory inductance plethysmography (RIP). Additionally, the shirt contains a 3-axis accelerometer. The presence of the 3-axis accelerometer and the ECG enables three different proxy methods for estimating V_E. To validate the Hexoskin shirt, HR, RR, V_T, and V_E measurements were compared to a 'gold standard' Vmax system in a physiology laboratory at Columbia University Medical Center. Participants carried out an incremental exercise test on a stationary bicycle to validate the shirt at rest, submaximal exercise (SM), and maximal exercise (M). Bland Altman analysis suggested good significant agreement for RR and HR measurements at all levels when excluding improper recordings, and resulted in strong correlations (R= 0.84 to 0.99). Correlation of V_E measurements at maximal exercise was high (R=0.85), but lower at rest (R= 0.49). V_T and V_E measurements taken by the Hexoskin shirt at SM and M exercise had a discrepancy of less than 12%. The Hexoskin shirt is advantageous as it provides three distinct metrics for V_E. In addition to the direct measurements, V_E can be estimated by HR or actigraphy recordings when the dual band RIP sensor does not provide a reliable measure.

Keywords: A-biomonitoring, A - exposure measurement, A-sensor technology

TH-PL-A2-634

Development and evaluation of hybrid LUR-CTM models in Canada

M. Johnson[1], A. Hakami[2], S. Saeednooran[2], J. Zhang[1], S. Zhao[2]; [1]Health Canada, Ottawa, Canada, [2]Carleton University, Ottawa, Canada

Withdrawn

TH-PL-A2-635

NO_2 Air Pollution Exposure Assessment in Urban Mysore, India

A. A. Nori-Sarma[1], R. K. Thimulappa[2], V. G. Venkatareddy[3], A. K. Fauzie[3], S. K. Dey[2], M. L. Bell[1]; [1]Yale University, New Haven, CT, [2]JSS Medical College, Mysore, India, [3]University of Mysore, Mysore, India

Abstract: *Introduction*: In India, rapid urbanization has lead to a decrease in urban air quality. Few ongoing studies apply spatially heterogeneous sampling to assess ambient air pollution levels in urban India. Understanding air pollution exposure differentials is particularly important in cities, where socio-economic disparity by neighborhood may increase health impacts of varying exposures. *Methods*: Seasonal sampling campaigns have been conducted at 150 sampling sites throughout urban Mysore, Karnataka, India. Sampling sites were selected using a 3-step combination of purposeful site selection, systematic random sampling, and purposeful site selection to fill in remaining gaps of exposure measurement. Nitrogen dioxide (NO_2) levels were assessed using passive Palmes tube and Ogawa badge technologies at these sites, to develop a spatial interpolation of ambient air pollution exposure in these cities. *Results*: Mysore, considered a "clean city" of India, has minimum air pollution levels during the post-monsoon season in excess of 20 ppb averaged over a 2-week period. Maximum air pollution levels measured in the city around 50-60ppb over a 2-week period, indicating that pollution levels may exceed the health-protective thresholds established by the World Health Organization in highly populated urban areas. Strong spatial and temporal patterns indicate a seasonal trend in air pollution levels, as well as the importance of point sources of pollution. *Conclusions*: While traffic pollution has been indicated in the past as a major contributor to ambient air pollution levels in urbanizing centers of Asia, our results indicate that air pollution levels do not follow major roadway patterns. Other pollution sources (e.g., building density), as well as highly localized environmental factors (such as waste burning and small trash fires), might be critically influencing air pollution exposure patterns in urban India.

Keywords: A - ambient monitoring, A - exposure measurement, A-exposure models, A-geospatial analysis/GIS, A-sampling methods

TH-PL-A2-636

Seasonal trend of fine particulate matter in indoor and outdoor environments of urban residential homes in Nanjing, China

Z. SHAO, Z. Ma, J. Bi; Nanjing University, Nanjing, China

Abstract: Most people spend more than 80% of their time indoors, hence indoor pollution level has a significant influence on individual's exposure to ambient air pollution. In this study, 48-h continuous real-time $PM_{2.5}$ concentrations in both indoor and outdoor air were simultaneously measured in 61 residential homes in Nanjing, China. The measurements were carried out using MicroPEM personal exposure monitors during spring, summer and winter seasons from March 2016 to February 2017. Information on building characteristics, occupants' activities and socioeconomic variables were collected through questionnaires during sampling period. The average indoor $PM_{2.5}$ concentrations were 50.97, 37.08 and 55.56 μg/m^3 in spring, summer and winter, and the mean outdoor concentrations were 61.62, 27.75 and 70.61 μg/m^3, respectively. Significant seasonal variations of indoor and outdoor $PM_{2.5}$ were observed ($p<0.001$). The correlation between 48-h average indoor and outdoor $PM_{2.5}$ concentrations in spring ($r=0.822$, $p<0.001$) was higher than in winter ($r=0.644$, $p<0.001$) and summer ($r=0.509$, $p<0.01$). Infiltration factors for $PM_{2.5}$ was found to be highest in spring due to more naturally ventilated by keeping windows opened in comparison with other seasons. Linear regression models were fitted to log-transformed $PM_{2.5}$ concentrations on potential influence factors. Concentration of $PM_{2.5}$ outdoors was found to be the major source for indoor $PM_{2.5}$ pollution, and dwelling ownership was associated with significantly decreased concentration of indoor $PM_{2.5}$ in three seasons. More frequent building ventilation could increase indoor $PM_{2.5}$ exposure especially in spring and winter when the outdoor air quality is poor. Other household activities like cooking without kitchen separated, indoor smoking were also found to be influence factors of higher indoor $PM_{2.5}$ concentrations, and the use of air cleaner was associated with reduced indoor $PM_{2.5}$, suggesting a potential method of reducing indoor $PM_{2.5}$ exposure.

Keywords: A-indoor environment, A - ambient monitoring, A-activity patterns

TH-PL-A2-637

Impacts of community design and commute behavior on exposures to traffic-related air pollution

A. L. Stuart, S. Gurram, H. Yu, R. Michael, A. Marroquin, M. Resto; University of South Florida, Tampa, FL

Abstract: Exposure to air pollution is an important environmental health and urban design challenge. To ensure improved health outcomes, knowledge is needed on the impacts of community planning alternatives and personal choices on exposures to air pollution. Effects on diverse population groups across multiple scales and pollutants particularly need to be studied. This presentation will discuss results from two urban areas, Tampa Florida and Fort Collins Colorado, where we are studying exposures to traffic-related air pollution. For Tampa, the focus pollutants are oxides of nitrogen, benzene, 1,3-butadiene, formaldehyde, and acetaldehyde. To diagnose and predict exposures under alternative urban design scenarios, we have applied passive field sampling, geospatial analyses, simulation of human and vehicular activity, modeling of emissions and dispersion of pollution, and analysis of population exposure. Characterizing differences and inequality in exposure between population groups has been an emphasis. For Fort Collins, we aim to understand the impacts of the travel route and mode (cycling versus driving) on commute exposures to carbon monoxide and fine particulate black carbon using modeling methods. For this presentation, we will discuss the integrated findings from these study areas, with an emphasis on the development and application of innovative exposure modeling approaches. We will describe the use of an activity-based travel demand model and a transportation route simulator to predict human and vehicular activity. We will also describe the development of a dispersion modeling system that applies a Bayesian Markov Chain Monte Carlo approach to assimilate measurement data into estimates of exposures. Results demonstrate that impacts of design and management choices on overall and group-level exposure differ between pollutants and between acute and chronic times scales. Healthy community decisions must balance multiple and conflicting outcomes.

Keywords: A - population exposure, C-air, A-exposure models, A-built environment, A-environmental justice

TH-PL-A2-638

Diesel exposure in urban environment: evidence based approach to understand health risks

A. Rashid[1], A. Naseem[2], A. Kamal[1]; [1]PMAS Arid Agriculture University, Rawalpindi, Pakistan, [2]National University of Sciences and Technology, Islamabad, Pakistan

Being presented in place of WE-PL-A2-453

TH-PL-B2: Environmental Exposure and Health

TH-PL-B2-639

Urinary Concentrations of Organophosphate Flame Retardants and Fertility Outcomes among Couples Undergoing in Vitro Fertilization

C. Carignan[1], L. Mínguez-Alarcón[1], J. Meeker[2], P. Williams[3], H. M. Stapleton[4], C. Butt[4], T. Toth[5], J. Ford[1], R. Hauser[1]; [1]Harvard T.H. Chan School of Public Health, Boston, MA, [2]University of Michigan, Ann Arbor, MI, [3]Harvard T.H. Chan School of Public Health, Boston, MA, [4]Duke University, Durham, NC, [5]Massachusetts General Hospital, Boston, MA

Abstract: Use of organophosphate flame retardants (PFRs) has increased over the past decade with the phase out of some brominated flame retardants. We recently reported associations of some urinary PFR metabolites with decreased proportions of fertilization, implantation, clinical pregnancy and live birth among women recruited from an academic fertility clinic. In this analysis, we report on urinary concentrations for the male partners, examine predictors of these concentrations, and examine associations between urinary concentrations of PFR metabolites and outcomes of *in vitro* fertilization in their partner. Our analysis included 209 couples enrolled in the Environment and Reproductive Health (EARTH) prospective cohort study (2005-2015). We measured five urinary PFR metabolites using negative electrospray ionization liquid chromatography tandem mass spectrometry. We used multivariable generalized linear mixed models to evaluate the association of demographic characteristics with the PFR metabolites and PFR metabolites with IVF outcomes, accounting for multiple IVF cycles per couple. Detection frequencies were high (>75%) for BDCIPP, DPHPand ip-PPP but low (<15%) for tb-PPP and BCIPP. Some PFR urinary metabolites were associated with race, body mass index, year of treatment cycle, and season. An 8% decline was observed for the highest compared to lowest quartile of urinary BDCIPP in adjusted means for cycles resulting in successful fertilization (adjusted absolute difference=0.06 (0.01, 0.12), p-trend=0.06), after adjusting for maternal exposure. We conclude that male partner urinary concentrations of BDCIPP may be associated with fertilization whereas *maternal* urinary DPHP, ip-PPP and ΣPFR account for negative associations previously observed for proportions of successful fertilization, implantation, clinical pregnancy and live birth. These results highlight potential reproductive effects of exposure to PFRs at levels currently common among the general population.

Keywords: A-biomonitoring, A-chemical alternatives, A-epidemiology, A-indoor environment, B-flame retardants

TH-PL-B2-640

Phthalates and organophosphate flame retardants in house dust and their relation to asllergies and oxidative stress marker 8-OHdG: the Hokkaido Study

A. Araki[1], Y. Ait Bamai[1], C. Miyashita[1], T. Kawai[2], T. Tsuboi[2], R. Kishi[1]; [1]Hokkaido University, Sapporo, Japan, [2]Japan Industrial Safety and Health Association, Osaka, Japan

Abstract: Abstract: Phthalates and organophosphate flame retardants (PFRs) are categorized as semi-volatile organic compounds. They are not chemically bond to the product, thus they leach out, and adhere to settled dust. Phthalates and PFRs are reported to associate with asthma and allergies, and oxidative stress is considered to be a mediator between exposure and allergies. The aim of this study is to find the association between phthalates and PFRs in dust and allergies, and 8-hydroxydeoxyguanosine (8-OHdG) as a systemic markers of oxidative stress. **Methods:** Participants are 7 years old children of the Hokkaido Study, who completed ISAAC questionnaire and submitted house dust and urine. According to the ISAAC criteria, 221 children with wheeze, eczema, and/or rhino-conjunctivitis are defined as cases and 221 were randomly selected from those without symptoms after sex stratification as references. We analyzed 8 phthalates and 11 PFRs in dust by GC/MS. Spot morning urine was used to measure 8-OHdG by ELISA. **Results:** The most detected phthalate was DEHP (median concentrations; 1422.0 µg/g), followed by DiNP (119.2 µg/g), and the most detected PFRs was TBEP (49.22 µg/g) followed by TICP (0.90 µg/g). When limited to eczema, 8-OHdG levels were marginally higher than those without eczema (p=0.072). For eczema, 2 fold increase of DINP showed OR (95%CI) =1.18 (1.04, 1.33), and for rhino-conjunctivitis, 2 fold increase of BBzP showed OR=1.11 (1.00, 1.23). Two fold increase of TPP and TCEP decreased log2 transformed 8-OHdG (β: 95%CI= -0.029: -0.058, 0.000 and -0.017: -0.033, -0.001, respectively). However, TPP and TCEP did not show any associations with allergies. **Conclusion:** We observed significant associations between DiNP and BBzP and allergies, however, our finding did not support the hypothesis of 8-OHdG as underlying the mechanisms. Limitation of the study is that levels in indoor dust do not reflect the exact exposure levels of each participants.

Keywords: A-biomarkers, A-epidemiology, D-children, B-phthalates, B-flame retardants

TH-PL-B2-641

Association between urinary metabolites of pesticides and biomarkers of oxidative stress/inflammation in children consuming an organic diet: preliminary findings

C. Konstantinou, S. Ioannou, K. C. Makris; Cyprus University of Technology, LImassol, Cyprus

Abstract: Introduction: The effectiveness of a systematic organic diet in improving key parameters of health in human studies has not been well studied. The objective of this study was to determine the intervention effectiveness of an organic diet in significantly reducing urinary pesticide metabolite concentrations and concomitantly affecting biomarkers of inflammation and oxidative stress in young children. **Methods**: A crossover study was designed for healthy children (n=150, 10-12 years old) with families historically consuming a conventional diet (>80% of weekly diet). The study period consisted of a 40-day conventional diet and a 40-day organic diet. Diaries were used to record the consumed diet. A total of five morning urine voids samples per child in select time points of both conventional and organic phases were collected. Select metabolites of neonicotinoid/pyrethroid pesticides were monitored in children's urine using tandem mass spectrometry. Biomarkers of oxidative stress/inflammation and biochemical parameters were measured in children urine (malondialdehyde, interleukin-6) using immunoassay kits. Linear mixed-effects models determined the main effects of organic diet accounting for both between- and within-subject variability. **Results**: The effect of organic diet on the magnitude and variability of pesticide metabolites accounting for known confounders will be analyzed and presented. The influence of organic diet upon the association between the biomarkers of oxidative stress/inflammation and those of the pesticide metabolites will be discussed. **Conclusions**: Never before the concomitant assessment of biomarkers of effect (oxidative stress/inflammation) along with biomarkers of exposure to pesticides has been performed in field trials. Larger-scale intervention studies are required to confirm these preliminary findings.

Keywords: A - population exposure, A-biomarkers, A-biomonitoring, B-pesticides

TH-PL-B2-642

Dietary and inhalation exposure to polycyclic aromatic hydrocarbons (PAHs) and monohydroxy metabolites in urine: A panel study for the elderly in Tianjin,

B. Han[1], P. Li[2], X. Qin[3], L. Zhang[3], T. Ni[4], J. Fan[3], N. Zhang[1], F. He[5], J. Xu[6], W. Yang[1], W. Zhang[1], X. Wang[1], Z. Bai[1]; [1]Chinese Research Academy of Environmental Sciences, Beijing, China, [2]Tianjin University of Technology, Tianjin, China, [3]Tianjin Medical University, Tianjin, China, [4]Tianjin University of Sport, Tianjin, China, [5]Hubei Meteorological Service Center, Wuhan, China, [6]University of Washington, Seattle, WA

Moved to Tuesday Poster Sessions

TH-PL-B2-643

Combining the exposome with the metabolome reveals associations between environmental chemicals and endogenous molecules involved in critical metabolic processes

V. Bessonneau[1], R. R. Gerona[2], R. Grashow[1], A. Wang[3], J. Trowbridge[4], T. Lin[2], R. Morello-Frosch[4], R. A. Rudel[1]; [1]Silent Spring Institute, Newton, MA, [2]University of California, San Francisco, San Francisco, CA, [3]University of California, San Francisco, San Francisco, CA, [4]University of California, Berkeley, Berkeley, CA

Abstract: A limitation of current biomonitoring for environmental chemicals is that targeted analytic methods require *a priori* selection of compounds to study. As a result, significant time and resources are expended to develop methods for chemicals that may or may not be present in biological specimens, and important exposures may be missed if not anticipated. One way to circumvent this issue is to perform non-targeted analysis of biospecimens using high-resolution mass spectrometry to measure environmental exposures (chemical exposome) and to assess which chemicals are associated with biological changes (metabolome). In this study, we computed partial correlations between the serum chemical exposome and metabolome - measured using liquid chromatography in combination with quadrupole time-of-flight mass spectrometry (LC-QTOF/MS) - in 89 California women workers from the Women Firefighter Collaborative Biomonitoring (WFBC) study to investigate the effects of environmental chemicals on biological processes. The network of correlation revealed many direct associations (FDR<0.1) between environmental chemicals and endogenous molecules, suggesting an effect of these exposures on critical metabolic processes. For example, we observed positive associations between perfluorooctane sulfonic acid (PFOS), mono-2-ethylhexyl phthalate, and ethyl paraben with inflammatory signaling molecules, between pyrethrin II and androgens, and negative associations between certain phthalates and phenols with neurosteroids. Several exposure-metabolite associations in our analysis were further confirmed using the National Health and Nutrition Examination Survey (NHANES). For example, exposure to PFOS was associated with a significant increase in serum C-reactive protein and lymphocyte count in women 19-84 years of age, after adjusting for age, race/ethnicity, socioeconomic status and body mass index, suggesting that PFOS might have effects on inflammation and immunity in women.

Keywords: A - exposure measurement, A-biomarkers, A-biomonitoring, D-occupational,

TH-PL-C2: Exposure & Health Effects to Perfluorinated Compounds

TH-PL-C2-644

Characterizing persistent EDC mixtures in a diverse pregnancy cohort

S. Mehta[1], K. Coleman-Phox[2], N. E. Adler[3], B. Laraia[4], E. Epel[3], A. R. Zota[1]; [1]The George Washington University, Washington, DC, [2]University of California, San Francisco, San Francisco, CA, [3]University of California, San Francisco, San Francisco, CA, [4]University of California, Berkeley, Berkeley, CA

Abstract: Background: Exposure to persistent halogenated endocrine disrupting chemicals (EDCs) during pregnancy can impact fetal and maternal health. Despite an overall phase-out in use, these synthetic compounds are still measured concurrently in pregnant women at detectable concentrations in the US. In our current study, we aim to characterize exposure to EDC mixtures during early pregnancy by utilizing multiple aggregation techniques. **Methods:** Single spot serum samples of polybrominated diphenyl ethers (PBDEs), hydroxylated PBDEs (OH-PBDEs), polychlorinated biphenyls (PCBs), and perfluorinated chemicals (PFCs) were measured in a cohort of 104 pregnant women living in Northern California. Cross-sectional associations between EDCs and predictors were examined using linear regression. To aggregate exposure, a chemical burden metric was calculated for each participant by counting the number of chemicals with concentrations in the 25th percentile of each chemical's distribution. Spearman correlation and principal components analysis were also used to understand grouping and to overcome the issue of multicollinearity. **Results:** 15 EDCs were detected in >50% of serum samples; PBDE-47, PFNA, and PFOS were detected in all participants. There was no consistency in the direction or magnitude of predictors in single pollutant and within-class models. Age was cross-sectionally associated with PCB-138, 153, and 180 (p<0.0001). Food insecurity was associated with maternal PFOS concentrations (p=0.03). Of the 15 EDCs, a range of 3-11 chemicals (median=6 chemicals) were detected at greater than the 75th percentile concentrations in all participants. Overall, strong with-in class and weak across-class correlations were seen. The first and second principal components, which explained 43% of the total variance, were highly correlated with PBDEs and PCBs, respectively. **Conclusions:** By utilizing various methods to aggregate exposure, we are better able to characterize EDC mixtures during pregnancy.

Keywords: A-aggregate exposure, B-mixtures, B-POPs, D-susceptible/vulnerable

TH-PL-C2-645

Fluorinated compounds are common in U.S. fast food packaging and potentially contribute to population-wide PFAS exposure

L. A. Schaider[1], S. Balan[2], A. Blum[3], H. Susmann[1], D. Andrews[4], M. Stryar[5], R. A. Rudel[1], R. Dodson[1], G. Peaslee[6]; [1]Silent Spring Institute, Newton, MA, [2]California Department of Toxic Substances Control, Sacramento, CA, [3]Green Science Policy Institute, Berkeley, CA, [4]Environmental Working Group, Washington, DC, [5]U.S. Environmental Protection Agency, Research Triangle Park, NC, [6]University of Notre Dame, Notre Dame, IN

Abstract: Per- and polyfluoroalkyl substances (PFASs) are highly persistent synthetic chemicals, some of which have been associated with cancer, developmental and immune toxicity, and other health effects. PFASs are used in some grease-resistant food packaging; prior studies found PFASs in packaging can leach into food. To evaluate potential contributions of fast food packaging to PFAS exposure, we screened food packaging samples for fluorinated chemicals and evaluated associations between reported fast food consumption and PFAS blood levels. In 2014-2015, we tested >400 samples of food contact papers, paperboard, and cups from fast food restaurants across the U.S. for total fluorine using particle-induced gamma-ray emission (PIGE) spectroscopy. PIGE can rapidly measure total fluorine in paper, a marker for PFASs. We detected fluorine (>16 nmol/cm^2) in 46% of food contact papers and 20% of paperboard samples. Liquid chromatography/high-resolution mass spectrometry analysis of a subset of 20 samples found 27 known PFASs and unidentified polyfluorinated compounds based on nontargeted analysis. Six of the 20 samples contained detectable levels of PFOA (C8), even though in 2011 U.S. food packaging manufacturers phased out C8 compounds. While production of long-chain PFASs has been phased out in the U.S., newer short-chain and polyether PFASs are similarly persistent and also raise health concerns. We explored associations between fast food consumption and PFAS blood serum levels in the general U.S. population using U.S. CDC's NHANES biomonitoring and health survey data. Number of fast food meals consumed in the prior week and consumption of fast food in the prior day were associated with higher levels of PFOS, PFOA, and several related PFASs after controlling for demographic variables, BMI, and NHANES cycle, with stronger associations in 2007-2010 data compared to 2011-2014 data. Our results suggest potentially significant contributions of fast food packaging to PFAS exposure.

Keywords: B-POPs, C-food, A - population exposure, A-biomonitoring

TH-PL-C2-646

Polyfluoroalkyl substance exposure in the Mid-Ohio River Valley, 1991-2012

R. L. Herrick[1], J. M. Buckholz[1], F. M. Biro[2], A. M. Calafat[3], X. Ye[3], C. Xie[1], S. M. Pinney[1]; [1]University of Cincinnati College of Medicine, Cincinnati, OH, [2]Cincinnati Children's Hospital Medical Center, Cincinnati, OH, [3]CDC, Atlanta, GA

Abstract: Background: Industrial discharges of perfluorooctanoic acid (PFOA) to the Ohio River, contaminating water systems near Parkersburg, WV, were previously associated with nearby residents' serum PFOA concentrations above US population medians. Ohio River PFOA concentrations downstream are elevated, suggesting Mid-Ohio River Valley (ORV) residents are exposed through drinking water. We quantified per- and polyfluoroalkyl substances (PFAS) in Mid-ORV resident sera collected between 1991 and 2013 and determined whether the Ohio River and Ohio River Aquifer were exposure sources. **Methods:** Using PFAS measurements in 1608 sera from 931 participants, we assessed water source associations using linear mixed-effects models. We estimated between-sample serum PFOA using one-compartment pharmacokinetics for participants with multiple samples. **Results:** In sera collected as early as 1991, PFOA (median=7.6ng/mL) was detected in 99.9%; 47% had concentrations greater than US population 95th percentiles. Five other PFAS were detected in > 82% of samples, with median concentrations similar to the US population. Serum PFOA was significantly

associated with water source, sampling year and age, tap water consumption, pregnancy, gravidity and breastfeeding, and was 40-60% lower with granular activated carbon (GAC) use. Repeated measurements and pharmacokinetics suggest serum PFOA peaked in 2000-2006 for participants using water without GAC treatment; where GAC was used, serum PFOA concentrations decreased from 1991 to 2012. **Conclusions:** Mid-ORV residents appear to have PFOA, but not other PFAS, serum concentrations above US population levels. Drinking water from the Ohio River and Ohio River Aquifer, primarily contaminated by industrial discharges 209-666 kilometers upstream, is likely the primary exposure source. GAC treatment of drinking water mitigates, but does not eliminate, PFOA exposure.
Funding: U01ES012770, U01ES019453, P30ES006096, R21ES017176,T32ES10957, R834788

Keywords: A-biomonitoring, A-exposure models, D-community, C-water, B-POPs

TH-PL-C2-647

Exposure and risk assessment of perfluorooctanoic acid in drinking water and ambient air for residents living close to a production plant in the Netherlands

B. Bokkers, P. Janssen, E. Jansen, M. Zeilmaker; National Institute of Public Health and the Environment (RIVM), Bilthoven, Netherlands

Abstract: People living in the neighborhood of the Dupont/Chemours factory in Dordrecht, the Netherlands, have been exposed to perfluorooctanoic acid (PFOA) by air and drinking water between the 1970 and 2012[1],[2]. Due to the geographic location of drinking water intakes and the predominant wind direction, two exposed populations can be distinguished. One exposed via air only and one exposed via water only. PFOA accumulates in the human body. Therefore, it is necessary to take the entire exposure period into regard, and perform the risk assessment based on the PFOA body burden. Based on emissions the ambient air concentrations in the vicinity of the factory was estimated. The water concentrations were obtained from historical information sampled at drinking water intakes. Subsequently, we calculated the PFOA blood concentrations using a kinetic model, which accounts for ventilation rate/consumption, absorption, volume of distribution and elimination rate. Exposure via drinking water did not lead to body burdens exceeding the RIVM limit value for chronic exposure of 89 ng/mL blood. It is likely that, in the past, locals have been exposed to PFOA in air resulting in body burdens chronically exceeding this limit value. Several scenarios for the emission were used to estimate the exposure. In the most unfavorable scenario, the limit value was exceeded for 25 years. The limit value was not exceeded anymore after the year 2002. PFOA concentration in blood samples of residents are measured to verify this finding. Furthermore, the added value of health surveys of the population living close to the factory is considered. [1] RIVM (2016) Risk assessment of the emission of PFOA: Location: Dupont/Chemours, Dordrecht, The Netherlands [in Dutch]. RIVM report 2016-0049. [2] RIVM (2016) Risk assessment of PFOA in drinking water at two locations [in Dutch]. http://www.rivm.nl/dsresource?objectid=2b34fa1a-cd9d-454b-9f6e-36a8e9e6aac6&type=pdf&disposition=inline

Keywords: A-exposure models, B-POPs, C-air, D-community, A-biomonitoring

TH-PL-C2-648

Perfluorooctanoate and Body Mass Index, Waist:Hip and Waist:Height Ratio in Young Girls in the Greater Cincinnati and San Francisco Bay Area

S. M. Pinney[1], G. C. Windham[2], R. L. Herrick[1], K. L. McWhorter[6], C. Xie[1], C. S. Fassler[1], R. A. Hiatt[3], L. Kushi[4], F. M. Biro[5]; [1]University of Cincinnati College of Medicine, Cincinnati, OH, [2]California Department of Public Health, Richmond, CA, [3]UCSF, San Francisco, CA, [4]Kaiser Permanente, Oakland, CA, [5]Cincinnati Children's Hospital Medical Center, Cincinnati, OH, [6]NIEHS, Research Triangle Park, NC

Abstract: Background: Perfluorooctanoate (PFOA) is an endocrine disrupting chemical with ubiquitous exposure in the US. Cross-sectional studies of the effect of PFOA on BMI have produced disparate findings possibly related to age. **Hypothesis:** Exposure to PFOA results in decreased BMI in young girls.

Methods: Through the NIH Breast Cancer and the Environmental Research Centers (BCERC), we conducted a study of polyfluoroalkyl compound (PFC) biomarkers in girls from Greater Cincinnati (CIN, N=353) and San Francisco (SFBA, N=351). PFOA concentration in the first serum sample was analyzed for 704 girls, age 6-8 years. Mixed effects models were used to analyze the effect of PFOA on BMI, waist-to-height and waist-to-hip ratios and PFOA concentration in this longitudinal cohort, including Age*PFOA interaction. **Results:** Median serum PFOA concentrations were 7.3 (CIN) and 5.8 (SFBA) ng/mL, above the US population median for children 12-19 years (3.8 ng/mL, NHANES, 2005-2006). Log-transformed serum PFOA had a strong inverse association with BMI in the CIN girls (p=0.0002) and the combined two-site data (p=0.0008) and the joint inverse effect of PFOA and Age*PFOA weakened at age 10-11 years. However, in the SFBA group alone, the relationship was not significant (p=0.1641) with no evidence of changing with age. **Conclusions:** PFOA is associated with decreased BMI in young girls but the strength of the relationship decreases with age. Site heterogeneity may be due to greater perinatal exposure in CIN. Funding: U01ES012770, U01ES012771, U01ES012800, U01ES012801, U01ES019435, U01ES019453, U01ES019457 P30ES006096, T32-ES10957 UL1RR024131, CSTA-UL1RR026314

Keywords: B-POPs, A-biomonitoring, D-children, A-epidemiology

TH-PL-D2: Tobacco Smoke Exposure

TH-PL-D2-649

Assessing dermal exposure to nicotine - an interdisciplinary approach

T. Salthammer[1], G. Bekö[2], G. Claussen[2], H. M. Koch[3], G. Morrison[4], T. Schripp[1], J. Toftum[2], C. Weschler[2]; [1]Fraunhofer WKI, Braunschweig, Germany, [2]Technical University of Denmark, Lyngby, Denmark, [3]Institute for Prevention and Occupational Medicine of the German Social Accident Insurance, Institute of the Ruhr-Universität Bochum (IPA), Bochum, Germany, [4]Missouri University of Science and Technology, Rolla, MO

Abstract: In a pilot study it was shown that dermal uptake of nicotine directly from air can be a significant exposure pathway. On the basis of these preliminary results, scientists from the fields of building sciences, human-biomonitoring, gas-phase analytics, physical chemistry and modeling collaborated to design a more detailed research plan, and experiments were performed in October 2016 at the Technical University of Denmark. During all of the exposure experiments, the volunteers breathed clean air from hoods they wore. Nicotine (dissolved in water) was delivered to the 55 m^3 exposure chamber from a step-motor driven syringe. In all experiments the average nicotine concentration in air was between 236 $\mu g/m^3$ and 324 $\mu g/m^3$. In week 1, four volunteers wearing only shorts and two volunteers wearing clean cotton clothes were exposed in the chamber for five hours. In week 2, two of the bare-skin participants were again exposed in the chamber for five hours, and then showered immediately after exposure rather than waiting at least 24 h as they had done in week 1. The two participants who wore clean clothes in week 1 were now exposed wearing a shirt, socks and gloves that had been exposed to nicotine at > 250 $\mu g/m^3$ for almost a month. They wore clean full-length pants that had not been exposed. Urine samples were collected before, during and after exposure and analyzed for nicotine, cotinine and 3OH-cotinine. The major results of the study can be summarized as follows: a) dermal uptake, directly from air, is a meaningful exposure pathway for nicotine - comparable to inhalation; b) clean clothing acts as a barrier to dermal exposure from air; c) clothing that has absorbed nicotine can promote its dermal uptake; c) skin is a reservoir - delivery continues after leaving chamber. The outcome emphasizes the advantage of interdisciplinary research design, which is helpful to undrstand exposure scenarios to indoor pollutants.

Keywords: A-indoor environment, B-VOCs, C-air, A-biomonitoring, A-exposure models

TH-PL-D2-650

Kinetics of dermal uptake of nicotine from air

G. Morrison[4], G. Bekö[1], G. Claussen[1], H. M. Koch[3], C. Paelmke[3], T. Salthammer[2], T. Schripp[2], J. Toftum[1], C. Weschler[1]; [1]Technical University of Denmark, Lyngby, Denmark, [2]Fraunhofer WKI, Braunschweig, Germany, [3]Institut der Ruhr-Universität Bochum (IPA), Bochum, Germany, [4]Missouri University of Science and Technology, Rolla, MO

Abstract: Recently, dermal uptake of nicotine from cigarette smoke has been observed. To study this process in more detail, six participants were exposed to nicotine in a chamber over a 5-hour period while breathing clean air through a breathing hood. During the first week four of the participants wore only shorts and two wore a set of clean clothes comprised of cotton, polyester and rayon (average air concentration of nicotine 240 µg/m^3). During the second week, two of the bare-skin participants were again exposed in the chamber (average nicotine concentration 290 µg/m^3). The two participants who wore clean clothes on week one, were now exposed wearing a shirt, socks and gloves that had been exposed to nicotine at an air concentration of ~500 µg/m^3 for 16 days, then ~250 µg/m^3 for 11 days. They wore full-length pants that had been laundered but not exposed. One urine sample was collected immediately before exposure and all urine was collected during the 84 hours after each exposure. Post-exposure urine samples were pooled; one pooled sample contained urine collected within the first 12 hours after exposure, the second, third and fourth pooled samples contained urine collected during the subsequent three 24-hour periods. All urine samples were analysed for nicotine and two of its metabolites, cotinine and 3-hydroxycotinine. For three of the participants, all individual (non-pooled) samples from both weeks were analysed. The average back-calculated absorbed nicotine for bare-skinned subjects was 590 mg, less than 55 mg for subjects wearing fresh clothes and 810 mg for subjects wearing exposed clothes. The average cotinine half-life was 35h, which is greater than the reported half-life for intravenous delivery (14h), smokers (16h) or non-smokers exposed to ETS (27h). These observations are consistent with the hypothesis that non-smokers exposed to ETS are absorbing some fraction of nicotine through their skin.

Keywords: A-indoor environment, A-biomonitoring, A-second-hand smoke, B-VOCs, C-air

TH-PL-D2-651

Occupational exposures in e-cigarette vape shops

K. Attfield[1], M. Zalay[3], E. Glassford[2], B. Materna[1], L. Zwack[2]; [1]California Department of Public Health, Richmond, CA, [2]National Institute of Occupational Safety and Health, Cincinnati, OH, [3]Public Health Institute, Oakland, CA

Abstract: The promise of lower exposures to carcinogenic combustion products from e-cigarettes as compared to conventional cigarettes has resulted in a rapid increase in e-cigarette use in California and across the country. Laboratory tests have detected a variety of compounds in e-liquids and e-cigarette emissions including nicotine, carrier components (propylene glycol and glycerin), formaldehyde, flavorings (including diacetyl), tobacco-specific nitrosamines, metals, and some volatile organic compounds (VOCs). These compounds have known health effects including acute nicotine poisoning, decreased lung function, severe lung disease, asthma, and carcinogenicity. In California, use of e-cigarettes in workplaces has only recently been limited by state regulations, and workplace vaping is especially prevalent in stores selling e-liquid and e-cigarette devices (vape shops). Methods: In late 2016 through early 2017, we began characterizing workplace exposures in a number of vape shops in the San Francisco Bay Area. Sampling included personal and area air monitoring of flavoring chemicals, formaldehyde, particulate matter, and VOCs. E-liquids vaped during the sampling day were purchased and headspace analysis was used to assess flavorings including alpha-diketones. Wipe samples were collected for metals on commonly touched surfaces. Results: Initial findings suggest that employee chemical handling practices need improvement. Personal air sampling showed detectable levels of formaldehyde (median 9.8 ppb, maximum 33.7 ppb) and flavoring compounds (4/16 detectable for diacetyl personal samples and 4/16 detectable for 2,3-pentanedione). E-liquid headspace analysis showed diacetyl in 17 of 26 samples and 2,3-pentandeione in 7 of 26 samples. Conclusions: With data from several sites, we will be developing a better understanding of this emerging industry and recommendations for workplace health and safety.

Keywords: A-workplace, B-VOCs, A - exposure measurement, A-indoor environment, A-industrial hygiene

TH-PL-D2-652

Chemical characteristic of source emission and inhalational carcinogenic risk of environmental tobacco smoking under real condition, China

N. Zhang, B. Han, Z. Bai, W. Yang, X. Wang, W. Zhang; Chinese Research Academy of Environmental Sciences, Beijing 100012, China, Beijing, China

Abstract: Nine sets of $PM_{2.5}$ samples were collected via volunteer smoking nine Chinese brands of cigarette, to describe chemical characteristic of source emissions and evaluate inhalational carcinogenic risk of environmental tobacco smoke (ETS). Source profile containing 14 elements, NO_3^-, SO_4^{2-}, organic carbon (OC), element carbon (EC) and 11 polycyclic aromatic hydrocarbons (PAHs) was conducted; source signatures were recognized; and carcinogenic risks of heavy metals/PAHs were assessed. Results showed that OC was the primary composition in ETS profile; Cr, Pb and Cd were the most abundant heavy metals; PAHs mainly distributed in MMW- and HMW-PAHs. BaA, BbF and CHR could separate ETS and outdoor stationary sources as source marker, but diagnostic ratios could hardly distinguish ETS and cooking/biomass combustion. The integrated carcinogenic risk of heavy metals and PAHs was 2.2×10^{-3} and 1.8×10^{-5}, respectively. The risks of Cr (VI) and Cd were over 1.0×10^{-4}; the value of CHR was beneath 1.0×10^{-6}; the risks of the other carcinogens ranged between these two limits. The total carcinogenic risk of heavy metals was higher by two orders of magnitude than that of PAHs, indicating that the adverse effect of inorganic compositions should not be neglected.

Keywords: A-indoor environment, A-risk assessment, A-second-hand smoke, B-particulate matter

TH-PL-D2-653

Acrylonitrile in the US Population and Exposures From Tobacco Smoke and Diet: NHANES 2005 - 2006, 2011 - 2012

R. deCastro, N. Geldner, U. Alwis, D. M. Chambers, B. Blount; CDC National Center for Environmental Health, Atlanta, GA

Abstract: Acrylonitrile is a byproduct in the combustion of biomass, and metabolic detoxification produces biomarkers in urine indicative of exposure to smoke, especially from tobacco, where acrylonitrile is a major toxicant. One major urinary metabolite specific to acrylonitrile is N-Acetyl-S-(2-cyanoethyl)-L-cysteine (CYMA). Analysis of data from the 2005 - 2006 and 2011 - 2012 cycles of the National Health and Nutrition Examination Survey (NHANES; N = 4,718 subjects ≥6 years old) revealed that serum cotinine, a specific biomarker of tobacco smoke exposure, was a significant predictor of urinary CYMA (4.05 ng/ml per ng cotinine/ml serum [95%CI: 1.81:6.29]) among non-users of tobacco products, but not among exclusive combustion tobacco users (0.32 [0.10:0.55]), controlling for potential confounders. Foods prepared by smoking or roasting were also significant sources of exposure to acrylonitrile. Consumption of smoked meats was a significant predictor of urinary CYMA among both non-users of tobacco products (16.07 ng/ml per kg consumed [28.20:60.35]) and exclusive combustion tobacco users (1314.48 [-915.06:3544.02]). Brewed coffee was a borderline significant predictor of urinary CYMA among non-users (2.63 ng/ml per kg consumed [-1.64:6.90]) and significant among exclusive users (10.80 [-11.91:33.52]). Urinary CYMA's sensitivity and specificity as a biomarker of self-reported exclusive combustion tobacco use was equivalent to serum cotinine. This study establishes that tobacco smoke is a significant route of exposure to acrylonitrile in a representative sample of the United States civilian, non-institutionalized population. Diet is also a significant exposure route.

Keywords: A-biomarkers, A-second-hand smoke, A-biomonitoring, B-VOCs

TH-PL-E2: Metal Exposures

TH-PL-E2-654

Investigation of Arsenic and co-occurring metals near abandoned mine wastes in Cheyenne River, South Dakota (CRST)

C. DeVore, L. Rodriguez Freire, J. Cerrato; University of New Mexico, Albuquerque, NM

Withdrawn

Keywords: B-metals

TH-PL-E2-655

Health and safety awareness training on lead exposure for Artisanal and Small-Scale gold miners in Anka and Bukkuyum Local Government Areas of Zamfara state, Nigeria

S. Kadiri; Zub Chord Tech ventures, Lagos, Nigeria

Abstract: Introduction Artisanal and small-scale gold mining (ASGM) has long been practiced in Nigeria. Mining often involves both occupational and community health and safety hazards that not only affect miners, but also their families and communities. In Zamfara, Nigeria where the gold bearing deposits contain unusually problematic concentrations of lead these are overwhelmed by the enormous effects of lead poisoning. In 2010, unregulated small-scale miners in Zamfara state, gave rise to an epidemic of childhood lead poisoning, with at least four hundred children under the age of five dying within a six-month period (a number that rose to over 700 by 2013). It was found out that a lack of training in Health and safety and support to the ASGM sector, and the need for the miners to make a living, even in a precarious environment contributed to this incidence. **Objectives** The objectives of this study are: To provide capacity building to help this group of workers learn concepts for improving work conditions and understand the risks in mining; To provide awareness on various approaches of workplace health and safety promotion as regards mining. **Methods** One hundred and thirteen (113) participants were randomly selected for this training. Modules Seven modules were used for this training. Hazard identification and risk assessment. First aid Mercury Exposure and related risks lead exposure and related risks Safer mining practice Personal Hygiene Personal protective equipment (PPE) used in mining **Results** The Health and safety knowledge of the ASGM workers were increased. The capacity building process enabled the workers to recognize risks associated with mining and therefore know how to implement safety measures by using PPE and by learning about safety improvement concepts.

Conclusion Our findings suggest that positive attitudes toward promoting safe working conditions and practices can be fostered among the ASGM workers.

Keywords: A-exposure factors, A - population exposure, D-occupational, A-activity patterns

TH-PL-E2-656

Health risks of heavy metals from drinking water exposure near a typical river basin area in China, after pollution control measurement

X. Zhao[1], S. Cao[1], X. Duan[2], B. Zheng[1], Y. Ma[1]; [1]Chinese Research Academy of Environmental Sciences, Beijing, China, [2]University of Science and Technology Beijing, Beijing, China

Abstract: The metal pollution still is a great concern due to the effects from urbanization and industrialization. While, the health risks from the toxic metal could decrease if strict pollution control measures were adopted. However, few studies to date investigate the health risks of heavy metal in a systematic river basin for the dependent residents, after taking pollution control measures. Thus, the contents of metal(loid)s (Cu, Pb, Zn, Cd, Mn, As) in surface water along a typical river basin were investigated in this study, and the potential non-carcinogenic and carcinogenic health risks posed to the residents were assessed. Although existing mining activity in the upstream, the soluble concentrations of all the metal were within the relevant thresholds. However, the closer to the mining area, the higher the pollution levels of metal. The total hazard index for non-carcinogenic risks of metal were far lower than the threshold (1) for the local population. Whereas, although the content of metal were low (such as As and Cd), they could pose relative higher non-carcinogenic health risks. The result illustrated that pollution levels, toxicity of the contaminants and exposure behavior patterns all could contribute to the potential detrimental health risks. Additionally, the non-carcinogenic and carcinogenic risks from ingestion exposure were ~2- ~4 orders of magnitude higher than those from dermal contact. The total carcinogenic risks were ~2- ~4 orders of magnitude lower than the maximum tolerable levels (1.0×10^{-4}), indicating carcinogenic risks could also be neglected. Among different population groups, heavy metal posed relative higher non-carcinogenic and carcinogenic risks to the children in 0~5 years old. Fortunately, the surface water in this basin is safe in usage for the local population and the health risks were acceptable in case exposed to the target metal, after the river basin was in the charge of strict pollution control measures.

Keywords: A - exposure measurement, A - population exposure, A-environmental policy, C-water

TH-PL-E2-657

Risk Assessment of Polluted Soils in Relation to Transfer of Metals to Human Food Chain

D. Golui, S. Datta, B. Dwivedi, M. Meena; ICAR-Indian Agricultural Research Institute, New Delhi, India

Abstract: The success of risk assessment of metal contaminated soils depends on how precisely one can predict the solubility of metals in soil and subsequent transfer of these metals from soil to human food chain via plant. The specific objectives of the present investigation were to predict the uptake of Zn, Cu, Ni, Cd and Pb by Indian spinach (*Beta Vulgaris* L. var. All green) grown on metal contaminated soil using free ion activity model (FIAM) and to assess the risk to human health from intake of these metals from consumption of this leafy vegetable. For this purpose, twenty eight bulk surface soil samples with diverse soil properties such as pH, organic carbon and metal content were collected from smelter effluent, industrial sewage, polluted river water, domestic sewage, solid waste and cycle factory effluent deposited sites across the country. Free metal ion activity was estimated in soil solution as extracted by rhizon samplers using version 7 of the 'Windermere Humic Aqueous Model' (WHAM-VII). Results indicated that 94, 70, 75, 81 and 91% variation in Zn, Cu, Ni, Pb and Cd content, respectively, of Indian spinach could be explained by free ion activity model based on WHAM VII (Model I). Predicted free ion activity derived from solubility model could also significantly predict Zn, Cu, Ni, Pb, and Cd content in spinach to the extent of 85, 64, 66, 78 and 95%, respectively. Risk assessment of metal contaminated soil in terms of hazard quotient (HQ) indicated that Indian spinach grown on solid waste and industrial effluent irrigated soils were not safe (>0.5) to be consumed by human being as far as Zn, Cd, Pb and Ni content in this leafy vegetable are concerned. A ready reckoner was developed for computing safe limit of extractable metal in soil based on the predicted HQ by Model I for intake of metal by human through consumption of spinach. Novel approach as used in assessing the risk of metal polluted soils in relation to human health proved to be useful and promising.

Keywords: A-risk assessment, B-metals, C-soil

TH-PL-E2-658

Relating Soil Geochemical Properties to Arsenic Bioaccessibility Through Hierarchical Modeling

C. M. Nelson[1], K. Li[2], D. Obenour[2], J. Misenheimer[1], K. Scheckel[3], A. Betts[4], A. Juhasz[5], D. Thomas[6], K. Bradham[1]; [1]US EPA, Research Triangle Park, NC, [2]NC State University, Raleigh, NC, [3]US EPA, Cincinnati, OH, [4]University of Delaware, Newark, DE, [5]University of South Australia, Mawson Lakes, Australia, [6]US EPA, Research Triangle Park, NC

Abstract: Arsenic (As) is the highest prioritized contaminant at U.S. Superfund sites. Estimating the site-specific bioavailability of As in soil, defined in this context as the fraction of ingested As that crosses the gastrointestinal epithelium and becomes available for distribution to internal tissues, improves the accuracy of exposure assessments to evaluate human health risks from As-contaminated soils. Soil geochemical properties influence As bioavailability. Interest in improved understanding of these relationships has motivated the use of regression models to evaluate the ability of soil properties to estimate the amount of As that dissolves from the soil matrix upon exposure to gastric-like conditions, a surrogate measure of bioavailability termed *bioaccessibility*. However, limits in the numbers and types of soils included in previous studies restrict the usefulness of previous models beyond the range of soil conditions evaluated, as evidenced by reduced predictive performance when models were applied to new data. In response, we developed a novel hierarchical model that can account for variability between contaminant sources and geographic locations to evaluate the ability of soil geochemical properties to estimate As bioaccessibility in 139 soils collected from 97 locations across three continents. This approach improved the estimation of As bioaccessibility in study soils and enabled the identification of a larger suite of elements as significant explanatory variables when compared to previous studies, reflecting the complexity of geochemical mechanisms that control As bioaccessibility across the range of soil properties and contaminant sources encountered in the environment. The hierarchical framework provides a promising approach for improved understanding of the relationships between soil geochemical properties and soil metal bioavailability, with applications for improved risk characterization and remediation of metal(loid)-contaminated soils.

Keywords: B-metals, C-soil, A-statistical methods, A-exposure factors, D-community

TH-PL-F2: Water Contaminants

TH-PL-F2-659

Disparities in exposures to nitrate in U.S. public drinking water supplies

L. A. Schaider[1], L. Swetschinski[1], C. Campbell[2], R. A. Rudel[1]; [1]Silent Spring Institute, Newton, MA, [2]Environmental Working Group, Washington, DC

Abstract: Nitrate, a common contaminant in U.S. drinking water, mainly originates from agricultural fertilizer as well as animal manure and human sewage. Nitrate is regulated in drinking water to protect infants from methemoglobinemia, and nitrate-contaminated drinking water has been linked with bladder and colorectal cancers, birth defects, and thyroid dysfunction. Small water supplies in poor rural areas are especially vulnerable due to proximity to agricultural and livestock production and limited financial and technical resources. A small number of regional drinking water studies have found higher nitrate concentrations in communities with higher proportions of minority and low-income residents. However, studies of environmental justice (EJ) indicators and water quality have been limited to local and regional scales. Our study—the first to examine disparities in drinking water nitrate at a national scale—evaluates relationships between nitrate concentrations and EJ indicators for about 40,000 community water systems serving 225 million Americans from 2010 to 2014. We combined water system data from U.S. EPA's Safe Drinking Water Information System, nitrate data from each state, and U.S. Census Bureau demographic data. Overall, 755 public water supplies, serving 5.9 million Americans, violated the nitrate Maximum Contaminant Level (10 mg/L NO_3-N) at least once over this five-year period; 3,512 public water supplies, serving 33.9 million Americans, had at least one sample >5 mg/L. We conducted regression analyses to assess relationships between nitrate and EJ indicators (median income, racial/ethnic minority populations, education). We also considered potential confounders, such as extent of nearby agricultural land-use. Our study provides new information about the degree to which EJ communities face elevated exposures to nitrate in drinking water, and our data analysis approaches can be extended to other drinking water contaminants such as pesticides and pharmaceuticals.

Keywords: A-environmental justice, C-water, D-community

TH-PL-F2-660

Inferring Instream Loading Rates of Organic Chemicals in United States Watersheds from Their Downstream Concentrations

P. Saranjampour[1], R. W. Setzer[1], C. Barber[2], K. Isaacs[2], P. P. Egeghy[2], K. Phillips[2], J. Wambaugh[1]; [1]U.S. Environmental Protection Agency, Durham, NC, [2]U.S. Environmental Protection Agency, Durham, NC

Withdrawn

Keywords: A-exposure models, A-ecological exposure, A-statistical methods, C-water, A - exposure measurement

TH-PL-F2-661

Carbon Nanotube Electrochemical Sensors for Quantifying Heavy Metal Exposure

N. Alvarez[1], D. Zhao[2], W. Heineman[2], E. N. Haynes[2], V. Shanov[2]; [1]University of Cincinnati, Cincinnati, OH, [2]University Cincinnati, Cincinnati, OH

Abstract: Heavy metal exposure concerns are no longer limited to underdeveloped countries, alarming levels of exposure have been reported in the USA. Rapid detection or continuous monitoring of metals, including lead, in drinking water is needed to ensure safe drinking water supplies. Detection of heavy metals in water is high priority for societies, particularly living in old cities where water infrastructure was developed more than 50 years ago. This talk will highlight the current progress toward the development of an electrochemical sensor based on carbon nanotubes from synthesis to heavy metal detection. The unique physical properties of individual Carbon Nanotubes (CNTs) surpass the properties of many advanced materials available today. Due to their large surface area, chemical stability, and electrical conductivity, they are the most promising candidates for a large number of electroanalytical applications. A material for sensor applications requires high purity CNT assemblies like fibers and films, that have good electrical conductivity, in some case additional insulation coatings. Polystyrene coated CNT fiber was used as the working electrode; bare CNT thread was used as the auxiliary electrode; and a pseudo-reference electrode was fabricated by electroplating CNT fiber with Ag that is subsequently anodized in chloride solution to form a layer of AgCl. This all-carbon CNT fiber three electrode cell is evaluated for simultaneous detection of trace levels of heavy metal ions by anodic stripping voltammetry (ASV). Hg^{2+}, Cu^{2+} and Pb^{2+} in H_2O were detected successfully, and the detection limits are 1.05 nM, 0.53 nM and 0.57 nM for Hg^{2+}, Cu^{2+} and Pb^{2+}, respectively. These electrodes significantly reduce the dimensions of the conventional three electrode electrochemical cell and have the potential to become low cost and disposable sensor.

Keywords: A-sensor technology, A-nanotechnology, A-analytical methods, B-metals, C-water

TH-PL-F2-662

Suspect screening analysis of drinking water using point-of-use filters

S. Newton[1], R. McMahen[2], J. R. Sobus[1], A. Williams[3], A. McEachran[2], M. Stryar[1]; [1]Environmental Protection Agency, Durham, NC, [2]Oak Ridge Institute for Science and Education, Raleigh, NC, [3]National Center for Computational Toxicology, Raleigh, NC

Abstract: Despite remarkable progress in recent years in the identification of unknown compounds using high resolution mass spectrometry (HRMS), monitoring programs for drinking water rely on targeted methods which cover a limited number of compounds. In an effort to advance suspect screening analysis (SSA) methodologies and more fully characterize the drinking water exposome, point-of-use water filtration devices (BritaTM filters) were employed to collect time-integrated (1-2 months) drinking water samples in a pilot study of nine central North Carolina homes. Filters were Soxhlet extracted and analyzed by HPLC-TOFMS. Formulas corresponding to observed molecular features (i.e., unknown compounds described by accurate mass, retention time, and mass spectra) were searched against the US EPA's CompTox Chemistry Dashboard, a recently developed web application and data hub for ~750,000 chemical substances. Out of 14,922 molecular features identified in the samples, 430 were assigned molecular formulas with a match score > 90, which mapped to 10,621 candidate compounds. These compounds were ranked by number of data sources and given a "ToxPi" score calculated using four elements: bioactivity data from EPA's ToxCast program, exposure estimates from EPA's ExpoCast program, detection frequency, and average peak area. Of all the candidate compounds, 91 were found as having the highest number of data sources, as well as the highest ToxPi ranking for their molecular formula, and thus grouped as the highest priority compounds. Of these, 15 were confirmed as being correctly identified using standards on hand. Product-use categories from EPA's CPCat database, again available via the CompTox Dashboard, revealed that the majority of priority compounds are associated with industrial processes which indicates that drinking water in central North Carolina may be impacted by local industries. Most of the priority compounds would not have been discovered without the use of SSA.

Keywords: A-analytical methods, C-water, A-chemical prioritization

TH-PL-F2-663

Influence of Hydrogeological Factors on Exposure to Emerging Contaminant in Karst Environments

I. Y. Padilla[1], N. I. Torres[1], V. L. Rivera[1], R. Macchiavelli[2], D. J. Vesper[3], J. Meeker[4], A. N. Alshawabkeh[5]; [1]University of Puerto Rico, Mayagüez, Mayagüez, PR, [2]University of Puerto Rico, Mayaguez, Mayagüez, PR, [3]West Virginia University, Morgantown, WV, [4]University of Michigan, Ann Arbor, MI, [5]Northeastern University, Boston, MA

Abstract: Providing 20-25% of the global population water needs, karst aquifers are extremely productive groundwater systems that develop in soluble geologic formations. The same characteristics that make karst aquifers highly productive and important freshwater resources for human consumption also impart a high vulnerability for contamination. As a consequence, these aquifers can serve as significant routes of contaminant exposure. Emerging contaminants, including phthalates, are of particular concern because they can easily enter karst groundwater and move toward areas of potential exposure to human. Many of these contaminants are known endocrine disruptors and potential precursors of adverse reproductive outcomes. Previous studies in the karst region of northern Puerto Rico have shown significant presence of phthalates in groundwater and tap water. This work evaluates potential correlations between hydrogeological factors, diethylhexyl phthalate (DEHP) contamination, and potential exposure at the tap water point of use. Geographic Information Systems technologies and statistical models are applied to attain these objectives. The analysis incorporates data gathered from regulatory agencies and current groundwater and tap water samples collected from homes. Results show widespread distribution of DEHP in both groundwater and tap water, suggesting that contamination comes from multiple sources. Spatial variability of DEHP in groundwater is significantly influenced by hydrogeological factors, reflecting the importance of groundwater transport from sources of contamination to potential points of exposure. Spatial correlation analysis indicates that association between tap water and groundwater contamination depends on spatial location and time. Though groundwater serves as route of contamination to tap water sources, additional sources of contamination from the water distribution system increases the presence of phthalates in tap water and the potential for exposure.

Keywords: A-geospatial analysis/GIS, B-phthalates, C-water, A-ecological exposure,

TH-PL-G2: Data Sharing and the Exposome

TH-PL-G2-664

Decentralized and HIPAA Compliant Geocoding to Characterize Community and Environmental Exposures for Multi-Site Studies

C. Brokamp, C. Wolfe, P. Ryan; Cincinnati Children's Hospital Medical Center, Cincinnati, OH

Abstract: Geocoding and geomarker assessment (characterizing study participants' communities and environmental exposures near their residence, school, etc.) are often used in epidemiological studies. Because addresses are identifiable private health information (PHI), this must be done in a HIPAA-compliant manner. This is a challenge for multi-site studies as current approaches include obtaining Institutional Review Board approval for a central site to conduct all analyses, a lengthy and sometimes unfeasible process, or allowing each site to conduct their own analyses which may yield inconsistencies across sites. To address these limitations, we created DeGAUSS (Decentralized Geomarker Assessment for Multi-Site Studies), a software application that can geocode and assign geomarkers per specific study needs. DeGAUSS is built on a containerization platform which guarantees that the process is reproducible regardless of its environment and ensures that PHI never leaves the user's machine. Geocoding 215,350 addresses in Cincinnati, Ohio, USA with DeGAUSS showed similar accuracy compared to Google and ArcGIS. The median distance between the geocoded and actual locations were 40, 7, and 22 m for DeGAUSS, Google, and ArcGIS, respectively. DeGAUSS, Google, and ArcGIS accurately assigned addresses to census tracts at a rate of 96%, 99%, and 95%, respectively. As a proof of concept, we applied DeGAUSS in a multi-site study, where each site independently geocoded and assigned median census tract level income and distance to nearest major roadway to their participants' addresses, stripped any PHI, and sent the de-identified data back. A total of 32,436 study participants at three different sites were geocoded with a mean distance to roadway of 9,659 m and a mean census tract income of $59,124. The open source DeGAUSS software not only solves the problem of using address data in multi-site studies, but also serves as a reproducible research tool for geocoding and geomarker assessment.

Keywords: A - exposure measurement, A-epidemiology, A-geospatial analysis/GIS

TH-PL-G2-665

Modeling the contribution of chemicals in building materials to population exposome

O. Jolliet, L. Huang; University of Michigan, Ann Arbor, MI

Abstract: Complementary to biomarker-oriented approaches, this presentation explores how recent databases on consumer products and modeling of releases from building materials enables us to estimate the contribution of building materials to chemical US exposome. Using the PHAROS database, we first created a list of 22 building product categories corresponding to homogeneous archetypes in term of material and application and with a minimum of 8 products in each category. We then identified the concentrations of the 467 different chemical ingredients present in each of the 676 final products, leading to a total of 10,391 chemical-product combinations. A systematic application of release models combined with gaseous, dust, inhalation and direct contact exposure route enable to systematically determine are used to determine the fraction of chemical released to indoor air and relevant exposures by gaseous dermal intake, dermal contact, and dust ingestion. Resulting product intake fractions (PiF) from chemicals in flooring range from 10^{-6} to 10^{-2} with inhalation as dominant pathways for VOCs and dust ingestion or gaseous dermal uptake dominating SVOCs exposures. Combining PiF with chemical content yields exposure doses that for phthalates range from 7 to max 190 µg/kg-d, which is of the same order of magnitude as the range of back-calculated exposure dose from the NHANES urine biomarker of DEHP in the US population of 6 to 21 µg/kg-d. The presentation will discuss the range and related hazard quotient for exposure to multiple organic chemicals in these building materials. The present approach demonstrates the feasibility to predict population and individual exposures to multiple chemicals in building products. It constitutes one component of a more comprehensive and systematic approach to predict the lifetime exposure of individuals to thousands of chemicals on the basis of the time evolution of their home location, economy sector workplace, consumption patterns and eating habits.

Keywords: A-exposure models, A-chemical prioritization, A - population exposure, A-biomarkers, A-built environment

TH-PL-G2-666

Promoting integration across diverse studies of environmental health: Development of a children's health and exposure ontology

S. L. Teitelbaum[1], J. A. Stingone[1], J. P. McCusker[2], P. Pinheiro[2], S. M. Rashid[2], Y. Liu[2], Z. Liang[2], D. L. McGuinness[2]; [1]Icahn School of Medicine at Mount Sinai, New York, NY, [2]Rensselaer Polytechnic Institute, Troy, NY

Abstract: Collective-science initiatives rely on the investigators' ability to harmonize data and methodologies across diverse studies in order to promote integration. The Children's Health Exposure Analysis Resource (CHEAR) provides the infrastructure to promote the inclusion of exposure science within diverse studies of children's health. In order to support the CHEAR Data Center's mission of providing support for the analysis, interpretation, curation and reuse of data within the CHEAR network, the CHEAR ontology was developed to provide a common vocabulary with unambiguous definitions of concepts and variables across studies. By integrating with existing ontologies and supplementing terms unique to exposure science and epidemiology, the CHEAR ontology supports chemical, biological and epidemiologic measurements that comprise the majority of data within environmental health studies. This includes distinctive features of environmental health research such as repeated measures of environmental biomarkers across time and parent-child linkages within study populations. Currently, the CHEAR ontology interfaces with thirteen foundational ontologies and contains 232 chemical analytes, 165 epidemiologic attributes, 21 biologic sample types, 15 classes of laboratory assays, and 150 laboratory instruments. Customizable templates have been created for researchers to accurately describe their data in a way that facilitates their semantic integration with other environmental health studies and existing ontologies. The template approach allows a wide range of users to input content while supporting dynamic growth of the ontology. The creation of this ontology will facilitate both the pooling of study data and the construction of searchable data warehouses with epidemiologic and environmental exposure data. The CHEAR ontology can be reused and expanded as needed to specify, curate and search data from studies within the broader fields of exposure science and environmental epidemiology.

Keywords: D-children, A-epidemiology, A-analytical methods,

TH-PL-G2-667

Creating a Pilot Web Portal to Facilitate Access to Consumer Exposure Science Methods, Databases, and Projects

R. Becker; American Chemistry Council, Washington, DC

Abstract: The Catalog of Consumer Exposure Initiatives (CCEI) website is a pilot project to explore the utility of creating a portal to help users find and access publicly available exposure tools, databases, and projects related to the field of consumer exposure science. Creation of the CCEI was driven, in part, by the recognition that information on consumer product exposures and use can be challenging to locate. The intended audiences are those seeking to learn more about new initiatives and current practices/tools—both new practitioners and those involved in emerging consumer risk assessment programs. The CCEI pilot web portal currently contains links to approximately 150 resources, including, for example: 1) links to consumer product ingredient disclosure and exposure initiatives by various product organizations or companies in the U.S. and Europe; 2) links to relevant activities by the USEPA and other government agencies; and 3) links to methods, models, and databases published in the scientific literature. CCEI has been recently updated to include additional details related to tools and models (e.g., user expertise needed, input, output, applicability domain), scientific meeting abstracts for current and emerging science, and standard operating procedures for web site postings. The presentation will include a demonstration of the CCEI pilot web portal, illustrating multiple ways of accessing cataloged information, including using an exposure framework graphical interface. As a portal, the CCEI is not intended to recommend or endorse specific data, models or methods; rather the goal is promote knowledge sharing that can maximize impact of ongoing activities and also better inform future research.

Keywords: C-consumer products, A-exposure models

TH-PL-G2-668

Harmonization of Sensor Metadata and Measurements to Support Exposomic Research

N. Burnett[1], R. Gouripeddi[2], J. Wen[2], P. Mo[3], R. Madsen[3], R. Butcher[3], K. A. Sward[4], J. Facelli[2]; [1]University of Utah, Salt Lake City, UT, [2]University of Utah, Salt Lake City, UT, [3]University of Utah, Salt Lake City, UT, [4]University of Utah, Salt Lake City, UT

Abstract: The concept of the exposome includes assessing exposures to different environmental factors and understanding endogenous processes within the body. Quantifying exposures and their effects therefore requires sensors that measure the general and personal environments for different physical, chemical and biological species. Using data from these different sensors requires them to be harmonized into a common data representation for ease of use in different data analytic approaches, and for integration for generation as comprehensive, high-resolution spatio-temporal profiles of exposomes. Towards developing such a common representation, we started with modeling air quality (AQ) data which has well-documented effects on health. We performed a literature review using PubMed with the search criterion “Pediatric Asthma Sensor Studies’. We manually extracted a list of metadata elements describing AQ sensors from the literature, and developed a first draft of a conceptual data model (CDM). Next we collected sample data from different research studies using different types of sensors including Environmental Protection Agency monitors, personal, in-home, outdoor and citizen networks. We evaluated the CDM with the collected and revised it with fields found in the data. We then met with AQ experts in Utah and The Pediatric Research using Integrated Sensor Monitoring Systems (PRISMS) group to review the model and modified it further based on their inputs. We now have a complete CDM that accommodate different types of sensors and is available for community review at goo.gl/DgvzGz. This model consists of Instrument, Deployment and Measurement Output domains that describe characteristics of a sensor device, its deployment in research settings, and its measurements respectively. We have deployed this model with different databases harmonizing vast amounts of air quality data from different sensors. This model will now be used in the PRISMS project to study and effects of AQ on health.

Keywords: C-air,

AUTHOR INDEX

Name/Abstract #

A

B

C

D

E

F

G

H

M

N

Q

T

International Society of Exposure Science (ISES)
c/o Infinity Conference Group
1035 Sterling Road, Suite 202
Herndon VA 20170
USA

ISBN 978-1-5108-9754-0

CURRAN ASSOCIATES INC. proceedings.com

7th International High Performance Buildings Conference at Purdue 2022

West Lafayette, Indiana, USA
10-14 July 2022

Herrick Laboratories
PURDUE UNIVERSITY™